Preface

College Algebra in Context is designed for a course in algebra that is based on data analysis, modeling, and real-life applications from the management, life, and social sciences. The text is intended to show students how to analyze, solve, and interpret problems in this course, in future courses, and in future careers. At the heart of this text is its emphasis on problem solving in meaningful contexts.

The text is application-driven and uses real-data problems that motivate interest in the skills and concepts of algebra. Modeling is introduced early, in the discussion of linear functions and in the discussion of quadratic and power functions. Additional models are introduced as exponential, logarithmic, logistic, cubic, and quartic functions are discussed. Mathematical concepts are introduced informally with an emphasis on applications. Each chapter contains real-data problems and extended application projects that can be solved by students working collaboratively.

The text features a constructive chapter-opening Algebra Toolbox, which reviews previously learned algebra concepts by presenting the prerequisite skills needed for successful completion of the chapter. In addition, a section on calculus preparation at the end of the text emphasizes how students can use their new knowledge in a variety of calculus courses.

Changes in the Fourth Edition

We have made a number of changes based on suggestions from users and reviewers of the third edition, as well as our own classroom experiences.

Chapter objectives are listed at the beginning of each chapter, and key objectives are given at the beginning of each section.

Appendix B has been expanded to include instructions for Excel 2007 and Excel 2010 as well as Excel 2003. Where a new technology procedure is introduced in the text, references to calculator and Excel instructions in Appendixes A and B have been added.

To keep the applications current, nearly all of the examples and exercises using real data have been updated or replaced. This is especially important in light of the numerous economic and financial changes that have occurred since publication of the previous edition.

Many new Section Previews and other applications provide motivation. Applications are referenced on the chapter-opening page in the order in which they will appear in the section.

To improve the exposition, the organization and content of some sections have been changed.

- Section 1.1 contains an expanded discussion of mathematical models.

- Section 1.3 provides a more complete introduction to revenue, cost, and profit functions.

- Section 2.1 includes expanded discussion of

- The term "Power Functions" has been incl   ght the increased emphasis on this topic in this

- Section 3.1 contains an expanded discussio

- Direct variation as an *n*th power and inverse variation have   3.3.

- Additional discussion of graphing and solving equations with Excel has been added to Chapter 3.

- Chapter 4 Toolbox has been revised to contain a library of functions.

- A more complete introduction to the average cost function now appears in Section 4.2.

- The discussion of one-to-one functions has been moved to Section 4.3 with inverse functions.

- Section 5.1 has been reorganized to complete the discussion of exponential functions before moving on to exponential growth and decay.

- Additional Skills Check problems have been added to ensure that every type of skill is well represented.

- Additional modeling questions involving more decision making and critical thinking have been added throughout the text.

- Sample homework exercises, chosen by the authors, are indicated with an underline in the Annotated Instructor's Edition.

Continued Features

Features of the text include the following:

- The development of algebra is motivated by the need to use algebra to find the solutions to **real data–based applications**.

    Real-life problems demonstrate the need for specific algebraic concepts and techniques. Each section begins with a motivational problem that presents a real-life setting. The problem is solved after the necessary skills have been presented in that section. The aim is to prepare students to solve problems of all types by first introducing them to various functions and then encouraging them to take advantage of available technology. Special business and finance models are included to demonstrate the applications of functions to the business world.

- **Technology** has been integrated into the text.

    The text discusses the use of graphing calculators and computers, but there are no specific technology requirements. When a new calculator or spreadsheet skill becomes useful in a section, students can find the required keystrokes or commands in the *Graphing Calculator and Excel® Manual* that accompanies the text, as well as in the appendixes mentioned above. The text indicates where calculators and spreadsheets can be used to solve problems. Technology is used to enhance and support learning when appropriate—not to supplant learning.

- The text contains two technology appendixes: a **Basic Calculator Guide** and a **Basic Guide to Excel 2003, Excel 2007, and Excel 2010.** Footnotes throughout the text refer students to these guides for a detailed exposition when a new use of technology is introduced. Additional Excel solution procedures have been added, but, as before, they can be omitted without loss of continuity in the text.

- Each chapter begins with an **Algebra Toolbox** section that provides the prerequisite skills needed for the successful completion of the chapter.

    Topics discussed in the Toolbox are topics that are prerequisite to a college algebra course (usually found in a Chapter R or appendix of a college algebra text). Key objectives are listed at the beginning of each Toolbox, and topics are introduced "just in time" to be used in the chapter under consideration.

- Many problems posed in the text are **multi-part and multi-level problems**.

    Many problems require thoughtful, real-world answers adapted to varying conditions, rather than numerical answers. Questions such as "When will this model no longer be valid?" "What additional limitations must be placed on your answer?" and "Interpret your answer in the context of the application" are commonplace in the text.

- Each chapter has a **Chapter Summary**, a **Chapter Skills Check**, and a **Chapter Review**.

  The Chapter Summary lists the key terms and formulas discussed in the chapter, with section references. Chapter Skills Check and Chapter Review exercises provide additional review problems.

- The text encourages **collaborative learning**.

  Each chapter ends with one or more Group Activities/Extended Applications that require students to solve multilevel problems involving real data or situations, making it desirable for students to collaborate in their solutions. These activities provide opportunities for students to work together to solve real problems that involve the use of technology and frequently require modeling.

- The text ends with a **Preparing for Calculus** section that shows how algebra skills from the first four chapters are used in a calculus context.

  Many students have difficulty in calculus because they have trouble applying algebra skills to calculus. This section reviews earlier topics and shows how they apply to the development and application of calculus.

- The text encourages students to improve **communication skills and research skills**.

  The Group Activities/Extended Applications require written reports and frequently require use of the internet or library. Some Extended Applications call for students to use literature or the internet to find a graph or table of discrete data describing an issue. They are then required to make a scatter plot of the data, determine the function type that is the best fit for the data, create the model, discuss how well the model fits the data, and discuss how it can be used to analyze the issue.

- **Answers to Selected Exercises** include answers to all Chapter Skills Checks and Chapter Reviews, so students have feedback regarding the exercises they work.

- **Supplements** are provided that will help students and instructors use technology to improve the learning and teaching experience. See the supplements list.

Acknowledgments

Many individuals contributed to the development of this textbook. We would like to thank the following reviewers, whose comments and suggestions were invaluable in preparing this text.

Jay Abramson, *Arizona State University*
Khadija Ahmed, *Monroe County Community College*
* Janet Arnold, *Indiana University—Southeast Campus*
Jamie Ashby, *Texarkana College*
Sohrab Bakhtyari, *St. Petersburg College*
Jean Bevis, *Georgia State University*
Thomas Bird, *Austin Community College*
Len Brin, *Southern Connecticut State University*
Marc Campbell, *Daytona Beach Community College*
Florence Chambers, *Southern Maine Community College*
Floyd Downs, *Arizona State University*
Aniekan Ebiefung, *University of Tennessee at Chattanooga*
Marjorie Fernandez Karwowski, *Valencia Community College*
Toni W. Fountain, *Chattanooga State Technical Community College*
John Gosselin, *University of Georgia*
David J. Graser, *Yavapai College*
Lee Graubner, *Valencia Community College East*
Linda Green, *Santa Fe Community College*
Richard Brent Griffin, *Georgia Highlands College*
Lee Hanna, *Clemson University*
Deborah Hanus, *Brookhaven College*
Steve Heath, *Southern Utah University*
Todd A. Hendricks, *Georgia Perimeter College*
Suzanne Hill, *New Mexico State University*
Sue Hitchcock, *Palm Beach Community College*

Sandee House, *Georgia Perimeter College*
Mary Hudacheck-Buswell, *Clayton State University*
* David Jabon, *DePaul University*
Arlene Kleinstein, *State University of New York—Farmingdale*
Danny Lau, *Kennesaw State University*
Ann H. Lawrence, *Wake Technical Community College*
Kit Lumley, *Columbus State University*
Antonio Magliaro, *Southern Connecticut State University*
Beverly K. Michael, *University of Pittsburgh*
* Phillip Miller, *Indiana University—Southeast Campus*
Nancy R. Moseley, *University of South Carolina Aiken*
Demetria Neal, *Gwinnett Technical College*
Malissa Peery, *University of Tennessee*
Ingrid Peterson, *University of Kansas*
* Beverly Reed, *Kent State University*
Jeri Rogers, *Seminole Community College—Oviedo*
Michael Rosenthal, *Florida International University*
Sharon Sanders, *Georgia Perimeter College*
Carolyn Spillman, *Georgia Perimeter College*
* Susan Staats, *University of Minnesota*
Jacqueline Underwood, *Chandler-Gilbert Community College*
* Erwin Walker, *Clemson University*
Denise Widup, *University of Wisconsin—Parkside*
Sandi Wilbur, *University of Tennessee—Knoxville*
* Jeffrey Winslow, *Yavapai College*

Many thanks to Helen Medley for checking the accuracy of this text. Special thanks go to the Pearson team for their assistance, encouragement, and direction throughout this project: Greg Tobin, Anne Kelly, Katie O'Connor, Elizabeth Bernardi, Tracy Patruno, Barbara Atkinson, Peggy Lucas, Justine Goulart, and Tracy Menoza.

Ronald J. Harshbarger
Lisa S. Yocco

* Denotes reviewers of the fourth edition.

List of Supplements

Student Supplements

Student's Solutions Manual

- By Lee Graubner, *Valencia Community College*.
- Provides selected solutions to the Skills Check problems and Exercises, as well as solutions to all Chapter Skills Checks, Review problems, Algebra Toolbox problems, and Extended Applications.
- ISBN-13: 978-0-321-78355-4; ISBN-10: 0-321-78355-7

A Review of Algebra

- By Heidi Howard, *Florida Community College at Jacksonville*.
- Provides additional support for those students needing further algebra review.
- ISBN-13: 978-0-201-77347-7; ISBN-10: 0-201-77347-3

Instructor Supplements

Annotated Instructor's Edition

- Provides answers to many text exercises right after the exercise and answers to all the exercises in the back of the book.
- Notes with an underline sample homework exercises, selected by the authors and supported by MathXL.
- ISBN-13: 978-0-321-78357-8; ISBN-10: 0-321-78357-3

Instructor's Solutions Manual

- By Lee Graubner, *Valencia Community College*.
- Provides complete solutions to all Algebra Toolbox problems, Skills Check problems, Exercises, Chapter Skills Checks, Review problems, and Extended Applications.
- ISBN-13: 978-0-321-78356-1; ISBN-10: 0-321-78356-5

Instructor's Testing Manual

- By Melanie Fulton.
- Contains three alternative forms of tests per chapter.
- Includes answer keys with more applications.
- Available for download from Pearson Education's online catalog.

TestGen®

- Enables instructors to build, edit, print, and administer tests.
- Features a computerized bank of questions developed to cover all text objectives.
- Algorithmically based, allowing instructors to create multiple but equivalent versions of the same question or test with the click of a button.
- Available for download from the Instructor Resource Center at pearsonhighered.com/irc.

Technology Resources

MathXL® Online Course (access code required)

MathXL is the homework and assessment engine that runs MyMathLab. (MyMathLab is MathXL plus a learning management system.) With MathXL, instructors can

- Create, edit, and assign online homework and tests using algorithmically generated exercises correlated at the objective level to the textbook.

- Create and assign their own online exercises and import TestGen tests for added flexibility.

- Maintain records of all student work tracked in MathXL's online gradebook.

With MathXL, students can

- Take chapter tests in MathXL and receive personalized study plans and/or personalized homework assignments based on their test results.

- Use the study plan and/or the homework to link directly to tutorial exercises for the objectives they need to study.

- Access supplemental animations and video clips directly from selected exercises.

MathXL is available to qualified adopters. For more information, visit our website at www.mathxl.com or contact your Pearson representative.

MyMathLab® Online Course (access code required)

MyMathLab delivers **proven results** in helping individual students succeed. It provides **engaging experiences** that personalize, stimulate, and measure learning for each student. And it comes from a **trusted partner** with educational expertise and an eye on the future.

- **Narrated Example Videos with subtitles** have been updated to reflect new content and current real data. Available to download from within MyMathLab®, the videos are correlated to the examples in the Graphing Calculator and Excel® Manual and walk students through algebraic solutions to pivotal examples and provide technological solutions where applicable.

- **NEW! Introductory Videos** contain a brief overview of the topics covered in each section and highlight objectives and key concepts. These videos give context to the Narrated Example Videos and are available in MyMathLab®.

- **NEW! Interactive Figures** enable you to manipulate figures to bring math concepts to life. They are assignable in MyMathLab and available in the Multimedia Library.

- **NEW! Animations** have been added to the Algebra Toolbox section to help students master the prerequisite skills needed to be successful in the chapter. Instructors can assign these multimedia learning aids as homework to help their students grasp the concepts.

To learn more about how MyMathLab combines proven learning applications with powerful assessment, visit www.mymathlab.com or contact your Pearson representative.

MyMathLab® Ready to Go Course (access code required)

These new Ready to Go courses provide students with all the same great MyMathLab features that you're used to, but make it easier for instructors to get started. Each course includes preassigned homeworks and quizzes to make creating your course even simpler. Ask your Pearson representative about the details for this particular course or to see a copy of this course.

College Algebra in Context was written to help you develop the math skills you need to model problems and analyze data—tasks that are required in many jobs in the fields of management, life science, and social science. As you read this text, you may be surprised by how many ways professionals use algebra—from predicting how a population will vote in an upcoming election to projecting sales of a particular good.

There are many ways in which this text will help you succeed in your algebra course, and you can be a partner in that success. Consider the following suggestions, which our own students have found helpful.

1. **Take careful notes in an organized notebook.** Good organization is essential in a math course so that you do not fall behind and so that you can quickly refer back to a topic when you need it.

   ■ **Separate your notebook into three sections:** class notes and examples, homework, and a problem log consisting of problems worked in class (which will provide a sample test).

   ■ **Begin each set of notes with identifying information:** the date, section of the book, page number from the book, and topic.

   ■ **Write explanations in words**, rather than just the steps to a problem, so that you will understand later what was done in each step. **Use abbreviations and short phrases**, rather than complete sentences, so that you can keep up with the explanation as you write.

   ■ **Write step-by-step instructions** for each process.

   ■ **Keep handouts and tests in your notebook,** using either a spiral notebook with pockets or a loose-leaf notebook.

2. **Read the textbook.** Reading a mathematics textbook is different from reading other textbooks. Some suggestions follow.

   ■ **Skim the material** to get a general idea of the major topics. As you skim the material, circle any words that you do not understand. Read the Chapter Summary and look at the Exercises at the end of each section.

   ■ **Make note cards** for terms, symbols, and formulas. Review these note cards often to retain the information.

   ■ **Read for explanation and study the steps to work a problem.** It is essential that you learn *how* to work a problem and *why* the process works rather than memorizing sample problems.

   ■ **Study any illustrations and other aids,** provided to help you understand a sample problem; then cover up the solution and try to work the problem on your own.

   ■ **Practice the process.** The more problems you do, the more confident you will become in your ability to do math and perform on tests. When doing your homework, don't give in to frustration. Put your homework aside for a while and come back to it later.

   ■ **Recite and review.** You should know and understand the example problems in your text well enough to be able to work similar problems on the test. Make note cards with example problems on one side and solutions on the other side.

3. **Work through the Algebra Toolbox.** The Algebra Toolbox will give you a great review of the skills needed for success in each chapter. Note the Key Objectives listed at the beginning of the Toolbox; reread them once you have completed the Toolbox Exercises.

4. **Practice with Skills Check exercises.** Skills Check exercises provide a way to practice your algebra skills before moving on to more applied problems.

5. **Work the Exercises carefully.** The examples and exercises in this book model ways in which mathematics is used in the world. Look for connections between the examples and what you are learning in your other classes. The examples will help you work through the applied exercises in each section.

6. **Prepare for your exams.**

   ■ **Make a study schedule.** Begin to study at least three days before the test. You should make a schedule, listing those sections of the book that you will study each day. Schedule a sample test to be taken upon completion of those sections. This sample test should be taken at least two days before the test date so that you have time to work on areas of difficulty.

   ■ **Rework problems.** You should actively prepare for a test. Do *more* than read your notes and the textbook. Do *more* than look over your homework. Review the note cards prepared from your class notes and text. Actually *rework* problems from each section of your book. Use the Chapter Summary at the end of each chapter to be sure you know and understand the key concepts and formulas. Then get some more practice with the Chapter Skills Checks and Review Exercises. Check your answers!

   ■ **Get help.** Do not leave questions unanswered. Remember to utilize all resources in getting the help you need. Some resources that you might consider using are your fellow classmates, tutors, MyMathLab, the *Student's Solutions Manual*, math videos, and even your teacher. Do not take the gamble that certain questions will not be on the test!

   ■ **Make a sample test.** Write a sample test by choosing a variety of problems from each section in the book. Then write the problems for your sample test in a different order than they appear in the book. (*Hint*: If you write each problem, with directions, on a separate index card and mix them up, you will have a good sample test.)

   ■ **Review and relax the night before the test.** The night before the test is best used *reviewing* the material. This may include working one problem from each section, reworking problems that have given you difficulty, or thinking about procedures you have used.

   ■ **Practice taking tests online.** Ask your professor if MathXL or MyMathLab is available at your school. Both provide online homework, tutorial, and assessment systems for unlimited practice exercises correlated to your textbook. (An access code is required to use these products.)

7. **Develop better math test-taking skills.**

   ■ **Do a memory download.** As soon as you receive your test, jot down formulas or rules that you will need but are likely to forget. If you get nervous later and forget this information, you merely have to refer to the memory cues that you have written down.

   ■ **Skip the difficult questions.** Come back to these later or try to work at least one step for partial credit.

   ■ **Keep a schedule.** The objective is to get the most points. Don't linger over one question very long.

   ■ **Review your work.** Check for careless errors and make sure your answers make sense.

   ■ **Use all the time given.** There are no bonus points for turning your test in early. Use extra time for checking your work.

8. **Have fun!** Look for mathematics all around you. Read the newspaper, look at data on government websites, and observe how professionals use mathematics to do their jobs and communicate information to the world.

We have enjoyed teaching this material to our students and watching their understanding grow. We wish you the very best this semester and in your future studies.

Ronald J. Harshbarger
Lisa S. Yocco

Functions, Graphs, and Models; Linear Functions

With digital TV becoming more affordable by the day, the demand for high-definition home entertainment is growing rapidly, with 115 million Americans having digital TV in 2010. The worldwide total was 517 million in 2010, with 40% of TV households having digital TV. More than 401 million digital TV homes are expected to be added between the end of 2010 and the end of 2015, and by 2015 more than 1 billion sets should be in use worldwide. Cell phone use is also on the rise, not just in the United States but throughout the world. By the middle of 2006, the number of subscribers to cell phone carriers had dramatically increased, and the number of total users had reached 4.6 billion by 2010. If the numbers continue to increase at a steady rate, the number of subscribers is expected to reach into the tens of billions over the next few years. These projections and others are made by collecting real-world data and creating mathematical models. The goal of this chapter and future chapters is to use real data and mathematical models to make predictions and solve meaningful problems.

section	objectives	applications
1.1 Functions and Models	Determine graphs, tables, and equations that represent functions; find domains and ranges; evaluate functions and mathematical models	Body temperature, personal computers, stock market, men in the workforce, public health expenditures
1.2 Graphs of Functions	Graph and evaluate functions with technology; graph mathematical models; align data; graph data points; scale data	Personal savings, cost-benefit, voting, U.S. executions, high school enrollment
1.3 Linear Functions	Identify and graph linear functions; find and interpret intercepts and slopes; find constant rates of change; model revenue, cost, and profit; find marginal revenue, marginal cost, and marginal profit; identify special linear functions	Hispanics in the United States, loan balances, revenue, cost, profit, marginal cost, marginal revenue, marginal profit
1.4 Equations of Lines	Write equations of lines; identify parallel and perpendicular lines; find average rates of change; model approximately linear data	Service call charges, blood alcohol percent, inmate population, hybrid vehicle sales, high school enrollment

Algebra TOOLBOX

KEY OBJECTIVES

- Write sets of numbers using description or elements
- Identify sets of real numbers as being integers, rational numbers, and/or irrational numbers
- Identify the coefficients of terms and constants in algebraic expressions
- Remove parentheses and simplify polynomials
- Express inequalities as intervals and graph inequalities
- Plot points on a coordinate system
- Use subscripts to represent fixed points

The Algebra Toolbox is designed to review prerequisite skills needed for success in each chapter. In this Toolbox, we discuss sets, the real numbers, the coordinate system, algebraic expressions, equations, inequalities, absolute values, and subscripts.

Sets

In this chapter we will use sets to write domains and ranges of functions, and in future chapters we will find solution sets to equations and inequalities. A **set** is a well-defined collection of objects, including but not limited to numbers. In this section, we will discuss sets of real numbers, including natural numbers, integers, and rational numbers, and later in the text we will discuss the set of complex numbers. There are two ways to define a set. One way is by listing the **elements** (or **members**) of the set (usually between braces). For example, we may say that a set A contains 2, 3, 5, and 7 by writing $A = \{2, 3, 5, 7\}$. To say that 5 is an element of the set A, we write $5 \in A$. To indicate that 6 is not an element of the set, we write $6 \notin A$. Domains of functions and solutions to equations are sometimes given in sets with the elements listed.

If all the elements of the set can be listed, the set is said to be a **finite set**. If all elements of a set cannot be listed, the set is called an **infinite set**. To indicate that a set continues with the established pattern, we use three dots. For example, $B = \{1, 2, 3, 4, 5, \ldots, 100\}$ describes the finite set of whole numbers from 1 through 100, and $N = \{1, 2, 3, 4, 5, \ldots\}$ describes the infinite set of all whole numbers beginning with 1. This set is called the **natural numbers**.

Another way to define a set is to give its description. For example, we may write $\{x \mid x \text{ is a math book}\}$ to define the set of math books. This is read as "the set of all x such that x is a math book." $N = \{x \mid x \text{ is a natural number}\}$ defines the set of natural numbers, which was also defined by $N = \{1, 2, 3, 4, 5, \ldots\}$ above.

The set that contains no elements is called the **empty set** and is denoted by $\varnothing$.

EXAMPLE 1 ▶

Write the following sets in two ways.

a. The set A containing the natural numbers less than 7.

b. The set B of natural numbers that are at least 7.

SOLUTION

a. $A = \{1, 2, 3, 4, 5, 6\}, A = \{x \mid x \in N, x < 7\}$

b. $B = \{7, 8, 9, 10, \ldots\}, B = \{x \mid x \in N, x \geq 7\}$

The relations that can exist between two sets follow.

Relations Between Sets

1. Sets X and Y are **equal** if they contain exactly the same elements.
2. Set A is called a **subset** of set B if each element of A is an element of B. This is denoted $A \subseteq B$.
3. If sets C and D have no elements in common, they are called **disjoint**.

EXAMPLE 2 ▶   For the sets $A = \{x \mid x \leq 9, x \text{ is a natural number}\}$, $B = \{2, 4, 6\}$, $C = \{3, 5, 8, 10\}$:

a. Which of the sets A, B, and C are subsets of A?

b. Which pairs of sets are disjoint?

c. Are any of these three sets equal?

SOLUTION

a. Every element of B is contained in A. Thus, set B is a subset of A. Because every element of A is contained in A, A is a subset of A.

b. Sets B and C have no elements in common, so they are disjoint.

c. None of these sets have exactly the same elements, so none are equal.

The Real Numbers

Because most of the mathematical applications you will encounter in an applied nontechnical setting use real numbers, the emphasis in this text is the **real number system**.* Real numbers can be rational or irrational. **Rational numbers** include integers, fractions containing only integers (with no 0 in a denominator), and decimals that either terminate or repeat. Some examples of rational numbers are

$$-9, \quad \frac{1}{2}, \quad 0, \quad 12, \quad -\frac{4}{7}, \quad 6.58, \quad -7.\overline{3}$$

Irrational numbers are real numbers that are not rational. Some examples of irrational numbers are π (a number familiar to us from the study of circles), $\sqrt{2}$, $\sqrt[3]{5}$, and $\sqrt[3]{-10}$.

The types of real numbers are described in Table 1.1.

Table 1.1

Types of Real Numbers	Descriptions
Natural numbers	1, 2, 3, 4, ...
Integers	Natural numbers, zero, and the negatives of the natural numbers: ..., $-3, -2, -1, 0, 1, 2, 3$, ...
Rational numbers	All numbers that can be written in the form $\frac{p}{q}$, where p and q are both integers with $q \neq 0$. Rational numbers can be written as terminating or repeating decimals.
Irrational numbers	All real numbers that are not rational numbers. Irrational numbers cannot be written as terminating or repeating decimals.

We can represent real numbers on a **real number line**. Exactly one real number is associated with each point on the line, and we say there is a one-to-one correspondence between the real numbers and the points on the line. That is, the real number line is a graph of the real numbers (see Figure 1.1).

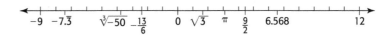

Figure 1.1

---

* The complex number system will be discussed in the Chapter 3 Toolbox.

Notice the number π on the real number line in Figure 1.1. This special number, which can be approximated by 3.14, results when the circumference of (distance around) any circle is divided by the diameter of the circle. Another special real number is e; it is denoted by

$$e \approx 2.71828$$

We will discuss this number, which is important in financial and biological applications, later in the text.

Inequalities and Intervals on the Number Line

In this chapter, we will sometimes use inequalities and interval notation to describe domains and ranges of functions. An **inequality** is a statement that one quantity is greater (or less) than another quantity. We say that a is less than b (written $a < b$) if the point representing a is to the left of the point representing b on the real number line. We may indicate that the number a is greater than or equal to b by writing $a \geq b$. The subset of real numbers x that lie between a and b (excluding a and b) can be denoted by the **double inequality** $a < x < b$ or by the **open interval** (a, b). This is called an open interval because neither of the endpoints is included in the interval. The **closed interval** $[a, b]$ represents the set of all real numbers satisfying $a \leq x \leq b$. Intervals containing one endpoint, such as $[a, b)$ or $(a, b]$, are called **half-open intervals**. We can represent the inequality $x \geq a$ by the interval $[a, \infty)$, and we can represent the inequality $x < a$ by the interval $(-\infty, a)$. Note that ∞ and $-\infty$ are not numbers, but ∞ is used in $[a, \infty)$ to represent the fact that x increases without bound and $-\infty$ is used in $(-\infty, a)$ to indicate that x decreases without bound. Table 1.2 shows the graphs of different types of intervals.

Table 1.2

Interval Notation	Inequality Notation	Verbal Description	Number Line Graph
(a, ∞)	$x > a$	x is greater than a	
$[a, \infty)$	$x \geq a$	x is greater than or equal to a	
$(-\infty, b)$	$x < b$	x is less than b	
$(-\infty, b]$	$x \leq b$	x is less than or equal to b	
(a, b)	$a < x < b$	x is between a and b, not including either a or b	
$[a, b)$	$a \leq x < b$	x is between a and b, including a but not including b	
$(a, b]$	$a < x \leq b$	x is between a and b, not including a but including b	
$[a, b]$	$a \leq x \leq b$	x is between a and b, including both a and b	

Note that open circles may be used instead of parentheses and solid circles may be used instead of brackets in the number line graphs.

EXAMPLE 3 ▶ Intervals

Write the interval corresponding to each of the inequalities in parts (a)–(e), and then graph the inequality.

a. $-1 \le x \le 2$ **b.** $2 < x < 4$ **c.** $-2 < x \le 3$ **d.** $x \ge 3$ **e.** $x < 5$

SOLUTION

a. $[-1, 2]$

b. $(2, 4)$

c. $(-2, 3]$

d. $[3, \infty)$

e. $(-\infty, 5)$

Algebraic Expressions

In algebra we deal with a combination of real numbers and letters. Generally, the letters are symbols used to represent unknown quantities or fixed but unspecified constants. Letters representing unknown quantities are usually called **variables**, and letters representing fixed but unspecified numbers are called **literal constants**. An expression created by performing additions, subtractions, or other arithmetic operations with one or more real numbers and variables is called an **algebraic expression**. Unless otherwise specified, the variables represent real numbers for which the algebraic expression is a real number. Examples of algebraic expressions include

$$5x - 2y, \quad \frac{3x - 5}{12 + 5y}, \quad \text{and} \quad 7z + 2$$

A **term** of an algebraic expression is the product of one or more variables and a real number; the real number is called a **numerical coefficient** or simply a **coefficient**. A constant is also considered a term of an algebraic expression and is called a **constant term**. For instance, the term $5yz$ is the product of the factors 5, y, and z; this term has coefficient 5.

Polynomials

An algebraic expression containing a finite number of additions, subtractions, and multiplications of constants and nonnegative integer powers of variables is called a **polynomial**. When simplified, a polynomial cannot contain negative powers of variables, fractional powers of variables, variables in a denominator, or variables inside a radical.

The expressions $5x - 2y$ and $7z^3 + 2y$ are polynomials, but $\frac{3x - 5}{12 + 5y}$ and $3x^2 - 6\sqrt{x}$ are not polynomials. If the only variable in the polynomial is x, then the polynomial is called a **polynomial in x**. The general form of a polynomial in x is

$$a_n x^n + a_{n-1} x^{n-1} + \cdots + a_1 x + a_0$$

where a_0 and each coefficient $a_n, a_{n-1}, \ldots$ are real numbers and each exponent $n, n - 1, \ldots$ is a positive integer.

For a polynomial in the single variable x, the power of x in each term is the **degree** of that term, with the degree of a constant term equal to 0. The term that has the highest power of x is called the **leading term** of the polynomial, the coefficient of this term is the **leading coefficient**, and the degree of this term is the **degree of the polynomial**. Thus, $5x^4 + 3x^2 - 6$ is a fourth-degree polynomial with leading coefficient 5. Polynomials with one term are called **monomials**, those with two terms are called **binomials**, and those with three terms are called **trinomials**. The right side of the equation $y = 4x + 3$ is a first-degree binomial, and the right side of $y = 6x^2 - 5x + 2$ is a second-degree trinomial.

EXAMPLE 4 ▶   For each polynomial, state the constant term, the leading coefficient, and the degree of the polynomial.

a. $5x^2 - 8x + 2x^4 - 3$ **b.** $5x^2 - 6x^3 + 3x^6 + 7$

SOLUTION

a. The constant term is -3; the term of highest degree is $2x^4$, so the leading coefficient is 2 and the degree of the polynomial is 4.

b. The constant term is 7; the term of highest degree is $3x^6$, so the leading coefficient is 3 and the degree of the polynomial is 6.

Terms that contain exactly the same variables with exactly the same exponents are called **like terms**. For example, $3x^2y$ and $7x^2y$ are like terms, but $3x^2y$ and $3xy$ are not. We can *simplify* an expression by adding or subtracting the coefficients of the like terms. For example, the simplified form of

$$3x + 4y - 8x + 2y \quad \text{is} \quad -5x + 6y$$

and the simplified form of

$$3x^2y + 7xy^2 + 6x^2y - 4xy^2 - 5xy \quad \text{is} \quad 9x^2y + 3xy^2 - 5xy$$

Removing Parentheses

We often need to remove parentheses when simplifying algebraic expressions and when solving equations. Removing parentheses frequently requires use of the **distributive property**, which says that for real numbers a, b, and c, $a(b + c) = ab + ac$. Care must be taken to avoid mistakes with signs when using the distributive property. Multiplying a sum in parentheses by a negative number changes the sign of each term in the parentheses. For example,

$$-3(x - 2y) = -3(x) + (-3)(-2y) = -3x + 6y$$

and

$$-(3xy - 5x^3) = -3xy + 5x^3$$

We add or subtract (**combine**) algebraic expressions by combining the like terms. For example, the sum of the expressions $5sx - 2y + 7z^3$ and $2y + 5sx - 4z^3$ is

$$(5sx - 2y + 7z^3) + (2y + 5sx - 4z^3) = 5sx - 2y + 7z^3 + 2y + 5sx - 4z^3$$
$$= 10sx + 3z^3$$

and the difference of these two expressions is

$$(5sx - 2y + 7z^3) - (2y + 5sx - 4z^3) = 5sx - 2y + 7z^3 - 2y - 5sx + 4z^3$$
$$= -4y + 11z^3$$

The Coordinate System

Much of our work in algebra involves graphing. To graph in two dimensions, we use a rectangular coordinate system, or **Cartesian coordinate system**. Such a system allows us to assign a unique point in a plane to each ordered pair of real numbers. We construct the coordinate system by drawing a horizontal number line and a vertical number line so that they intersect at their origins (Figure 1.2). The point of intersection is called the **origin** of the system, the number lines are called the coordinate **axes**, and the plane is divided into four parts called **quadrants**. In Figure 1.3, we call the horizontal axis the *x*-axis and the vertical axis the *y*-axis, and we denote any point in the plane as the ordered pair (x, y).

The ordered pair (a, b) represents the point P that is $|a|$ units from the *y*-axis (right if a is positive, left if a is negative) and $|b|$ units from the *x*-axis (up if b is positive, down if b is negative). The values of a and b are called the **rectangular coordinates** of the point. Figure 1.3 shows point P with coordinates (a, b). The point is in the second quadrant, where $a < 0$ and $b > 0$.

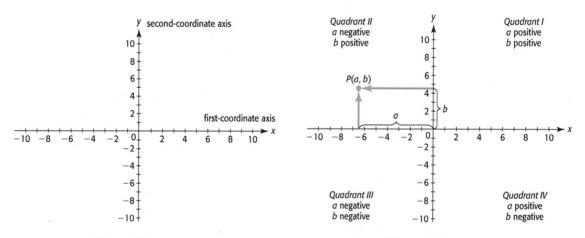

Figure 1.2   Figure 1.3

Subscripts

We sometimes need to distinguish between two *y*-values and/or *x*-values in the same problem or on the same graph, or to designate literal constants. It is often convenient to do this by using **subscripts**. For example, if we have two fixed but unidentified points on a graph, we can represent one point as (x_1, y_1) and the other as (x_2, y_2). Subscripts also can be used to designate different equations entered in graphing utilities; for example, $y = 2x - 5$ may appear as $y_1 = 2x - 5$ when entered in the equation editor of a graphing calculator.

Toolbox EXERCISES

1. Write "the set of all natural numbers N less than 9" in two different ways.

2. Is it true that $3 \in \{1, 3, 4, 6, 8, 9, 10\}$?

3. Is A a subset of B if $A = \{2, 3, 5, 7, 8, 9, 10\}$ and $B = \{3, 5, 8, 9\}$?

4. Is it true that $\frac{1}{2} \in N$ if N is the set of natural numbers?

5. Is the set of integers a subset of the set of rational numbers?

6. Are sets of rational numbers and irrational numbers disjoint sets?

Identify the sets of numbers in Exercises 7–9 as containing one or more of the following: integers, rational numbers, and/or irrational numbers.

7. $\{5, 2, 5, 8, -6\}$ 8. $\left\{\frac{1}{2}, -4.1, \frac{5}{3}, 1\frac{2}{3}\right\}$

9. $\left\{\sqrt{3}, \pi, \frac{\sqrt[3]{2}}{4}, \sqrt{5}\right\}$

In Exercises 10–12, express each interval or graph as an inequality.

10. x
    -3

11. $[-3, 3]$ 12. $(-\infty, 3]$

In Exercises 13–15, express each inequality or graph in interval notation.

13. $x \leq 7$ 14. $3 < x \leq 7$

15. x
    4

In Exercises 16–18, graph the inequality or interval on a real number line.

16. $(-2, \infty)$ 17. $5 > x \geq 2$

18. $x < 3$

In Exercises 19–21, plot the points on a coordinate system.

19. $(-1, 3)$ 20. $(4, -2)$

21. $(-4, 3)$

22. Plot the points $(-1, 2)$, $(3, -1)$, $(4, 2)$, and $(-2, -3)$ on the same coordinate system.

23. Plot the points (x_1, y_1) and (x_2, y_2) on a coordinate system if

$$x_1 = 2, y_1 = -1, x_2 = -3, y_2 = -5$$

Determine if each expression in Exercises 24–27 is a polynomial. If it is, state the degree of the polynomial.

24. $14x^4 - 6x^3 + 9x - 7$

25. $\dfrac{5x - 8}{3x + 2}$

26. $10x - \sqrt{y}$

27. $-12x^4 + 5x^6$

For each algebraic expression in Exercises 28 and 29, give the coefficient of each term and give the constant term.

28. $-3x^2 - 4x + 8$ 29. $5x^4 + 7x^3 - 3$

30. Find the sum of $z^4 - 15z^2 + 20z - 6$ and $2z^4 + 4z^3 - 12z^2 - 5$.

31. Simplify the expression

$$3x + 2y^4 - 2x^3y^4 - 119 - 5x - 3y^2 + 5y^4 + 110$$

Remove the parentheses and simplify in Exercises 32–37.

32. $4(p + d)$

33. $-2(3x - 7y)$

34. $-a(b + 8c)$

35. $4(x - y) - (3x + 2y)$

36. $4(2x - y) + 4xy - 5(y - xy) - (2x - 4y)$

37. $2x(4yz - 4) - (5xyz - 3x)$

1.1 | Functions and Models

KEY OBJECTIVES

- Determine if a table, graph, or equation defines a function
- Find the domains and ranges of functions
- Create a scatter plot of a set of ordered pairs
- Use function notation to evaluate functions
- Apply real-world information using a mathematical model

SECTION PREVIEW Body Temperatures

One indication of illness in children is elevated body temperature. There are two common measures of temperature, Fahrenheit (°F) and Celsius (°C), with temperature measured in Fahrenheit degrees in the United States and in Celsius degrees in many other countries of the world. Suppose we know that a child's normal body temperature is 98.6°F and that his or her temperature is now 37°C. Does this indicate that the child is ill? To help decide this, we could find the Fahrenheit temperature that corresponds to 37°C. We could do this easily if we knew the relationship between Fahrenheit and Celsius temperature scales. In this section, we will see that the relationship between these measurements can be defined by a **function** and that functions can be defined numerically, graphically, verbally, or by an equation. We also explore how this and other functions can be applied to help solve problems that occur in real-world situations. ∎

Function Definitions

There are several techniques to show how Fahrenheit degree measurements are related to Celsius degree measurements.

One way to show the relationship between Celsius and Fahrenheit degree measurements is by listing some Celsius measurements and the corresponding Fahrenheit measurements. These measurements, and any other real-world information collected in numerical form, are called **data**. These temperature measurements can be shown in a table (Table 1.3).

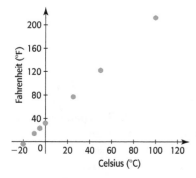

Figure 1.4

Table 1.3

Celsius Degrees (°C)	−20	−10	−5	0	25	50	100
Fahrenheit Degrees (°F)	−4	14	23	32	77	122	212

This relationship is also defined by the set of ordered pairs

$$\{(-20, -4), (-10, 14), (-5, 23), (0, 32), (25, 77), (50, 122), (100, 212)\}$$

We can picture the relationship between the measurements with a graph. Figure 1.4 shows a **scatter plot** of the data—that is, a graph of the ordered pairs as points. Table 1.3, the set of ordered pairs below the table, and the graph in Figure 1.4 define a **function** with a set of Celsius temperature **inputs** (called the **domain** of the function) and a set of corresponding Fahrenheit **outputs** (called the **range** of the function). A function that will give the Fahrenheit temperature measurement F that corresponds to *any* Celsius temperature measurement C between −20°C and 100°C is the equation

$$F = \frac{9}{5}C + 32$$

This equation defines F as a function of C because each input C results in exactly one output F. Its graph, shown in Figure 1.5, is a line that contains the points on the scatter plot in Figure 1.4 as well as other points. If we consider only Celsius temperatures from −20 to 100, then the domain of this function defined by the equation above is $-20 \leq C \leq 100$ and the resulting range is $-4 \leq F \leq 212$.

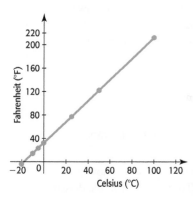

Figure 1.5

Function

A function is a rule or correspondence that assigns to each element of one set (called the domain) exactly one element of a second set (called the range).

The function may be defined by a set of ordered pairs, a table, a graph, an equation, or a verbal description.

EXAMPLE 1 ▶ **Body Temperature**

Suppose a child's normal temperature is 98.6°F and the only thermometer available, which is Celsius, indicates that the child's temperature is 37°C. Does this reading indicate that the child's temperature is normal?

SOLUTION

We now have a function that will give the output temperature F that corresponds to the input temperature C. Substituting 37 for C in $F = \dfrac{9}{5}C + 32$ gives $F = \dfrac{9}{5}(37) + 32 = 98.6$. This indicates that the child's temperature is normal.

Domains and Ranges

How a function is defined determines its domain and range. For instance, the domain of the function defined by Table 1.3 or by the scatter plot in Figure 1.4 is the finite set $\{-20, -10, -5, 0, 25, 50, 100\}$ with all values measured in degrees Celsius, and the range is the set $\{-4, 14, 23, 32, 77, 122, 212\}$ with all values measured in degrees Fahrenheit. This function has a finite number of inputs in its domain.

The function defined by $F = \dfrac{9}{5}C + 32$ and graphed in Figure 1.5 had the inputs (domain) and outputs (range) restricted to $-20 \le C \le 100$ and $-4 \le F \le 212$, respectively. Functions defined by equations can also be restricted by the context in which they are used. For example, if the function $F = \dfrac{9}{5}C + 32$ is used in measuring the temperature of water, its domain is limited to real numbers from 0 to 100 and its range is limited to real numbers from 32 to 212, because water changes state with other temperatures.

If x represents any element in the domain, then x is called the **independent variable**, and if y represents an output of the function from an input x, then y is called the **dependent variable**. The figure at left shows a general "function machine" in which the input is called x, the rule is denoted by f, and the output is symbolized by $f(x)$. The symbol $f(x)$ is read "f of x."

If the domain of a function is not specified or restricted by the context in which the function is used, it is assumed that the domain consists of all real number inputs that result in real number outputs in the range, and that the range is a subset of the real numbers. Two special cases where the domain of a function may be limited follow.

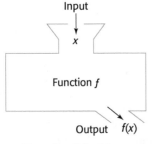

Input
x

Function f

Output $f(x)$

Function Machine

Domains and Ranges

1. Functions with variables in the denominator may have input values that give 0 in the denominator. To find values *not in* the domain:

   Set the denominator equal to 0 and solve for the variable
   (see Example 2(b)).

2. Functions with variables inside an even root may have one or more input values that give negative values inside the even root. To find values *in* the domain:

   Set the expression inside the even root greater than or equal to 0
   and solve for the variable (see Example 2(c)).

We can use the graph of a function to find or to verify its domain and range. We can usually see the interval(s) on which the graph exists and therefore agree on the subset of the real numbers for which the function is defined. This set is the domain of the function. We can also usually determine if the outputs of the graph form the set of all real numbers or some subset of real numbers. This set is the range of the function.

EXAMPLE 2 ▶ Domains and Ranges

For each of the following functions, determine the domain. Determine the range of the function in parts (a) and (c).

a. $y = 4x^2$ **b.** $y = 1 + \dfrac{1}{x - 2}$ **c.** $y = \sqrt{4 - x}$

SOLUTION

a. Because any real number input for x, when squared and multiplied by 4, results in a real number output for y, we conclude that the domain is the set of all real numbers. Because

$$y = 4x^2$$

cannot be negative for any value of x that is input, the range is the set of all non-negative real numbers ($y \geq 0$). This can be confirmed by looking at the graph of this function, shown in Figure 1.6.

b. Because the denominator of the fractional part of this function will be 0 when $x = 2$, and the outputs for every other value of x are real numbers, the domain of this function contains all real numbers except 2. The graph in Figure 1.7 confirms this conclusion.

c. Because

$$y = \sqrt{4 - x}$$

cannot be a real number if $4 - x$ is negative, the only values of x that give real outputs to the function are values that satisfy $4 - x \geq 0$, or $4 \geq x$, so the domain is $x \leq 4$. Because $\sqrt{4 - x}$ (the principal square root) can never be negative, the range is $y \geq 0$. Selected inputs and their outputs are shown in Table 1.4, and the graph of this function is shown in Figure 1.8. They confirm that the domain is $(-\infty, 4]$ and the range is $[0, \infty)$.

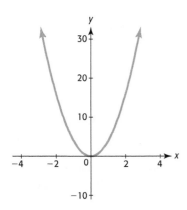

Figure 1.6

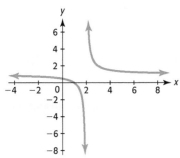

Figure 1.7

Table 1.4

x	y
2	$\sqrt{2}$
3	1
4	0
4.001	Undefined (not real)

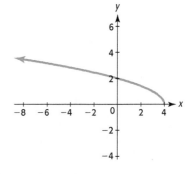

Figure 1.8

Arrow diagrams that show how each individual input results in exactly one output can also represent functions. Each arrow in Figure 1.9(a) and in Figure 1.9(c) goes from an input to exactly one output, so each of these diagrams defines a function. On the other hand, the arrow diagram in Figure 1.9(b) does not define a function because one input, 8, goes to two different outputs, 6 and 9.

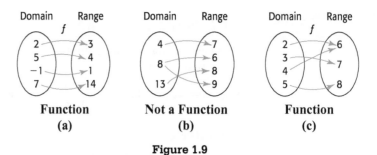

Figure 1.9

Tests for Functions

Functions play an important role in the solution of mathematical problems. Also, to graph a relationship using graphing calculators and computer software, it is usually necessary to express the association between the variables in the form of a function. It is therefore essential for you to recognize when a relationship is a function. Recall that a function is a rule or correspondence that determines exactly one output for each input.

EXAMPLE 3 ▶ **Recognizing Functions**

For each of the following, determine whether or not the indicated relationship represents a function. Explain your reasoning. For each function that is defined, give the domain and range.

a. The amount N of U.S. sales of personal computers, in millions of dollars, determined by the year x, as defined in Table 1.5. Is N a function of x?

Table 1.5

Year, x	U.S. Sales of Personal Computers, N ($ millions)
1997	15,950
2000	16,400
2004	18,233
2008	23,412
2009	21,174
2010	26,060

(Source: Consumer Electronics Association)

b. The daily profit P (in dollars) from the sale of x pounds of candy, as shown in Figure 1.10. Is P a function of x?

c. The number of tons x of coal sold determined by the profit P that is made from the sale of the product, as shown in Table 1.6. Is x a function of P?

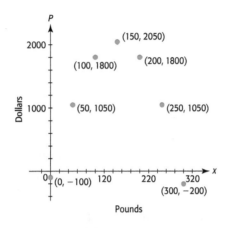

Figure 1.10

Table 1.6

x (tons)	P ($)
0	−100,000
500	109,000
1000	480,000
1500	505,000
2000	480,000
2500	109,000
3000	−100,000

d. W is a person's weight in pounds during the nth week of a diet for $n = 1$ and $n = 2$. Is W a function of n?

SOLUTION

a. For each year (input) listed in Table 1.5, only one value is given for the amount of sales (output), so Table 1.5 represents N as a function of x. The set $\{1997, 2000, 2004, 2008, 2009, 2010\}$ is the domain, and the range is the set $\{15,950, 16,400, 18,233, 23,412, 21,174, 26,060\}$ million dollars.

b. Each input x corresponds to only one daily profit P, so this scatter plot represents P as a function of x. The domain is $\{0, 50, 100, 150, 200, 250, 300\}$ pounds, and the range is $\{-200, -100, 1050, 1800, 2050\}$ dollars. For this function, x is the independent variable and P is the dependent variable.

c. The number of tons x of coal sold is not a function of the profit P that is made, because some values of P result in two values of x. For example, a profit of $480,000 corresponds to both 1000 tons of coal and 2000 tons of coal.

d. A person's weight varies during any week; for example, a woman may weigh 122 lb on Tuesday and 121 lb on Friday. Thus, there is more than one output (weight) for each input (week), and this relationship is not a function.

EXAMPLE 4 ▶   **Functions**

a. Does the equation $y^2 = 3x - 3$ define y as a function of x?

b. Does the equation $y = -x^2 + 4x$ define y as a function of x?

c. Does the graph in Figure 1.11 give the price of Home Depot, Inc., stock as a function of the day for three months in 2007?

Figure 1.11

SOLUTION

a. This indicated relationship between x and y is not a function because there can be more than one output for each input. For instance, the rule $y^2 = 3x - 3$ determines both $y = 3$ and $y = -3$ for the input $x = 4$. Note that if we solve this equation for y, we get $y = \pm\sqrt{3x - 3}$, so two values of y will result for any value of $x > 1$. In general, if y raised to an even power is contained in an equation, y cannot be solved for uniquely and thus y cannot be a function of another variable.

b. Because each value of x results in exactly one value of y, this equation defines y as a function of x.

c. The graph in Figure 1.11 gives the stock price of Home Depot, Inc. for each business day for ten months. The graph shows that the price of a share of stock during each day of these months is not a function. The vertical bar above each day shows that the stock has many prices between its daily high and low. Because of the fluctuation in price during each day of the month, the price of Home Depot stock during these months is not a function of the day.

Vertical Line Test

Another way to determine whether an equation defines a function is to inspect its graph. If y is a function of x, no two distinct points on the graph of $y = f(x)$ can have the same first coordinate. There are two points, $(4, 3)$ and $(4, -3)$, on the graph of $y^2 = 3x - 3$ shown in Figure 1.12, so the equation does not represent y as a function of x (as we concluded in Example 4). In general, no two points on its graph can lie on the same vertical line if the relationship is a function.

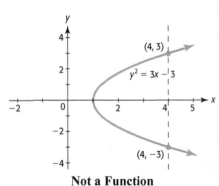

Not a Function

Figure 1.12

Vertical Line Test

A set of points in a coordinate plane is the graph of a function if and only if no vertical line intersects the graph in more than one point.

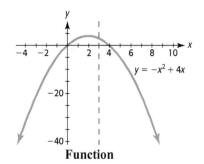

Function

Figure 1.13

When we look at the graph of $y = -x^2 + 4x$ in Figure 1.13, we can see that any vertical line will intersect the graph in at most one point for the portion of the graph that is visible. If we know that the graph will extend indefinitely in the same pattern, we can conclude that no vertical line will intersect the graph in two points and so the equation $y = -x^2 + 4x$ represents y as a function of x.

Function Notation

We can use the function notation $y = f(x)$, read "y equals f of x," to indicate that the variable y is a function of the variable x. For specific values of x, $f(x)$ represents the resulting outputs, or y-values. In particular, the point $(a, f(a))$ lies on the graph of $y = f(x)$ for any number a in the domain of the function. We can also say that $f(a)$ is $f(x)$ evaluated at $x = a$. Thus, if

$$f(x) = 4x^2 - 2x + 3$$

then

$$f(3) = 4(3)^2 - 2(3) + 3 = 33$$
$$f(-1) = 4(-1)^2 - 2(-1) + 3 = 9$$

This means that $(3, 33)$ and $(-1, 9)$ are points on the graph of $f(x) = 4x^2 - 2x + 3$. We can find function values using an equation, values from a table, or points on a graph. For example, because N is a function of x in Table 1.7, we can write $N = f(x)$ and see that $f(1991) = 129.4$ and $f(2008) = 1190.1$.

Table 1.7

Year, x	Worldwide Personal Computers, N (millions)
1991	129.4
1992	150.8
1993	177.4
1994	208.0
1995	245.0
2000	535.6
2005	903.9
2008	1190.1
2009	1299.0

(Source: *The Time Almanac*)

EXAMPLE 5 ▶   **Function Notation**

Figure 1.14 shows the graph of

$$f(x) = 2x^3 + 5x^2 - 28x - 15$$

a. Use the points shown on the graph to find $f(-2)$ and $f(4)$.

b. Use the equation to find $f(-2)$ and $f(4)$.

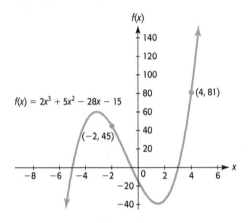

Figure 1.14

SOLUTION

a. By observing Figure 1.14, we see that the point $(-2, 45)$ is on the graph of $f(x) = 2x^3 + 5x^2 - 28x - 15$, so $f(-2) = 45$. We also see that the point $(4, 81)$ is on the graph, so $f(4) = 81$.

b. $f(-2) = 2(-2)^3 + 5(-2)^2 - 28(-2) - 15 = 2(-8) + 5(4) - 28(-2) - 15 = 45$
    $f(4) = 2(4)^3 + 5(4)^2 - 28(4) - 15 = 2(64) + 5(16) - 28(4) - 15 = 81$

Note that the values found by substitution agree with the y-coordinates of the points on the graph.

EXAMPLE 6 ▶   **Men in the Workforce**

The points on the graph in Figure 1.15 give the number of men in the workforce (in millions) as a function g of the year t for selected years from 1890 to 2009.

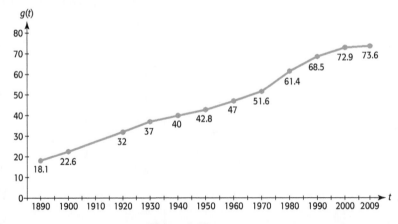

Figure 1.15

(Source: *World Almanac*)

a. Find and interpret $g(1940)$.

b. What is the input t if the output is $g(t) = 51.6$ million men?

c. What can be said about the number of men in the workforce during 1890–2009?

d. What is the maximum number of men in the workforce during the period shown on the graph?

SOLUTION

a. The point above $t = 1940$ has coordinates $(1940, 40)$, so $g(1940) = 40$. This means that there were 40 million men in the workforce in 1940.

b. The point $(1970, 51.6)$ occurs on the graph, so $g(1970) = 51.6$ and $t = 1970$ is the input when the output is $g(t) = 51.6$.

c. The number of men in the workforce increased during the period 1890–2009.

d. The maximum number of men in the workforce during the period was 73.6 million, in 2009.

Mathematical Models

The process of translating real-world information into a mathematical form so that it can be applied and then interpreted in the real-world setting is called **modeling**. As we most often use the term in this text, a **mathematical model** is a functional relationship (usually in the form of an equation) that includes not only the function rule but also descriptions of all involved variables and their units of measure. For example, the function $F = \dfrac{9}{5}C + 32$ was used to describe the relationship between temperature scales at the beginning of the section. The **model** that describes how to convert from one temperature scale to another must include the equation $\left(F = \dfrac{9}{5}C + 32 \right)$ and a description of the variables (F is the temperature in degrees Fahrenheit and C is the temperature in degrees Celsius). A mathematical model can sometimes provide an exact description of a real situation (such as the Celsius/Fahrenheit model), but a model frequently provides only an approximate description of a real-world situation.

Consider Table 1.8 and Figure 1.16(a), which show the number of drinks and resulting blood alcohol percent for a 90-pound woman. (One drink is equal to 1.25 oz of 80-proof liquor, 12 oz of regular beer, or 5 oz of table wine; many states have set 0.08%

Table 1.8

Number of Drinks	0	1	2	3	4	5	6	7	8	9
Blood Alcohol Percent	0	0.05	0.10	0.15	0.20	0.25	0.30	0.35	0.40	0.45

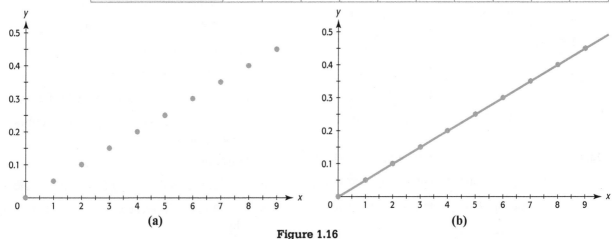

(a) **(b)**

Figure 1.16

as the legal limit for driving under the influence.) As Figure 1.16(b) shows, the graph of the function $f(x) = 0.05x$ lies on every point of the graph in Figure 1.16(a). Thus, every point in Table 1.8 can be found using this function, and it is a model that gives the blood alcohol percent for a 90-pound woman as a function of the number of drinks.

The model above fits the data points exactly, but many models are approximate fits to data points. We will see how to fit functions to data points later in the text.

EXAMPLE 7 ▶ Public Health Care Expenditures

Public health care expenditures for the period 1990–2012 can be modeled (that is, accurately approximated) by the function $E(t) = 738.1(1.065)^t$, where $E(t)$ is in billions of dollars and t is the number of years after 1990.

a. What value of t represents 2010?

b. Approximate the public health care expenditure for 2010.

c. Use the model to estimate the public health care expenditure for 2015. Can we be sure this estimate is accurate?

(Source: U.S. Department of Health and Human Services)

SOLUTION

a. Since t is the number of years after 1990, $t = 20$ represents 2010.

b. Substituting 20 for t in $E(t) = 738.1(1.065)^t$ gives a public health care expenditure of $738.1(1.065)^{20}$, or approximately 2600.8 billion dollars, for 2010.

c. Substituting 25 for t in $E(t) = 738.1(1.065)^t$ gives a public health care expenditure of $738.1(1.065)^{25}$, or 3563.3 billion dollars, for 2015. Note that this is a prediction, and many factors, including the Affordable Care Act of 2010, could make this prediction inaccurate.

Skills CHECK 1.1

Use the tables below in Exercises 1–6.

Table A

x	−9	−5	−7	6	12	17	20
y = f(x)	5	6	7	4	9	9	10

Table B

x	−4	−1	0	1	3	7	12
y = g(x)	5	7	3	15	8	9	10

1. Table A gives y as a function of x, with $y = f(x)$.

   a. Is −5 an input or an output of this function?

   b. Is $f(-5)$ an input or an output of this function?

   c. State the domain and range of this function.

   d. Explain why this relationship describes y as a function of x.

2. Table B gives y as a function of x, with $y = g(x)$.

   a. Is 0 an input or an output of this function?

   b. Is $g(7)$ an input or an output of this function?

   c. State the domain and range of this function.

   d. Explain why this relationship describes y as a function of x.

3. Use Table A to find $y = f(-9)$ and $y = f(17)$.

4. Use Table B to find $y = g(-4)$ and $y = g(3)$.

5. Does Table A describe x as a function of y? Why or why not?

6. Does Table B describe x as a function of y? Why or why not?

7. For each of the functions $y = f(x)$ described below, find $f(2)$.

a.

x	−1	0	1	2	3
f(x)	5	7	2	−1	−8

b. $y = 10 - 3x^2$

c.

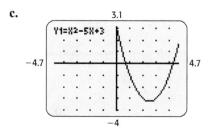

8. For each of the functions $y = f(x)$ described below, find $f(-1)$.

a.

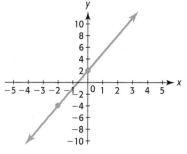

b.

x	f(x)
−3	12
−1	−8
0	5
3	16

c. $f(x) = x^2 + 3x + 8$

In Exercises 9 and 10, refer to the graph of the function $y = f(x)$ to complete the table.

9.

x	y
0	
−2	

10.

x	y
4	
−4	

11. If $R(x) = 5x + 8$, find (a) $R(-3)$, (b) $R(-1)$, and (c) $R(2)$.

12. If $C(s) = 16 - 2s^2$, find (a) $C(3)$, (b) $C(-2)$, and (c) $C(-1)$.

13. Does the table below define y as a function of x? If so, give the domain and range of f. If not, state why not.

x	−1	0	1	2	3
y	5	7	2	−1	−8

14. Does the table below define y as a function of x? If so, give the domain and range of f. If not, state why not.

x	−3	2	0	3	2
y	12	−3	5	16	4

Determine if each graph in Exercises 15–18 indicates that y is a function of x.

15.

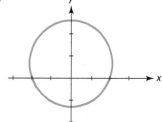

16.

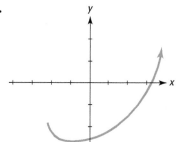

17.

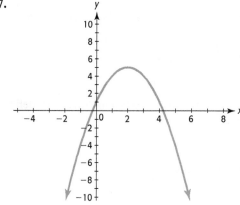

18.

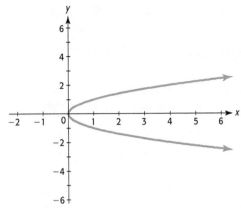

19. Determine if the graph in the figure represents y as a function of x. Explain your reasoning.

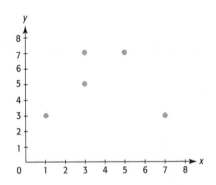

20. Determine if the graph below represents y as a function of x. Explain your reasoning.

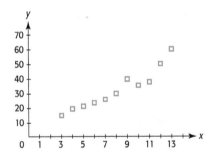

21. Which of the following sets of ordered pairs defines a function?

  a. $\{(1, 6), (4, 12), (4, 8), (3, 3)\}$

  b. $\{(2, 4), (3, -2), (1, -2), (7, 7)\}$

22. Which of the following sets of ordered pairs defines a function?

  a. $\{(1, 3), (-2, 4), (3, 5), (4, 3)\}$

  b. $\{(3, 4), (-2, 5), (4, 6), (3, 6)\}$

23. Which of the following arrow diagrams defines a function?

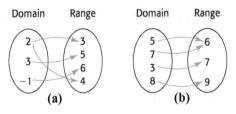

(a)   (b)

24. Which of the following arrow diagrams defines a function?

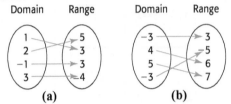

(a)   (b)

In Exercises 25–28, find the domain and range for the function shown in the graph.

25.

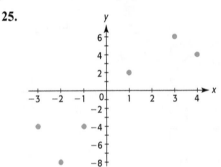

26.

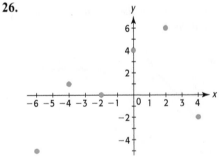

27.

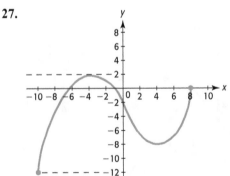

28.

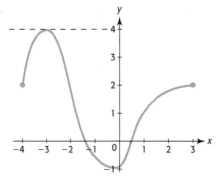

31. $y = 2 - \dfrac{5}{x + 4}$ **32.** $y = 4 + \dfrac{8}{2x - 6}$

33. Does $x^2 + y^2 = 4$ describe y as a function of x?

34. Does $x^2 + 2y = 9$ describe y as a function of x?

35. Write an equation to represent the function described by this statement: "The circumference C of a circle is found by multiplying 2π times the radius r."

36. Write a verbal statement to represent the function $D = 3E^2 - 5$.

In Exercises 29–32, find the domain of each function.

29. $y = \sqrt{3x - 6}$ **30.** $y = \sqrt{2x - 8}$

EXERCISES 1.1

In Exercises 37–41, determine whether the given relationship defines a function. If so, identify the independent and dependent variable and why the relationship is a function.

37. *Stock Prices*
  a. The price p at which IBM stock can be bought on a given day x.

  b. The closing price p of IBM stock on a given day x.

38. *Odometer*
  a. The odometer reading s when m miles are traveled.

  b. The miles traveled m when the odometer reads s.

39. *Life Insurance*
  a. The monthly premium p for a $100,000 life insurance policy determined by age a for males aged 26–32 (as shown in the table).

  b. Ages a that get a $100,000 life insurance policy for a monthly premium of $11.81 (from the table).

Age (years)	26	27	28	29	30	31	32
Premium (dollars per month)	11.72	11.81	11.81	11.81	11.81	11.81	11.81

40. *Earnings* The mean earnings M in dollars by level of education d for females in 2007, as defined by the graph in the figure.

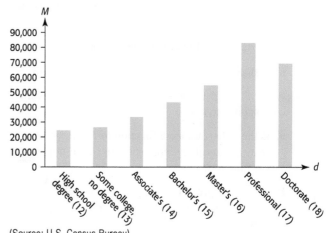

(Source: U.S. Census Bureau)

41. *Unemployment Rate* The U.S. unemployment rate r for month m during June–November 2010, defined by the table.

Month	Unemployment Rate
6	9.5
7	9.5
8	9.6
9	9.6
10	9.6
11	9.8

(Source: Bureau of Labor Statistics)

42. *Temperature* The figure below shows the graph of $T = 0.43m + 76.8$, which gives the temperature T (in degrees Fahrenheit) inside a concert hall m minutes after a 40-minute power outage during a summer rock concert. State what the variables m and T represent, and explain why T is a function of m.

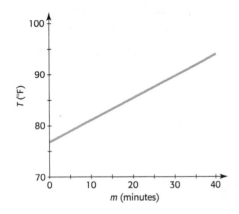

43. *Barcodes* A grocery scanner at Safeway connects the barcode number on a grocery item with the corresponding price.

  a. Is the price a function of the barcode? Explain.

  b. Is the barcode a function of the price? Explain.

44. *Piano* A child's piano has 12 keys, each of which corresponds to a note.

  a. Is the note sounded a function of the key pressed? If it is, how many elements are in the domain of the function?

  b. Is the key pressed a function of the note desired? If it is, how many elements are in the range of the function?

45. *Depreciation* A business property valued at $300,000 is depreciated over 30 years by the straight-line method, so its value x years after the depreciation began is

$$V = 300,000 - 10,000x$$

Explain why the value of the property is a function of the number of years.

46. *Seawater Pressure* In seawater, the pressure p is related to the depth d according to the model

$$p = \frac{18d + 496}{33}$$

where d is the depth in feet and p is in pounds per square inch. Is p a function of d? Why or why not?

47. *Weight* During the first two weeks in May, a man weighs himself daily (in pounds) and records the data in the table below.

  a. Does the table define weight as a function of the day in May?

  b. What is the domain of this function?

  c. What is the range?

  d. During what day(s) did he weigh most?

  e. During what day(s) did he weigh least?

  f. He claimed to be on a diet. What is the longest period of time during which his weight decreased?

May	1	2	3	4	5	6	7
Weight (lb)	178	177	178	177	176	176	175

May	8	9	10	11	12	13	14
Weight (lb)	176	175	174	173	173	172	171

48. *Test Scores*

  a. Is the average score on the final exam in a course a function of the average score on a placement test for the course, if the table below defines the relationship?

  b. Is the average score on a placement test a function of the average score on the final exam in a course, if the table below defines the relationship?

Average Score on Math Placement Test (%)	81	75	60	90	75
Average Score on Final Exam in Algebra Course (%)	86	70	58	95	81

49. *Car Financing* A couple wants to buy a $35,000 car and can borrow the money for the purchase at 8%, paying it off in 3, 4, or 5 years. The table below gives the monthly payment and total cost of the purchase (including the loan) for each of the payment plans.

t (years)	Monthly Payment ($)	Total Cost ($)
3	1096.78	39,484.08
4	854.46	41,014.08
5	709.68	42,580.80

(Source: Sky Financial Mortgage Tables)

Suppose that when the payment is over t years, $P(t)$ represents the monthly payment and $C(t)$ represents the total cost for the car and loan.

a. Find $P(3)$ and write a sentence that explains its meaning.

b. What is the total cost of the purchase if it is financed over 5 years? Write the answer using function notation.

c. What is t if $C(t) = 41,014.08$?

d. How much money will the couple save if they finance the car for 3 years rather than 5 years?

50. *Mortgage* A couple can afford $800 per month to purchase a home. As indicated in the table, if they can get an interest rate of 7.5%, the number of years t that it will take to pay off the mortgage is a function of the dollar amount A of the mortgage for the home they purchase.

Amount A ($)	t (years)
40,000	5
69,000	10
89,000	15
103,000	20
120,000	30

(Source: Comprehensive Mortgage Tables [Publication No. 492], Financial Publishing Co.)

a. If the couple wishes to finance $103,000, for how long must they make payments? Write this correspondence in function form if $t = f(A)$.

b. What is $f(120,000)$? Write a sentence that explains its meaning.

c. What is $f(3 \cdot 40,000)$?

d. What value of A makes $f(A) = 5$ true?

e. Does $f(3 \cdot 40,000) = 3 \cdot f(40,000)$? Explain your reasoning.

51. *Women in the Workforce* From 1930 and projected to 2015, the number of women in the workforce increased steadily. The points on the figure that follows give the number (in millions) of women in the workforce as a function f of the year for selected years.

a. Approximately how many women were in the workforce in 1960?

b. Estimate $f(1930)$ and write a sentence that explains its meaning.

c. What is the domain of this function if we consider only the indicated points?

d. How does the graph reflect the statement that the number of women in the workforce increases?

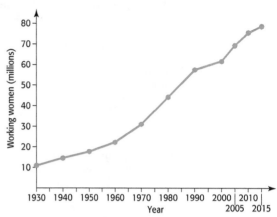

(Source: *World Almanac*)

52. *Working Age* The projected ratio of the working-age population (25- to 64-year-olds) to the elderly shown in the figure below defines the ratio as a function of the year shown. If this function is defined as $y = f(t)$ where t is the year, use the graph to answer the following:

a. What is the projected ratio of the working-age population to the elderly population in 2020?

b. Estimate $f(2005)$ and write a sentence that explains its meaning.

c. What is the domain of this function?

d. Is the projected ratio of the working-age population to the elderly increasing or decreasing over the domain shown in the figure?

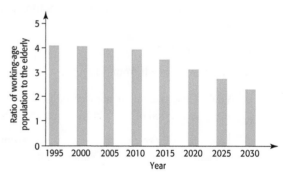

(Source: *Newsweek*)

53. *Age at First Marriage* The table on the following page gives the U.S. median age at first marriage for selected years from 1890 to 2009.

a. If the function f is the median age at first marriage for men, find $f(1890)$ and $f(2004)$.

b. If the function g is the median age at first marriage for women, find g(1940) and g(2000).

c. For what value of x is f(x) = 24.7? Write the result in a sentence.

d. Did f(x) increase or decrease from 1960 to 2009?

Median Age at First Marriage

Year	Men	Women
1890	26.1	22.0
1900	25.9	21.9
1910	25.1	21.6
1920	24.6	21.2
1930	24.3	21.3
1940	24.3	21.5
1950	22.8	20.3
1960	22.8	20.3
1970	23.2	20.8
1980	24.7	22.0
1990	26.1	23.9
2000	26.8	25.1
2004	27.1	25.8
2009	28.1	25.9

(Source: U.S. Census Bureau)

54. *Gun Crime* The table below gives the total number of nonfatal firearm incidents (crimes) as a function f of the year t.

a. What is f(2005)?

b. Interpret the result from (a).

c. What is the maximum number of firearm crimes during this period of time? In what year did it occur?

Year	Firearm Incidents	Year	Firearm Incidents
1993	1,054,820	2000	428,670
1994	1,060,800	2001	467,880
1995	902,680	2002	353,880
1996	845,220	2003	366,840
1997	680,900	2004	280,890
1998	557,200	2005	419,640
1999	457,150		

(Source: National Crime Victimization Survey)

55. *Internet Use* The following figure gives the number of millions of U.S. homes using the Internet for the years 1996–2008. If the number of millions of U.S. homes is the function f(x), where x is years, use the graph to answer the following:

a. How many homes used the Internet during 2000?

b. Find f(2008) and explain its meaning.

c. In what year did 26 million homes use the Internet?

d. Is this function increasing or decreasing? What do you think has happened to home use of the Internet since 2008?

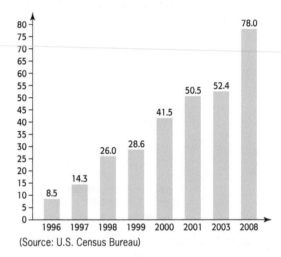

(Source: U.S. Census Bureau)

56. *Farms* The points on the following figure show the number N of U.S. farms (in millions) for selected years t.

a. Is the number of U.S. farms a function of the year?

b. What is f(1940) if N = f(t)?

c. What is t if f(t) = 2.4?

d. Write a sentence explaining the meaning of f(2009) = 2.2.

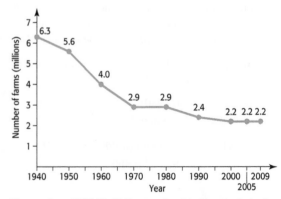

(Source: From *2011 World Almanac*. Reprinted by permission.)

57. *Teen Birth Rate* The birth rate in 2009 for U.S. girls ages 15 to 19 was the lowest since the government began tracking the statistic in 1940. The recent sharp drop is attributed to the recession. Answer the following questions using the figure below.

   a. Find the output when the input is 1995, and explain its meaning.

   b. For what year was the rate 40.5?

   c. For what year on the graph was the birth rate at its maximum?

   d. Despite the significant and steady decline in teen birth rates in recent decades, the decrease appears to have slowed recently. How would you describe the rates from 2005 to 2009?

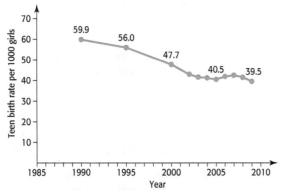

(Source: Centers for Disease Control and Prevention)

58. *Social Security Funding* Social Security benefits are funded by individuals who are currently employed. The following graph, based on known data and projections into the future, defines a function that gives the number of workers n supporting each retiree as a function of time t (given by the calendar year). Denote this function by $n = f(t)$.

   a. Find $f(1990)$ and explain its meaning.

   b. In what year will the number of workers supporting each retiree equal 2?

   c. What does this function tell us about Social Security in the future?

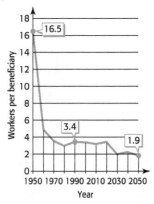

(Source: Social Security Administration)

59. *Revenue* The revenue from the sale of specialty golf hats is given by the function $R(x) = 32x$ dollars, where x is the number of hats sold.

   a. What is $R(200)$? Interpret this result.

   b. What is the revenue from the sale of 2500 hats? Write this in function notation.

60. *Cost* The cost of producing specialty golf hats is given by the function $C(x) = 4000 + 12x$, where x is the number of hats produced.

   a. What is $C(200)$? Interpret this result.

   b. What is the cost of producing 2500 hats? Write this in function notation.

61. *Utilities* An electric utility company determines the monthly bill by charging 85.7 cents per kilowatt-hour (kWh) used plus a base charge of $19.35 per month. Thus, the monthly charge is given by the function

$$f(W) = 0.857W + 19.35 \text{ dollars}$$

where W is the number of kilowatt-hours.

   a. Find $f(1000)$ and explain what it means.

   b. What is the monthly charge if 1500 kWh are used?

62. *Profit* The profit from the production and sale of iPod players is given by the function $P(x) = 450x - 0.1x^2 - 2000$, where x is the number of units produced and sold.

   a. What is $P(500)$? Interpret this result.

   b. What is the profit from the production and sale of 4000 units? Write this in function notation.

63. *Profit* The daily profit from producing and selling Blue Chief bicycles is given by

$$P(x) = 32x - 0.1x^2 - 1000$$

where x is the number produced and sold and $P(x)$ is in dollars.

   a. Find $P(100)$ and explain what it means.

   b. Find the daily profit from producing and selling 160 bicycles.

64. *Projectiles* Suppose a ball thrown into the air has its height (in feet) given by the function

$$h(t) = 6 + 96t - 16t^2$$

where t is the number of seconds after the ball is thrown.

   a. Find $h(1)$ and explain what it means.

b. Find the height of the ball 3 seconds after it is thrown.

c. Test other values of $h(t)$ to decide if the ball eventually falls. When does the ball stop climbing?

65. *Test Reliability* If a test that has reliability 0.7 has the number of questions increased by a factor n, the reliability R of the new test is given by

$$R(n) = \frac{0.7n}{0.3 + 0.7n}$$

a. What is the domain of the function defined by this equation?

b. If the application used requires that the size of the test be increased, what values of n make sense in the application?

66. *Body-Heat Loss* The description of body-heat loss due to convection involves a coefficient of convection K_c, which depends on wind speed s according to the equation $K_c = 4\sqrt{4s + 1}$.

a. Is K_c a function of s?

b. What is the domain of the function defined by this equation?

c. What restrictions does the physical nature of the model put on the domain?

67. *Cost-Benefit* Suppose that the cost C (in dollars) of removing $p\%$ of the particulate pollution from the smokestack of a power plant is given by

$$C(p) = \frac{237{,}000p}{100 - p}$$

a. Use the fact that percent is measured between 0 and 100 and find the domain of this function.

b. Evaluate $C(60)$ and $C(90)$.

68. *Demand* Suppose the number of units q of a product that is demanded by consumers is given as a function of p by

$$q = \frac{100}{\sqrt{2p + 1}}$$

where p is the price charged per unit.

a. What is the domain of the function defined by this equation?

b. What should the domain and range of this function be to make sense in the application?

69. *Postal Restrictions* Some postal restrictions say that in order for a box to be shipped, its length (longest side) plus its girth (distance around the box in the other two dimensions) must be equal to no more than 108 inches.

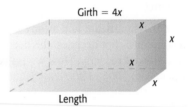

Girth = $4x$

Length

If the box with length plus girth equal to 108 inches has a square cross section that is x inches on each side, then the volume of the box is given by $V = x^2(108 - 4x)$ cubic inches.

a. Find $V(12)$ and $V(18)$.

b. What restrictions must be placed on x to satisfy the conditions of this model?

c. Create a table of function values to investigate the value of x that maximizes the volume. What are the dimensions of the box that has the maximum volume?

70. *Height of a Bullet* The height of a bullet shot into the air is given by $S(t) = -4.9t^2 + 98t + 2$, where t is the number of seconds after it is shot and $S(t)$ is in meters.

a. Find $S(0)$ and interpret it.

b. Find $S(9)$, $S(10)$, and $S(11)$.

c. What appears to be happening to the bullet at 10 seconds? Evaluate the function at some additional times near 10 seconds to confirm your conclusion.

Graphs of Functions

KEY OBJECTIVES

- Graph equations using the point-plotting method
- Graph equations using graphing calculators
- Graph equations using Excel spreadsheets
- Align inputs and scale outputs to model data
- Graph data points

SECTION PREVIEW **Personal Savings**

Using data from 1960 through 2009, a model can be created that gives the U.S. personal savings rate as a percent of disposable income. If the data are **aligned** so that the input values (x) represent the number of years after 1960, the personal savings rate can be modeled by the function

$$y = 0.000469x^3 - 0.0387x^2 + 0.714x + 6.787$$

(Source: U.S. Census Bureau)

To graph this function for values of x representing 1960 through 2009, we use x-values from 0 through 49 in the function to find the corresponding y-values, which represent percents. These points can be used to sketch the graph of the function. If technology is used to graph the function, the **viewing window** on which the graph is shown can be determined by these values of x and y. (See Example 6.)

In this section, we will graph functions by point plotting and with technology. We will graph application functions on windows determined by the context of the applications, and we will align data so that smaller inputs can be used in models. We will also graph data points. ■

Graphs of Functions

If an equation defines y as a function of x, we can sketch the graph of the function by plotting enough points to determine the shape of the graph and then drawing a line or curve through the points. This is called the **point-plotting method** of sketching a graph.

EXAMPLE 1 ▶ Graphing an Equation by Plotting Points

a. Graph the equation $y = x^2$ by drawing a smooth curve through points determined by integer values of x between 0 and 3.

b. Graph the equation $y = x^2$ by drawing a smooth curve through points determined by integer values of x between -3 and 3.

SOLUTION

a. We use the values in Table 1.9 to determine the points. The graph of the function drawn through these points is shown in Figure 1.17(a).

Table 1.9

x	0	1	2	3
$y = x^2$	0	1	4	9
Points	(0, 0)	(1, 1)	(2, 4)	(3, 9)

b. We use Table 1.10 to find the additional points. The graph of the function drawn through these points is shown in Figure 1.17(b).

Table 1.10

x	−3	−2	−1	0	1	2	3
$y = x^2$	9	4	1	0	1	4	9
Points	(−3, 9)	(−2, 4)	(−1, 1)	(0, 0)	(1, 1)	(2, 4)	(3, 9)

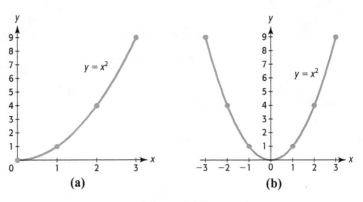

Figure 1.17

Although neither graph in Figure 1.17 shows all of the points satisfying the equation $y = x^2$, the graph in Figure 1.17(b) is a much better representation of the function (as you will learn later). When enough points are connected to determine the shape of the graph and the important parts of the graph and to suggest what the unseen parts of the graph look like, the graph is called **complete**.

Complete Graph

A graph is a complete graph if it shows the basic shape of the graph and important points on the graph (including points where the graph crosses the axes and points where the graph turns)* and suggests what the unseen portions of the graph will be.

EXAMPLE 2 ▶ Graphing a Complete Graph

Sketch the complete graph of the equation $f(x) = x^3 - 3x$, using the fact that the graph has at most two turning points.

SOLUTION

We use the values in Table 1.11 to determine some points. The graph of the function drawn through these points is shown in Figure 1.18.

Table 1.11

x	−3	−2	−1	0	1	2	3
$y = x^3 - 3x$	−18	−2	2	0	−2	2	18
Points	(−3, −18)	(−2, −2)	(−1, 2)	(0, 0)	(1, −2)	(2, 2)	(3, 18)

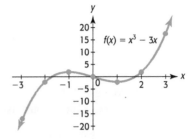

Figure 1.18

Note that this graph has two turning points, so it is a complete graph.

---

* Points where graphs turn from rising to falling or from falling to rising are called turning points.

The ability to draw complete graphs, with or without the aid of technology, improves with experience and requires knowledge of the shape of the basic function(s) involved in the graph. As you learn more about types of functions, you will be better able to determine when a graph is complete.

Graphing with Technology

Obtaining a sufficient number of points to sketch a graph by hand can be time consuming. Computers and graphing calculators have graphing utilities that can be used to plot many points quickly, thereby producing the graph with minimal effort. However, graphing with technology involves much more than pushing a few buttons. Unless you know what to enter into the calculator or computer (that is, the domain of the function and sometimes the range), you cannot see or use the graph that is drawn. We suggest the following steps for graphing most functions with a graphing calculator.

Using a Graphing Calculator to Draw a Graph

1. Write the function with x representing the independent variable and y representing the dependent variable. Solve for y, if necessary.

2. Enter the function in the equation editor of the graphing utility. Use parentheses* as needed to ensure the mathematical correctness of the expression.

3. Activate the graph by pressing the $\boxed{\text{ZOOM}}$ or $\boxed{\text{GRAPH}}$ key. Most graphing utilities have several preset **viewing windows** under $\boxed{\text{ZOOM}}$, including the **standard viewing window** that gives the graph of a function on a coordinate system in which the x-values range from -10 to 10 and the y-values range from -10 to 10.[†]

4. To see parts of the graph of a function other than those shown in a standard window, press $\boxed{\text{WINDOW}}$ to set the x- and y-boundaries of the viewing window before pressing $\boxed{\text{GRAPH}}$. Viewing window boundaries are discussed below and in the technology supplement. As you gain more knowledge of graphs of functions, determining viewing windows that give complete graphs will become less complicated.

Although the standard viewing window is a convenient window to use, it may not show the desired graph. To see parts of the graph of a function other than those that might be shown in a standard window, we change the x- and y-boundaries of the viewing window. The values that define the viewing window can be set manually under $\boxed{\text{WINDOW}}$ or by using the $\boxed{\text{ZOOM}}$ key. The boundaries of a viewing window are

$x_{\min}$: the smallest value on the x-axis (the left boundary of the window)

$y_{\min}$: the smallest value on the y-axis (the bottom boundary of the window)

$x_{\max}$: the largest value on the x-axis (the right boundary of the window)

$y_{\max}$: the largest value on the y-axis (the top boundary of the window)

x_{scl}: the distance between ticks on the x-axis (helps visually find x-intercepts and other points)

y_{scl}: the distance between ticks on the y-axis (helps visually find the y-intercept and other points)

---

* Parentheses should be placed around numerators and/or denominators of fractions, fractions that are multiplied by a variable, exponents consisting of more than one symbol, and in other places where the order of operations needs to be indicated.

[†] For more information on viewing windows and graphing, see Appendix A, page 618.

When showing viewing window boundaries on calculator graphs in this text, we write them in the form

$$[x_{min}, x_{max}] \text{ by } [y_{min}, y_{max}]$$

EXAMPLE 3 ▶   Graphing a Complete Graph

Sketch the graph of $y = x^3 - 3x^2 - 13$

a. using the standard window.

b. using the window $x_{min} = -10, x_{max} = 10, y_{min} = -25, y_{max} = 10$.

Which graph gives a better view of the graph of the function?

SOLUTION

a. A graph of this function in the standard window appears to be a line (Figure 1.19(a)).

b. By setting the window with $x_{min} = -10, x_{max} = 10, y_{min} = -25, y_{max} = 10$, we obtain the graph in Figure 1.19(b). This window gives a better view of the graph of this equation. As you learn more about functions, you will see that the graph in Figure 1.19(b) is a complete graph of this function.

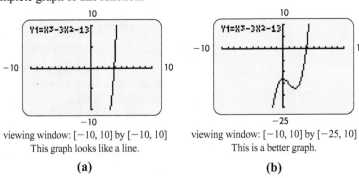

viewing window: $[-10, 10]$ by $[-10, 10]$ viewing window: $[-10, 10]$ by $[-25, 10]$
This graph looks like a line.   This is a better graph.

(a)   **(b)**

Figure 1.19

Different viewing windows give different views of a graph. There are usually many different viewing windows that give complete graphs for a particular function, but some viewing windows do not show all the important parts of a graph.

EXAMPLE 4 ▶   Cost-Benefit

Suppose that the cost C of removing $p\%$ of the pollution from drinking water is given by the model

$$C = \frac{5350p}{100 - p} \text{ dollars}$$

a. Use the restriction on p to determine the limitations on the horizontal-axis values (which are the x-values on a calculator).

b. Graph the function on the viewing window $[0, 100]$ by $[0, 50,000]$. Why is it reasonable to graph this model on a viewing window with the limitation $C \geq 0$?

c. Find the point on the graph that corresponds to $p = 90$. Interpret the coordinates of this point.

SOLUTION

a. Because p represents the percent of pollution removed, it is limited to values from 0 to 100. However, $p = 100$ makes C undefined in this model, so p is restricted to $0 \leq p < 100$ for this model.

b. The graph of the function is shown in Figure 1.20(a), with x representing p and y representing C. The interval [0, 100] contains all the possible values of p. The value of C is bounded below by 0 because C represents the cost of removing the pollution, which cannot be negative.

c. We can find (or estimate) the output of a function $y = f(x)$ at specific inputs with a graphing utility. We do this with TRACE or TABLE .* Using TRACE with the x-value 90 gives the point (90, 48,150) (see Figure 1.20(b)). Because the values of p are represented by x-values and the values of C are represented by y, the coordinates of the point tell us that the cost of removing 90% of the pollution from the drinking water is $48,150. Figure 1.20(c) shows the value of y for $x = 90$ and other values in a calculator table.

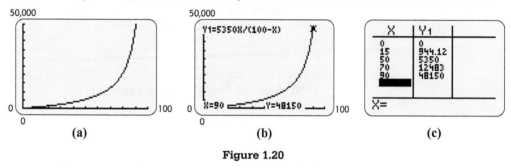

(a)   (b)   (c)

Figure 1.20

Spreadsheet ▸ SOLUTION We have briefly described how to use a graphing calculator to graph. Computer software of several types can be used to create more accurate and better-looking graphs. Software such as Scientific Notebook, Maple, and Mathematica can be used to create graphs, and Excel **spreadsheets** can also be used to generate graphs.

Table 1.12 shows a spreadsheet with inputs and outputs for the function $C = \dfrac{5350p}{100 - p}$ given in Example 4. When evaluating a function with a spreadsheet, we use the cell location of the data to represent the variable. Thus, to evaluate $C = \dfrac{5350p}{100 - p}$ at $p = 0$, we type 0 in cell A2 and $= 5350*A2/(100 - A2)$ in cell B2. Typing 10 in cell A3 and using the fill-down capacity of the spreadsheet gives the outputs of this function for $p = 0$ to $p = 90$ in increments of 10 (see Table 1.12). These values can be used to create the graph of this function in Excel (Figure 1.21).†

Table 1.12

	A	B
1	p	C = 5350p/(100 − p)
2	0	0
3	10	594.4444
4	20	1337.5
5	30	2292.857
6	40	3566.667
7	50	5350
8	60	8025
9	70	12483.33
10	80	21400
11	90	48150

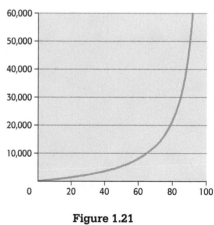

Figure 1.21

---

* For more details, see Appendix A, page 619.
† See the Excel Guide in Appendix B, pages 636–640, for details.

Aligning Data

When finding a model to fit a set of data, it is often easier to use aligned inputs rather than the actual data values. **Aligned inputs** are simply input values that have been converted to smaller numbers by subtracting the same number from each input. For instance, instead of using x as the actual year in the following example, it is more convenient to use an aligned input that is the number of years after 1950.

EXAMPLE 5 ▶ ## Voting

Between 1950 and 2008, the percent of the voting population who voted in presidential elections (during election years) is given by

$$f(x) = 0.0155x^2 - 1.213x + 75.26$$

where x is the number of years after 1950. That is, the input variable $x = 0$ represents 1950. (Source: Federal Election Commission)

a. What are the values of x that correspond to the years 1960 and 2008?

b. Find $f(10)$ and explain its meaning.

c. If this model is accurate for 2008, find the percent of the voting population who voted in the 2008 presidential election.

SOLUTION

a. Because 1960 is 10 years after 1950, $x = 10$ corresponds to the year 1960 in the aligned data. By similar reasoning, 2008 corresponds to $x = 58$.

b. An input value of 10 represents 10 years after 1950, which is the year 1960. To find $f(10)$, we substitute 10 for x to obtain $f(10) = 0.0155(10)^2 - 1.213(10) + 75.26 = 64.7$. Here is one possible explanation of this answer: Approximately 65% of the voting population voted in the 1960 presidential election.

c. To find the percent in 2008, we first calculate the aligned input for the function. Because 2008 is 58 years after 1950, we use $x = 58$. Thus, we find

$$f(58) = 0.0155(58)^2 - 1.213(58) + 75.26 = 57.0$$

So, if this model is accurate for 2008, approximately 57% of the voting population voted in the 2008 presidential election.

Determining Viewing Windows

Finding the functional values (y-values) for selected inputs (x-values) can be useful when setting viewing windows for graphing utilities.

Technology Note

Once the input values for a viewing window have been selected, $\boxed{\text{TRACE}}$ or $\boxed{\text{TABLE}}$ can be used to find enough output values to determine a y-view that gives a complete graph.*

---

* It is occasionally necessary to make more than one attempt to find a window that gives a complete graph.

EXAMPLE 6 ▶   Personal Savings

Using data from 1960 to 2009, the personal savings rate (as a percent of disposable income) of Americans can be modeled by the function

$$y = 0.000469x^3 - 0.0387x^2 + 0.714x + 6.787$$

where x is the number of years after 1960. (Source: U.S. Census Bureau)

a. Choose an appropriate window and graph the function with a graphing calculator.

b. Use the model to estimate the personal savings rate in 2013.

c. Use the graph to estimate the year in which the personal savings rate is a maximum.

SOLUTION

a. The viewing window should include values of x that are equivalent to the years 1960 to 2009, so the x-view should include the aligned values $x = 0$ to $x = 49$. The y-view should include the outputs obtained from the inputs between 0 and 49. We can use TRACE or TABLE with some or all of the integers 0 through 49 to find corresponding y-values (regardless of how the y-view is set when we are evaluating). See Figure 1.22(a). These evaluations indicate that the y-view should include values from about –1 through about 11. Choosing the window $x_{min} = 0$, $x_{max} = 54$, $y_{min} = -2$, and $y_{max} = 12$ gives the graph of the function shown in Figure 1.22(b).

b. The year 2013 is represented by $x = 53$. Evaluating the function with TRACE on the graph (see Figure 1.22(c)) gives an output of approximately 5.7 when $x = 53$. Figure 1.22(a) shows the value of the function at $x = 53$ using TABLE. Thus, we estimate the personal savings rate in 2013 to be 5.7%.

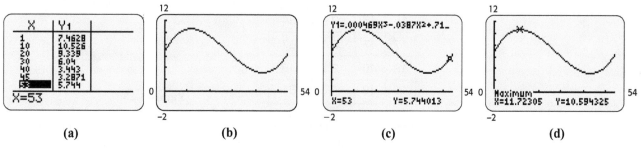

Figure 1.22

c. We can use the "maximum" feature located under 2ND TRACE to find the maximum point on the graph.* Figure 1.22(d) shows that the maximum point is (11.72, 10.59). Because x represents the number of years after 1960, an x-value of 11.72 means the year 1972, if we assume data are collected and reported at the end of the year.† Thus, the maximum personal savings rate was 10.6% in 1972 (see Figure 1.22(d)).

---

* For more information on computing minimum and maximum points, see the Calculator Guide in Appendix A, page 621.

†This topic will be explored further in Section 2.2.

Spreadsheet
▸ SOLUTION
Excel can also be used to graph and evaluate the function $y = 0.000469x^3 - 0.0387x^2 + 0.714x + 6.787$, discussed in Example 6. The Excel graph is shown in Figure 1.23, and an Excel spreadsheet with values of the function at selected values of x is shown in Table 1.13.

Table 1.13

	A	B
	x	$f(x)$
1	x	$f(x)$
2	0	6.787
3	5	9.448125
4	10	10.526
5	15	10.37238
6	20	9.339
7	25	7.777625
8	30	6.04
9	35	4.477875
10	40	3.443
11	41	3.330249
12	42	3.255472
13	43	3.221483
14	44	3.231096
15	45	3.287125
16	46	3.392384
17	47	3.549687
18	48	3.761848
19	49	4.031681

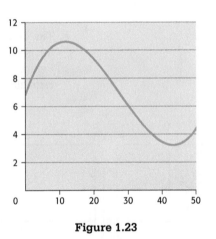

Figure 1.23

To use Excel to find the year in which the personal savings rate is a maximum, invoke Solver from the Data tab. Click on Max in the Solver dialog box. Set the Target Cell to B2 and the Changing Cells to A2. Then click Solve in the dialog box and click OK to accept the solution.* See Table 1.14.

Table 1.14

1	x	$f(x)$
2	11.72305	10.59433

Graphing Data Points

Graphing utilities can also be used to create lists of numbers and to create graphs of the data stored in the lists. To see how lists and scatter plots are created, consider the following example.

---

* For more information on computing maximum and minimum points, see the Excel Guide in Appendix B, page 641.

EXAMPLE 7 ▶ U.S. Executions

Table 1.15 gives the number of executions in the United States for selected years from 1984 to 2008. (Source: "The Death Penalty in the U.S.," www.clarkprosecuter.org)

Table 1.15

Year	Number of Executions
1984	21
1987	25
1990	23
1993	38
1996	45
1999	98
2002	71
2005	60
2007	42
2008	37

a. Align the data so that x = the number of years after 1980, and enter these x-values in list L1 of your graphing calculator. Enter the number of executions in L2.

b. Use a graphing command* to create the scatter plot of these data points.

SOLUTION

a. Figure 1.24(a) shows the first 7 aligned inputs in L1 and the first 7 outputs from the data in L2.

b. The scatter plot is shown in Figure 1.24(b), with each entry in L1 represented by an x-coordinate of a point on the graph and the corresponding entry in L2 represented by the y-coordinate of that point on the graph. The window can be set automatically or manually to include x-values from L1 and y-values from L2.

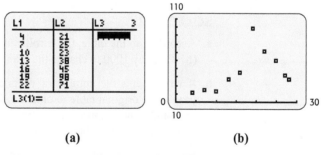

(a)   (b)

Figure 1.24

---

* Many calculators have a Stat Plot command, and Excel has an XY(Scatter) command. See Appendix A, page 618, and Appendix B, page 642.

Table 1.16 shows the values from Table 1.15 in an Excel spreadsheet, and Figure 1.25 shows the scatter plot (graph) of the data.

Table 1.16

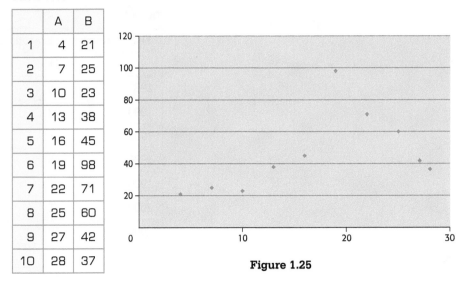

	A	B
1	4	21
2	7	25
3	10	23
4	13	38
5	16	45
6	19	98
7	22	71
8	25	60
9	27	42
10	28	37

Figure 1.25

EXAMPLE 8 ▶   **High School Enrollment**

Table 1.17 shows the enrollment (in thousands) in grades 9–12 at U.S. public and private schools for the years 1990–2008.

a. According to the table, what was the enrollment in 1998?

b. Create a new table with x representing the number of years after 1990 and y representing the enrollment in millions.

c. According to this new table, what was the enrollment in 1998?

d. Use a graphing utility to graph the new data as a scatter plot.

SOLUTION

a. According to the table, the enrollment in 1998 was 14,428 thousand, or 14,428,000. (Multiply by 1000 to change the number of thousands to a number in standard form.)

b. If x represents the number of years after 1990, we must subtract 1990 from each year in the original table to obtain x. The enrollment in the original table is given in thousands, and because one million is 1000 times one thousand, we must divide each number by 1000 to obtain y. The result is shown in Table 1.18.

c. Because x represents the number of years after 1990, we subtract $1998 - 1990$ to obtain an x-value of 8. Table 1.18 indicates that in year 8 the population was 14.428 million, or (multiplying by 1000) 14,428 thousand. Note that this is the same result as in part (a).

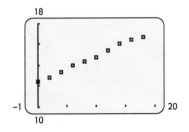

Figure 1.26

d. A scatter plot of the data is shown in Figure 1.26.

Table 1.17

Year	Enrollment (thousands)
1990	12,488
1992	12,882
1994	13,376
1996	14,060
1998	14,428
2000	14,802
2002	15,426
2004	16,048
2006	16,498
2008	16,715

(Source: U.S. Department of Education)

Table 1.18

Number of Years after 1990, x	Enrollment, y (millions)
0	12.488
2	12.882
4	13.376
6	14.060
8	14.428
10	14.802
12	15.426
14	16.048
16	16.498
18	16.715

Alignment and Scaling of Data

Recall that replacing the input so that x represents the number of years after a certain year is called *alignment of the data*. Replacing the outputs by dividing or multiplying each number by a given number is called **scaling the data**. In Example 8, we aligned the data by replacing the years so that x represents the number of years after 1990, and we scaled the data by replacing the enrollment in thousands so that y represents the enrollment in millions.

Skills CHECK 1.2

1. a. Complete the table of values for the function $y = x^3$, plot the values of x and y as points in a coordinate plane, and draw a smooth curve through the points.

x	−3	−2	−1	0	1	2	3
y							

b. Graph the function $y = x^3$ with a graphing calculator, using the viewing window $x_{min} = -4$, $x_{max} = 4$, $y_{min} = -30$, $y_{max} = 30$.

c. Compare the graphs in parts (a) and (b).

2. a. Complete the table of values for the function $y = 2x^2 + 1$, plot the values of x and y as points in a coordinate plane, and draw a smooth curve through the points.

x	−3	−2	−1	0	1	2	3
y							

b. Graph the function $y = 2x^2 + 1$ with a graphing calculator, using the viewing window $x_{min} = -5$, $x_{max} = 5$, $y_{min} = -1$, $y_{max} = 20$.

c. Compare the graphs in parts (a) and (b).

In Exercises 3–8, make a table of values for each function and then plot the points to graph the function by hand.

3. $f(x) = 3x - 1$

4. $f(x) = 2x - 5$

5. $f(x) = \dfrac{1}{2}x^2$

6. $f(x) = 3x^2$

7. $f(x) = \dfrac{1}{x - 2}$

8. $f(x) = \dfrac{x}{x + 3}$

In Exercises 9–14, graph the functions with a graphing calculator using a standard viewing window. State whether the graph has a turning point in this window.

9. $y = x^2 - 5$

10. $y = 4 - x^2$

11. $y = x^3 - 3x^2$

12. $y = x^3 - 3x^2 + 4$

13. $y = \dfrac{9}{x^2 + 1}$

14. $y = \dfrac{40}{x^2 + 4}$

For Exercises 15–18, graph the given function with a graphing calculator using (a) the standard viewing window and (b) the specified window. Which window gives a better view of the graph of the function?

15. $y = x + 20$ with $x_{min} = -10$, $x_{max} = 10$, $y_{min} = -10$, $y_{max} = 30$

16. $y = x^3 - 3x + 13$ with $x_{min} = -5$, $x_{max} = 5$, $y_{min} = -10$, $y_{max} = 30$

17. $y = \dfrac{0.04(x - 0.1)}{x^2 + 300}$ on $[-20, 20]$ by $[-0.002, 0.002]$

18. $y = -x^2 + 20x - 20$ on $[-10, 20]$ by $[-20, 90]$

For Exercises 19–22, find an appropriate viewing window for the function, using the given x-values. Then graph the function. Find the coordinates of any turning points.

19. $y = x^2 + 50$, for x-values between -8 and 8.

20. $y = x^2 + 60x + 30$, for x-values between -60 and 0.

21. $y = x^3 + 3x^2 - 45x$, for x-values between -10 and 10.

22. $y = (x - 28)^3$, for x-values between 25 and 31.

23. Find a complete graph of $y = 10x^2 - 90x + 300$. (A complete graph of this function shows one turning point.)

24. Find a complete graph of $y = -x^2 + 34x - 120$. (A complete graph of this function shows one turning point.)

25. Use a calculator or a spreadsheet to find $S(t)$ for the values of t given in the following table.

t	S(t) = 5.2t − 10.5
12	
16	
28	
43	

26. Use a calculator or a spreadsheet to find $f(q)$ for the values of q given in the following table.

q	f(q) = 3q² − 5q + 8
−8	
−5	
24	
43	

27. Enter the data below into lists and graph the scatter plot of the data, using the window $[0, 100]$ by $[0, 500]$.

x	20	30	40	50	60	70	80	90	100
y	500	320	276	80	350	270	120	225	250

28. Enter the data below into lists and graph the scatter plot of the data, using the window $[-20, 120]$ by $[0, 80]$.

x	−15	−1	25	40	71	110	116
y	12	16	10	32	43	62	74

29. Use the table below in parts (a)–(c).

x	−3	−1	1	5	7	9	11
y	−42	−18	6	54	78	102	126

a. Use a graphing utility to graph the points from the table.

b. Use a graphing utility to graph the equation $f(x) = 12x - 6$ on the same set of axes as the data in part (a).

c. Do the points fit on the graph of the equation? Do the respective values of x and y in the table satisfy the equation in part (b)?

30. Use the table below in parts (a)–(c).

x	1	3	6	8	10
y	−10	0	15	25	35

a. Use a graphing utility to graph the points from the table.

b. Use a graphing utility to graph the equation $f(x) = 5x - 15$ on the same set of axes as the data in part (a).

c. Do the points fit on the graph of the equation? Do the respective values of x and y in the table satisfy the equation in part (b)?

31. Suppose $f(x) = x^2 - 5x$ million dollars are earned, where x is the number of years after 2000.

    a. What is $f(20)$?

    b. The answer to part (a) gives the number of millions of dollars earned for what year?

32. Suppose $f(x) = 100x^2 - 5x$ thousand units are produced, where x is the number of years after 2000.

    a. What is $f(10)$?

    b. How many units are produced in 2010, according to this function?

EXERCISES 1.2

33. *Women in the Workforce* The number y (in millions) of women in the workforce is given by the function

    $$y = 0.006x^2 - 0.018x + 5.607$$

    where x is the number of years after 1900.

    a. Find the value of y when $x = 44$. Explain what this means.

    b. Use the model to find the number of women in the workforce in 2010.
    (Source: U.S. Department of Labor)

34. *Unemployment Rates* The unemployment rates for Canada for selected years from 1970 to 2008 can be modeled by

    $$y = -0.009x^2 + 0.321x + 5.676$$

    where x is the number of years after 1970.

    a. What are the values of x that correspond to the years 1990 and 2005?

    b. Find the value of y when $x = 25$. Explain what this means.

    c. What was the unemployment rate in Canada in 2008, according to the model?
    (Source: Bureau of Labor Statistics, U.S. Department of Labor)

35. *Internet Access* The function $P = 5.8t + 7.13$ gives the percent of households with Internet access as a function of t, the number of years after 1995.

    a. What are the values of t that correspond to the years 1996 and 2014?

    b. $P = f(10)$ gives the value of P for what year? What is $f(10)$?

    c. What x_{min} and x_{max} should be used to set the viewing window so that t represents 1995–2015?

36. *State Lotteries* The total cost of prizes and expenses of state lotteries is given by $P = 35t^2 + 740t + 1207$ million dollars, with t equal to the number of years after 1980.

    a. What are the values of t that correspond to the years 1988, 2000, and 2012?

    b. $P = f(14)$ gives the value of P for what year? What is $f(14)$?

    c. What x_{min} and x_{max} should be used to set a viewing window so that t represents 1980–2007?

37. *Height of a Ball* If a ball is thrown into the air at 64 feet per second from the top of a 100-foot-tall building, its height can be modeled by the function $S = 100 + 64t - 16t^2$, where S is in feet and t is in seconds.

    a. Graph this function on a viewing window [0, 6] by [0, 200].

    b. Find the height of the ball 1 second after it is thrown and 3 seconds after it is thrown. How can these values be equal?

    c. Find the maximum height the ball will reach.

38. *Depreciation* A business property valued at $600,000 is depreciated over 30 years by the straight-line method, so its value x years after the depreciation began is

    $$V = 600{,}000 - 20{,}000x$$

    a. Graph this function on a viewing window [0, 30] by [0, 600,000].

    b. What is the value 10 years after the depreciation is started?

39. *Earnings and Gender* A model that relates the median annual salary (in thousands of dollars) of females, F, and males, M, in the United States is given by $F = 0.78M - 1.05$.

    a. Use a graphing utility to graph this function on the viewing window [0, 100] by [0, 80].

    b. Use the graphing utility to find the median female salary that corresponds to a male salary of $63,000.
    (Source: U.S. Census Bureau)

40. *Medical School* The number of students (in thousands) of osteopathic medicine in the United States can be described by

$$S = 0.027t^2 - 4.85t + 218.93$$

where t is the number of years after 1980.

a. Graph this function on the viewing window [0, 17] by [0, 300].

b. Use technology to find S when t is 15.

c. Use the model to estimate the number of osteopathic students in 2005.
(Source: *Statistical Abstract of the United States*)

41. *Advertising Expenditures* The amounts of money spent on advertising in the Yellow Pages, either national or local, for the years 1990 to 2008 can be modeled by

$$A = -1.751x^2 + 259.910x + 8635.242$$

million dollars, where x is the number of years after 1990.

a. Graph this function on the viewing window [0, 30] by [8000, 15,000].

b. Determine the amount of money spent on advertising in the Yellow Pages in 2004 and 2008.

c. Does this model predict that the amount spent will continue to increase through 2020?
(Source: *Statistical Abstract of the United States*)

42. *State Lotteries* The cost (in millions of dollars) of prizes and expenses for state lotteries can be described by $L = 35.3t^2 + 740.2t + 1207.2$, where t is the number of years after 1980.

a. Graph this function on the viewing window [0, 27] by [1200, 45,000].

b. Use technology to find L when t is 26.

c. What was the cost of prizes and expenses for state lotteries in 2006?
(Source: *Statistical Abstract of the United States*)

43. *Crime* The rate (number per 100,000 people) of arrests for violent crimes is given by

$$f(x) = -0.711x^2 + 14.244x + 581.178$$

where x is the number of years after 1980.
(Source: Federal Bureau of Investigation)

a. Use technology to graph this model on the viewing window [0, 30] by [0, 800].

b. This viewing window shows the graph for what time period?

c. Did the number of arrests per 100,000 people increase or decrease after 1990?

d. Use the model to find the number of arrests per 100,000 in 1994 and in 2009. Did the number increase or decrease over this period?

e. Is your answer to part (c) consistent with the answer to part (d)?

44. *Tax Burden* Using data from the Internal Revenue Service, the per capita tax burden B (in hundreds of dollars) can be described by $B(x) = 1.37x^2 + 26.6x + 152$, where x is the number of years after 1960.

a. Graph this function with technology using a viewing window with $x \geq 0$ and $B(x) \geq 0$.

b. Did the tax burden increase or decrease?
(Source: Internal Revenue Service)

45. *Cost* Suppose the cost of the production and sale of x Electra dishwashers is $C(x) = 15,000 + 100x + 0.1x^2$ dollars. Graph this function on a viewing window with x between 0 and 50.

46. *Revenue* Suppose the revenue from the sale of x coffee makers is given by $R(x) = 52x - 0.1x^2$. Graph this function on a viewing window with x between 0 and 100.

47. *Profit* The profit from the production and sale of x laser printers is given by the function $P(x) = 200x - 0.01x^2 - 5000$, where x is the number of units produced and sold. Graph this function on a viewing window with x between 0 and 1000.

48. *Profit* The profit from the production and sale of x digital cameras is given by the function $P(x) = 1500x - 8000 - 0.01x^2$, where x is the number of units produced and sold. Graph this function on a viewing window with x between 0 and 500.

49. *Teacher Salaries* The average salary of a U.S. classroom teacher is given by $f(t) = 1216.879t + 31,148.869$, where t is the number of years from 1990.

a. What inputs correspond to the years 1990–2009?

b. What outputs correspond to the inputs determined for 1990 and 2009?

c. Use the answers to parts (a) and (b) and the fact that the function increases to find an appropriate viewing window and graph this function.
(Source: National Center for Education Statistics)

50. *Cocaine Use* The percent of high school seniors during the years 1975–2010 who have ever used cocaine can be described by

$$y = 0.0152x^2 - 0.9514x + 21.5818$$

where x is the number of years after 1975.

a. What inputs correspond to the years 1975 through 2010?

b. What outputs for y could be used to estimate the percent of seniors who have ever used cocaine?

c. Based on your answers to parts (a) and (b), choose an appropriate window and graph the equation on a graphing utility.

d. Graph the function again with a new window that gives a graph nearer the center of the screen.

e. Use this function to estimate the percent in 2013.
(Source: monitoringthefuture.org)

51. *U.S. Population* The projected population of the United States for selected years from 2000 to 2060 is shown in the table below, with the population given in millions.

a. According to this table, what should the U.S. population have been in 2010?

b. Create a new table with x representing the number of years after 2000 and y representing the number of millions.

c. Use a graphing utility to graph the data from the new table as a scatter plot.

Year	Population (millions)
2000	275.3
2010	299.9
2020	324.9
2030	351.1
2040	377.4
2050	403.7
2060	432.0

(Source: U.S. Census Bureau)

52. *Hotel Values* The following table gives the annual rental value of an average hotel room for the years 2000–2005.

a. Let x represent the number of years after 2000 and y represent the value per room in thousands of dollars and sketch the scatter plot of the data.

b. Graph the equation $y = 0.973x^2 - 4.667x + 73.950$ on the same axes as the scatter plot.

Year	Value per Room
2000	$73,978
2001	$70,358
2002	$68,377
2003	$68,192
2004	$71,691
2005	$74,584

(Source: Pennsylvania State University)

53. *Social Security Benefits* The maximum monthly benefits payable to individuals who retired at age 65 are given in the table below. Use a graphing utility to graph the data, with x representing the number of years after 1990 when the retiree reached age 65 and y representing the monthly benefits.

Year Attaining Age 65	Benefit Payable at Retirement ($)
1990	975
1995	1199
1996	1248
1997	1326
1998	1342
1999	1373
2000	1434
2001	1538
2002	1660
2003	1721
2004	1784
2005	1874
2006	1961
2007	1998
2008	2030
2009	2172

(Source: Social Security Administration)

54. *Expenditures for U.S. Health Care* The cost per person per year for health care, with projections to 2018, is shown in the following table.

Year	$ per Person	Year	$ per Person
2000	4789	2010	8465
2002	5563	2012	9275
2004	6331	2014	10,289
2006	7091	2016	11,520
2008	7826	2018	12,994

(Source: U.S. Medicare and Medicaid Services)

a. Let x represent the number of years after 2000 and y represent the annual cost and sketch the scatter plot of the data.

b. Graph the function $y = 1.149x^3 - 21.564x^2 + 473.575x + 4749.394$ on the same axes as the scatter plot.

c. Is the function a good visual fit to the data?

55. *Unemployment Rate* The U.S. civilian unemployment rate (as a percent) is given by the table.

a. According to this table, what was the unemployment rate in 2003?

b. Graph the data from this table as a scatter plot, using the number of years after 1990 as x.

c. Graph the equation $y = 0.0085x^4 - 0.4385x^3 + 8.2396x^2 - 66.627x + 199.962$ on the same axes as the scatter plot.

Year	Unemployment (%)
1998	4.5
1999	4.2
2000	4.0
2001	4.7
2002	5.8
2003	6.0
2004	5.5
2005	5.1
2006	4.6
2007	4.6
2008	5.8
2009	9.3

(Source: Bureau of Labor Statistics)

56. *Dropout Rates* The table gives the dropout rates (as percents) for students ages 16 to 24 during given years.

a. What is the dropout rate in 2004, according to the data?

b. Graph the data from this table as a scatter plot, using the number of years after 1980 as x.

c. On the same axes as the scatter plot, graph the equation $y = -0.0008x^3 + 0.030x^2 - 0.413x + 14.081$.

Year	Dropout (%)
1980	14.1
1985	12.6
1990	12.1
1995	12.0
2000	10.9
2004	10.3
2008	8.0

(Source: U.S. Department of Education)

KEY OBJECTIVES

- Identify linear functions
- Find the intercepts and slopes of graphs of linear functions
- Graph linear functions
- Find the rate of change of a linear function
- Identify identity and constant functions
- Apply linear revenue, cost, and profit functions
- Find marginal revenue and marginal profit from linear revenue and linear profit functions

SECTION PREVIEW **Hispanics in the United States**

Using data and projections from 1990 through 2050, the percent of Hispanics in the U.S. population can be modeled by

$$H(x) = 0.224x + 9.01$$

with x equal to the number of years after 1990. The graph of this model is shown in Figure 1.27. (Source: U.S. Census Bureau)

Because the graph of this function is a line, it is called a linear function. The rate at which the Hispanic population is increasing is constant in this model, and it is equal to the slope of the line shown in Figure 1.27. (See Example 6.) In this section, we investigate linear functions and discuss slope, constant rates of change, intercepts, revenue, cost, and profit.

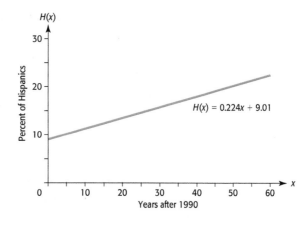

Figure 1.27

Linear Functions

A function whose graph is a line is a **linear function**.

Linear Function

A linear function is a function that can be written in the form $f(x) = ax + b$, where a and b are constants.

If x and y are in separate terms in an equation and each appears to the first power (and not in a denominator), we can rewrite the equation relating them in the form

$$y = ax + b$$

for some constants a and b. So the original equation is a linear equation that represents a linear function. If no restrictions are stated or implied by the context of the problem situation and if its graph is not a horizontal line, both the domain and the range of a linear function consist of the set of all real numbers. Note that an equation of the form $x = d$, where d is a constant, is not a function; its graph is a vertical line.

Recall that in the form $y = ax + b$, a and b represent constants and the variables x and y can represent any variables, with x representing the independent (input) variable and y representing the dependent (output) variable. For example, the function $5q + p = 400$ can be written in the form $p = -5q + 400$, so we can say that p is a linear function of q.

EXAMPLE 1 ▶   ## Linear Functions

Determine whether each equation represents a linear function. If so, give the domain and range.

a. $0 = 2t - s + 1$ **b.** $y = 5$ **c.** $xy = 2$

SOLUTION

a. The equation

$$0 = 2t - s + 1$$

does represent a linear function because each of the variables t and s appears to the first power and each is in a separate term. We can solve this equation for s, getting

$$s = 2t + 1$$

so s is a linear function of t. Because any real number can be multiplied by 2 and increased by 1 and the result is a real number, both the domain and the range consist of the set of all real numbers. (Note that we could also solve the equation for t so that t is a linear function of s.)

b. The equation $y = 5$ is in the form $y = ax + b$, where $a = 0$ and $b = 5$, so it represents a linear function. (It is in fact a constant function, which is a special linear function.) The domain is the set of all real numbers (because $y = 5$ regardless of what x we choose), and the range is the set containing 5.

c. The equation

$$xy = 2$$

does not represent a linear function because x and y are not in separate terms and the equation cannot be written in the form $y = ax + b$.

Intercepts

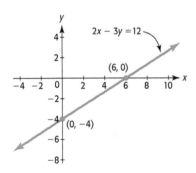

Figure 1.28

The points where a graph crosses or touches the x-axis and the y-axis are called the **x-intercepts** and **y-intercepts**, respectively, of the graph. For example, Figure 1.28 shows that the graph of the linear function $2x - 3y = 12$ crosses the x-axis at $(6, 0)$, so the x-intercept is $(6, 0)$. The graph crosses the y-axis at $(0, -4)$, so the y-intercept is $(0, -4)$. In this text, we will use the widely accepted convention that the x-coordinate of the x-intercept may also be called the x-intercept, and the y-coordinate of the y-intercept may also be called the y-intercept.* The procedure for finding intercepts is a direct result of these definitions.

Finding Intercepts Algebraically

To find the y-intercept of a graph of $y = f(x)$, set $x = 0$ in the equation and solve for y. If the solution is b, we say the y-intercept is b and the graph intersects the y-axis at the point $(0, b)$.

To find the x-intercept(s) of the graph of $y = f(x)$, set $y = 0$ in the equation and solve for x. If the solution is a, we say the x-intercept is a and the graph intersects the x-axis at the point $(a, 0)$.

---

* We usually call the horizontal axis intercept(s) the x-intercept(s) and the vertical axis intercept(s) the y-intercept(s), but realize that symbols other than x and y can be used to represent the input and output. For example, if $p = f(q)$, the vertical intercept is called the p-intercept and the horizontal intercept is called the q-intercept.

EXAMPLE 2 ▶ Finding Intercepts Algebraically

Find the x-intercept and the y-intercept of the graph of $2x - 3y = 12$ algebraically.

SOLUTION

The y-intercept can be found by substituting 0 for x in the equation and solving for y.

$$2(0) - 3y = 12$$
$$-3y = 12$$
$$y = -4$$

Thus, the y-intercept is -4, and the graph crosses the y-axis at the point $(0, -4)$.

Similarly, the x-intercept can be found by substituting 0 for y in the equation and solving for x.

$$2x - 3(0) = 12$$
$$2x = 12$$
$$x = 6$$

Thus, the x-intercept is 6, and the graph crosses the x-axis at the point $(6, 0)$.

Finding Intercepts Graphically

To find the intercept(s) of a graph of $y = f(x)$, first graph the function in a window that shows all intercepts.

To find the y-intercept, ⃞TRACE to $x = 0$ and the y-intercept will be displayed. To find the x-intercept(s) of the graph of $y = f(x)$, use the ⃞ZERO command under the ⃞CALC menu (accessed by ⃞2ND ⃞TRACE).*

The graph of a linear function has one y-intercept and one x-intercept unless the graph is a horizontal line. The intercepts of the graph of a linear function are often easy to calculate. If the intercepts are distinct, then plotting these two points and connecting them with a line gives the graph.

Technology Note

When graphing with a graphing utility, finding or estimating the intercepts can help you set the viewing window for the graph of a linear equation.

EXAMPLE 3 ▶ Loan Balance

A business property is purchased with a promise to pay off a \$60,000 loan plus the \$16,500 interest on this loan by making 60 monthly payments of \$1275. The amount of money, y, remaining to be paid on \$76,500 (the loan plus interest) is reduced by \$1275 each month. Although the amount of money remaining to be paid changes every month, it can be modeled by the linear function

$$y = 76,500 - 1275x$$

* For more details, see Appendix A, page 620.

where x is the number of monthly payments made. We recognize that only integer values of x from 0 to 60 apply to this application.

a. Find the x-intercept and the y-intercept of the graph of this linear equation.

b. Interpret the intercepts in the context of this problem situation.

c. How should x and y be limited in this model so that they make sense in the application?

d. Use the intercepts and the results of part (c) to sketch the graph of the given equation.

SOLUTION

a. To find the x-intercept, set $y = 0$ and solve for x.

$$0 = 76{,}500 - 1275x$$

$$1275x = 76{,}500$$

$$x = \frac{76{,}500}{1275} = 60$$

Thus, 60 is the x-intercept.

To find the y-intercept, set $x = 0$ and solve for y.

$$y = 76{,}500 - 1275(0)$$

$$y = 76{,}500$$

Thus, 76,500 is the y-intercept.

b. The x-intercept corresponds to the number of months that must pass before the amount owed is \$0. Therefore, a possible interpretation of the x-intercept is "The loan is paid off in 60 months." The y-intercept corresponds to the total (loan plus interest) that must be repaid 0 months after purchase—that is, when the purchase is made. Thus, the y-intercept tells us "A total of \$76,500 must be repaid."

c. We know that the total time to pay the loan is 60 months. A value of x larger than 60 will result in a negative value of y, which makes no sense in the application, so the values of x vary from 0 to 60. The output, y, is the total amount owed at any time during the loan. The amount owed cannot be less than 0, and the value of the loan plus interest will be at its maximum, 76,500, when time is 0. Thus, the values of y vary from 0 to 76,500.

d. The graph intersects the horizontal axis at (60, 0) and intersects the y-axis at (0, 76,500), as indicated in Figure 1.29. Because we know that the graph of this equation is a line, we can simply connect the two points to obtain this first-quadrant graph.

Figure 1.29 shows the graph of the function on a viewing window determined by the context of the application.

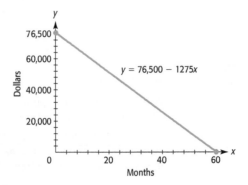

Figure 1.29

Slope of a Line

Consider two stairways: One goes from the park entrance to the shogun shrine at Nikko, Japan, and rises vertically 1 foot for every 1 foot of horizontal increase, and the second goes from the Tokyo subway to the street and rises vertically 20 centimeters for every 25 centimeters of horizontal increase. To see which set of stairs would be easier to climb, we can find the steepness, or **slope**, of each stairway. If a board is placed along the steps of each stairway, its slope is a measure of the incline of the stairway.

$$\text{shrine stairway steepness} = \frac{\text{vertical increase}}{\text{horizontal increase}} = \frac{1 \text{ foot}}{1 \text{ foot}} = 1$$

$$\text{subway stairway steepness} = \frac{\text{vertical increase}}{\text{horizontal increase}} = \frac{20 \text{ cm}}{25 \text{ cm}} = 0.8$$

The shrine stairway has a slope that is larger than that of the subway stairway, so it is steeper than the subway stairway. Thus, the subway steps would be easier to climb. In general, we define the slope of a line as follows.

Slope of a Line

The slope of a line is defined as

$$\text{slope} = \frac{\text{vertical change}}{\text{horizontal change}} = \frac{\text{rise}}{\text{run}}$$

The slope can be found by using any two points on the line (see Figure 1.30). If a nonvertical line passes through the two points, P_1 with coordinates (x_1, y_1) and P_2 with coordinates (x_2, y_2), its slope, denoted by m, is found by using

$$m = \frac{y_2 - y_1}{x_2 - x_1}$$

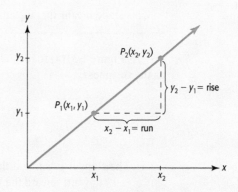

Figure 1.30

The slope of a vertical line is undefined because $x_2 - x_1 = 0$ and division by 0 is undefined.

The slope of any given nonvertical line is a constant. Thus, the same slope will result regardless of which two points on the line are used in its calculation.

EXAMPLE 4 ▶ **Calculating the Slope of a Line**

a. Find the slope of the line passing through the points $(-3, 2)$ and $(5, -4)$. What does the slope mean?

b. Find the slope of the line joining the x-intercept point and y-intercept point in the loan situation of Example 3.

SOLUTION

a. We choose one point as P_1 and the other point as P_2. Although it does not matter which point is chosen to be P_1, it is important to keep the correct order of the terms in the numerator and denominator of the slope formula. Letting $P_1 = (-3, 2)$ and $P_2 = (5, -4)$ and substituting in the slope formula gives

$$m = \frac{-4 - 2}{5 - (-3)} = \frac{-6}{8} = -\frac{3}{4}$$

Note that letting $P_1 = (5, -4)$ and $P_2 = (-3, 2)$ gives the same slope:

$$m = \frac{2 - (-4)}{-3 - 5} = \frac{6}{-8} = -\frac{3}{4}$$

A slope of $-\dfrac{3}{4}$ means that, from a given point on the line, by moving 3 units down and 4 units to the right or by moving 3 units up and 4 units to the left, we arrive at another point on the line.

b. Because $x = 60$ is the x-intercept in part (b) of Example 3, $(60, 0)$ is a point on the graph. The y-intercept is $y = 76{,}500$, so $(0, 76{,}500)$ is a point on the graph. Recall that x is measured in months and that y has units of dollars in the real-world setting (that is, the *context*) of Example 3. Substituting in the slope formula, we obtain

$$m = \frac{76{,}500 - 0}{0 - 60} = \frac{76{,}500}{-60} = -1275$$

This slope means that the amount owed decreases by \$1275 each month.

As Figure 1.31(a) to (d) indicates, the slope describes the direction of a line as well as the steepness.

The Relation Between Orientation of a Line and Its Slope

1. The slope is *positive* if the line *rises* upward toward the right.

2. The slope is *negative* if the line *falls* downward toward the right.

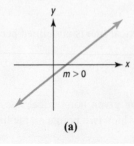

(a)

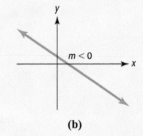

(b)

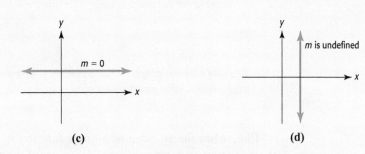

Figure 1.31

3. The slope of a *horizontal line* is 0 because a horizontal line has a vertical change (rise) of 0 between any two points on the line.

4. The slope of a *vertical line* does not exist because a vertical line has a horizontal change (run) of 0 between any two points on the line.

Remembering how the orientation of a line is related to its slope can help you check that you have the correct order of the points in the slope formula. For instance, if you find that the slope of a line is positive and you see that the line falls downward from left to right when you observe its graph, you know that there is a mistake either in the slope calculation or in the graph that you are viewing.

Slope and y-Intercept of a Line

There is an important connection between the slope of the graph of a linear equation and its equation when written in the form $y = f(x)$. To investigate this connection, we can graph the equation

$$y = 3x + 2$$

by plotting points or by using a graphing utility. By creating a table for values of x equal to 0, 1, 2, 3, and 4 (see Table 1.19), we can see that each time x increases by 1, y increases by 3 (see Figure 1.32). Thus, the slope is

$$\frac{\text{change in } y}{\text{change in } x} = \frac{3}{1} = 3$$

From this table, we see that the y-intercept is 2, the same value as the constant term of the equation. Observe also that the slope of the graph of $y = 3x + 2$ is the same as the coefficient of x.

Table 1.19

x	y
0	2
1	5
2	8
3	11
4	14

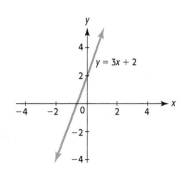

Figure 1.32

Because we denote the slope of a line by *m*, we have the following.

Slope and *y*-Intercept of a Line

The slope of the graph of the equation $y = mx + b$ is *m* and the *y*-intercept of the graph is *b*, so the graph crosses the *y*-axis at $(0, b)$.

Thus, when the equation of a linear function is written in the form $y = mx + b$ or $f(x) = mx + b$, we can "read" the values of the slope and *y*-intercept of its graph.

EXAMPLE 5 ▶   Loan Balance

As we saw in Example 3, the amount of money *y* remaining to be paid on the loan of $60,000 with $16,500 interest is

$$y = 76,500 - 1275x$$

where *x* is the number of monthly payments that have been made.

a. What are the slope and *y*-intercept of the graph of this function?

b. How does the amount owed on the loan change as the number of months increases?

SOLUTION

a. Writing this equation in the form $y = mx + b$ gives $y = -1275x + 76,500$. The coefficient of *x* is -1275, so the slope is $m = -1275$. The constant term is 76,500, so the *y*-intercept is $b = 76,500$.

b. The slope of the line indicates that the amount owed decreases by $1275 each month. This can be verified from the graph of the function in Figure 1.33 and the sample inputs and outputs in Table 1.20.

Table 1.20

Months (x)	Amount Owed (y)
1	75,225
2	73,950
3	72,675
4	71,400
5	70,125
6	68,850
7	67,575
8	66,300
9	65,025
10	63,750

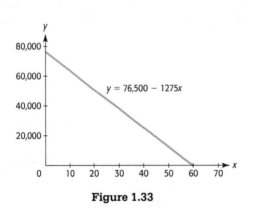

Figure 1.33

Constant Rate of Change

The function $y = 76,500 - 1275x$, whose graph is shown in Figure 1.33, gives the amount owed as a function of the months remaining on the loan. The coefficient of x, -1275, indicates that for each additional month, the value of y changes by -1275 (see Table 1.20). That is, the amount owed decreases at the **constant rate** of -1275 each month. Note that the **constant rate of change** of this linear function is equal to the **slope** of its graph. This is true for all linear functions.

Constant Rate of Change

The rate of change of the linear function $y = mx + b$ is the constant m, the slope of the graph of the function.

Note: The rate of change in an applied context should include appropriate units of measure. The rate of change describes by how much the output changes (increases or decreases) for every input unit.

EXAMPLE 6 ▶

Hispanics in the United States

Using data and projections from 1990 through 2050, the percent of Hispanics in the U.S. population can be modeled by

$$H(x) = 0.224x + 9.01$$

with x equal to the number of years after 1990. The graph of this model is shown in Figure 1.34. (Source: U.S. Census Bureau)

a. What is the slope of the graph of this function?

b. What does this slope tell us about the annual rate of change in the percent of Hispanics in the United States?

SOLUTION

a. The coefficient of x in the linear function $H(x) = 0.224x + 9.01$ is 0.224, so the slope of the line that is its graph is 0.224.

b. This slope tells us that the percent of Hispanics in the U.S. population is increasing by 0.224 each year since 1990. We say that the percent has increased by 0.224 *percentage point.*

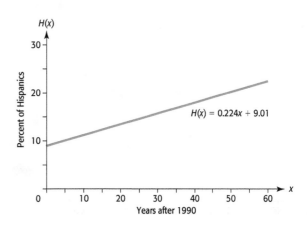

Figure 1.34

Revenue, Cost, and Profit

The **profit** that a company makes on its product is the difference between the amount received from sales (revenue) and the production and sales costs. If x units are produced and sold, we can write

$$P(x) = R(x) - C(x)$$

where

$$P(x) = \text{profit from sale of } x \text{ units}$$
$$R(x) = \text{total revenue from sale of } x \text{ units}$$
$$C(x) = \text{total cost of production and sale of } x \text{ units}$$

In general, **revenue** is found by using the equation

$$\text{revenue} = (\text{price per unit})(\text{number of units})$$

The **total cost** is composed of two parts: fixed costs and variable costs. **Fixed costs** (FC), such as depreciation, rent, and utilities, remain constant regardless of the number of units produced. **Variable costs** (VC) are those directly related to the number of units produced. Thus, the total cost, often simply called the cost, is found by using the equation

$$\text{cost} = \text{variable costs} + \text{fixed costs}$$

EXAMPLE 7 ▶ Cost, Revenue, and Profit

Suppose that a company manufactures 50-inch 3D plasma TVs and sells them for $1800 each. The costs incurred in the production and sale of the TVs are $400,000 plus $1000 for each TV produced and sold. Write the profit function for the production and sale of x TVs.

SOLUTION

The total revenue for x TVs is $1800x$, so the revenue function is $R(x) = 1800x$. The fixed costs are $400,000, so the total cost for x TVs is $1000x + 400,000$. Hence, $C(x) = 1000x + 400,000$. The profit function is given by $P(x) = R(x) - C(x)$, so

$$P(x) = 1800x - (1000x + 400,000)$$
$$P(x) = 800x - 400,000$$

Figure 1.35 shows the graphs of the three linear functions: $R(x)$, $C(x)$, and $P(x)$.

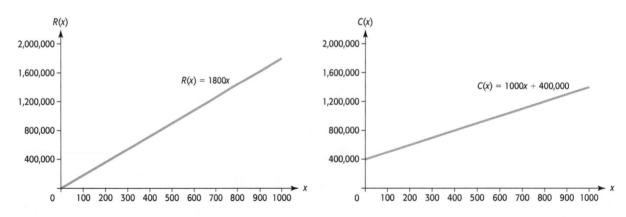

Figure 1.35

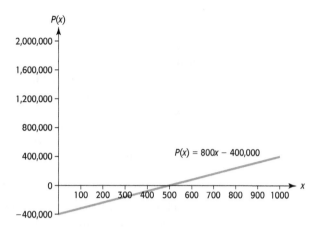

Figure 1.35 (Continued)

Marginal Cost, Revenue, and Profit

For total cost, total revenue, and profit functions* that are linear, the rates of change are called **marginal cost, marginal revenue**, and **marginal profit**, respectively. Suppose that the cost to produce and sell a product is $C(x) = 54.36x + 6790$ dollars, where x is the number of units produced and sold. This is a linear function, and its graph is a line with slope 54.36. Thus, the *rate of change* of this cost function, called the **marginal cost**, is \$54.36 per unit produced and sold. This means that the production and sale of each additional unit will cost an additional \$54.36.

EXAMPLE 8 ▶   **Marginal Revenue and Marginal Profit**

A company produces and sells a BlackBerry smartphone with revenue given by $R(x) = 89.50x$ dollars and cost given by $C(x) = 54.36x + 6790$ dollars, where x is the number of BlackBerries produced and sold.

a. What is the marginal revenue for this BlackBerry, and what does it mean?

b. Find the profit function.

c. What is the marginal profit for this BlackBerry, and what does it mean?

SOLUTION

a. The marginal revenue for this BlackBerry is the rate of change of the revenue function, which is the slope of its graph. Thus, the marginal revenue is \$89.50 per unit sold. This means that the sale of each additional BlackBerry will result in additional revenue of \$89.50.

b. To find the profit function, we subtract the cost function from the revenue function.

$$P(x) = 89.50x - (54.36x + 6790) = 35.14x - 6790$$

c. The marginal profit for this BlackBerry is the rate of change of the profit function, which is the slope of its graph. Thus, the marginal profit is \$35.14 per BlackBerry sold. This means that the production and sale of each additional BlackBerry will result in an additional profit of \$35.14.

---

* In this text, we frequently use "total cost" and "total revenue" interchangeably with "cost" and "revenue," respectively.

Special Linear Functions

A special linear function that has the form $y = 0x + b$, or $y = b$, where b is a real number, is called a **constant function**. The graph of the constant function $y = 3$ is shown in Figure 1.36(a). The temperature inside a sealed case containing an Egyptian mummy in a museum is a constant function of time because the temperature inside the case never changes. Notice that even though the range of a constant function consists of a single value, the input of a constant function is any real number or any real number that makes sense in the context of an applied problem. Another special linear function is the **identity function**

$$y = 1x + 0, \quad \text{or} \quad y = x$$

which is a linear function of the form $y = mx + b$ with slope $m = 1$ and y-intercept $b = 0$. For the general identity function $f(x) = x$, the domain and range are each the set of all real numbers. A graph of the identity function f is shown in Figure 1.36(b).

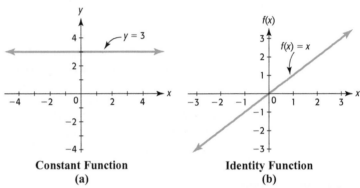

Constant Function	Identity Function
(a)	(b)

Figure 1.36

Skills CHECK 1.3

1. Which of the following functions are linear?

   a. $y = 3x^2 + 2$ **b.** $3x + 2y = 12$ **c.** $y = \dfrac{1}{x} + 2$

2. Is the graph in the figure below a function?

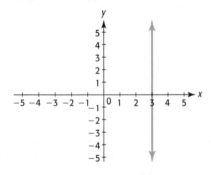

3. Find the slope of the line through $(4, 6)$ and $(28, -6)$.

4. Find the slope of the line through $(8, -10)$ and $(8, 4)$.

5. Find the slope of the line in the graph that follows.

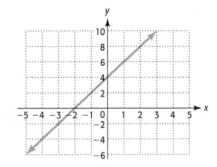

6. Find the slope of the line in the graph below.

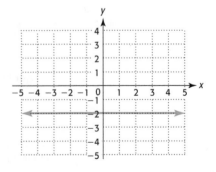

In Exercises 7–10, (a) find the x- and y-intercepts of the graph of the given equation, if they exist, and (b) graph the equation.

7. $5x - 3y = 15$ **8.** $x + 5y = 17$

9. $3y = 9 - 6x$ **10.** $y = 9x$

11. If a line is horizontal, then its slope is ____. If a line is vertical, then its slope is ____.

12. Describe the line whose slope was determined in Exercise 4.

For Exercises 13–14, determine whether the slope of the graph of the line is positive, negative, 0, or undefined.

13. a.

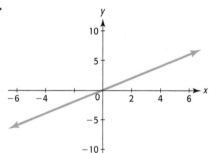

b.

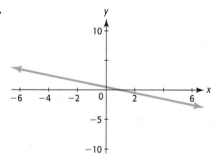

14. a.

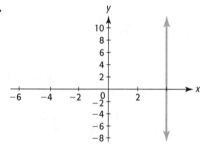

b.

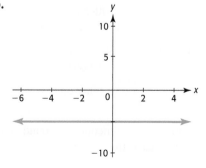

For Exercises 15–18, (a) give the slope of the line (if it exists) and the y-intercept (if it exists) and (b) graph the line.

15. $y = 4x + 8$ **16.** $3x + 2y = 7$

17. $5y = 2$ **18.** $x = 6$

For each of the functions in Exercises 19–21, do the following:
  a. *Find the slope and y-intercept (if possible) of the graph of the function.*
  b. *Determine if the graph is rising or falling.*
  c. *Graph each function on a window with the given x-range and a y-range that shows a complete graph.*

19. $y = 4x + 5; [-5, 5]$

20. $y = 0.001x - 0.03; [-100, 100]$

21. $y = 50,000 - 100x; [0, 500]$

22. Rank the functions in Exercises 19–21 in order of increasing steepness.

For each of the functions in Exercises 23–26, find the rate of change.

23. $y = 4x - 3$ **24.** $y = \dfrac{1}{3}x + 2$

25. $y = 300 - 15x$ **26.** $y = 300x - 15$

27. If a linear function has the points $(-1, 3)$ and $(4, -7)$ on its graph, what is the rate of change of the function?

28. If a linear function has the points $(2, 1)$ and $(6, 3)$ on its graph, what is the rate of change of the function?

29. a. Does graph (i) or graph (ii) represent the identity function?

  b. Does graph (i) or graph (ii) represent a constant function?

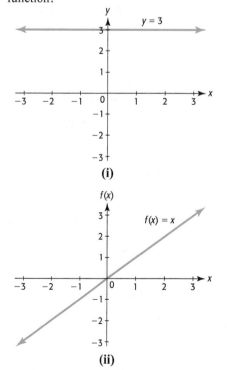

(i)

(ii)

30. What is the slope of the identity function?

31. a. What is the slope of the constant function $y = k$?

b. What is the rate of change of a constant function?

32. What is the rate of change of the identity function?

EXERCISES 1.3

33. *College Enrollment* The total fall enrollment in 4-year public institutions for the years 1990 through 2008 is given by $y = 0.014x + 2.290$, where x is the number of years after 1990 and y is millions of students. Is this a linear function? Why or why not? What is the independent variable?
(Source: U.S. Department of Education, National Center for Education Statistics)

34. *Women in the Workforce* Using data and projections from 1950 through 2050 gives the number y (in thousands) of women in the workforce as the function $y = -0.005x^2 + 1.278x + 13.332$, where x is the number of years from 1950. Is this a linear function? Why or why not?
(Source: U.S. Census Bureau, U.S. Dept. of Commerce)

35. *Marriage Rate* The marriage rate per 1000 population for the years 1987–2009 is given by $M(x) = -0.146x + 11.074$, where x is the number of years after 1980.

a. Why is this a linear function, with $y = M(x)$?

b. What is the slope? What does this slope tell you about the number of unmarried women who get married?
(Source: *National Vital Statistics Report*)

36. *Prescription Drug Sales* Retail prescription drug sales for the years 1995–2005, in billions of dollars, can be modeled by the function

$$y = 14.232x + 8.073$$

where x is the number of years after 1990.

a. Why is this a linear function?

b. What is the slope of the graph of the function?

c. What is the rate at which the sales grew during this period?
(Source: U.S. Census Bureau)

37. *Marijuana Use* The percent p of high school seniors who ever used marijuana can be related to x, the number of years after 2000, by the equation $25p + 21x = 1215$.

a. Find the x-intercept of the graph of this function.

b. Find and interpret the p-intercept of the graph of this function.

c. Graph the function, using the intercepts. What values of x on the graph represent years 2000 and after?

38. *Depreciation* An $828,000 building is depreciated for tax purposes by its owner, using the straight-line depreciation method. The value of the building after x months of use is given by $y = 828,000 - 2300x$ dollars.

a. Find and interpret the y-intercept of the graph of this function.

b. Find and interpret the x-intercept of the graph of this function.

c. Use the intercepts to graph the function for non-negative x- and y-values.

39. *Asparagus Cultivation* In Canada, the cultivation area (in hectares) of asparagus between 1999 and 2008 is shown in the table.

Year	1999	2000	2001	2002	2003
Cultivation Area (hectares)	1200	1200	1200	1200	1200

Year	2004	2005	2006	2007	2008
Cultivation Area (hectares)	1200	1200	1200	1200	1200

(Source: Sustainablog.org)

a. Sketch the data as a scatter plot, with y equal to the cultivation area and x equal to the year.

b. Could the data in the table be modeled by a constant function or the identity function?

c. Write the equation of a function that fits the data points.

d. Sketch a graph of the function you found in part (c) on the same axes as the scatter plot.

40. *Life Insurance* The monthly rates for a $100,000 life insurance policy for males aged 27–32 are shown in the table.

Age (years), x	27	28	29
Premium (dollars per month), y	11.81	11.81	11.81

Age (years), x	30	31	32
Premium (dollars per month), y	11.81	11.81	11.81

a. Could the data in this table be modeled by a constant function or the identity function?

b. Write an equation whose graph contains the data points in the table.

c. What is the slope of the graph of the function found in part (b)?

d. What is the rate of change of the data in the table?

41. *Cigarette Use* For the years 1965–2009, the percent p of adults who have tried cigarettes can be modeled by $p = 43.3 - 0.504t$, where t is the number of years after 1960.

a. Is the rate of change of the percent positive or negative?

b. How fast was the percent of adults who tried cigarettes during this period changing? Use the units in the problem in your answer.
(Source: monitoringthefuture.org)

42. *Crickets* The number of times per minute n that a cricket chirps can be modeled as a function of the Fahrenheit temperature T. The data can be approximated by the function

$$n = \frac{12T}{7} - \frac{52}{7}$$

a. Is the rate of change of the number of chirps positive or negative?

b. What does this tell us about the relationship between temperature and the number of chirps?

43. *Twitter* Between 2007 and 2010, "tweeting" became increasingly popular. The number of tweets (in thousands) reported by the microblogging company Twitter can be modeled by $T(x) = 16,665x - 116,650$, where x is the number of years after 2000.

a. What is the slope of the graph of this function?

b. Interpret the slope as a rate of change.

c. According to Computerworld, the number of tweets grew by 1400% from 2009 to 2010. Do your results from (b) confirm this?

44. *Diabetes* The figure below shows the projected percent of U.S. adults with diabetes for the years 2010 through 2050. What is the average rate of growth over this period of time?

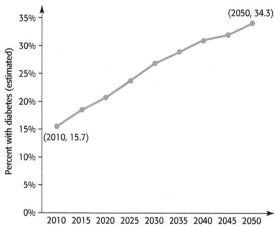

(Source: Centers for Disease Control and Prevention)

45. *Earnings and Minorities* According to the U.S. Equal Employment Opportunity Commission, the relationship between the median annual salaries of minorities and whites can be modeled by the function $M = 0.959W - 1.226$, where M and W represent the median annual salary (in thousands of dollars) for minorities and whites, respectively.

a. Is this function a linear function?

b. What is the slope of the graph of this function?

c. Interpret the slope as a rate of change.
(Source: *Statistical Abstract of the United States*)

46. *Marijuana Use* The percent p of high school seniors using marijuana daily can be modeled by $30p - 19x = 30$, where x is the number of years after 1990.

a. Use this model to determine the slope of the graph of this function if x is the independent variable.

b. What is the rate of change of the percent of high school seniors using marijuana per year?
(Source: Index of Leading Cultural Indicators)

47. *Black Population* Using data and projections from 1990 through 2050, the percent of the U.S. population that is black can be modeled by $B(x) = 0.057x + 12.3$, where x is the number of years after 1990.

a. What is the slope of the graph of this function?

b. Interpret the slope as a rate of change.
(Source: U.S. Census Bureau)

48. *Seawater Pressure* In seawater, the pressure p is related to the depth d according to the model $33p - 18d = 496$, where d is the depth in feet and p is in pounds per square inch.

a. What is the slope of the graph of this function?

b. Interpret the slope as a rate of change.

49. *Advertising Impact* An advertising agency has found that when it promotes a new product in a city, the weekly rate of change R of the number of people who are aware of it x weeks after it is introduced is given by $R = 3500 - 70x$. Find the x- and R-intercepts and then graph the function on a viewing window that is meaningful in the application.

50. *Internet Users* The percent of the U.S. population with Internet access can be modeled by

$$y = 5.0x - 6.5$$

where x is the number of years after 1995.

a. Find the slope and the y-intercept of the graph of this equation.

b. What interpretation could be given to the slope? (Source: Jupiter Media Metrix)

51. *Call Centers* Call centers are booming in the Philippines as multinational companies increasingly outsource these operations. The annual revenue for call centers in the Philippines from 2006 to 2010, in billions of dollars, can be modeled by $R(x) = 0.975x - 3.45$, where x is the number of years after 2000.

a. According to the model, what was the rate of change of revenue for call centers in the Philippines?

b. According to the model, what was the revenue for call centers in the Philippines in 2010?

c. Would this model be valid to estimate the revenue in 2000? Why or why not? (Source: Business Processing Association of the Philippines)

52. *Wireless Service Spending* The total amount spent in the United States for wireless communication services S (in billions of dollars) can be modeled by the function

$$S = 6.205 + 11.23t$$

where t is the number of years after 1995.

a. Find the slope and the y-intercept of the graph of this equation.

b. What interpretation could be given to the y-intercept?

c. What interpretation could be given to the slope? (Source: Cellular Telecommunications and Internet Association)

53. *Depreciation* Suppose the cost of a business property is $1,920,000 and a company depreciates it with the straight-line method. Suppose V is the value of the property after x years and the line representing the value as a function of years passes through the points (10, 1,310,000) and (20, 700,000).

a. What is the slope of the line through these points?

b. What is the annual rate of change of the value of the property?

54. *Men in the Workforce* The number of men in the workforce (in millions) for the years from 1890 to 2008 can be approximated by the linear model determined by connecting the points (1890, 18.1) and (2008, 67.8).

a. Find the annual rate of change of the model whose graph is the line connecting these points.

b. What does this tell us about men in the workforce?

55. *Profit* A company charting its profits notices that the relationship between the number of units sold x and the profit P is linear. If 300 units sold results in $4650 profit and 375 units sold results in $9000 profit, find the marginal profit, which is the rate of change of the profit.

56. *Cost* A company buys and retails baseball caps, and the total cost function is linear. The total cost for 200 caps is $2690, and the total cost for 500 caps is $3530. What is the marginal cost, which is the rate of change of the function?

57. *Marginal Cost* Suppose the monthly total cost for the manufacture of golf balls is $C(x) = 3450 + 0.56x$, where x is the number of balls produced each month.

a. What is the slope of the graph of the total cost function?

b. What is the marginal cost (rate of change of the cost function) for the product?

c. What is the cost of each additional ball that is produced in a month?

58. *Marginal Cost* Suppose the monthly total cost for the manufacture of 19-inch television sets is $C(x) = 2546 + 98x$, where x is the number of TVs produced each month.

a. What is the slope of the graph of the total cost function?

b. What is the marginal cost for the product?

c. Interpret the marginal cost for this product.

59. *Marginal Revenue* Suppose the monthly total revenue for the sale of golf balls is $R(x) = 1.60x$, where x is the number of balls sold each month.

a. What is the slope of the graph of the total revenue function?

b. What is the marginal revenue for the product?

c. Interpret the marginal revenue for this product.

60. *Marginal Revenue* Suppose the monthly total revenue from the sale of 19-inch television sets is $R(x) = 198x$, where x is the number of TVs sold each month.

    a. What is the slope of the graph of the total revenue function?

    b. What is the marginal revenue for the product?

    c. Interpret the marginal revenue for this product.

61. *Profit* The profit for a product is given by $P(x) = 19x - 5060$, where x is the number of units produced and sold. Find the marginal profit for the product.

62. *Profit* The profit for a product is given by the function $P(x) = 939x - 12,207$, where x is the number of units produced and sold. Find the marginal profit for the product.

1.4   Equations of Lines

KEY OBJECTIVES

- Write equations of lines using the slope-intercept form and the point-slope form
- Write equations of horizontal and vertical lines
- Write the equations of lines parallel or perpendicular to given lines
- Find the average rate of change over an interval for nonlinear functions
- Find the slope of the secant line between two points on a graph
- Find the difference quotient from $(x, f(x))$ to $(x + h, f(x + h))$
- Find average rates of change for approximately linear data

SECTION PREVIEW   **Blood Alcohol Percent**

Suppose that the blood alcohol percent for a 180-lb man is 0.11% if he has 5 drinks and that the percent increases by 0.02% for each additional drink. We can use this information to write the linear equation that models the blood alcohol percent p as a function of the number of drinks he has, because the rate of change is constant. (See Example 3.) In this section, we see how to write the equation of a linear function from information about the line, such as the slope and a point on the line or two points on the line. We also discuss **average rates of change** for data that can be modeled by nonlinear functions, slopes of secant lines, and how to create linear models that approximate data that are nearly linear. ■

Writing Equations of Lines

Creating a linear model from data involves writing a linear equation that describes the mathematical situation. If we know two points on a line or the slope of the line and one point, we can write the equation of the line.

Recall that if a linear equation has the form $y = mx + b$, then the coefficient of x is the slope of the line that is the graph of the equation, and the constant b is the y-intercept of the line. Thus, if we know the slope and the y-intercept of a line, we can write the equation of this line.

Slope-Intercept Form of the Equation of a Line

The slope-intercept form of the equation of a line with slope m and y-intercept b is

$$y = mx + b$$

In an applied context, m is the rate of change and b is the initial value (when $x = 0$).

EXAMPLE 1 ▶ Appliance Repair

An appliance repairman charges $60 for a service call plus $25 per hour for each hour spent on the repair. Assuming his service call charges can be modeled by a linear function of the number of hours spent on the repair, write the equation of the function.

SOLUTION

Let x represent the number of hours spent on the appliance repair and let y be the service call charge in dollars. The slope of the line is the amount that the charge increases for every hour of work done, so the slope is 25. Because $60 is the basic charge before any time is spent on the repair, 60 is the y-intercept of the line. Substituting these values in the slope-intercept form of the equation gives the equation of the linear function modeling this situation. When the repairman works x hours on the service call, the charge is

$$y = 25x + 60 \text{ dollars}$$

We next consider a form of an equation of a line that can be written if we know the slope and a point. If the slope of a line is m, then the slope between a fixed point (x_1, y_1) and any other point (x, y) on the line is also m. That is,

$$m = \frac{y - y_1}{x - x_1}$$

Solving for $y - y_1$ (that is, multiplying both sides of this equation by $x - x_1$) gives the **point-slope form** of the equation of a line.

Point-Slope Form of the Equation of a Line

The equation of the line with slope m that passes through a known point (x_1, y_1) is

$$y - y_1 = m(x - x_1)$$

EXAMPLE 2 ▶ Using a Point and Slope to Write an Equation of a Line

Write an equation for the line that passes through the point $(-1, 5)$ and has slope $\frac{3}{4}$.

SOLUTION

We are given $m = \frac{3}{4}$, $x_1 = -1$, and $y_1 = 5$. Substituting in the point-slope form, we obtain

$$y - 5 = \frac{3}{4}[x - (-1)]$$

$$y - 5 = \frac{3}{4}(x + 1)$$

$$y - 5 = \frac{3}{4}x + \frac{3}{4}$$

$$y = \frac{3}{4}x + \frac{23}{4}$$

Figure 1.37 shows the graph of this line.

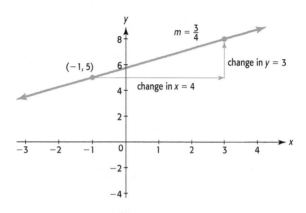

Figure 1.37

EXAMPLE 3 ▶ **Blood Alcohol Percent**

Table 1.21 gives the number of drinks and the resulting blood alcohol percent for a 180-pound man. (One drink is equal to 1.25 oz of 80-proof liquor, 12 oz of regular beer, or 5 oz of table wine; many states have set 0.08% as the legal limit for driving under the influence.)

a. Is the rate of change of the blood alcohol percent for a 180-pound man a constant? What is it?

b. Write the equation of the function that models the blood alcohol percent as a function of the number of drinks.

Table 1.21

Number of Drinks (x)	3	4	5	6	7	8	9	10
Blood Alcohol Percent (y)	0.07	0.09	0.11	0.13	0.15	0.17	0.19	0.21

SOLUTION

a. Yes. Each additional drink increases the blood alcohol percent by 0.02, so the rate of change is 0.02 percentage point per drink.

b. The rate of change is constant, so the function is linear, with its rate of change (and slope of its graph) equal to 0.02. Using this value and any point, like (5, 0.11), on the line gives us the equation we seek, with x equal to the number of drinks and y the blood alcohol percent.

$$y - 0.11 = 0.02(x - 5), \quad \text{or} \quad y = 0.02x + 0.01$$

If we know two points on a line, we can find the slope of the line and use either of the points and this slope to write the equation of the line.

EXAMPLE 4 ▶   **Using Two Points to Write an Equation of a Line**

Write the equation of the line that passes through the points $(-1, 5)$ and $(2, 4)$.

SOLUTION

Because we know two points on the line, we can find the slope of the line.

$$m = \frac{4 - 5}{2 - (-1)} = \frac{-1}{3} = -\frac{1}{3}$$

We can now substitute one of the points and the slope in the point-slope form to write the equation. Using the point $(-1, 5)$ gives

$$y - 5 = -\frac{1}{3}[x - (-1)], \quad \text{or} \quad y - 5 = -\frac{1}{3}x - \frac{1}{3}, \quad \text{so} \quad y = -\frac{1}{3}x + \frac{14}{3}$$

Using the point $(2, 4)$ gives

$$y - 4 = -\frac{1}{3}(x - 2)$$

$$y - 4 = -\frac{1}{3}x + \frac{2}{3}$$

$$y = -\frac{1}{3}x + \frac{14}{3}$$

Notice that the equations are the same, regardless of which of the two given points is used in the point-slope form.

If the rate of change of the outputs with respect to the inputs is a constant and we know two points that satisfy the conditions of the application, then we can write the equation of the linear function that models the application.

EXAMPLE 5 ▶   **Inmate Population**

The number of people (in millions) in U.S. prisons or jails grew at a constant rate from 2001 to 2009, with 1.345 million people incarcerated in 2001 and 1.617 million incarcerated in 2009.

a. What is the rate of growth of people incarcerated from 2001 to 2009?

b. Write the linear equation that models the number N of prisoners as a function of the year x.

c. The Pew Center on the States projected that 1.7 million people would be incarcerated in 2011. Does our model agree with this projection?

SOLUTION

a. The rate of growth is constant, so the points fit on a line and a linear function can be used to find the model for the number of incarcerated people as a function of the year. The rate of change (and slope of the line) is given by

$$m = \frac{1.617 - 1.345}{2009 - 2001} = 0.034$$

b. Substituting in the point-slope form of the equation of a line (with either point) gives the equation of the line that contains the two points and thus models the application.

$$N - 1.345 = 0.034(x - 2001), \quad \text{or} \quad N = 0.034x - 66.689$$

c. Substituting 2011 for x in the function gives

$$N(2011) = 0.034(2011) - 66.689 = 1.685$$

Thus, our estimate that 1.685 million people would be incarcerated in 2011 is close to the Pew projection.

EXAMPLE 6 ▶ **Writing Equations of Horizontal and Vertical Lines**

Write equations for the lines that pass through the point $(-1, 5)$ and have

a. slope 0.   **b.** undefined slope.

SOLUTION

a. If $m = 0$, the point-slope form gives us the equation

$$y - 5 = 0\left[x - (-1)\right]$$
$$y = 5$$

Because the output is always the same value, the graph of this linear function is a horizontal line. See Figure 1.38(a).

b. Because m is undefined, we cannot use the point-slope form to write the equation of this line. Lines with undefined slope are vertical lines. Every point on the vertical line through $(-1, 5)$ has an x-coordinate of -1. Thus, the equation of the line is $x = -1$. Note that this equation does not represent a function. See Figure 1.38(b).

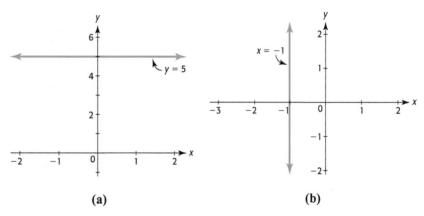

(a)   (b)

Figure 1.38

As we saw in Example 6, parts (a) and (b), there are special forms when the lines are horizontal or vertical.

Vertical and Horizontal Lines

A vertical line has the form $x = a$, where a is a constant and a is the x-coordinate of any point on the line.
A horizontal line has the form $y = b$, where b is a constant and b is the y-coordinate of any point on the line.

Parallel and Perpendicular Lines

Clearly horizontal lines are perpendicular to vertical lines. Two distinct nonvertical lines that have the same slope are **parallel**, and conversely. If a line has slope $m \neq 0$, any line **perpendicular** to it will have slope $-\dfrac{1}{m}$. That is, the slopes of perpendicular lines are negative reciprocals of each other if neither line is horizontal.

EXAMPLE 7 ▶   Parallel and Perpendicular Lines

Write the equation of the line through $(4, 5)$ and

a. parallel to the line with equation $3x + 2y = -1$.

b. perpendicular to the line with equation $3x + 2y = -1$.

SOLUTION

a. To find the slope of the line with equation $3x + 2y = -1$, we solve for y.

$$3x + 2y = -1$$
$$2y = -3x - 1$$
$$y = -\frac{3}{2}x - \frac{1}{2}$$

The line through $(4, 5)$ parallel to this line has the same slope, $-\dfrac{3}{2}$, and its equation is

$$y - 5 = -\frac{3}{2}(x - 4)$$

$$y = -\frac{3}{2}x + 11$$

b. The line through $(4, 5)$ perpendicular to $3x + 2y = -1$ has slope $\dfrac{2}{3}$, and its equation is

$$y - 5 = \frac{2}{3}(x - 4)$$

$$y = \frac{2}{3}x + \frac{7}{3}$$

In Example 7(b) we found that the equation of the line was

$$y = \frac{2}{3}x + \frac{7}{3}$$

By multiplying both sides of this equation by 3 and writing the new equation with x and y on the same side of the equation, we have a new form of the equation:

$$3y = 2x + 7, \quad \text{or} \quad 2x - 3y = -7$$

This equation is called the **general form** of the equation of the line.

General Form of the Equation of a Line

The general form of the equation of a line is $ax + by = c$, where a, b, and c are real numbers, with a and b not both equal to 0.

In summary, these are the forms we have discussed for the equation of a line.

Forms of Linear Equations

General form:	$ax + by = c$	where a, b, and c are real numbers, with a and b not both equal to 0.
Point-slope form:	$y - y_1 = m(x - x_1)$	where m is the slope of the line and (x_1, y_1) is a point on the line.
Slope-intercept form:	$y = mx + b$	where m is the slope of the line and b is the y-intercept.
Vertical line:	$x = a$	where a is a constant, and a is the x-coordinate of any point on the line. The slope is undefined.
Horizontal line:	$y = b$	where b is a constant, and b is the y-coordinate of any point on the line. The slope is 0.

Average Rate of Change

If the graph of a function is not a line, the function is nonlinear and the slope of a line joining two points on the curve may change as different points are chosen on the curve. The calculation and interpretation of the slope of a curve are topics you will study if you take a calculus course. However, there is a quantity called the **average rate of change** that can be applied to any function relating two variables.

In general, we can find the average rate of change of a function between two input values if we know how much the function outputs change between the two input values.

Average Rate of Change

The average rate of change of $f(x)$ with respect to x over the interval from $x = a$ to $x = b$ (where $a < b$) is calculated as

$$\text{average rate of change} = \frac{\text{change in } f(x) \text{ values}}{\text{corresponding change in } x \text{ values}} = \frac{f(b) - f(a)}{b - a}$$

How is the average rate of change over some interval of points related to the slope of a line connecting the points? For any function, the average rate of change between two points on its graph is the slope of the line joining the two points. Such a line is called a **secant line**.

EXAMPLE 8 ▶   Worldwide Hybrid Vehicle Sales

The worldwide hybrid vehicle sales for the years from 2004 to 2009 are shown in Figure 1.39(a). The number of sales can be modeled by the equation

$$S(x) = 8.348x^3 - 159.010x^2 + 1071.294x - 2116.822$$

where $S(x)$ is in thousands and x is the number of years after 2000. A graph of $S(x)$ is shown in Figure 1.39(b).

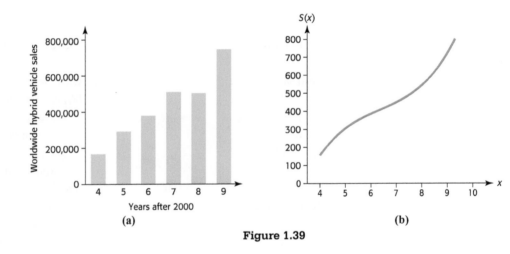

Figure 1.39

a. Use the model to find the average rate of change of hybrid vehicle sales between 2004 and 2009.

b. Interpret your answer to part (a).

c. What is the relationship between the slope of the secant line joining the points (4, 158.5) and (9, 730.7) and the answer to part (a)?

SOLUTION

a. We find the average rate of change between the two points to be

$$\frac{S(9) - S(4)}{9 - 4} = \frac{730.7 - 158.5}{9 - 4} = \frac{572.2}{5} = 114.44$$

or 114,440 units per year.

b. One possible interpretation is that, on average between 2004 and 2009, sales of hybrid vehicles increased by 114,440 per year.

c. The slope of this line, shown in Figure 1.40, is numerically the same as the average rate of change of the sales between 2004 and 2009 found in part (a).

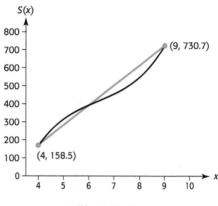

Figure 1.40

If we have points $(x, f(x))$ and $(x + h, f(x + h))$ on the graph of $y = f(x)$, we can find the average rate of change of the function from x to $x + h$, called the **difference quotient**, as follows.

Difference Quotient

The average rate of change of the function $f(x)$ from x to $x + h$ is

$$\frac{f(x + h) - f(x)}{x + h - x} = \frac{f(x + h) - f(x)}{h}$$

EXAMPLE 9 ▶ **Average Rate of Change**

For the function $f(x) = x^2 + 1$, whose graph is shown in Figure 1.41, find

a. $f(x + h)$.

b. $f(x + h) - f(x)$.

c. the average rate of change $\dfrac{f(x + h) - f(x)}{h}$.

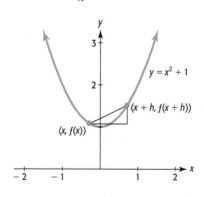

Figure 1.41

SOLUTION

a. $f(x + h)$ is found by substituting $x + h$ for x in $f(x) = x^2 + 1$:

$$f(x + h) = (x + h)^2 + 1 = (x^2 + 2xh + h^2) + 1$$

b. $f(x + h) - f(x) = (x^2 + 2xh + h^2) + 1 - (x^2 + 1)$

$$= x^2 + 2xh + h^2 + 1 - x^2 - 1 = 2xh + h^2$$

c. The average rate of change is $\dfrac{f(x + h) - f(x)}{h} = \dfrac{2xh + h^2}{h} = 2x + h.$

Approximately Linear Data

Real-life data are rarely perfectly linear, but some sets of real data points lie sufficiently close to a line that two points can be used to create a linear function that models the data. Consider the following example.

EXAMPLE 10 ▶   **High School Enrollment**

Table 1.22 and Figure 1.42 show the enrollment (in thousands) in grades 9–12 at U.S. public and private schools for the even years 1990–2008.

a. Create a scatter plot of the data. Does a line fit the data points exactly?

b. Find the average rate of change of the high school enrollment between 1990 and 2008.

c. Write the equation of the line determined by this rate of change and one of the two points.

Table 1.22

Year, x	Enrollment, y (thousands)
1990	12,488
1992	12,882
1994	13,376
1996	14,060
1998	14,428
2000	14,802
2002	15,332
2004	16,048
2006	16,498
2008	16,715

(Source: U.S. National Center for Education Statistics)

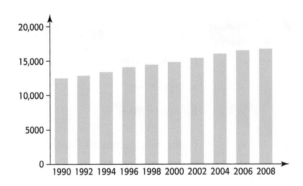

Figure 1.42

SOLUTION

a. A scatter plot is shown in Figure 1.43. The graph shows that the data do not fit a linear function exactly, but that a linear function could be used as an approximate model for the data.

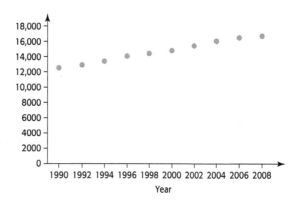

Figure 1.43

b. To find the average rate of change between 1990 and 2008, we use the points (1990, 12,488) and (2008, 16,715).

$$\frac{16{,}715 - 12{,}488}{2008 - 1990} = \frac{4227}{18} \approx 235$$

This means that the high school enrollments have grown at an average rate of 235 thousand per year (because the outputs are in thousands).

c. Because the enrollment growth is approximately linear, the average rate of change can be used as the slope of the graph of a linear function describing the enrollments from 1990 through 2008. The graph connects the points (1990, 12,488) and (2008, 16,715), so we can use either of these two points to write the equation of the line. Using the point (1990, 12,488) and $m = 235$ gives the equation

$$y - 12{,}488 = 235(x - 1990)$$
$$y - 12{,}488 = 235x - 467{,}650$$
$$y = 235x - 455{,}162$$

This equation is a model of the enrollment, in thousands, as a function of the year.

The equation in Example 10 approximates a model for the data, but it is not the best possible model for the data. In Section 2.2, we will see how technology can be used to find the linear function that is the best fit for a set of data of this type.

Skills CHECK 1.4

For Exercises 1–18, write the equation of the line with the given conditions.

1. Slope 4 and *y*-intercept $\dfrac{1}{2}$

2. Slope 5 and *y*-intercept $\dfrac{1}{3}$

3. Slope $\dfrac{1}{3}$ and *y*-intercept 3

4. Slope $-\dfrac{1}{2}$ and *y*-intercept -8

5. Through the point $(4, -6)$ with slope $-\dfrac{3}{4}$

6. Through the point $(-4, 3)$ with slope $-\dfrac{1}{2}$

7. Vertical line, through the point $(9, -10)$

8. Horizontal line, through the point $(9, -10)$

9. Passing through $(-2, 1)$ and $(4, 7)$

10. Passing through $(-1, 3)$ and $(2, 6)$

11. Passing through $(5, 2)$ and $(-3, 2)$

12. Passing through $(9, 2)$ and $(9, 5)$

13. x-intercept -5 and y-intercept 4

14. x-intercept 4 and y-intercept -5

15. Passing through $(4, -6)$ and parallel to the line with equation $3x + y = 4$

16. Passing through $(5, -3)$ and parallel to the line with equation $2x + y = -3$

17. Passing through $(-3, 7)$ and perpendicular to the line with equation $2x + 3y = 7$

18. Passing through $(-4, 5)$ and perpendicular to the line with equation $3x + 2y = -8$

For Exercises 19 and 20, write the equation of the line whose graph is shown.

19.

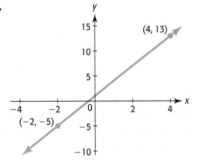

20.

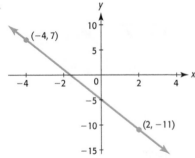

21. Write the equation of a function if its rate of change is -15 and $y = 12$ when $x = 0$.

22. Write the equation of a function if its rate of change is -8 and $y = -7$ when $x = 0$.

23. For the function $y = x^2$, compute the average rate of change between $x = -1$ and $x = 2$.

24. For the function $y = x^3$, compute the average rate of change between $x = -1$ and $x = 2$.

25. For the function shown in the figure, find the average rate of change from $(-2, 7)$ to $(1, -2)$.

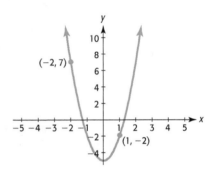

26. For the function shown in the figure, find the average rate of change from $(-1, 2)$ to $(2, -4)$.

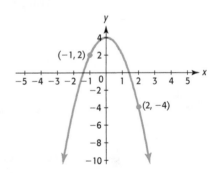

For the functions given in Exercises 27–30, find

$$\frac{f(x + h) - f(x)}{h}$$

27. $f(x) = 45 - 15x$

28. $f(x) = 32x + 12$

29. $f(x) = 2x^2 + 4$

30. $f(x) = 3x^2 + 1$

In Exercises 31–32, use the tables on the next page, which give a set of input values x and the corresponding outputs y that satisfy a function.

31. **a.** Would you say that a linear function could be used to model the data? Explain.

b. Write the equation of the linear function that fits the data.

x	10	20	30	40	50
y	585	615	645	675	705

32. a. Verify that the values satisfy a linear function.

b. Write the equation of the linear function that fits the data.

x	1	7	13	19
y	−0.5	8.5	17.5	26.5

EXERCISES 1.4

33. *Utility Charges* Palmetto Electric determines its monthly bills for residential customers by charging a base price of $12.00 plus an energy charge of 10.34 cents for each kilowatt-hour (kWh) used. Write an equation for the monthly charge y (in dollars) as a function of x, the number of kWh used.

34. *Phone Bills* For interstate calls, AT&T charges 10 cents per minute plus a base charge of $2.99 each month. Write an equation for the monthly charge y as a function of the number of minutes of use.

35. *Depreciation* A business uses straight-line depreciation to determine the value y of a piece of machinery over a 10-year period. Suppose the original value (when $t = 0$) is equal to $36,000 and the value is reduced by $3600 each year. Write the linear equation that models the value y of this machinery at the end of year t.

36. *Sleep* Each day a young person should sleep 8 hours plus $\frac{1}{4}$ hour for each year the person is under 18 years of age.

a. Based on this information, how much sleep does a 10-year-old need?

b. Based on this information, how much sleep does a 14-year-old need?

c. Use the answers from parts (a) and (b) to write a linear equation relating hours of sleep y to age x, for $6 \leq x \leq 18$.

d. Use your equation from part (c) to verify that an 18-year-old needs 8 hours of sleep.

37. *Internet Advertising* The amount spent on Internet advertising was $22.7 billion in 2009 and is expected to grow at a rate of $2.25 billion per year for the next five years.

a. Write an equation for the amount of Internet advertising spending as a function of the number of years after 2009.

b. Use the function to estimate the amount that will be spent on Internet advertising in 2015.
(Source: emarketer.com)

38. *Magazine Advertising* The amount spent on magazine advertising was $15.5 billion in 2009 and is expected to decrease at a rate of $0.65 billion per year for the next five years.

a. Write an equation for the amount of magazine advertising spending as a function of the number of years after 2009.

b. Use the function to estimate the amount that will be spent on magazine advertising in 2015.
(Source: emarketer.com)

39. *Depreciation* A business uses straight-line depreciation to determine the value y of an automobile over a 5-year period. Suppose the original value (when $t = 0$) is equal to $26,000 and the salvage value (when $t = 5$) is equal to $1000.

a. By how much has the automobile depreciated over the 5 years?

b. By how much is the value of the automobile reduced at the end of each of the 5 years?

c. Write the linear equation that models the value s of this automobile at the end of year t.

40. *Retirement* For Pennsylvania state employees for whom the average of the three best yearly salaries is $75,000, the retirement plan gives an annual pension of 2.5% of 75,000, multiplied by the number of years of service. Write the linear function that models the pension P in terms of the number of years of service y.
(Source: Pennsylvania State Retirement Fund)

41. *Patrol Cars* The Beaufort County Sheriff's office assigns a patrol car to each of its deputy sheriffs, who keeps the car 24 hours a day. As the population of the county grew, the number of deputies and the number of patrol cars also grew. The data in the following table could describe how many patrol cars were necessary as the number of deputies increased. Write the function that models the relationship between the number of deputies and the number of patrol cars.

Number of Deputies	10	50	75	125	180	200	250
Number of Patrol Cars in Use	10	50	75	125	180	200	250

42. *Search Market Share* Data from comScore shows that Google's share of the U.S. search market has remained flat, as shown in the table below.

Date	April 2010	May 2010	June 2010	July 2010	August 2010	Sept 2010
Market Share (%)	66	66	66	66	66	66

Date	Oct 2010	Nov 2010	Dec 2010	Jan 2011	Feb 2011	March 2011
Market Share (%)	66	66	66	66	66	66

a. Write the linear model that gives Google's market share as a function of the number of months after April 2010.

b. What is the rate of change of Google's market share for this time period?

43. *Profit* A company charting its profits notices that the relationship between the number of units sold, x, and the profit, P, is linear. If 300 units sold results in $4650 profit and 375 units sold results in $9000 profit, write the equation that models its profit.

44. *Cost* A company buys and retails baseball caps. The total cost function is linear, the total cost for 200 caps is $2680, and the total cost for 500 caps is $3580. Write the equation that models this cost function.

45. *Depreciation* Suppose the cost of a business property is $1,920,000 and a company depreciates it with the straight-line method. If V is the value of the property after x years and the line representing the value as a function of years passes through the points

(10, 1,310,000) and (20, 700,000), write the equation that gives the annual value of the property.

46. *Depreciation* Suppose the cost of a business property is $860,000 and a company depreciates it with the straight-line method. Suppose y is the value of the property after t years.

a. What is the value at the beginning of the depreciation (when $t = 0$)?

b. If the property is completely depreciated ($y = 0$) in 25 years, write the equation of the line representing the value as a function of years.

47. *Cigarette Use* The percent of adults who smoke cigarettes can be modeled by a linear function $p = f(t)$, where t is the number of years after 1960. If two points on the graph of this function are (25, 30.7) and (50, 18.1), write the equation of this function.

48. *Earnings and Race* Data from 2003 for various age groups show that for each $100 increase in median weekly income for whites, the median weekly income for blacks increases by $61.90. For these workers, the median weekly income for whites was $676 and for blacks was $527. Write the equation that gives the median weekly income for blacks as a function of the median weekly income for whites. (Source: U.S. Department of Labor)

49. *Drinking and Driving* The following table gives the number of drinks and the resulting blood alcohol percent for a 100-lb woman legally considered to be driving under the influence (DUI). (One drink is equal to 1.25 oz of 80-proof liquor, 12 oz of regular beer, or 5 oz of table wine; many states have set 0.08% as the legal limit for driving under the influence.)

a. The average rate of change of the blood alcohol percent with respect to the number of drinks is a constant. What is it?

b. Use the rate of change and one point determined by a number of drinks and the resulting blood alcohol percent to write the equation of a linear model for this data.

Number of Drinks	2	3	4	5	6	7
Blood Alcohol Percent	0.09	0.135	0.18	0.225	0.27	0.315

(Source: Pennsylvania Liquor Control Board)

50. *Blood Alcohol Percent* The table on the next page gives the number of drinks and the resulting blood alcohol percent for a 220-lb man.

a. The rate of change in blood alcohol percent per drink for a 220-lb man is a constant. What is it?

b. Write the equation of the function that models the blood alcohol percent as a function of the number of drinks.

Number of Drinks	0	1	2	3	4
Blood Alcohol Percent	0	0.017	0.034	0.051	0.068

Number of Drinks	5	6	7	8	9
Blood Alcohol Percent	0.085	0.102	0.119	0.136	0.153

(Source: Pennsylvania Liquor Control Board)

51. *Men in the Workforce* The number of men in the workforce (in millions) for selected decades from 1890 to 2010 is shown in the figure. The decade is defined by the year at the beginning of the decade, and $g(t)$ is defined by the average number of men (in millions) in the workforce during the decade (indicated by the point on the graph within the decade). The data can be approximated by the linear model determined by the line connecting (1890, 18.1) and (1990, 68.5).

a. Write the equation of the line connecting these two points to find a linear model for the data.

b. Does this line appear to be a reasonable fit to the data points?

c. How does the slope of this line compare with the average rate of change in the function during this period?

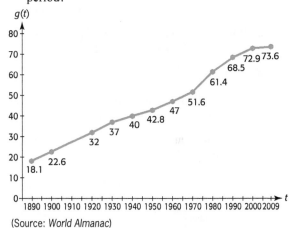

(Source: *World Almanac*)

52. *Farm Workers* The figure below shows the percent p of U.S. workers in farm occupations for selected years t.

a. Write the equation of a line connecting the points (1820, 71.8) and (2005, 1.5), with values rounded to two decimal places.

b. Does this line appear to be a reasonable fit to the data points?

c. Interpret the slope of this line as a rate of change of the percent of farm workers.

d. Can the percent p of U.S. workers in farm occupations continue to fall at this rate for 20 more years? Why or why not?

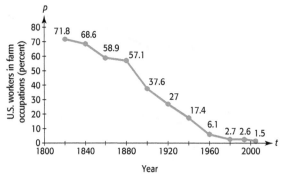

(Source: Data from World Almanac)

53. *Education Spending* Personal expenditures for higher education rose dramatically from 1990 to 2008. The figure shows the personal expenditures for higher education in billions of dollars for selected years during 1990–2008.

a. If a line were drawn connecting the points (1990, 34.7) and (2008, 135), what would be the slope of the line?

b. What is the average rate of change of expenditures between 1990 and 2008?

c. Is the rate of change of expenditures the same each year?

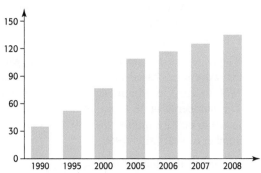

(Source: Bureau of Economic Analysis, U.S. Department of Commerce)

54. *Enrollment Projection* The figure on the next page shows enrollment projections for three schools. (Outputs are measured in students.)

a. What is the slope of the line joining the two points in the figure that show the Beaufort enrollment projections?

b. Find the average rate of change in projected enrollment for students in Beaufort schools between 2000 and 2005.

c. How should this information be used when planning future school building in Beaufort?

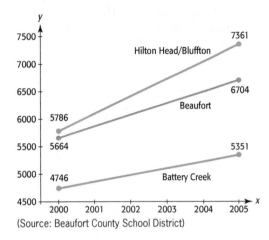

(Source: Beaufort County School District)

55. *Teen Birth Rate* The birth rate in 2009 for U.S. girls ages 15 to 19 was the lowest since the government began tracking the statistic in 1940. The highest birth rate for U.S. teens was 96.3 per 1000 girls in 1957. In 2009, the birth rate was lowest at 39.1.

a. What is the slope of the line joining the points (7, 96.3) and (59, 39.1)?

b. What is the average rate of change in the birth rate over this period?

c. What does this average rate of change tell about the birth rate?

d. Use the slope from part (a) and the birth rate from 2009 to write the equation of the line. Let x represent the number of years after 1950.
(Source: Centers for Disease Control and Prevention)

56. *Women in the Workforce* The number of women in the workforce, based on data and projections from 1950 to 2050, can be modeled by a linear function. The number was 18.4 million in 1950 and is projected to be 81.6 million in 2030. Let x represent the number of years after 1950.

a. What is the slope of the line through (0, 18.4) and (80, 81.6)?

b. What is the average rate of change in the number of women in the workforce during this time period?

c. Use the slope from part (a) and the number of millions of women in the workforce in 1950 to write the equation of the line.
(Source: U.S. Department of Labor)

57. *Prison Population* The graph showing the total number of prisoners in state and federal prisons for the years 1960 through 2009 is shown in the figure. There were 212,953 prisoners in 1960 and 1,617,478 in 2009.

a. What is the average rate of growth of the prison population from 1960 to 2009?

b. What is the slope of the line connecting the points satisfying the conditions above?

c. Write the equation of the secant line joining these two points on the curve.

d. Can this secant line be used to make a good estimate of total number of prisoners in 2015?

e. If a line were drawn between points associated with 1990 and 2004, would the equation of this line give a better estimate of total number of prisoners in 2015?

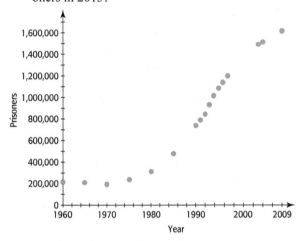

(Source: U.S. Bureau of Justice Statistics)

58. *Investment* The graph of the future value of an investment of $1000 for x years earning interest at a rate of 8% compounded continuously is shown in the following figure. The $1000 investment is worth approximately $1083 after 1 year and about $1492 after 5 years.

a. What is the average rate of change in the future value over the 4-year period?

b. Interpret this average rate of change.

c. What is the slope of the line connecting the points satisfying the conditions above?

d. Write the equation of the secant line joining the two given points.

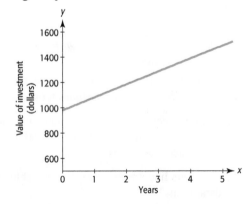

59. *Women in the Workforce* The number of women in the workforce (in millions) for selected years from 1890 to 2010 is shown in the following figure.

a. Would the data in the scatter plot be modeled well by a linear function? Why or why not?

b. The number of women in the workforce was 10.519 million in 1930 and 16.443 million in 1950. What is the average rate of change in the number of women in the workforce during this period?

c. If the number of women in the workforce was 16.443 million in 1950 and 75.500 million in 2010, what is the average rate of change in the number of women in the workforce during this period?

d. Is it reasonable that these two average rates of change are different? How can you tell this from the graph?

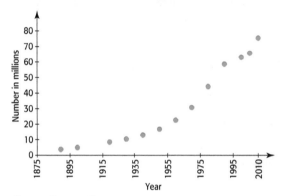

(Source: *Newsweek*)

60. *Working Age* The scatter plot below projects the ratio of the working-age population to the elderly.

a. Do the data appear to fit a linear function?

b. The data points shown in the scatter plot from 2010 to 2030 are projections made from a mathematical model. Do those projections appear to be made with a linear model? Explain.

c. If the ratio is 3.9 in 2010 and projected to be 2.2 in 2030, what is the average annual rate of change of the data over this period of time?

d. Write the equation of the line joining these two points.

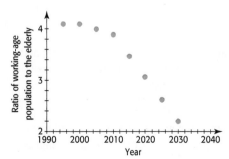

(Source: *Newsweek*)

61. *U.S. Population* The total U.S. population for selected years from 1950 to 2009 is shown in the table below, with the population given in thousands. Use x to represent the number of years from 1950 and y to represent the number of thousands of people.

a. Find the average annual rate of change in population during 1950–2009, with the appropriate units.

b. Use the slope from part (a) and the population in 1950 to write the equation of the line associated with 1950 and 2009.

c. Use the equation to estimate the population in 1975. Does it agree with the population shown in the table?

d. Why might the values be different?

Year	Population (thousands)	Year	Population (thousands)
1950	152,271	1985	238,466
1955	165,931	1990	249,948
1960	180,671	1995	263,044
1965	194,303	2000	281,422
1970	205,052	2005	296,410
1975	215,973	2009	307,007
1980	227,726		

(Source: U.S. Census Bureau)

62. *Social Agency* A social agency provides emergency food and shelter to two groups of clients. The first group has x clients who need an average of $300 for emergencies, and the second group has y clients who need an average of $200 for emergencies. The agency has $100,000 to spend for these two groups.

a. Write the equation that gives the number of clients who can be served in each group.

b. Find the y-intercept and the slope of the graph of this equation. Interpret each value.

c. If 10 clients are added from the first group, what happens to the number served in the second group?

chapter 1 ▶ SUMMARY

In this chapter, we studied the basic concept of a function and how to recognize functions with tables, graphs, and equations. We studied linear functions in particular—graphing, finding slope, model formation, and equation writing. We used graphing utilities to graph functions and to evaluate functions. We solved business and economics problems involving linear functions.

Key Concepts and Formulas

1.1 Functions and Models

Scatter plot of data	A scatter plot is a graph of pairs of values that represent real-world information collected in numerical form.
Function definition	A function is a rule or correspondence that determines exactly one output for each input. The function may be defined by a set of ordered pairs, a table, a graph, or an equation.
Domain and range	The set of possible inputs for a function is called its domain, and the set of possible outputs is called its range. In general, if the domain of a function is not specified, we assume that it includes all real numbers except • values that result in a denominator of 0 • values that result in an even root of a negative number
Independent variable and dependent variable	If x represents any element in the domain, then x is called the independent variable; if y represents an output of the function from an input x, then y is called the dependent variable.
Tests for functions	We can test for a function graphically, numerically, and analytically.
Vertical line test	A set of points in a coordinate plane is the graph of a function if and only if no vertical line intersects the graph in more than one point.
Function notation	We denote that y is a function of x using function notation when we write $y = f(x)$.
Modeling	The process of translating real-world information into a mathematical form so that it can be applied and then interpreted in the real-world setting is called modeling.
Mathematical model	A mathematical model is a functional relationship (usually in the form of an equation) that includes not only the function rule but also descriptions of all involved variables and their units of measure.

1.2 Graphs of Functions

Point-plotting method	The point-plotting method of sketching a graph involves sketching the graph of a function by plotting enough points to determine the shape of the graph and then drawing a smooth curve through the points.
Complete graph	A graph is a complete graph if it shows the basic shape of the graph and important points on the graph (including points where the graph crosses the axes and points where the graph turns) and suggests what the unseen portions of the graph will be.
Using a calculator to draw a graph	After writing the function with x representing the independent variable and y representing the dependent variable, enter the function in the equation editor of the graphing utility. Activate the graph with ZOOM or GRAPH . (Set the x- and y-boundaries of the viewing window before pressing GRAPH .)

Viewing windows	The values that define the viewing window can be set manually or by using the $\boxed{\text{ZOOM}}$ keys. The boundaries of a viewing window are $x_{\min}$: the smallest value on the x-axis (the left boundary of the window) $x_{\max}$: the largest value on the x-axis (the right boundary of the window) $y_{\min}$: the smallest value on the y-axis (the bottom boundary of the window) $y_{\max}$: the largest value on the y-axis (the top boundary of the window)
Spreadsheets	Spreadsheets like Excel can be used to create accurate graphs, sometimes better looking than those created with graphing calculators.
Aligning data	Using aligned inputs (input values that have been shifted horizontally to smaller numbers) rather than the actual data values results in smaller coefficients in models and less involved computations.
Evaluating functions with a calculator	We can find (or estimate) the output of a function $y = f(x)$ at specific inputs with a graphing calculator by using $\boxed{\text{TRACE}}$ and moving the cursor to (or close to) the value of the independent variable.
Graphing data points	Graphing utilities can be used to create lists of numbers and to create graphs of the data stored in the lists.
Scaling data	Replacing the outputs of a set of data by dividing or multiplying each number by a given number results in models with less involved computations.

1.3 Linear Functions

Linear functions	A linear function is a function of the form $f(x) = ax + b$, where a and b are constants. The graph of a linear function is a line.
Intercepts	A point (or the x-coordinate of the point) where a graph crosses or touches the horizontal axis is called an x-intercept, and a point (or the y-coordinate of the point) where a graph crosses or touches the vertical axis is called a y-intercept. To find the x-intercept(s) of the graph of an equation, set $y = 0$ in the equation and solve for x. To find the y-intercept(s), set $x = 0$ and solve for y.
Slope of a line	The slope of a line is a measure of its steepness and direction. The slope is defined as $$\text{slope} = \frac{\text{vertical change}}{\text{horizontal change}} = \frac{\text{rise}}{\text{run}}$$ If (x_1, y_1) and (x_2, y_2) are two points on a line, then the slope of the line is $$m = \frac{y_2 - y_1}{x_2 - x_1}$$
Slope and y-intercept of a line	The slope of the graph of the equation $y = mx + b$ is m, and the y-intercept of the graph is b. (This is called slope-intercept form.)
Constant rate of change	If a model is linear, the rate of change of the outputs with respect to the inputs will be constant, and the rate of change equals the slope of the line that is the graph of a linear function. Thus, we can determine if a linear model fits a set of real data by determining if the rate of change remains constant for the data.

Revenue, cost, and profit	If a company sells x units of a product for p dollars per unit, then the total revenue for this product can be modeled by the linear function $R(x) = px$. $$\text{profit} = \text{revenue} - \text{cost}, \quad \text{or} \quad P(x) = R(x) - C(x)$$
Marginal cost, marginal revenue, and marginal profit	For total cost, revenue, and profit functions that are linear, the rates of change are called marginal cost, marginal revenue, and marginal profit, respectively.
Special linear functions	
• **Constant function**	A special linear function that has the form $y = c$, where c is a real number, is called a constant function.
• **Identity function**	The identity function $y = x$ is a linear function of the form $y = mx + b$ with slope $m = 1$ and y-intercept $b = 0$.

1.4 Equations of Lines

Writing equations of lines	
• **Slope-intercept form**	The slope-intercept form of the equation of a line with slope m and y-intercept b is $$y = mx + b$$
• **Point-slope form**	The point-slope form of the equation of the line with slope m and passing through a known point (x_1, y_1) is $$y - y_1 = m(x - x_1)$$
• **Horizontal line**	The equation of a horizontal line is $y = b$, where b is a constant.
• **Vertical line**	The equation of a vertical line is $x = a$, where a is a constant.
• **General form**	The general form of the equation of a line is $ax + by = c$, where a, b, and c are constants.
Parallel and perpendicular lines	Two distinct nonvertical lines are parallel if they have the same slope. Two nonvertical and nonhorizontal lines are perpendicular if their slopes are negative reciprocals.
Average rate of change	The average rate of change of a quantity over an interval describes how a change in the output of a function describing that quantity $f(x)$ is related to a change in the input x over that interval. The average rate of change of $f(x)$ with respect to x over the interval from $x = a$ to $x = b$ (where $a < b$) is calculated as $$\text{average rate of change} = \frac{\text{change in } f(x) \text{ values}}{\text{corresponding change in } x \text{ values}} = \frac{f(b) - f(a)}{b - a}$$
Secant line	When a function is not linear, the average rate of change between two points is the slope of the line joining two points on the curve, which is called a secant line.
Difference quotient	The average rate of change of a function from a point $(x, f(x))$ to the point $(x + h, f(x + h))$ is the difference quotient $\dfrac{f(x + h) - f(x)}{h}$.
Approximately linear data	Some sets of real data lie sufficiently close to a line that two points can be used to create a linear function that models the data.

chapter 1 ▸ SKILLS CHECK

Use the values in the table below in Exercises 1–4.

x	−3	−1	1	3	5	7	9	11	13
y	9	6	3	0	−3	−6	−9	−12	−15

1. Explain why the relationship shown by the table describes y as a function of x.

2. State the domain and range of the function.

3. If the function defined by the table is denoted by f, so that $y = f(x)$, what is $f(3)$?

4. Do the outputs in this table indicate that a linear function fits the data? If so, write the equation of the line.

5. If $C(s) = 16 - 2s^2$, find

   a. $C(3)$ b. $C(-2)$ c. $C(-1)$

6. For each of the functions $y = f(x)$ described below, find $f(-3)$.

   a.   b.

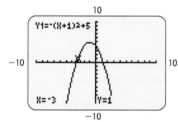   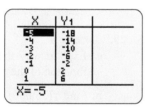

7. Graph the function $f(x) = -2x^3 + 5x$.

8. Graph the function $y = 3x^2$.

9. Graph $y = -10x^2 + 400x + 10$ on a standard window and on a window with $x_{min} = 0$, $x_{max} = 40$, $y_{min} = 0$, $y_{max} = 5000$. Which window gives a better view of the graph of the function?

10. Use a graphing utility to graph the points (x, y) from the table.

x	10	15	20	25	30	35	40
y	−8	−6	−3	0	5	8	10

11. Find the domain of each function.

   a. $y = \sqrt{2x - 8}$ b. $f(x) = \dfrac{x + 2}{x - 6}$

12. One line passes through the points $(-12, 16)$ and $(-1, 38)$, and a second line has equation $2y + x = 23$. Are the lines parallel, perpendicular, or neither?

13. A line passes through $(-1, 4)$ and $(5, -3)$. Find the slope of a line parallel to this line and the slope of a line perpendicular to this line.

14. Find the slope of the line through $(-4, 6)$ and $(8, -16)$.

15. Given the equation $2x - 3y = 12$, (a) find the x- and y-intercepts of the graph and (b) graph the equation.

16. What is the slope of the graph of the function given in Exercise 15?

17. Find the slope and y-intercept of the graph of $y = -6x + 3$.

18. Find the rate of change of the function whose equation is given in Exercise 17.

19. Write the equation of the line that has slope $\dfrac{1}{3}$ and y-intercept 3.

20. Write the equation of a line that has slope $\dfrac{-3}{4}$ and passes through $(4, -6)$.

21. Write the equation of the line that passes through $(-1, 3)$ and $(2, 6)$.

22. For the function $y = x^2$, compute the average rate of change between $x = 0$ and $x = 3$.

For the functions given in Exercises 23 and 24, find

   a. $f(x + h)$ b. $f(x + h) - f(x)$

   c. $\dfrac{f(x + h) - f(x)}{h}$

23. $f(x) = 5 - 4x$ 24. $f(x) = 10x - 50$

25. *Voters* It wasn't until Harry Truman garnered 77% of the black vote in 1948 that a majority of blacks reported that they considered themselves Democrats. The table gives the percent p of black voters who supported Democratic candidates for president for the years 1960–2008.

a. Is the percent p a function of the year y?

b. Let $p = f(y)$ denote that p is a function of y. Find $f(1992)$ and explain what it means.

c. What is y if $f(y) = 94$? What does this mean?

Year	Democrat (%)	GOP (%)
1960	68	32
1964	94	6
1968	85	15
1972	87	13
1976	85	15
1980	86	12
1984	89	9
1992	82	11
1996	84	12
2000	93	7
2004	74	26
2008	95	5

(Source: Joint Center for Political and Economic Studies)

26. *Voters*

a. What is the domain of the function defined by the table in Exercise 25?

b. Is this function defined for 1982? Why is this value not included in the table?

27. *Voters* Graph the function defined by the table in Exercise 25 on the window [1960, 2008] by [60, 100].

28. *Voters*

a. Find the slope of the line joining points (1968, 85) and (2008, 95).

b. Use the answer from part (a) to find the average annual rate of change of the percent of black voters who voted for Democratic candidates for president between the years 1968 and 2008, inclusive.

c. Is the average annual rate of change from 1968 to 1980 equal to the average annual rate of change from 1968 to 2008?

When money is borrowed to purchase an automobile, the amount borrowed A determines the monthly payment P. In particular, if a dealership offers a 5-year loan at 2.9% interest, then the amount borrowed for the car determines the payment according to the following table. Use the table to define the function $P = f(A)$ in Exercises 29–31.

Amount Borrowed ($)	Monthly Payment ($)
10,000	179.25
15,000	268.87
20,000	358.49
25,000	448.11
30,000	537.73

(Source: Sky Financial)

29. *Car Loans*

a. Explain why the monthly payment P is a linear function of the amount borrowed A.

b. Find $f(25,000)$ and interpret its meaning.

c. If $f(A) = 358.49$, what is A?

30. *Car Loans*

a. What are the domain and range of the function $P = f(A)$ defined by the table?

b. Is this function $P = f(A)$ defined for a $12,000 loan?

31. *Car Loans* The equation of the line that fits the data points in the table is $f(A) = 0.017924A + 0.01$.

a. Use the equation to find $f(28,000)$ and explain what it means.

b. Can the function f be used to find the monthly payment for any dollar amount A of a loan if the interest rate and length of loan are unchanged?

32. *Life Expectancy* The figure gives the number of years the average woman is estimated to live beyond age 65 during selected years between 1950 and 2030. Let x represent the year and let y represent the expected number of years a woman will live past age 65. Write $y = f(x)$ and answer the following questions:

a. What is $f(1960)$ and what does it mean?

b. What is the life expectancy for the average woman in 2010?

c. In what year was the average woman expected to live 19 years past age 65?

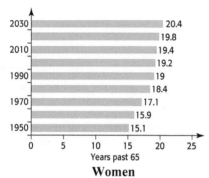

Women
(Source: National Center for Health Statistics)

33. *Life Expectancy* The figure gives the number of years the average man is estimated to live beyond age 65 during selected years between 1950 and 2030. Let x represent the year and let y represent the expected number of years a man will live past age 65. Write $y = g(x)$ and answer the following questions:

a. What is $g(2020)$ and what does it mean?

b. What was the life expectancy for the average man in 1950?

c. Write a function expression that indicates that the average man in 1990 has a life expectancy of 80 years.

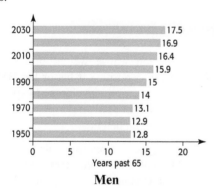

Men
(Source: National Center for Health Statistics)

34. *Teacher Salaries* The average salary of a classroom teacher in the United States is given by $f(t) = 982.06t + 32,903.77$, where t is the number of years from 1990.

a. What was the average salary in 2000? Write this in function notation.

b. Find $f(15)$ and interpret it.

c. Are the salaries increasing or decreasing?
(Source: www.ors2.state.sc.us/abstract)

35. *Drug Users* According to the Substance Abuse and Mental Health Service Administration's *2008 National Survey on Drug Use and Health*, an estimated 117,325,000 Americans 12 years and older had used an illicit drug at least once in their lifetime. In 2003, the number was 110,205,000.

a. Find the slope of the line connecting the points (2003, 110.205) and (2008, 117.325).

b. Find the average annual rate of change in number of drug users between 2003 and 2008.

36. *Teacher Salaries* The average salary of a classroom teacher in the United States is given by the function $f(t) = 982.06t + 32,903.77$, where t is the number of years from 1990.

a. Graph this function for values of t between 0 and 15.

b. What years do these values represent?
(Source: www.ors2.state.sc.us/abstract)

37. *Fuel* The table below shows data for the number of gallons of gas purchased each day x of a certain week by the 150 taxis owned by the Inner City Transportation Taxi Company. Write the equation that models the data.

Days from First Day	0	1	2	3	4	5	6
Gas Used (gal)	4500	4500	4500	4500	4500	4500	4500

38. *Work Hours* The average weekly hours worked by employees at PriceLo Company are given by the data in the table below.

Month and Year	Average Weekly Hours	Month and Year	Average Weekly Hours
June 2011	33.8	Oct. 2011	33.8
July 2011	33.8	Nov. 2011	33.8
Aug. 2011	33.8	Dec. 2011	33.8
Sept. 2011	33.8	Jan. 2011	33.8
		Feb. 2011	33.8

a. Write the equation of a function that describes the average weekly hours using an input x equal to the number of months past May 2011.

b. What type of function is this?

39. *Revenue* A company has revenue given by $R(x) = 564x$ dollars and total costs given by $C(x) = 40,000 + 64x$ dollars, where x is the number of units produced and sold.

a. What is the revenue when 120 units are produced?

b. What is the cost when 120 units are produced?

c. What is the marginal cost and what is the marginal revenue for this product?

d. What is the slope of the graph of $C(x) = 40,000 + 64x$?

e. Graph $R(x)$ and $C(x)$ on the same set of axes.

40. *Profit* A company has revenue given by $R(x) = 564x$ dollars and total cost given by $C(x) = 40,000 + 64x$ dollars, where x is the number of units produced and sold. The profit can be found by forming the function $P(x) = R(x) - C(x)$.

a. Write the profit function.

b. Find the profit when 120 units are produced and sold.

c. How many units give break-even?

d. What is the marginal profit for this product?

e. How is the marginal profit related to the marginal revenue and the marginal cost?

41. *Depreciation* A business property can be depreciated for tax purposes by using the formula $y + 3000x = 300,000$, where y is the value of the property x years after it was purchased.

a. Find the y-intercept of the graph of this function. Interpret this value.

b. Find the x-intercept. Interpret this value.

42. *Marginal Profit* A company has determined that its profit for a product can be described by a linear function. The profit from the production and sale of 150 units is $455, and the profit from 250 units is $895.

a. What is the average rate of change of the profit for this product when between 150 and 250 units are sold?

b. What is the slope of the graph of this profit function?

c. Write the equation of the profit function for this product.

d. What is the marginal profit for this product?

e. How many units give break-even for this product?

Group Activities
▶ EXTENDED APPLICATIONS

1. Body Mass Index

Obesity is a risk factor for the development of medical problems, including high blood pressure, high cholesterol, heart disease, and diabetes. Of course, how much a person can safely weigh depends on his or her height. One way of comparing weights that account for height is the *body mass index (BMI)*. The table on the next page gives the BMI for a variety of heights and weights of people. Roche Pharmaceuticals state that a BMI of 30 or greater can create an increased risk of developing medical problems associated with obesity.

Describe how to assist a group of people in using the information. Some things you would want to include in your description follow.

1. How a person uses the table to determine his or her BMI.
2. How a person determines the weight that will put him or her at medical risk.

3. How a person whose weight or height is not in the table can determine if his or her BMI is 30. To answer this question, develop and explain a formula to find the weight that would give a BMI of 30 for a person of a given height:

a. Pick the points from the table that correspond to a BMI of 30 and create a table of these heights and weights. Change the heights to inches to simplify the data.

b. Create a scatter plot of the data.

c. Using the two points with the smallest and largest heights, write a linear equation that models the data.

d. Graph the linear equation from part (c) with the scatter plot and discuss the fit.

e. Explain how to use the model to test for obesity.

Body Mass Index for Specified Height (ft/in.) and Weight (lb)

Height/Weight	120	130	140	150	160	170	180	190	200	210	220	230	240	250
5'0"	23	25	27	29	31	33	35	37	39	41	43	45	47	49
5'1"	23	25	27	28	30	32	34	36	38	40	42	44	45	47
5'2"	22	24	26	27	29	31	33	35	37	38	40	42	44	46
5'3"	21	23	25	27	28	30	32	34	36	37	39	41	43	44
5'4"	21	22	24	26	28	29	31	33	34	36	38	40	41	43
5'5"	20	22	23	25	27	28	30	32	33	35	37	38	40	42
5'6"	19	21	23	24	26	27	29	31	32	34	36	37	39	40
5'7"	19	20	23	24	25	27	28	30	31	33	35	36	38	39
5'8"	18	20	21	23	24	26	27	29	30	32	34	35	37	38
5'9"	18	19	21	22	24	25	27	28	30	31	33	34	36	37
5'10"	17	19	20	22	23	24	25	27	28	29	31	33	35	36
5'11"	17	18	20	21	22	24	25	27	28	29	31	32	34	35
6'0"	16	18	19	20	22	23	24	26	27	29	30	31	33	34
6'1"	16	17	19	20	21	22	24	25	26	28	29	30	32	33

(Source: Roche Pharmaceuticals)

2. Total Revenue, Total Cost, and Profit

The total revenue is the amount a company receives from the sales of its products. It can be found by multiplying the selling price per unit times the number of units sold. That is, the revenue is

$$R(x) = p \cdot x$$

where p is the price per unit and x is the number of units sold.

The total cost comprises two costs, the fixed costs and the variable costs. Fixed costs (FC), such as depreciation, rent, and utilities, remain constant regardless of the number of units produced. Variable costs (VC) are those directly related to the number of units produced. The variable cost is the cost per unit (c) times the number of units produced (VC $= c \cdot x$) and the fixed cost is constant (FC $= k$), so the total cost is found by using the equation

$$C(x) = c \cdot x + k$$

where c is the cost per unit and x is the number of units produced.

Also, as discussed in Section 1.3, the profit a company makes on x units of a product is the difference between the revenue and the cost from production and sale of x units.

$$P(x) = R(x) - C(x)$$

Suppose a company manufactures MP3 players and sells them to retailers for $98 each. It has fixed costs of $262,500 related to the production of the MP3 players, and the cost per unit for production is $23.

1. What is the total revenue function?
2. What is the marginal revenue for this product?
3. What is the total cost function?
4. What is the marginal cost for this product? Is the marginal cost equal to the variable cost or the fixed cost?
5. What is the profit function for this product? What is the marginal profit?
6. What are the cost, revenue, and profit if 0 units are produced?
7. Graph the total cost and total revenue functions on the same axes and estimate where the graphs intersect.
8. Graph the profit function and estimate where the graph intersects the x-axis.
9. What do the points of intersection in Question 7 and Question 8 give?

Linear Models, Equations, and Inequalities

Annual data can be used to find average annual salaries by gender and educational attainment. These salaries can be compared by creating a linear equation that gives female annual earnings as a function of male annual earnings. We can then investigate if female salaries are approaching male salaries. We can also find when the manufacturing sector of China reaches and surpasses that of the United States, by finding linear models and solving them simultaneously. Market equilibrium occurs when the number of units of a product demanded equals the number of units supplied. In this chapter, we solve these and other problems by creating linear models, solving linear equations and inequalities, and solving systems of equations.

section	objectives	applications
2.1 Algebraic and Graphical Solution of Linear Equations	Find solutions algebraically; relate solutions, zeros, and x-intercepts; solve linear equations; find functional forms of equations in two variables; solve literal equations; apply direct variation	Prison sentence length, credit card debt, stock market, simple interest, blood alcohol percent
2.2 Fitting Lines to Data Points; Modeling Linear Functions	Fit lines to data points with linear regression; model with linear functions; apply linear models; evaluate goodness of fit	Earnings and gender, retirement, health service employment, U.S. population
2.3 Systems of Linear Equations in Two Variables	Solve systems of linear equations in two variables with graphing, substitution, and elimination; find break-even and market equilibrium; find nonunique solutions if they exist	China's manufacturing, break-even, market equilibrium, investments, medication
2.4 Solutions of Linear Inequalities	Solve linear inequalities with analytical and graphical methods; solve double inequalities	Profit, body temperature, apparent temperature, course grades, expected prison sentences

Algebra TOOLBOX

KEY OBJECTIVES

- Use properties of equations to solve linear equations

- Determine if a linear equation is a conditional equation, an identity, or a contradiction

- Use properties of inequalities to solve linear inequalities

Properties of Equations

In this chapter, we will solve equations. To solve an equation means to find the value(s) of the variable(s) that makes the equation a true statement. For example, the equation $3x = 6$ is true when $x = 2$, so 2 is a solution of this equation. Equations of this type are sometimes called **conditional equations** because they are true only for certain values of the variable. Equations that are true for all values of the variables for which both sides are defined are called **identities**. For example, the equation $7x - 4x = 5x - 2x$ is an identity. Equations that are not true for any value of the variable are called **contradictions**. For example, the equation $2(x + 3) = 2x - 1$ is a contradiction.

We frequently can solve an equation for a given variable by rewriting the equation in an equivalent form whose solution is easy to find. Two equations are **equivalent** if and only if they have the same solutions. The following operations give equivalent equations:

Properties of Equations

1. **Addition Property** Adding the same number to both sides of an equation gives an equivalent equation. For example, $x - 5 = 2$ is equivalent to $x - 5 + 5 = 2 + 5$, or to $x = 7$.

2. **Subtraction Property** Subtracting the same number from both sides of an equation gives an equivalent equation. For example, $z + 12 = -9$ is equivalent to $z + 12 - 12 = -9 - 12$, or to $z = -21$.

3. **Multiplication Property** Multiplying both sides of an equation by the same nonzero number gives an equivalent equation. For example, $\dfrac{y}{6} = 5$ is equivalent to $6\left(\dfrac{y}{6}\right) = 6(5)$, or to $y = 30$.

4. **Division Property** Dividing both sides of an equation by the same nonzero number gives an equivalent equation. For example, $17x = -34$ is equivalent to $\dfrac{17x}{17} = \dfrac{-34}{17}$, or to $x = -2$.

5. **Substitution Property** The equation formed by substituting one expression for an equal expression is equivalent to the original equation. For example, if $y = 3x$, then $x + y = 8$ is equivalent to $x + 3x = 8$, so $4x = 8$ and $x = 2$.

EXAMPLE 1 ▶ **Properties of Equations**

State the property (or properties) of equations that can be used to solve each of the following equations, and then use the property (or properties) to solve the equation.

a. $3x = 6$ **b.** $\dfrac{x}{5} = 12$ **c.** $3x - 5 = 17$ **d.** $4x - 5 = 7 + 2x$

SOLUTION

a. Division Property. Dividing both sides of the equation by 3 gives the solution to the equation.

$$3x = 6$$

$$\frac{3x}{3} = \frac{6}{3}$$

$$x = 2$$

Thus, $x = 2$ is the solution to the original equation.

b. Multiplication Property. Multiplying both sides of the equation by 5 gives the solution to the equation.

$$\frac{x}{5} = 12$$

$$5\left(\frac{x}{5}\right) = 5(12)$$

$$x = 60$$

Thus, $x = 60$ is the solution to the original equation.

c. Addition Property and Division Property. Adding 5 to both sides of the equation and dividing both sides by 3 gives an equivalent equation.

$$3x - 5 = 17$$

$$3x - 5 + 5 = 17 + 5$$

$$3x = 22$$

$$\frac{3x}{3} = \frac{22}{3}$$

$$x = \frac{22}{3}$$

Thus, $x = \dfrac{22}{3}$ is the solution to the original equation.

d. Addition Property, Subtraction Property, and Division Property. Adding 5 to both sides of the equation and subtracting $2x$ from both sides of the equation gives an equivalent equation.

$$4x - 5 = 7 + 2x$$

$$4x - 5 + 5 = 7 + 2x + 5$$

$$4x = 12 + 2x$$

$$4x - 2x = 12 + 2x - 2x$$

$$2x = 12$$

Dividing both sides by 2 gives the solution to the original equation.

$$\frac{2x}{2} = \frac{12}{2}$$

$$x = 6$$

Thus, $x = 6$ is the solution to the original equation.

EXAMPLE 2 ▶ Equations

Determine whether each equation is a conditional equation, an identity, or a contradiction.

a. $3(x - 5) = 2x - 7$ **b.** $5x - 6(x + 1) = -x - 9$

c. $-3(2x - 4) = 2x + 12 - 8x$

SOLUTION

a. To solve $3(x - 5) = 2x - 7$, we first use the Distributive Property, then the Addition Property.

$$3(x - 5) = 2x - 7$$
$$3x - 15 = 2x - 7$$
$$3x - 15 - 2x = 2x - 7 - 2x$$
$$x - 15 + 15 = -7 + 15$$
$$x = 8$$

Because this equation is true for only the value $x = 8$, it is a conditional equation.

b. To solve $5x - 6(x + 1) = -x - 9$, we first use the Distributive Property, then the Addition Property.

$$5x - 6(x + 1) = -x - 9$$
$$5x - 6x - 6 = -x - 9$$
$$-x - 6 = -x - 9$$
$$-x - 6 + x = -x - 9 + x$$
$$-6 = -9$$

Because $-6 = -9$ is false, the original equation is a contradiction.

c. To solve $-3(2x - 4) = 2x + 12 - 8x$, we first use the Distributive Property, then the Addition Property.

$$-3(2x - 4) = 2x + 12 - 8x$$
$$-6x + 12 = 12 - 6x$$
$$-6x + 12 + 6x = 12 - 6x + 6x$$
$$12 = 12$$

Because $12 = 12$ is a true statement, the equation is true for all real numbers and thus is an identity.

Properties of Inequalities

We will solve linear inequalities in this chapter. As with equations, we can find solutions to inequalities by finding equivalent inequalities from which the solutions can be easily seen. We use the following properties to reduce an inequality to a simple equivalent inequality.

Properties of Inequalities	Examples	
Substitution Property The inequality formed by substituting one expression for an equal expression is equivalent to the original inequality.	$7x - 6x \geq 8$ $x \geq 8$	$3(x - 1) < 8$ $3x - 3 < 8$
Addition and Subtraction Properties The inequality formed by adding the same quantity to (or subtracting the same quantity from) both sides of an inequality is equivalent to the original inequality.	$x - 6 < 12$ $x - 6 + 6 < 12 + 6$ $x < 18$	$3x > 13 + 2x$ $3x - 2x > 13 + 2x - 2x$ $x > 13$
Multiplication Property I The inequality formed by multiplying (or dividing) both sides of an inequality by the same *positive* quantity is equivalent to the original inequality.	$\frac{1}{3}x \leq 6$ $3\left(\frac{1}{3}x\right) \leq 3(6)$ $x \leq 18$	$3x > 6$ $\dfrac{3x}{3} > \dfrac{6}{3}$ $x > 2$
Multiplication Property II The inequality formed by multiplying (or dividing) both sides of an inequality by the same *negative* number and reversing the inequality symbol is equivalent to the original inequality.	$-x > 7$ $-1(-x) < -1(7)$ $x < -7$	$-4x \geq 12$ $\dfrac{-4x}{-4} \leq \dfrac{12}{-4}$ $x \leq -3$

Of course, these properties can be used in combination to solve an inequality. This means that the steps used to solve a linear inequality are the same as those used to solve a linear equation, except that the inequality symbol is reversed if both sides are multiplied (or divided) by a negative number.

EXAMPLE 3 ▶ **Properties of Inequalities**

State the property (or properties) of inequalities that can be used to solve each of the following inequalities, and then use the property (or properties) to solve the inequality.

a. $\dfrac{x}{2} \geq 13$ **b.** $-3x < 6$ **c.** $5 - \dfrac{x}{3} \geq -4$ **d.** $-2(x + 3) < 4$

SOLUTION

a. Multiplication Property I. Multiplying both sides by 2 gives the solution to the inequality.

$$\frac{x}{2} \geq 13$$

$$2\left(\frac{x}{2}\right) \geq 2(13)$$

$$x \geq 26$$

The solution to the inequality is $x \geq 26$.

b. Multiplication Property II. Dividing both sides by -3 and reversing the inequality symbol gives the solution to the inequality.

$$-3x < 6$$

$$\frac{-3x}{-3} > \frac{6}{-3}$$

$$x > -2$$

The solution to the inequality is $x > -2$.

c. Subtraction Property, Multiplication Property II. First we subtract 5 from both sides of the inequality.

$$5 - \frac{x}{3} \geq -4$$

$$5 - \frac{x}{3} - 5 \geq -4 - 5$$

$$-\frac{x}{3} \geq -9$$

Multiplying both sides by -3 and reversing the inequality symbol gives the solution to the inequality.

$$-\frac{x}{3} \geq -9$$

$$-3\left(-\frac{x}{3}\right) \leq -3(-9)$$

$$x \leq 27$$

The solution to the inequality is $x \leq 27$.

d. Substitution Property, Addition Property, Multiplication Property II. First we distribute -2 on the left side of the inequality.

$$-2(x + 3) < 4$$

$$-2x - 6 < 4$$

Adding 6 to both sides and then dividing both sides by -2 and reversing the inequality symbol gives the solution to the inequality.

$$-2x - 6 + 6 < 4 + 6$$

$$-2x < 10$$

$$\frac{-2x}{-2} > \frac{10}{-2}$$

$$x > -5$$

The solution to the inequality is $x > -5$.

Toolbox EXERCISES

In Exercises 1–8, state the property (or properties) of equations that can be used to solve each of the following equations; then use the property (or properties) to solve the equation.

1. $3x = 6$

2. $x - 7 = 11$

3. $x + 3 = 8$

4. $x - 5 = -2$

5. $\dfrac{x}{3} = 6$

6. $-5x = 10$

7. $2x + 8 = -12$

8. $\dfrac{x}{4} - 3 = 5$

Solve the equations in Exercises 9–16.

9. $4x - 3 = 6 + x$

10. $3x - 2 = 4 - 7x$

11. $\dfrac{3x}{4} = 12$

12. $\dfrac{5x}{2} = -10$

13. $3(x - 5) = -2x - 5$

14. $-2(3x - 1) = 4x - 8$

15. $2x - 7 = -4\left(4x - \dfrac{1}{2}\right)$

16. $-2(2x - 6) = 3\left(3x - \dfrac{1}{3}\right)$

In Exercises 17–20, use the Substitution Property of Equations to solve the equation.

17. Solve for x if $y = 2x$ and $x + y = 12$

18. Solve for x if $y = 4x$ and $x + y = 25$

19. Solve for x if $y = 3x$ and $2x + 4y = 42$

20. Solve for x if $y = 6x$ and $3x + 2y = 75$

In Exercises 21–24, determine whether the equation is a conditional equation, an identity, or a contradiction.

21. $3x - 5x = 2x + 7$ **22.** $3(x + 1) = 3x - 7$

23. $9x - 2(x - 5) = 3x + 10 + 4x$

24. $\dfrac{x}{2} - 5 = \dfrac{x}{4} + 2$

In Exercises 25–32, solve the inequalities.

25. $5x + 1 > -5$ **26.** $1 - 3x \geq 7$

27. $\dfrac{x}{4} > -3$ **28.** $\dfrac{x}{6} > -2$

29. $\dfrac{x}{4} - 2 > 5x$ **30.** $\dfrac{x}{2} + 3 > 6x$

31. $-3(x - 5) < -4$ **32.** $-\dfrac{1}{2}(x + 4) < 6$

2.1 Algebraic and Graphical Solution of Linear Equations

KEY OBJECTIVES

- Solve linear equations algebraically

- Solve real-world application problems

- Compare solutions of equations with zeros and x-intercepts of graphs of functions

- Solve linear equations graphically using the x-intercept and intersection methods

- Solve literal equations for a specified variable

- Solve direct variation problems

SECTION PREVIEW Prison Sentences

The average sentence length and the average time served in state prison for various crimes are shown in Table 2.1. These data can be used to create a linear function that is an approximate model for the data:

$$y = 0.55x - 2.886$$

Table 2.1

Average Sentence (months)	Average Time Served (months)
62	30
85	45
180	95
116	66
92	46
61	33
56	26

(Source: U.S. Department of Justice)

This function describes the mean time y served in prison for a crime as a function of the mean sentence length x, where x and y are each measured in months. Assuming that this model remains valid, it predicts that the mean time served on a 5-year (60 months) sentence is $0.55(60) - 2.886 \approx 30$ months, or approximately $2\dfrac{1}{2}$ years.

To find the sentence that would give an expected time served of 10 years (120 months), we use algebraic or graphical methods to solve the equation

$$120 = 0.55x - 2.866$$

for x. (See Example 6.) In this section, we use additional algebraic methods and graphical methods to solve linear equations in one variable. ■

Algebraic Solution of Linear Equations

We can use algebraic, graphical, or a combination of algebraic and graphical methods to solve linear equations. Sometimes it is more convenient to use algebraic solution methods rather than graphical solution methods, especially if an exact solution is desired. The steps used to solve a linear equation in one variable algebraically follow.

Steps for Solving a Linear Equation in One Variable

1. If a linear equation contains fractions, multiply both sides of the equation by a number that will remove all denominators from the equation. If there are two or more fractions, use the least common denominator (LCD) of the fractions.
2. Perform any multiplications or divisions to remove any parentheses or other symbols of grouping.
3. Perform any additions or subtractions to get all terms containing the variable on one side and all other terms on the other side of the equation. Combine like terms.
4. Divide both sides of the equation by the coefficient of the variable.
5. Check the solution by substitution in the original equation. If a real-world solution is desired, check the algebraic solution for reasonableness in the real-world situation.

EXAMPLE 1 ▶ **Algebraic Solutions**

a. Solve for x: $\dfrac{2x - 3}{4} = \dfrac{x}{3} + 1$

b. Solve for y: $\dfrac{y}{5} - \dfrac{1}{2}\left(\dfrac{y}{2}\right) = \dfrac{7}{20}$

c. Solve for x if $y = 0.72$: $y = 1.312x - 2.56$

SOLUTION

a. $\dfrac{2x - 3}{4} = \dfrac{x}{3} + 1$

$12\left(\dfrac{2x - 3}{4}\right) = 12\left(\dfrac{x}{3} + 1\right)$ Multiply both sides by the LCD, 12.

$3(2x - 3) = 12\left(\dfrac{x}{3} + 1\right)$ Simplify the fraction $\dfrac{12}{4}$ on the left.

$6x - 9 = 4x + 12$ Remove parentheses using the distributive property.

$2x = 21$ Subtract $4x$ from both sides and add 9 to both sides.

$x = \dfrac{21}{2}$ Divide both sides by 2.

Check the result: $\dfrac{2\left(\dfrac{21}{2}\right) - 3}{4} \overset{?}{=} \dfrac{\dfrac{21}{2}}{3} + 1 \Rightarrow \dfrac{9}{2} = \dfrac{9}{2}$

b. $\dfrac{y}{5} - \dfrac{1}{2}\left(\dfrac{y}{2}\right) = \dfrac{7}{20}$

$$\dfrac{y}{5} - \dfrac{y}{4} = \dfrac{7}{20} \qquad \text{Remove parentheses first to find the LCD.}$$

$$20\left(\dfrac{y}{5} - \dfrac{y}{4}\right) = 20\left(\dfrac{7}{20}\right) \qquad \text{Multiply both sides by the LCD, 20.}$$

$$4y - 5y = 7 \qquad \text{Remove parentheses using the distributive property.}$$

$$-y = 7 \qquad \text{Combine like terms.}$$

$$y = -7 \qquad \text{Multiply both sides by } -1.$$

Check the result: $\dfrac{-7}{5} - \dfrac{1}{2}\left(\dfrac{-7}{2}\right) \overset{?}{=} \dfrac{7}{20} \Rightarrow \dfrac{-7}{5} + \dfrac{7}{4} \overset{?}{=} \dfrac{7}{20} \Rightarrow \dfrac{7}{20} = \dfrac{7}{20}$

c. $y = 1.312x - 2.56$

$$0.72 = 1.312x - 2.56 \qquad \text{Substitute 0.72 for } y.$$

$$3.28 = 1.312x \qquad \text{Add 2.56 to both sides.}$$

$$x = 2.5 \qquad \text{Divide both sides by 1.312.}$$

Check the result: $0.72 \overset{?}{=} 1.312(2.5) - 2.56 \Rightarrow 0.72 = 0.72$

As the next example illustrates, we solve an application problem that is set in a real-world context by using the same solution methods. However, you must remember to include units of measure with your answer and check that your answer makes sense in the problem situation.

EXAMPLE 2 ▶ Credit Card Debt

It is hard for some people to pay off credit card debt in a reasonable period of time because of high interest rates. The interest paid on a \$10,000 debt over 3 years is approximated by

$$y = 175.393x - 116.287 \text{ dollars}$$

when the interest rate is $x\%$. What is the interest rate if the interest is \$1637.60? (Source: Consumer Federation of America)

SOLUTION

To answer this question, we solve the linear equation

$$1637.60 = 175.393x - 116.287$$

$$1753.887 = 175.393x$$

$$x = 9.9998$$

Thus, if the interest rate is approximately 10%, the interest is \$1637.60. (Note that if you check the approximate answer, you are checking only for the reasonableness of the estimate.)

EXAMPLE 3 ▶ **Stock Market**

For a period of time, a man is very successful speculating on an Internet stock, with its value growing to $100,000. However, the stock drops rapidly until its value has been reduced by 40%. What percent increase will have to occur before the latest value returns to $100,000?

SOLUTION

The value of the stock after the 40% loss is $100,000 - 0.40(100,000) = 60,000$. To find the percent p of increase that is necessary to return the value to 100,000, we solve

$$60,000 + 60,000p = 100,000$$
$$60,000p = 40,000$$
$$p = \frac{40,000}{60,000} = \frac{2}{3} = 66\frac{2}{3}\%$$

Thus, the stock value must increase by $66\frac{2}{3}\%$ to return to a value of $100,000.

Solutions, Zeros, and x-Intercepts

Because x-intercepts are x values that make the output of the function equal to 0, these intercepts are also called **zeros** of the function.

Zero of a Function

Any number a for which $f(a) = 0$ is called a **zero** of the function $f(x)$. If a is real, a is an x-intercept of the graph of the function.

The zeros of the function are values that make the function equal to 0, so they are also solutions to the equation $f(x) = 0$.

The following three concepts are numerically the same:

The x-intercepts of the graph of $y = f(x)$

The real zeros of the function f

The real solutions to the equation $f(x) = 0$

The following example illustrates the relationships that exist among x-intercepts of the graph of a function, zeros of the function, and solutions to associated equations.

EXAMPLE 4 ▶ **Relationships Among x-Intercepts, Zeros, and Solutions**

For the function $f(x) = 13x - 39$, find

a. $f(3)$ **b.** The zero of $f(x) = 13x - 39$

c. The x-intercept of the graph of $y = 13x - 39$

d. The solution to the equation $13x - 39 = 0$

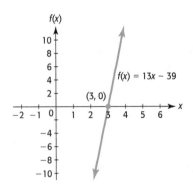

Figure 2.1

Table 2.2

	A	B
1	x	y = 13x − 39
2	3	0

SOLUTION

a. Evaluate the output $f(3)$ by substituting the input 3 for x in the function:

$$f(3) = 13(3) - 39 = 0$$

b. Because $f(x) = 0$ when $x = 3$, we say that 3 is a zero of the function.

c. The x-intercept of a graph occurs at the value of x where $y = 0$, so the x-intercept is $x = 3$. The only x-intercept is $x = 3$ because the graph of $f(x) = 13x - 39$ is a line that crosses the x-axis at only one point (Figure 2.1).

d. The solution to the equation $13x - 39 = 0$ is $x = 3$ because $13x - 39 = 0$ gives $13x = 39$, or $x = 3$.

We can use TRACE or TABLE on a graphing utility to check the reasonableness of a solution. The graph of $y = 13x - 39$ given in Figure 2.1 confirms that $x = 3$ is the x-intercept. We can also confirm that $f(3) = 0$ by using an Excel spreadsheet (Table 2.2.). When we enter 3 in cell A2 and the formula "$= 13 * A2 - 39$" in cell B2, the function $f(x) = 13x - 39$ is evaluated at $x = 3$, with a result of 0.*

Graphical Solution of Linear Equations

We can also view the graph or use TRACE on a graphing utility to obtain a quick estimate of an answer. TRACE may or may not provide the exact solution to an equation, but it will provide an estimate of the solution. If your calculator or computer has a graphical or numerical solver, the solver can be used to obtain the solution to the equation $f(x) = 0$. Recall that an x-intercept of the graph of $y = f(x)$, a real zero of $f(x)$, and the real solution to the equation $f(x) = 0$ are all different names for the same input value. If the graph of the linear function $y = f(x)$ is not a horizontal line, we can find the one solution to the linear equation $f(x) = 0$ as described below. We call this solution method the **x-intercept method**.

Solving a Linear Equation Using the x-Intercept Method with Graphing Utilities

1. Rewrite the equation to be solved with 0 (and nothing else) on one side of the equation.

2. Enter the nonzero side of the equation found in the previous step in the equation editor of your graphing utility and graph the line in an appropriate viewing window. Be certain that you can see the line cross the horizontal axis on your graph.

3. Find the x-intercept by using ZERO . The x-intercept is the value of x that makes the equation equal to zero, so it is the solution to the equation. The value of x displayed by this method is sometimes a decimal approximation of the exact solution rather than the exact solution. *Note:* Using MATH ▶ FRAC will often convert a decimal solution of a linear equation (approximated on the display) to the exact solution.†

---

*See Appendix B, page 638.

†See Appendix A, pages 621–622.

EXAMPLE 5 ▶   Graphical Solution

Solve $\dfrac{2x-3}{4} = \dfrac{x}{3} + 1$ for x using the x-intercept method.

SOLUTION

To solve the equation using the x-intercept method, first rewrite the equation with 0 on one side.

$$0 = \frac{x}{3} + 1 - \frac{2x-3}{4}$$

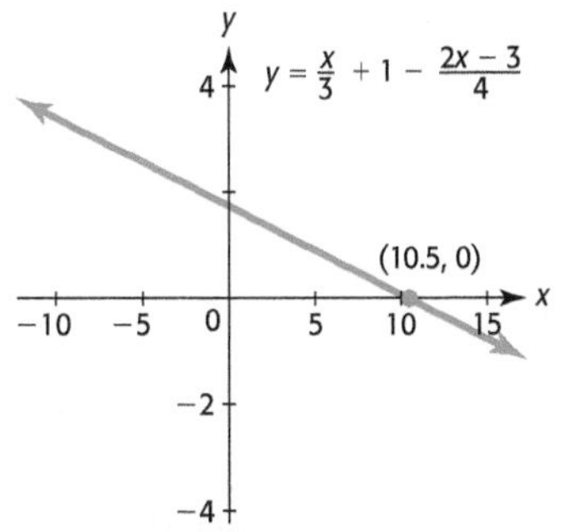

Next, enter the right side of the equation,

$$\frac{x}{3} + 1 - \frac{2x-3}{4}$$

in the equation editor of your graphing utility as

$$y_1 = (x/3) + 1 - (2x - 3)/4$$

and graph this function. Use parentheses as needed to preserve the order of operations. You should obtain a graph similar to the one seen in Figure 2.2, which was obtained using the viewing window $[-10, 15]$ by $[-4, 4]$. However, any graph in which you can see the x-intercept will do. Using $\boxed{\text{ZERO}}$ gives the x-intercept of the graph, 10.5. (See Figure 2.2.) This is the value of x that makes $y = 0$, and thus the original equation true, so it is the solution to this linear equation.

You can determine if $x = 10.5 = \dfrac{21}{2}$, found graphically, is the exact answer by substituting this value into the original equation. Both sides of the original equation are equal when $x = 10.5$, so it is the exact solution.

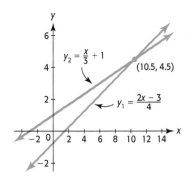

Figure 2.2

We can also use the intersection method described below to solve linear equations:

Solving a Linear Equation Using the Intersection Method

1. Enter the left side of the equation as y_1 and the right side as y_2. Graph both of these equations on a window that shows their point of intersection.

2. Find the point of intersection of the two graphs with $\boxed{\text{INTERSECT}}$.* This is the point where $y_1 = y_2$. The x-value of this point is the value of x that makes the two sides of the equation equal, so it is the solution to the original equation.

The equation in Example 5,

$$\frac{2x-3}{4} = \frac{x}{3} + 1$$

can be solved using the intersection method. Enter the left side as $y_1 = (2x - 3)/4$ and the right side as $y_2 = (x/3) + 1$. Graphing these two functions gives the graphs in Figure 2.3. The point of intersection is found to be $(10.5, 4.5)$. So the solution is $x = 10.5$ (as we found in Example 5).

Figure 2.3

---

*See Appendix A, page 622.

EXAMPLE 6 ▶   Criminal Sentences

The function $y = 0.55x - 2.886$ describes the mean time y served in prison for a crime as a function of the mean sentence length x, where x and y are each measured in months. To find the sentence for a crime that would give an expected time served of 10 years, write an equation and solve it by using (a) the x-intercept method and (b) the intersection method.

SOLUTION

a. Note that 10 years is 120 months. We solve the linear equation $120 = 0.55x - 2.886$ by using the x-intercept method as follows:

$$0 = 0.55x - 2.886 - 120 \qquad \text{Rewrite the equation in a form with 0 on one side.}$$

$$0 = 0.55x - 122.886 \qquad \text{Combine the constant terms.}$$

Enter $y_1 = 0.55x - 122.886$ and graph this equation (Figure 2.4). Using ZERO gives an x-intercept of approximately 223 (Figure 2.4). Thus, if a prisoner receives a sentence of 223 months, we would expect him or her to serve 120 months, or 10 years.

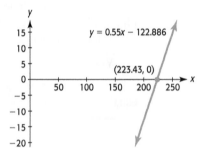

Figure 2.4

b. To use the intersection method, enter $y_1 = 0.55x - 2.886$ and $y_2 = 120$, graph these equations, and find the intersection of the lines with INTERSECT . Figure 2.5 shows the point of intersection, and again we see that x is approximately 223.

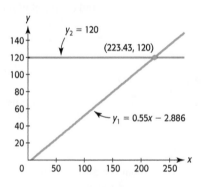

Figure 2.5

Spreadsheet ▸ SOLUTION   We can use the **Goal Seek** feature of Excel to find the x-intercept of the graph of a function—that is, the zeros of the function. To solve

$$120 = 0.55x - 2.886$$

we enter the nonzero side of $0 = 0.55x - 122.886$ as shown in Table 2.3. Using **Goal Seek** with cell B2 set to value 0 gives the value of x that solves the equation, as shown in Table 2.4.*

Table 2.3

	A	B
1	x	y
2		= 0.55*A2 − 122.886

Table 2.4

	A	B
1	x	y
2	223.43	0

Literal Equations; Solving an Equation for a Specified Linear Variable

An equation that contains two or more letters that represent constants or variables is called a **literal equation**. Formulas are examples of literal equations. If one of two or more of the variables in an equation is present only to the first power, we can solve for that variable by treating the other variables as constants and using the same steps that we used to solve a linear equation in one variable. This is often useful because it is necessary to get equations in the proper form to enter them in a graphing utility.

EXAMPLE 7 ▶   Simple Interest

The formula for the future value of an investment of P dollars at simple interest rate r for t years is $A = P(1 + rt)$. Solve the formula for r, the interest rate.

SOLUTION

$$A = P(1 + rt)$$
$$A = P + Prt \qquad \text{Multiply to remove parentheses.}$$
$$A - P = Prt \qquad \text{Get the term containing } r \text{ by itself on one side of the equation.}$$
$$r = \frac{A - P}{Pt} \qquad \text{Divide both sides by } Pt.$$

EXAMPLE 8 ▶   Solving an Equation for a Specified Variable

Solve the equation $2(2x - b) = \dfrac{5cx}{3}$ for x.

SOLUTION

We solve the equation for x by treating the other variables as constants:

$$2(2x - b) = \frac{5cx}{3}$$
$$6(2x - b) = 5cx \qquad \text{Clear the equation of fractions by multiplying by the LCD, 3.}$$

---

*See Appendix B, page 645.

$$12x - 6b = 5cx$$ Multiply to remove parentheses.

$$12x - 5cx = 6b$$ Get all terms containing x on one side and all other terms on the other side.

$$x(12 - 5c) = 6b$$ Factor x from the expression. The remaining factor is the coefficient of x.

$$\frac{x(12 - 5c)}{(12 - 5c)} = \frac{6b}{12 - 5c}$$ Divide both sides by the coefficient of x.

$$x = \frac{6b}{12 - 5c}$$

EXAMPLE 9 ▶   **Writing an Equation in Functional Form and Graphing Equations**

Solve each of the following equations for y so that y is expressed as a function of x. Then graph the equation on a graphing utility with a standard viewing window.

a. $2x - 3y = 12$

b. $x^2 + 4y = 4$

SOLUTION

a. Because y is to the first power in the equation, we solve the equation for y using linear equation solution methods:

$$2x - 3y = 12$$

$$-3y = -2x + 12$$ Isolate the term involving y by subtracting $2x$ from both sides.

$$y = \frac{2x}{3} - 4$$ Divide both sides by -3, the coefficient of y.

The graph of this equation is shown in Figure 2.6.

b. The variable y is to the first power in the equation

$$x^2 + 4y = 4$$

so we solve for y by using linear solution methods. (Note that to solve this equation for x would be more difficult; this method will be discussed in Chapter 3.)

$$4y = -x^2 + 4$$ Isolate the term containing y by subtracting x^2 from both sides.

$$y = \frac{-x^2}{4} + 1$$ Divide both sides by 4, the coefficient of y.

The graph of this equation is shown in Figure 2.7.

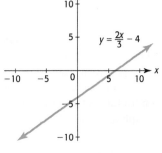

Figure 2.6

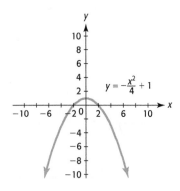

Figure 2.7

Direct Variation

Often in mathematics we need to express relationships between quantities. One relationship that is frequently used in applied mathematics occurs when two quantities are proportional. Two variables x and y are proportional to each other (or vary directly) if their quotient is a constant. That is, y is **directly proportional** to x, or y **varies directly** with x, if y and x are related by

$$\frac{y}{x} = k \quad \text{or, equivalently,} \quad y = kx$$

where k is called the **constant of proportionality** or the **constant of variation**.

As we saw earlier, the revenue from the sale of x units at p dollars per unit is $R(x) = px$, so we can say that the revenue is directly proportional to the number of units, with the constant of proportionality equal to p.

We know that blood alcohol percent is dependent on the number of drinks consumed and that excess consumption can lead to impairment of the drinker's judgment and motor skills (and sometimes to a DUI arrest). The blood alcohol percent p of a 130-pound man is directly proportional to the number of drinks x, where a drink is defined as 1.25 oz of 80-proof liquor, 12 oz of regular beer, or 5 oz of table wine.

EXAMPLE 10 ▶ **Blood Alcohol Percent**

The blood alcohol percent of a 130-pound man is directly proportional to the number of drinks consumed, and 3 drinks give a blood alcohol percent of 0.087. Find the constant of proportionality and the blood alcohol percent resulting from 5 drinks.

SOLUTION

The equation representing blood alcohol percent is $0.087 = k(3)$. Solving for k in $0.087 = k(3)$ gives $k = 0.029$, so the constant of proportionality is 0.029. Five drinks would result in a blood alcohol percent of $5(0.029) = 0.145$.

Skills CHECK 2.1

In Exercises 1–12, solve the equations.

1. $5x - 14 = 23 + 7x$ **2.** $3x - 2 = 7x - 24$

3. $3(x - 7) = 19 - x$ **4.** $5(y - 6) = 18 - 2y$

5. $x - \dfrac{5}{6} = 3x + \dfrac{1}{4}$ **6.** $3x - \dfrac{1}{3} = 5x + \dfrac{3}{4}$

7. $\dfrac{5(x - 3)}{6} - x = 1 - \dfrac{x}{9}$

8. $\dfrac{4(y - 2)}{5} - y = 6 - \dfrac{y}{3}$

9. $5.92t = 1.78t - 4.14$

10. $0.023x + 0.8 = 0.36x - 5.266$

11. $\dfrac{3}{4} + \dfrac{1}{5}x - \dfrac{1}{3} = \dfrac{4}{5}x$

12. $\dfrac{2}{3}x - \dfrac{6}{5} = \dfrac{1}{2} + \dfrac{5}{6}x$

For Exercises 13–16, find (a) the solution to the equation $f(x) = 0$, (b) the x-intercept of the graph of $y = f(x)$, and (c) the zero of $f(x)$.

13. $f(x) = 32 + 1.6x$ **14.** $f(x) = 15x - 60$

15. $f(x) = \dfrac{3}{2}x - 6$ **16.** $f(x) = \dfrac{x - 5}{4}$

In Exercises 17 and 18, you are given a table showing input and output values for a given function $y_1 = f(x)$. Using this table, find (if possible) (a) the x-intercept of the graph of $y = f(x)$, (b) the y-intercept of the graph of $y = f(x)$, and (c) the solution to the equation $f(x) = 0$.

17. **18.**

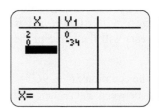

In Exercises 19 and 20, you are given the graph of a certain function $y = f(x)$ and the zero of that function. Using this graph, find (a) the x-intercept of the graph of $y = f(x)$ and (b) the solution to the equation $f(x) = 0$.

19.

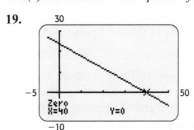

20.

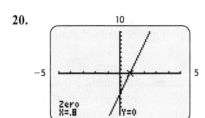

In Exercises 21–24, you are given the equation of a function. For each function, (a) find the zero of the function, (b) find the x-intercept of the graph of the function, and (c) solve the equation $f(x) = 0$.

21. $f(x) = 4x - 100$ **22.** $f(x) = 6x - 120$

23. $f(x) = 330 + 40x$ **24.** $f(x) = 250 + 45x$

In Exercises 25–32, solve the equations using graphical methods.

25. $14x - 24 = 27 - 3x$ **26.** $3x - 8 = 15x + 4$

27. $3(s - 8) = 5(s - 4) + 6$

28. $5(2x + 1) + 5 = 5(x - 2)$

29. $\dfrac{3t}{4} - 2 = \dfrac{5t - 1}{3} + 2$

30. $4 - \dfrac{x}{6} = \dfrac{3(x - 2)}{4}$ **31.** $\dfrac{t}{3} - \dfrac{1}{2} = \dfrac{t + 4}{9}$

32. $\dfrac{x - 5}{4} + x = \dfrac{x}{2} + \dfrac{1}{3}$

33. Solve $A = P(1 + rt)$ for r.

34. Solve $V = \dfrac{1}{3}\pi r^2 h$ for h.

35. Solve $5F - 9C = 160$ for F.

36. Solve $4(a - 2x) = 5x + \dfrac{c}{3}$ for x.

37. Solve $\dfrac{P}{2} + A = 5m - 2n$ for n.

38. Solve $y - y_1 = m(x - x_1)$ for x.

In Exercises 39–42, solve the equations for y and graph them with a standard window on a graphing utility.

39. $5x - 3y = 5$ **40.** $3x + 2y = 6$

41. $x^2 + 2y = 6$ **42.** $4x^2 + 2y = 8$

EXERCISES 2.1

43. *Depreciation* An $828,000 building is depreciated for tax purposes by its owner using the straight-line depreciation method. The value of the building y, after x months of use, is given by $y = 828{,}000 - 2300x$ dollars. After how many years will the value of the building be $690,000?

44. *Temperature Conversion* The equation $5F - 9C = 160$ gives the relationship between Fahrenheit and Celsius temperature measurements. What Fahrenheit measure is equivalent to a Celsius measurement of $20°$?

45. *Investments* The future value of a simple interest investment is given by $S = P(1 + rt)$. What principal P must be invested for $t = 5$ years at the simple interest rate $r = 10\%$ so that the future value grows to $9000?

46. *Temperature–Humidity Index* The temperature–humidity index I is $I = t + 0.55(1 - h)(t - 58)$, where t is the air temperature in degrees Fahrenheit and h is the relative humidity expressed as a decimal. If the air temperature is $80°$, find the humidity h (as a percent) that gives an index value of 102.

47. *Earnings and Minorities* The median annual salary (in thousands of dollars) of minorities M is related to the median salary (in thousands of dollars) of whites W by $M = 0.819W - 3.214$. What is the median annual salary for whites when the median annual salary for minorities is $41,520?
(Source: *Statistical Abstract of the United States*)

48. *Game Show Question* The following question was worth $32,000 on the game show *Who Wants to Be a Millionaire?*: "At what temperature are the Fahrenheit and Celsius temperature scales the same?" Answer this question. Recall that Fahrenheit and Celsius temperatures are related by $5F - 9C = 160$.

49. *Cell Phone Bills* The average monthly bill for wireless telephone subscribers from 1985 to 2008 can be modeled by $B(x) = -1.871x + 95.793$, where x is the number of years after 1980. If this model remains valid, in what year will the average monthly bill be $20.95?

50. *Reading Score* The average reading score on the National Assessment of Progress tests is given by $y = 0.155x + 255.37$, where x is the number of years past 1970. In what year would the average reading score be 259.4 if this model is accurate?

51. *Banks* The number of banks in the United States for selected years from 1980 to 2009 is given by $y = -380.9x + 18,483$, where x is the number of years after 1980. If this model is accurate after 2009, in what year would the number of banks be 7056?
(Source: Federal Deposit Insurance Corporation)

52. *Profit* The profit from the production and sale of specialty golf hats is given by the function $P(x) = 20x - 4000$, where x is the number of hats produced and sold.

   a. Producing and selling how many units will give a profit of $8000?

   b. How many units must be produced and sold to avoid a loss?

53. *World Forest Area* When x is the number of years after 1990, the world forest area (natural forest or planted stands) as a percent of land area is given by $F(x) = -0.065x + 31.39$. In what year will the percent be 29.44 if this model is accurate?
(Source: The World Bank)

54. *Marriage Rate* The marriage rate per 1000 population for the years 1987–2009 can be modeled by $y = -0.146x + 11.074$, where x is the number of years after 1980. During what year does the model indicate the marriage rate to be 5.8?
(Source: *National Vital Statistics Report 2010*)

55. *U.S. Population* The U.S. population can be modeled for the years 1960–2060 by the function $p = 2.6x + 177$, where p is in millions of people and x is in years from 1960. During what year does the model estimate the population to be 320,000,000?
(Source: www.census.gov/statab)

56. *Inmates* The total number of inmates in custody between 1990 and 2005 in state and federal prisons is given approximately by $y = 76x + 115$ thousand prisoners, where x is the number of years after 1990. If the model remains accurate, in what year should the number of inmates be 1,787,000?
(Source: Bureau of Justice Statistics)

57. *Twitter* Since 2007, "tweeting" has become increasingly popular. The number of tweets (in thousands) reported by the microblogging company Twitter can be modeled by $T(x) = 16,665x - 116,650$, where x is the number of years after 2000. If this model remains accurate, in what year will the number of tweets be 133,325 thousand, according to the model?

58. *Marijuana Use* The percent p of high school seniors using marijuana daily can be related to x, the number of years after 1990, by the equation $30p - 19x = 1$. Assuming the model remains accurate, during what

year should the percent using marijuana daily equal 12.7%?
(Source: MonitoringtheFuture.org)

59. *Hispanic Population* Using data and projections from 1980 through 2050, the number (in millions) of Hispanics in the U.S. civilian noninstitutional population is given by $y = 0.876x + 6.084$, where x is the number of years after 1980. During what year was the number of Hispanics 14.8 million, if this model is accurate?
(Source: U.S. Census Bureau)

60. *Personal Income* U.S. personal income increased between 1980 and 2005 according to the model $I(x) = 386.17x + 524.32$ billion dollars, where x is the number of years after 1980. In what year would U.S. personal income be estimated to reach $10,951 billion if this model is accurate beyond 2005?
(Source: *Statistical Abstract of the United States*)

61. *Cigarette Ads* The FTC's annual report on cigarette sales and advertising shows that the major cigarette manufacturers spent $15.15 billion on advertising and promotional expenditures in 2003, an increase of $2.68 billion from 2002 and the most ever reported to the Commission.

   a. Write the linear equation that represents the expenditure as a function of the number of years after 2000.

   b. Use the model to determine in what year the expenditures are expected to exceed 60 billion dollars.

62. *Cell Phone Subscribers* The number of cell phone subscribers (in millions) between 2001 and 2009 can be modeled by $S(x) = 21x + 101.7$, where x is the number of years after 1995. In what year does this model indicate that there were 311,700,000 subscribers?
(Source: Semiannual CTIA Wireless Survey)

63. *Grades* To earn an A in a course, a student must get an average score of at least 90 on five tests. If her first four test scores are 92, 86, 79, and 96, what score does she need on the last test to obtain a 90 average?

64. *Grades* To earn an A in a course, a student must get an average score of at least 90 on three tests and a final exam. If his final exam score is higher than his lowest score, then the lowest score is removed and the final exam score counts double. If his first three test scores are 86, 79, and 96, what is the lowest score he can get on the last test and still obtain a 90 average?

65. *Tobacco Judgment* As part of the largest-to-date damage award in a jury trial in history, the July 2000 penalty handed down against the tobacco industry included a $74 billion judgment against Philip Morris. If this amount was 94% of this company's 1999 revenue, how much was Philip Morris's 1999 revenue?

(Source: *Newsweek*, July 24, 2000)

66. *Tobacco Judgment* As part of the largest-to-date damage award in a jury trial in history, the July 2000 penalty handed down against the tobacco industry included a $36 billion judgment against R. J. Reynolds. If this amount was 479% of this company's 1999 revenue, how much was R. J. Reynolds' 1999 revenue?

(Source: *Newsweek*, July 24, 2000)

67. *Sales Commission* A salesman earns $50,000 in commission in 1 year and then has his commission reduced by 20% the next year. What percent increase in commission over the second year will give him $50,000 in the third year?

68. *Salaries* A man earning $100,000 per year has his salary reduced by 5% because of a reduction in his company's market. A year later, he receives $104,500 in salary. What percent raise from the reduced salary does this represent?

69. *Sales Tax* The total cost of a new automobile, including a 6% sales tax on the price of the automobile, is $29,998. How much of the total cost of this new automobile is sales tax?

70. *Wildlife Management* In wildlife management, the capture–mark–recapture technique is used to estimate the population of certain types of fish or animals. To estimate the population, we enter information in the equation

$$\frac{\text{total in population}}{\text{total number marked}} = \frac{\text{total number in second capture}}{\text{number found marked in second capture}}$$

Suppose that 50 sharks are caught along a certain shoreline, marked, and then released. If a second capture of 50 sharks yields 20 sharks that have been marked, what is the resulting population estimate?

71. *Investment* The formula for the future value A of a simple interest investment is $A = P + Prt$, where P is the original investment, r is the annual interest rate, and t is the time in years. Solve this formula for t.

72. *Investment* The formula for the future value A of a simple interest investment is $A = P + Prt$, where P is the original investment, r is the annual interest rate, and t is the time in years. Solve this formula for P.

73. *Investment* If P dollars are invested for t years at simple interest rate r, the future value of the investment is $A = P + Prt$. If $2000 invested for 6 years gives a future value of $3200, what is the simple interest rate of this investment?

74. *Investment* If an investment at 7% simple interest has a future value of $5888 in 12 years, what is the original investment?

75. *Investment* The simple interest earned in 9 years is directly proportional to the interest rate r. If the interest is $920 when r is 12%, what is the amount of interest earned in 9 years at 8%?

76. *Investment* The interest earned at 9% simple interest is directly proportional to the number of years the money is invested. If the interest is $4903.65 in 5 years, in how many years will the interest earned at 9% be $7845.84?

77. *Circles* Does the circumference of a circle vary directly with the radius of the circle? If so, what is the constant of variation?

78. *Calories* The amount of heat produced in the human body by burning protein is directly proportional to the amount of protein burned. If burning 1 gram of protein produces 32 calories of heat, how much protein should be burned to produce 180 calories?

79. *Body Mass Index* Body mass index (BMI) is a measure that helps determine obesity, with a BMI of 30 or greater indicating that the person is obese. A BMI table for heights in inches and weights in pounds is shown on page 83. BMI was originally defined in the metric system of measure, and the BMI is directly proportional to the weight of a person of a given height.

a. If the BMI of a person who is 1.5 meters tall is 20 when the person weighs 45 kilograms, what is the constant of variation?

b. If a woman of this height has a BMI of 32, what does she weigh?

80. *Land Cost* The cost of land in Savannah is directly proportional to the size of the land. If a 2500-square-foot piece of land costs $172,800, what is the cost of a piece of land that is 5500 square feet?

Fitting Lines to Data Points: Modeling Linear Functions

KEY OBJECTIVES

- Find exact linear models for data

- Determine if a set of data can be modeled exactly or approximately

- Create scatter plots for sets of data

- Find approximate linear models for data

- Visually determine if a linear model is a "good" fit for data

- Solve problems using linear models

SECTION PREVIEW   Earnings and Gender

Table 2.5 shows the earnings of year-round full-time workers by gender and educational attainment. Figure 2.8 shows the scatter plot of the data, and it appears that a line would approximately fit along these data points. We can determine the relationship between the two sets of earnings by creating a linear equation that gives female annual earnings as a function of male annual earnings. This model can be used to estimate an annual salary for a female from a given annual salary for a male. (See Example 3.) In this section, we will create linear models from data points, use graphing utilities to create linear models, and use the model to answer questions about the data.

Table 2.5

Educational Attainment	Average Annual Earnings of Males ($ thousand)	Average Annual Earnings of Females ($ thousand)
Less than ninth grade	21.659	17.659
Some high school	26.277	19.162
High school graduate	35.725	26.029
Some college	41.875	30.816
Associate's degree	44.404	33.481
Bachelor's degree	57.220	41.681
Master's degree	71.530	51.316
Doctorate degree	82.401	68.875
Professional degree	100.000	75.036

(Source: U.S. Census Bureau)

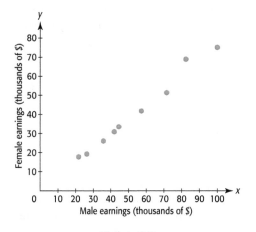

Figure 2.8

Exact and Approximate Linear Models

We have seen that if data points fit exactly on a line, we can use two of the points to model the linear function (write the equation of the line). We can determine that the data points fit exactly on a line by determining that the changes in output values are equal for equal changes in the input values. In this case, we say that the inputs are **uniform** and the **first differences** are constant.

- If the first differences of data outputs are constant for uniform inputs, the rate of change is constant and a linear function can be found that fits the data exactly.

- If the first differences are "nearly constant," a linear function can be found that is an approximate fit for the data.

If the first differences of data outputs are constant for uniform inputs, we can use two of the points to write the linear equation that models the data. If the first differences of data outputs are constant for inputs differing by 1, this constant difference is the rate of change of the function, which is the slope of the line fitting the points exactly.

EXAMPLE 1 ▶ Retirement

Table 2.6 gives the annual retirement payment to a 62-year-old retiree with 21 or more years of service at Clarion State University as a function of the number of years of service and the first differences of the outputs.

Table 2.6

Year	21	22	23	24	25
Retirement Payment	40,950	42,900	44,850	46,800	48,750
First Differences		1950	1950	1950	1950

Because the first differences of the outputs of this function are constant (equal to 1950) for each unit change of the input (years), the rate of change is the constant 1950. Using this rate of change and a point gives the equation of the line that contains all the points. Representing the annual retirement payment by y and the years of service by x and using the point (21, 40,950), we obtain the equation

$$y - 40,950 = 1950(x - 21)$$
$$y = 1950x$$

Note that the values in Table 2.6 represent points satisfying a **discrete function** (a function with a finite number of inputs), with each input representing the number of years of service. Although only points with integer inputs represent the annual retirement payments, we can model the application with the **continuous function** $y = 1950x$, whose graph is a line that passes through the 5 data points. Informally, a continuous function can be defined as a function whose graph can be drawn over its domain without lifting the pen from the paper.

In the context of this application, we must give a discrete interpretation to the model. This is because the only inputs of the function that make sense in this case are nonnegative integers representing the number of years of service. The graph of the discrete function defined in Table 2.6 is the scatter plot shown in Figure 2.9(a) on the next page, and the continuous function that fits these data points, $y = 1950x$, is shown in Figure 2.9(b).

When a scatter plot of data can be approximately fitted by a line, we attempt to find the graph that visually gives the best fit for the data and find its equation. For now, we informally define the "best-fit" line as the one that appears to come closest to all the data points.

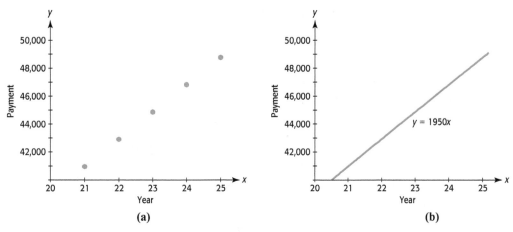

Figure 2.9

EXAMPLE 2 ▶	Health Service Employment

Table 2.7 gives the number of full- and part-time employees in offices and clinics of dentists for selected years between 1990 and 2005. (Source: U.S. Bureau of Labor Statistics)

Table 2.7

Year	1990	1995	2000	2001	2002	2003	2004	2005
Employees (in thousands)	513	592	688	705	725	744	760	771

a. Draw a scatter plot of the data with the x-value of each point representing the number of years after 1990 and the y-value representing the number of dental employees (in thousands) corresponding to that year.

b. Graph the equation $y = 16x + 510$ on the same graph as the scatter plot and determine if the line appears to be a good fit.

c. Draw a visual fit line that fits the data well (a piece of spaghetti or pencil lead over your calculator screen works well) and select two points on that line (use the free-moving cursor). Use these two points to write an equation of the visual fit line. Determine whether this line or the one from part (b) is the better fit.

SOLUTION

a. By using x as the number of years after 1990, we have aligned the data with $x = 0$ representing 1990, $x = 5$ representing 1995, and so forth. We enter into the lists of a graphing utility (or an Excel spreadsheet) the aligned input data representing the years in Table 2.7 and the output data representing the numbers of employees shown in the second row of Table 2.7. Table 2.8 shows the lists containing the data. The graph of these data points is shown in Figure 2.10. The window for this scatter plot can be set manually or with a command on the graphing utility such as ZoomStat, which is used to automatically set the window and display the graph.

Table 2.8

L1	L2
0	513
5	592
10	688
11	705
12	725
13	744
14	760
15	771

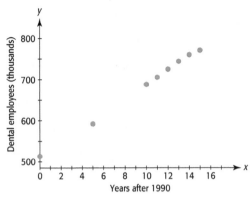

Figure 2.10

b. The graph of the equation $y = 16x + 510$ and the scatter plot of the data are shown in Figure 2.11. The line does not appear to be the best possible fit to the data points.

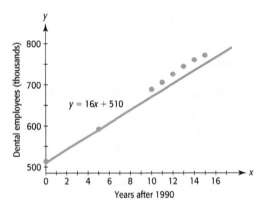

Figure 2.11

c. Figure 2.12(a) shows one example of a visual fit line that fits the data obtained by placing a piece of spaghetti on the calculator screen close to the data points. By using the free-moving cursor (press the right, left, up, and down arrows), we obtain two points that lie close to the visual fit line. Two points that lie on the line are (3.096, 563.934) and (12.479, 731.218) (rounded to three decimal places), and they are shown in Figure 2.12(b) and (c).*

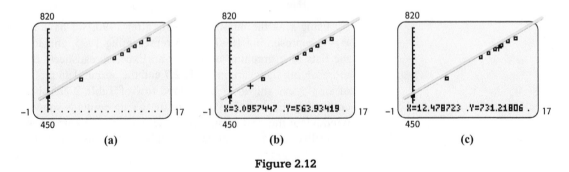

Figure 2.12

*Many other points are possible.

To write the equation of the line, we must first find the slope of the line between these two points.

$$m = \frac{731.218 - 563.934}{12.479 - 3.096} \approx 17.83$$

The equation of our visual fit line is

$$y - 563.934 = 17.83(x - 3.096)$$
$$y = 17.830x + 508.732$$

The graphs of the data points and the visual fit line are shown in Figure 2.13.

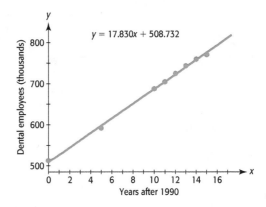

Figure 2.13

Figures 2.11 and 2.13 show that the visual fit line, with equation $y = 17.830x + 508.732$, appears to be a better fit to the data.

Is either of the lines in Figures 2.11 and 2.13 the best-fitting of all possible lines that could be drawn on the scatter plot? How do we determine the best-fit line? We now discuss the answers to these questions.

Fitting Lines to Data Points; Linear Regression

The points determined by the data in Table 2.7 on page 105 do not all lie on a line, but we can determine the equation of the line that is the "best fit" for these points by using a procedure called **linear regression**. This procedure defines the *best-fit line* as the line for which the sum of the squares of the vertical distances from the data points to the line is a minimum. For this reason, the linear regression procedure is also called the **least squares method**.

The vertical distance between a data point and the corresponding point on a line is simply the amount by which the line misses going through the point—that is, the difference in the y-values of the data point and the point on the line. If we call this difference in outputs d_i (where i takes on the values from 1 to n for n data points), the least squares method* requires that

$$d_1^2 + d_2^2 + d_3^2 + \cdots + d_n^2$$

be as small as possible. The line for which this sum of squared differences is as small as possible is called the *linear regression line* or *least squares line* and is the one that we consider to be the **best-fit line** for the data.

---

*The sum of the squared differences is often called SSE, the *sum of squared errors*. The development of the equations that lead to the minimum SSE is a calculus topic.

For an illustration of the linear regression process, consider again the two lines in Figures 2.11 and 2.13 and the data points in Table 2.7. Figures 2.14 and 2.15 indicate the vertical distances for each of these lines. To the right of each figure is the calculation of the sum of squared differences for the line.

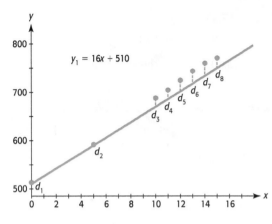

$$d_1{}^2 + d_2{}^2 + d_3{}^2 + \cdots + d_8{}^2$$
$$= (513 - 510)^2 + (592 - 590)^2 + (688 - 670)^2$$
$$+ (705 - 686)^2 + (725 - 702)^2 + (744 - 718)^2$$
$$+ (760 - 734)^2 + (771 - 750)^2$$
$$= 9 + 4 + 324 + 361 + 529 + 676 + 676 + 441$$
$$= 3020$$

Figure 2.14

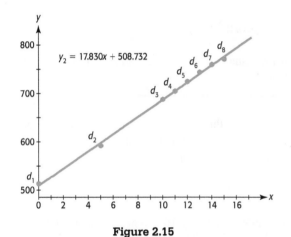

$$d_1{}^2 + d_2{}^2 + d_3{}^2 + \cdots + d_8{}^2$$
$$= (513 - 508.73)^2 + (592 - 597.88)^2 + (688 - 687.03)^2$$
$$+ (705 - 704.86)^2 + (725 - 722.69)^2 + (744 - 740.52)^2$$
$$+ (760 - 758.35)^2 + (771 - 776.18)^2$$
$$= 100.77$$

Figure 2.15

We see numerically from the sum of squared differences calculations, as well as visually from Figures 2.14 and 2.15, that the line

$$y_2 = 17.830x + 508.732$$

is a better model for the data than the line y_1. However, is this line the best-fit line for these data? Remember that the best-fit line must have the smallest sum of squared differences of *all* lines that can be drawn through the data! We will see that the line which is the best fit for these data points, with coefficients rounded to three decimal places, is

$$y = 17.733x + 509.917$$

The sum of squared differences for the regression line is approximately 99, which is slightly smaller than the sum 100.77, which was found for $y_2 = 17.830x + 508.732$.

Development of the formulas that give the best-fit line for a set of data is beyond the scope of this text, but graphing calculators, computer programs, and spreadsheets have built-in formulas or programs that give the equation of the best-fit line. That is, we can use technology to find the best linear model for the data. The calculation of the sum of squared differences for the dental employees data was given only to illustrate what the best-fit line means. You will not be asked to calculate—nor do we use from this point on—the value of the sum of squared differences in this text.

We illustrate the use of technology to find the regression line by returning to the data of Example 2. To find the equation that gives the number of employees in dental offices as a function of the years after 1990, we use the following steps.

Modeling Data

Step 1: Enter the data into lists of a graphing utility.

Step 2: Create a scatter plot of the data to see if a linear model is reasonable. The data should appear to follow a linear pattern with no distinct curvature.

Step 3: Use the graphing utility to obtain the linear equation that is the best fit for the data. Figure 2.16(a) shows the equation for Example 2, which can be approximated by $y = 17.733x + 509.917$.*

Step 4: Graph the linear function (unrounded) and the data points on the same graph to see how well the function fits the data. (The equation entered into an equation editor and the graph of the data and the best-fit line for Example 2 are shown in Figure 2.16(b) and Figure 2.16(c), respectively.)

Step 5: Report the function and/or numerical results in a way that makes sense in the context of the problem, with the appropriate units and with the variables identified. Unless otherwise indicated, report functions with coefficients rounded to three decimal places.

Figure 2.16(c) shows that

$$y = 17.733x + 509.917$$

is a good model of the number of full- and part-time employees in dentists' offices and clinics, where y is in thousands of employees and x is the number of years after 1990.

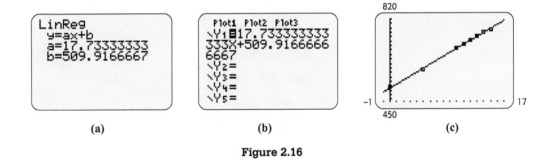

(a) (b) (c)

Figure 2.16

The screens for the graphing utility that you use may vary slightly from those given in this text. Also, the regression line you obtain is dependent on your particular technology and may have some decimal places slightly different from those shown here.

In modeling a set of data, it is important to be careful when rounding coefficients in equations and when rounding during calculations. We will use the following guidelines in this text.

---

*See Appendix A, page 625.

> **Technology Note**
>
> After a model for a data set has been found, it can be rounded for reporting purposes. However, do not use a rounded model in calculations, and do not round answers during the calculation process unless instructed to do so. When the model is used to find numerical answers, the answers should be rounded in a way that agrees with the context of the problem.

EXAMPLE 3 ▶ **Earnings and Gender**

Table 2.9 shows the earnings of year-round full-time workers by gender and educational attainment.

a. Let x represent earnings for males, let y represent earnings for females, and create a scatter plot of the data.

b. Create a linear model that expresses female annual earnings as a function of male annual earnings.

c. Graph the linear function and the data points on the same graph, and discuss how well the function models the data.

Table 2.9

Educational Attainment	Average Annual Earnings of Males ($ thousand)	Average Annual Earnings of Females ($ thousand)
Less than ninth grade	21.659	17.659
Some high school	26.277	19.162
High school graduate	35.725	26.029
Some college	41.875	30.816
Associate's degree	44.404	33.481
Bachelor's degree	57.220	41.681
Master's degree	71.530	51.316
Doctorate degree	82.401	68.875
Professional degree	100.000	75.036

(Source: U.S. Census Bureau)

SOLUTION

a. Enter the data from Table 2.9 in the lists of a graphing utility. Figure 2.17(a) shows a partial list of the data points. The scatter plot of all the data points from Table 2.9 is shown in Figure 2.17(b).

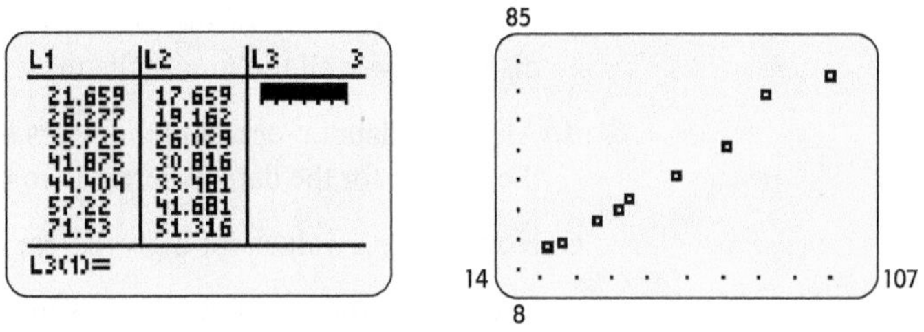

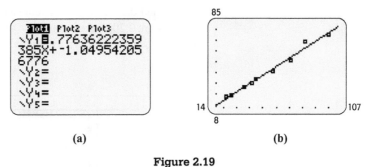

Figure 2.17

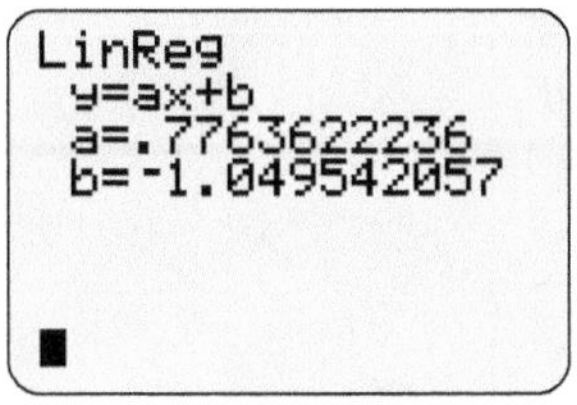

Figure 2.18

b. The points in the scatter plot in Figure 2.17(b) exhibit a nearly linear pattern, so a linear function could be used to model these data. The equation of the line that is the best fit for the data points can be found with a graphing calculator (Figure 2.18). The equation, rounded to three decimal places, is

$$y = 0.776x - 1.050$$

Remember that a *model* gives not only the equation but also a description of the variables and their units of measure. The rounded model

$$y = 0.776x - 1.050$$

where x and y are in thousands of dollars, expresses female annual earnings y as a function of male annual earnings x.

c. Using the unrounded function in the equation editor (Figure 2.19(a)) and graphing it along with the data points,* we observe that the line follows the general trend indicated by the data (Figure 2.19(b)). Note that not all points lie on the graph of the equation, even though this is the line that is the best fit for the data.

(a)   **(b)**

Figure 2.19

EXAMPLE 4 ▶   **U.S. Population**

The total U.S. population for selected years beginning in 1960 and projected to 2050 is shown in Table 2.10, with the population given in millions.

a. Align the data to represent the number of years after 1960, and draw a scatter plot of the data.

b. Create the linear equation that is the best fit for these data, where y is in millions and x is the number of years after 1960.

---

*Even though we write rounded models found from data given in this text, all graphs and calculations use the unrounded model found by the graphing utility.

c. Graph the equation of the linear model on the same graph with the scatter plot and discuss how well the model fits the data.

d. Align the data to represent the years after 1950 and create the linear equation that is the best fit for the data, where y is in millions.

e. How do the x-values for a given year differ?

f. Use both unrounded models to estimate the population in 2000 and in 2010. Are the estimates equal?

Table 2.10

Year	Population (millions)	Year	Population (millions)
1960	180.671	1995	263.044
1965	194.303	1998	270.561
1970	205.052	2000	281.422
1975	215.973	2003	294.043
1980	227.726	2025	358.030
1985	238.466	2050	408.695
1990	249.948		

(Source: U.S. Census Bureau)

SOLUTION

a. The aligned data have $x = 0$ representing 1960, $x = 5$ representing 1965, and so forth. Figure 2.20(a) shows the first seven entries using the aligned data. The scatter plot of these data is shown in Figure 2.20(b).

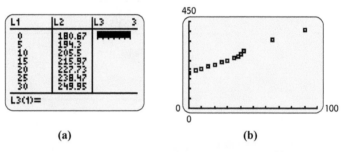

(a)   (b)

Figure 2.20

b. The equation of the best-fit line is found by using linear regression with a graphing calculator. With the decimals rounded to three places, the linear model for the U.S. population is

$$y = 2.607x + 177.195 \text{ million}$$

where x is the number of years after 1960.

c. Using the unrounded function in the equation editor (Figure 2.21(a)) and graphing it along with the scatter plot shows that the graph of the best-fit line is very close to the data points (Figure 2.21(b)). However, the points do not all fit the line because the U.S. population did not increase by exactly the same amount each year.

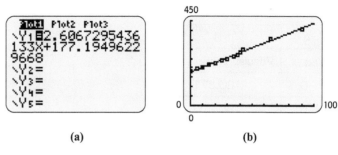

(a)   (b)

Figure 2.21

d. If we align the data to represent the years after 1950, then $x = 10$ corresponds to 1960, $x = 15$ corresponds to 1965, and so forth. Figure 2.22(a) shows the first seven entries using the aligned data, and Figure 2.22(b) shows the scatter plot. The equation that best fits the data, found using linear regression with a calculator, is

$$y = 2.607x + 151.128$$

where x is the number of years from 1950 and y is in millions. Figure 2.22(c) shows the regression equation.

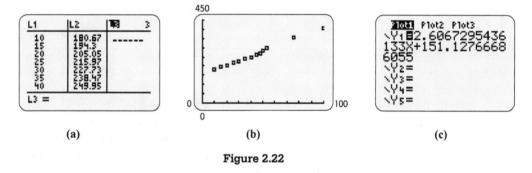

(a)   (b)   (c)

Figure 2.22

e. The x-values for a given year are 10 more with this model than with the first model.

f. Both models estimate the population to be 281.464 million in 2000 and to be 307.531 million in 2010. They are equal estimates. In fact, if you substitute $x - 10$ for x in the first model, you will get the second model.

Spreadsheet ▸ SOLUTION   We can also use spreadsheets to find the linear function that is the best fit for the data. Table 2.11 shows the Excel spreadsheet for the data of Example 4, with x equal to the years after 1960. Selecting the cells containing the data, using **Chart Wizard** to get the scatter plot of the data, and selecting **Add Trendline** gives the equation of the linear function that is the best fit for the data, along with the scatter plot and the graph of the best-fitting line. The equation and the graph of the line are shown in Figure 2.23.*

---

*See Appendix B, page 643.

Table 2.11

	A	B
	Year x	Population (millions) y
1	0	180.671
2	5	194.303
3	10	205.052
4	15	215.973
5	20	227.726
6	25	238.466
7	30	249.948
8	35	263.044
9	38	270.561
10	40	281.422
11	43	294.043
12	65	358.03
13	90	408.695

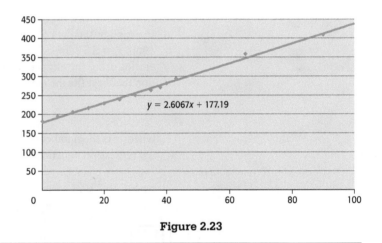

Figure 2.23

Applying Models

Because 1997 is a year between two given values in the table of Example 4, using the model to find the population in 1997 is called **interpolation**. When a model is evaluated for a *prediction* using input(s) outside the given data points, the process is called **extrapolation**.

A question arises whenever data that involve time are used. If we label the *x*-coordinate of a point on the input axis as 1999, what time during 1999 do we mean? Does 1999 refer to the beginning of the year, the middle, the end, or some other time? Because most data from which the functions in applications are derived represent end-of-year totals, we adopt the following convention when modeling, unless otherwise specified:

A point on the input axis indicating a time refers to the end of the time period.

For instance, the point representing the year 1999 means "at the end of 1999." Notice that this instant in time also represents the beginning of the year 2000. If, for example, data are aligned with *x* equal to the number of years after 1990, then any *x*-value greater than 9 and less than or equal to 10 represents some time in the year 2000. If a point represents anything other than the end of the period, this information will be clearly indicated. Also, if *a* and *b* are points in time, we use the phrases "from *a* to *b*" and "between *a* and *b*" to indicate the same time interval.

Goodness of Fit

Consider again Example 4, where we modeled the U.S. population for selected years. Looking at Table 2.12, notice that for uniform inputs the first differences of the outputs appear to be relatively close to the same constant, especially compared to the size of the population. (If these differences were closer to a constant, the fit would be better.)

Table 2.12

Uniform Inputs (years)	Outputs Population (millions)	First Differences in Output (difference in population)
1960	180.671	
1965	194.303	**13.632**
1970	205.052	**10.749**
1975	215.973	**10.921**
1980	227.726	**11.753**
1985	238.466	**10.740**
1990	249.948	**11.482**
1995	263.044	**13.096**
2000	281.422	**18.378**

So how "good" is the fit of the linear model $y = 2.6067295436133x + 177.19496229668$ to the data in Example 4? Based on observation of the graph of the line and the data points on the same set of axes (see Figure 2.23), it is reasonable to say that the regression line provides a very good fit, but not an exact fit, to the data.

The goodness of fit of a line to a set of data points can be observed from the graph of the line and the data points on the same set of axes, and it can be measured if your graphing utility computes the **correlation coefficient**.* The correlation coefficient is a number r, $-1 \le r \le 1$, that measures the strength of the linear relationship that exists between the two variables. The closer $|r|$ is to 1, the more closely the data points fit the linear regression line. (There is no linear relationship between the two variables if $r = 0$.) Positive values of r indicate that the output variable increases as the input variable increases, and negative values of r indicate that the output variable decreases as the input variable increases. But the *strength* of the relationship is indicated by how close $|r|$ is to 1. For the data of Example 4, computing the correlation coefficient gives $r = .997$ (Figure 2.24), which means that the linear relationship is strong and that the linear model is an excellent fit for the data. When a calculator feature or computer program is used to fit a linear model to data, the resulting equation can be considered the best linear fit for the data. As we will see later in this text, other mathematical models may be better fits to some sets of data, especially if the first differences of the outputs are not close to being constant.

```
LinReg
y=ax+b
a=2.606729544
b=177.1949623
r²=.9940145759
r=.9970027964
```

Figure 2.24

*See Appendix A, page 625.

Report models to three decimal places unless otherwise specified. Use unrounded models to graph and calculate unless otherwise specified.

Discuss whether the data shown in the scatter plots in the figures for Exercises 1 and 2 should be modeled by a linear function.

1.

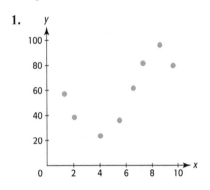

2.

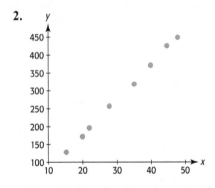

Discuss whether the data shown in the scatter plots in the figures for Exercises 3 and 4 should be modeled by a linear function exactly or approximately.

3.

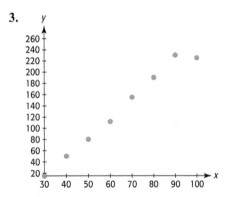

4.

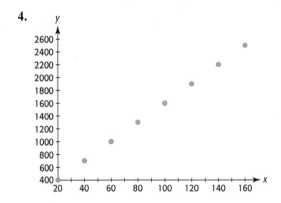

Create a scatter plot for each of the sets of data in Exercises 5 and 6.

5.

x	1	3	5	7	9
y	4	7	10	13	16

6.

x	1	2	3	5	7	9	12
y	1	3	6	1	9	2	6

7. Can the scatter plot in Exercise 5 be fit exactly or only approximately by a linear function? How do you know?

8. Can the scatter plot in Exercise 6 be fit exactly or only approximately by a linear function? How do you know?

9. Find the linear function that is the best fit for the data in Exercise 5.

10. Find the linear function that is the best fit for the data in Exercise 6.

Use the data in the following table for Exercises 11–14.

x	5	8	11	14	17	20
y	7	14	20	28	36	43

11. Construct a scatter plot of the data in the table.

12. Determine if the points plotted in Exercise 11 appear to lie near some line.

13. Create a linear model for the data in the table.

14. Use the function $y = f(x)$ created in Exercise 13 to evaluate $f(3)$ and $f(5)$.

Use the data in the table for Exercises 15–18.

x	2	5	8	9	10	12	16
y	5	10	14	16	18	21	27

15. Construct a scatter plot of the data in the table.

16. Determine if the points plotted in Exercise 15 appear to lie near some line.

17. Create a linear model for the data in the table.

18. Use the rounded function $y = f(x)$ that was found in Exercise 17 to evaluate $f(3)$ and $f(5)$.

19. Determine which of the equations, $y = -2x + 8$ or $y = -1.5x + 8$, is the better fit for the data points $(0, 8)$, $(1, 6)$, $(2, 5)$, $(3, 3)$.

20. Determine which of the equations, $y = 2.3x + 4$ or $y = 2.1x + 6$, is the better fit for the data points $(20, 50)$, $(30, 73)$, $(40, 96)$, $(50, 119)$, $(60, 142)$.

21. Without graphing, determine which of the following data sets are exactly linear, approximately linear, or nonlinear.

a.

x	y
1	5
2	8
3	11
4	14
5	17

b.

x	y
1	2
2	5
3	10
4	5
5	2

c.

x	y
1	6
2	8
3	12
4	14
5	18

22. Why can't first differences be used to tell if the following data are linear? $(1, 3)$, $(4, 5)$, $(5, 7)$, $(7, 9)$

EXERCISES 2.2

Report models to three decimal places unless otherwise specified. Use unrounded models to graph and calculate unless otherwise specified.

23. *Women in the Workforce* The number of women in the workforce for selected years from 1930 through 2015 is shown in the following figures.

   a. Do the points on the graph in the figure define the number of working women as a discrete or continuous function of the year?

   b. Does the graph of $y = W(x)$ shown in the figure define the number of working women as a discrete or continuous function of the year?

   c. Would the data in the scatter plot be better modeled by a linear function than by the nonlinear function $y = W(x)$? Why or why not?

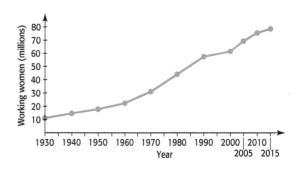

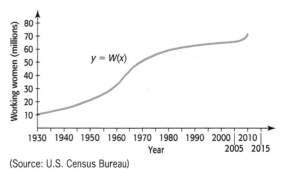

(Source: U.S. Census Bureau)

24. *Working Age* The scatter plot in the figure projects the ratio of the working-age population to the elderly population.

   a. Does the scatter plot in the figure represent discrete or continuous data?

   b. Would a linear function be a good model for the data points shown in the figure? Explain.

   c. Could the data shown for the years beyond 2010 be modeled by a linear function?

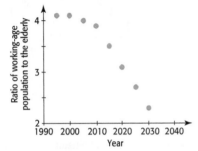

(Source: *Newsweek*)

25. *Education Spending* Personal expenditures for higher education rose dramatically from 1990 to 2008. The chart shows the personal expenditures for higher education in billions of dollars for selected years during 1990–2008. Can the data be modeled exactly by a linear function? Explain.

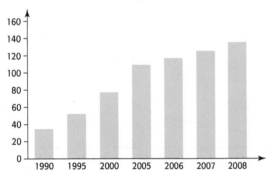

(Source: Bureau of Economic Analysis, U.S. Department of Commerce)

26. *Future Value of an Investment* If $1000 is invested at 6% simple interest, the initial value and the future value *S* at the end of each of 5 years are shown in the table that follows.

   a. Can a linear function model exactly the points from the table? Explain.

   b. If so, find a linear function $S = f(t)$ that models the points.

   c. Use the model to find the future value of this investment at the end of the 7th year. Is this an interpolation or an extrapolation from the data?

   d. Should this model be interpreted discretely or continuously?

Year (t)	0	1	2	3	4	5
Future Value (S)	1000	1060	1120	1180	1240	1300

27. *Taxes* The table below shows some sample incomes and the income tax due for each taxable income.

   a. Can a linear function model exactly the points from the table? Explain.

   b. If so, find a linear function $T = f(x)$ that models the points.

   c. Verify that the model fits the data by evaluating the function at $x = 30,100$ and $x = 30,300$ and comparing the resulting *T*-values with the income tax due for these taxable incomes.

   d. If the model can be interpreted continuously, use it to find the tax due on taxable income of $30,125. Is this an interpolation or an extrapolation from the data?

   e. Can this model be used to compute all tax due on taxable income?

Taxable Income	Income Tax Due
$30,000	$3665.00
30,050	3672.50
30,100	3680.00
30,150	3687.50
30,200	3695.00
30,250	3702.50
30,300	3710.00

(Source: U.S. Federal tax table for 2009)

28. *Farms* The figure gives the number of farms (in millions) for selected years from 1940 to 2009. Can the data be modeled exactly by a linear function? Explain.

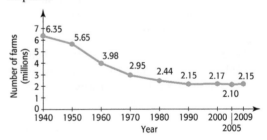

29. *Diabetes* As the following table shows, projections indicate that the percent of U.S. adults with diabetes could dramatically increase.

   a. Find a linear model that fits the data in the table, with $x = 0$ for 2000.

b. Use the model to predict the percent of U.S. adults with diabetes in 2018.

c. In what year does this model predict the percent to be 25.05%?

Year	Percent	Year	Percent	Year	Percent
2010	15.7	2025	24.2	2040	31.4
2015	18.9	2030	27.2	2045	32.1
2020	21.1	2035	29.0	2050	34.3

30. *High School Enrollment* The table below gives the enrollment (in thousands) in grades 9–12 at U.S. public and private schools for the years 1990–2009.

   a. Create a scatter plot of the data with x equal to the year after 1990 and y equal to enrollment in thousands.

   b. Find the linear model that is the best fit for the data.

   c. Graph the unrounded model and the scatter plot on the same axes. Is the model a "good" fit?

Year, x	Enrollment, y (thousands)	Year, x	Enrollment, y (thousands)
1990	12,488	2000	14,802
1991	12,703	2001	15,058
1992	12,882	2002	15,332
1993	13,093	2003	15,721
1994	13,376	2004	16,048
1995	13,697	2005	16,328
1996	14,060	2006	16,498
1997	14,272	2007	16,451
1998	14,428	2008	16,322
1999	14,623	2009	16,175

(Source: U.S. Department of Education)

31. *U.S. Domestic Travel* The graph that follows gives the number of millions of persons who took trips of 50 miles or more for the years 2000 through 2008.

   a. Find the equation of the line which is the best fit for these data, with x equal to the number of years after 2000 and y equal to the millions of persons.

   b. Use the model to estimate the number of persons who took trips in 2010.

   c. In what year does the model estimate the number of people taking trips as 1737.3 million?

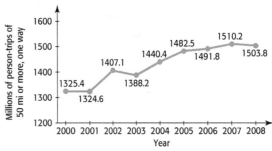

U.S. Domestic Leisure Travel Volume

(Source: From *2011 World Almanac*. Reprinted by permission.)

32. *Earnings and Race* The table gives the median household income for whites and blacks in various years.

   a. Let x represent the median household income for whites and y represent the corresponding median household income for blacks, and make a scatter plot of these data.

   b. Find a linear model that expresses the median household income for blacks as a function of the median household income for whites.

   c. Find the slope of the linear model in (b) and write a sentence that interprets it.

Median Household Income

Year	Whites	Blacks
1981	38,954	21,859
1985	40,614	24,163
1990	42,622	25,488
1995	42,871	26,842
2000	46,910	31,690
2001	46,261	30,625
2002	46,119	29,691
2003	45,631	29,645
2005	48,554	30,858

(Source: U.S. Census Bureau)

33. *Poverty* The table shows the number of millions of people in the United States who lived below the poverty level for selected years.

   a. Find a linear model that approximately fits the data, using x as the number of years after 1970.

   b. Use a graph of the model and a scatter plot to determine if the model is nearly an exact fit for the data.

Year	Persons Living Below the Poverty Level (millions)
1970	25.4
1975	25.9
1980	29.3
1986	32.4
1990	33.6
1995	36.4
2000	31.1
2004	37.0
2005	37.0
2006	36.5
2007	37.2
2008	39.8

(Source: U.S. Census Bureau, U.S. Department of Commerce)

34. *Consumer Price Index* Prices as measured by the U.S. Consumer Price Index (CPI) have risen steadily since World War II. The data in the table give the CPI for selected years between 1970 and 2010. The CPI in this table has 1984 as a reference year; that is, what cost $1 in 1984 cost about $1.31 in 1990 and $2.18 in 2010.

Year	CPI
1970	38.8
1980	82.4
1990	130.7
2000	172.2
2005	195.3
2010	217.5

(Source: Bureau of Labor Statistics)

a. Align the input data as the number of years after 1970 and find a linear model for the data rounded to three decimal places.

b. Use the model to estimate when the CPI will be 262.39.

35. *Personal Consumption* The sum of the personal consumption expenditures in the United States, in billions of dollars, for selected years from 1990 through 2009 is shown in the table that follows.

a. Make a scatter plot of the data, with x equal to the number of years past 1990 and y equal to the billions of dollars spent.

b. Does it appear that a line will be a reasonable fit for the data?

c. Find the linear model which is the best fit for the data.

d. Use the unrounded model to estimate the U.S. personal consumption for 2012.

Year	Personal Consumption ($ billions)
1990	3835.5
1995	4987.3
2000	6830.4
2005	8819.0
2007	9806.3
2008	10,104.5
2009	10,001.3

(Source: U.S. Department of Commerce)

36. *U.S. Population* The following table gives projections of the U.S. population from 2000 to 2100.

a. Find a linear function that models the data, with x equal to the number of years after 2000 and $f(x)$ equal to the population in millions.

b. Find $f(65)$ and state what it means.

c. What does this model predict the population to be in 2080? How does this compare with the value for 2080 in the table?

Year	Population (millions)	Year	Population (millions)
2000	275.3	2060	432.0
2010	299.9	2070	463.6
2020	324.9	2080	497.8
2030	351.1	2090	533.6
2040	377.4	2100	571.0
2050	403.7		

(Source: www.census.gov/population/projections)

37. *Gross Domestic Product* The table on the next page gives the gross domestic product (the value of all goods and services, in billions of dollars) of the United States for selected years from 1970 to 2009.

a. Create a scatter plot of the data, with y representing the GDP in billions of dollars and x representing the number of years after 1970.

b. Find the linear function that best fits the data, with *x* equal to the number of years after 1970.

c. Graph the model with the scatter plot to see if the line is a good fit for the data.

Year	Gross Domestic Product
1970	1038.3
1980	2788.1
1990	5800.5
2000	9951.5
2005	12,638.4
2007	14,061.8
2008	14,369.1
2009	14,119.0

(Source: U.S. Bureau of Economic Analysis)

38. *Smoking* The table gives the percent of adults aged 18 and over in the United States who reported smoking for selected years.

a. Write the equation that is the best fit for the data, with *x* equal to the number of years after 1990.

b. What does the model estimate the percent to be in 2012?

c. When will the percent be 16, according to the model?

Year	Smoking
1999	23.5
2000	23.2
2001	22.7
2002	22.4
2003	21.6
2004	20.9
2005	20.9
2006	20.8
2007	19.7
2008	20.5
2009	20.6

(Source: Centers for Disease Control)

39. *Online Marketing* Online marketing is emerging as an increasingly effective way for Hollywood to reach its target audience. In the past five years, Hollywood has placed more emphasis on new media channels, as shown in the table below.

Year	Online Ad Spending ($ millions)
2006	259
2007	370
2008	508
2009	650
2010	760
2011	857

(Source: eMarketer)

a. Write the linear equation that models online ad spending as a function of the number of years after 2000.

b. According to the model, what is the annual increase in the amount of online spending?

c. According to the model, what is the percent increase in the amount of online spending from 2009 to 2010?

d. What is the average rate of change of spending from 2006 to 2011?

40. *Prison Sentences* The proportion of those convicted in federal court who are imprisoned has been increasing. The table shows the number of those imprisoned for selected years from 1980 to 2007.

Year	Number Imprisoned
1980	13,766
1985	20,605
1990	28,659
1995	31,805
2000	50,451
2005	61,151
2006	63,699
2007	62,883

(Source: Bureau of Justice)

a. Use the data to create a linear equation that models the number of those imprisoned as a function of the number of years after 1980.

b. Use the model to estimate the number imprisoned in 2002. Is this interpolation or extrapolation?

c. Use the model to determine in what year the number will be 78,750.

**Defendants in Cases Concluded
in U.S. District Courts**

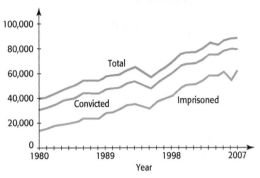

(Source: Bureau of Justice Statistics)

41. *Drug Doses* The table below shows the usual dosage for a certain prescription drug that combats bacterial infections for a person's weight.

Weight (lb)	Usual Dosage (mg)
88	40
99	45
110	50
121	55
132	60
143	65
154	70
165	75
176	80
187	85
198	90

a. Find a linear function $D = f(W)$ that models the dosage given in the table as a function of the patient's weight.

b. Compare the outputs of the model with the data outputs from the table for several values in the table. How well does the model fit the data?

c. What does the model give as the dosage for a 150-pound person?

d. Should this model be interpreted discretely or continuously?

42. *Parcel Post Postal Rates* The table that follows gives U.S. postage rates for local parcel post mail for zone 3. Each given weight refers to the largest weight package that can be mailed for the corresponding postage.

a. Find a linear function $P = f(W)$ that models the postage in the table as a function of the weight in the table.

b. Compare the outputs of the model with the data outputs from the table for several values in the table. Is the model a perfect fit?

Weight not over:	Parcel Post Rate
1 lb	$5.15
2	5.38
3	6.39
4	7.14
5	8.28

(Source: USPS)

43. *Box Office Revenue* Worldwide box office revenue for all films reached $29.9 billion in 2009, up 7.6% over the 2008 total. U.S./Canada and international box office revenues in U.S. dollars were both up significantly over the 2005 total.

Worldwide Box Office Revenue (US$ billions)

	2005	2006	2007	2008	2009
U.S./Canada	8.8	9.2	9.1	9.6	10.6
International	14.3	16.3	16.6	18.1	19.3
Worldwide	23.1	25.5	26.3	27.8	29.9

(Source: Motion Picture Association of America)

a. Let x = the number of years after 2000 and draw a scatter plot of the U.S./Canada data.

b. Find and graph the linear function that is the best fit for the U.S./Canada data.

c. Let x = the number of years after 2000 and draw a scatter plot of the international data.

d. Find and graph the linear function that is the best fit for the international data.

e. If the models are accurate, will U.S./Canada box office revenue ever reach the level of international box office revenue?

44. *U.S. Households with Internet Access* The following table gives the percentage of U.S. households with Internet access in various years.

Year	1996	1997	1998	1999	2000
Percent	8.5	14.3	26.2	28.6	41.5

Year	2001	2003	2007	2008
Percent	50.5	52.4	61.7	78.0

(Source: U.S. Census Bureau)

a. Create a scatter plot of the data, with x equal to the number of years from 1995.

b. Create a linear equation that models the data.

c. Graph the function and the data on the same graph, to see how well the function models (fits) the data.

45. *Marriage Rate* The marriage rate per 1000 population for selected years from 1991 to 2009 is shown in the table.

a. Create a scatter plot of the data, where x is the number of years after 1990.

b. Create a linear function that models the data, with x equal to the number of years after 1990.

c. Graph the function and the data on the same axes, to see how well the function models the data.

d. In what year is the marriage rate expected to be 6.5, according to the model?

Year	Marriage Rate per 1000 Population	Year	Marriage Rate per 1000 Population
1991	9.4	2000	8.5
1992	9.3	2001	8.2
1993	9.0	2002	7.9
1994	9.1	2003	7.9
1995	8.9	2004	7.6
1996	8.8	2005	7.5
1997	8.9	2007	7.3
1998	8.4	2008	7.1
1999	8.6	2009	6.8

(Source: National Vital Statistics Report 2010)

2.3 | Systems of Linear Equations in Two Variables

KEY OBJECTIVES

- Solve systems of linear equations graphically
- Solve systems of linear equations algebraically with the substitution method
- Solve systems of linear equations algebraically by elimination
- Model systems of equations to solve problems
- Determine if a system of linear equations is inconsistent or dependent

SECTION PREVIEW China's Manufacturing

Figure 2.25 shows that the size of the manufacturing sector of China will exceed that of the United States in this decade. If we find the linear functions that model these graphs, with x representing the number of years past 2000 and y representing the sizes of the manufacturing sector in trillions of 2005 dollars, the point of intersection of the graphs of these functions will represent the *simultaneous* solution of the two equations because both equations will be satisfied by the coordinates of the point. (See Example 5.)

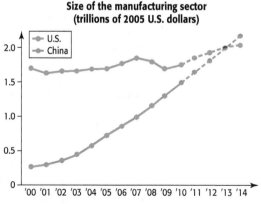

Size of the manufacturing sector (trillions of 2005 U.S. dollars)

Note: Figures starting in 2010 are forecasts

Figure 2.25

(Source: Data from IHS Global Insight, in the *Wall Street Journal*, June 2010)

In this section, we solve systems of linear equations in two variables graphically, by substitution, and by the elimination method. ∎

System of Equations

A **system of linear equations** is a collection of linear equations containing the same set of variables. A system of equations can have exactly one solution, no solution, or infinitely many solutions. A *solution to a system of equations* in two variables is an ordered pair that satisfies both equations in the system. We will solve systems of two equations in two variables by graphing, by substitution, and by the elimination method.

Graphical Solution of Systems

In Section 2.1, we used the intersection method to solve a linear equation by first graphing functions representing the expressions on each side of the equation and then finding the intersection of these graphs. For example, to solve

$$3000x - 7200 = 5800x - 8600$$

we can graph

$$y_1 = 3000x - 7200 \quad \text{and} \quad y_2 = 5800x - 8600$$

and find the point of intersection to be $(0.5, -5700)$. The x-coordinate of the point of intersection of the lines is the value of x that satisfies the original equation, $3000x - 7200 = 5800x - 8600$. Thus, the solution to this equation is $x = 0.5$ (Figure 2.26). In this example, we were actually using a graphical method to solve a **system of two equations in two variables** denoted by

$$\begin{cases} y = 3000x - 7200 \\ y = 5800x - 8600 \end{cases}$$

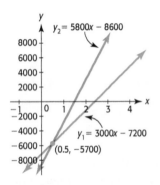

Figure 2.26

The coordinates of the point of intersection of the two graphs give the x- and y-values that satisfy both equations **simultaneously**, and these values are called the **solution** to the system. The following example uses the graphical method to solve a system of equations in two variables.*

EXAMPLE 1 ▶   **Break-Even**

A company is said to **break even** from the production and sale of a product if the total revenue equals the total cost—that is, if $R(x) = C(x)$. Because profit $P(x) = R(x) - C(x)$, we can also say that the company breaks even if the profit for the product is zero.

Suppose a company has its total revenue for a product given by $R = 5585x$ and its total cost given by $C = 61{,}740 + 440x$, where x is the number of thousands of tons of the product that are produced and sold per year. The company is said to break even when the total revenue equals the total cost—that is, when $R = C$. Find the number of thousands of tons of the product that gives break-even and how much the revenue and cost are at that level of production.

SOLUTION

We graph the revenue function as $y_1 = 5585x$ and the cost function as $y_2 = 61{,}740 + 440x$ (Figure 2.27(a)). We can find break-even with the **intersection method** on a calculator by graphing the two equations, $y_1 = 5585x$ and $y_2 = 61{,}740 + 440x$, on a window that contains the point of intersection and then finding the point of intersection,

---
*See Appendix A, page 623.

which is the point where the y-values are equal. This point, which gives break-even, is (12, 67,020) (Figure 2.27(b)).

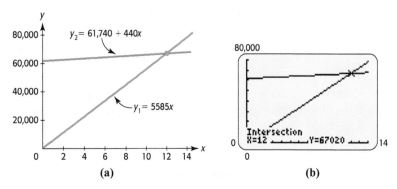

(a)   (b)

Figure 2.27

Thus, the company will break even on this product if 12 thousand tons of the product are sold, when both the cost and revenue equal $67,020.

It is frequently necessary to solve each equation for a variable so that the equation can be graphed with a graphing utility. It is also necessary to find a viewing window that contains the point of intersection. Consider the following example.

EXAMPLE 2 ▶ Solving a System of Linear Equations

Solve the system

$$\begin{cases} 3x - 4y = 21 \\ 2x + 5y = -9 \end{cases}$$

SOLUTION

To solve this system with a graphing utility, we first solve both equations for y:

$$3x - 4y = 21 \qquad\qquad 2x + 5y = -9$$
$$-4y = 21 - 3x \qquad\qquad 5y = -9 - 2x$$
$$y = \frac{21 - 3x}{-4} = \frac{3x - 21}{4} \qquad\qquad y = \frac{-9 - 2x}{5}$$

Graphing these equations with a window that contains the point of intersection (Figure 2.28 (a)) and finding the point of intersection (Figure 2.28(b)) gives $x = 3$, $y = -3$, so the solution is $(3, -3)$.*

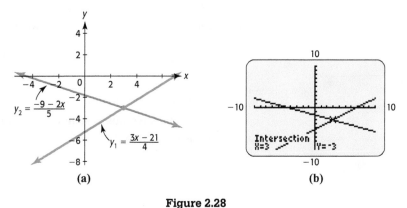

(a)   (b)

Figure 2.28

---

*For a discussion of solving systems of equations using Excel, see Appendix B, page 647.

Solution by Substitution

Graphing with a graphing utility is not always the easiest method to use to solve a system of equations because the equations must be solved for y to be entered in the utility and an appropriate window must be found. A second solution method for a system of linear equations is the **substitution method**, where one equation is solved for a variable and that variable is replaced by the equivalent expression in the other equation.

The substitution method is illustrated by the following example, which finds market equilibrium. The quantity of a product that is demanded by consumers is called the **demand** for the product, and the quantity that is supplied is called the **supply**. In a free economy, both demand and supply are related to the price, and the price where the number of units demanded equals the number of units supplied is called the **equilibrium price**.

EXAMPLE 3 ▶   **Market Equilibrium**

Suppose that the daily demand for a product is given by $p = 200 - 2q$, where q is the number of units demanded and p is the price per unit in dollars, and that the daily supply is given by $p = 60 + 5q$, where q is the number of units supplied and p is the price in dollars. If a price results in more units being supplied than demanded, we say there is a *surplus*, and if the price results in fewer units being supplied than demanded, we say there is a *shortfall*. **Market equilibrium** occurs when the supply quantity equals the demand quantity (and when the prices are equal)—that is, when q and p both satisfy the system

$$\begin{cases} p = 200 - 2q \\ p = 60 + 5q \end{cases}$$

a. If the price is \$140, how many units are supplied and how many are demanded?

b. Does this price give a surplus or a shortfall of the product?

c. What price gives market equilibrium?

SOLUTION

a. If the price is \$140, the number of units demanded satisfies $140 = 200 - 2q$, or $q = 30$, and the number of units supplied satisfies $140 = 60 + 5q$, or $q = 16$.

b. At this price, the quantity supplied is less than the quantity demanded, so a shortfall occurs.

c. Because market equilibrium occurs where q and p both satisfy the system

$$\begin{cases} p = 200 - 2q \\ p = 60 + 5q \end{cases}$$

we seek the solution to this system.

We can solve this system by substitution. Substituting $60 + 5q$ for p in the first equation gives

$$60 + 5q = 200 - 2q$$

Solving this equation gives

$$60 + 5q = 200 - 2q$$
$$7q = 140$$
$$q = 20$$

Thus, market equilibrium occurs when the number of units is 20, and the equilibrium price is

$$p = 200 - 2(20) = 60 + 5(20) = 160 \text{ dollars per unit}$$

The substitution in Example 3 was not difficult because both equations were solved for p. In general, we use the following steps to solve systems of two equations in two variables by substitution.

Solution of Systems of Equations by Substitution

1. Solve one of the equations for one of the variables in terms of the other variable.
2. Substitute the expression from step 1 into the other equation to give an equation in one variable.
3. Solve the linear equation for the variable.
4. Substitute this solution into the equation from step 1 or into one of the original equations and solve this equation for the second variable.
5. Check the solution in both original equations or check graphically.

EXAMPLE 4 ▶ **Solution by Substitution**

Solve the system $\begin{cases} 3x + 4y = 10 \\ 4x - 2y = 6 \end{cases}$ by substitution.

SOLUTION

To solve this system by substitution, we can solve either equation for either variable and substitute the resulting expression into the other equation. Solving the second equation for y gives

$$4x - 2y = 6$$
$$-2y = -4x + 6$$
$$y = 2x - 3$$

Substituting this expression for y in the first equation gives

$$3x + 4(2x - 3) = 10$$

Solving this equation gives

$$3x + 8x - 12 = 10$$
$$11x = 22$$
$$x = 2$$

Substituting $x = 2$ into $y = 2x - 3$ gives $y = 2(2) - 3 = 1$, so the solution to the system is $x = 2, y = 1$, or $(2, 1)$.

Checking shows that this solution satisfies both original equations.

EXAMPLE 5 ▶ China's Manufacturing

Figure 2.29 shows the size of the manufacturing sector of China and that of the United States in this century. Suppose these graphs for China and the United States can be modeled by the functions

$$y = 0.158x - 0.0949 \quad \text{and} \quad y = 0.037x + 1.477$$

respectively, with x representing the years after 2000 and y representing the sizes of the manufacturing sector in trillions of 2005 dollars. Find the year during which China reaches the United States on its way to becoming the world's largest manufacturer.

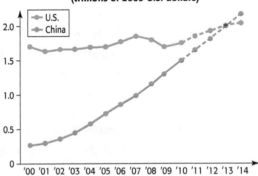

**Size of the manufacturing sector
(trillions of 2005 U.S. dollars)**

Note: Figures starting in 2010 are forecasts

Figure 2.29

(Source: Data from IHS Global Insight, in the *Wall Street Journal*, June 2010)

SOLUTION

China's manufacturing sector reaches that of the United States when the y-values of the two models are equal, so we solve the following system by substitution:

$$\begin{cases} y = 0.158x - 0.0949 \\ y = 0.037x + 1.477 \end{cases}$$

$$0.158x - 0.0949 = 0.037x + 1.477$$

$$0.121x = 1.5719$$

$$x = 12.99 \approx 13$$

Thus, the size of China's manufacturing sector is estimated to reach that of the United States in 2013. We can check the solution graphically by graphing the two functions and using INTERSECT on a graphing calculator. (See Figure 2.30.)

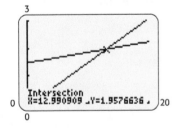

Figure 2.30

Solution by Elimination

A second algebraic method, called the **elimination method**, is frequently an easier method to use to solve a system of linear equations. The elimination method is based on rewriting one or both of the equations in an equivalent form that allows us to eliminate one of the variables by adding or subtracting the equations.

Solving a System of Two Equations in Two Variables by Elimination

1. If necessary, multiply one or both equations by a nonzero number that will make the coefficients of one of the variables in the equations equal, except perhaps for sign.

2. Add or subtract the equations to eliminate one of the variables.

3. Solve for the variable in the resulting equation.

4. Substitute the solution from step 3 into one of the original equations and solve for the second variable.

5. Check the solutions in the remaining original equation, or check graphically.

EXAMPLE 6 ▶ **Solution by Elimination**

Use the elimination method to solve the system

$$\begin{cases} 3x + 4y = 10 \\ 4x - 2y = 6 \end{cases}$$

and check the solution graphically.

SOLUTION

The goal is to convert one of the equations into an equivalent equation of a form such that addition of the two equations will eliminate one of the variables. Notice that the coefficient of y in the second equation, -2, is a factor of the coefficient of y in the first equation, 4. If we multiply both sides of the second equation by 2 and add the two equations, this will eliminate the y-variable.

$$\begin{cases} 3x + 4y = 10 & (1) \\ 4x - 2y = 6 & (2) \end{cases}$$

$$\begin{cases} 3x + 4y = 10 & (1) \quad \text{Multiply 2 times Equation (2), getting equivalent} \\ 8x - 4y = 12 & (3) \quad \text{Equation (3)} \end{cases}$$

$$\begin{aligned} 11x \quad\;\; &= 22 \qquad \text{Add Equations (1) and (3) to eliminate } y. \\ x \quad\;\; &= 2 \qquad \text{Solve the new equation for } x. \end{aligned}$$

Substituting $x = 2$ in the first equation gives $3(2) + 4y = 10$, or $y = 1$. Thus, the solution to the system is $x = 2$, $y = 1$, or $(2, 1)$.

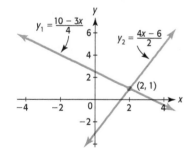

To check graphically, we solve both equations for y, getting $y_1 = \dfrac{10 - 3x}{4}$ and $y_2 = \dfrac{4x - 6}{2}$, then graph the equations and find the point of intersection to be $(2, 1)$ (Figure 2.31).

Figure 2.31

Modeling Systems of Linear Equations

Solving some real problems requires us to create two or more equations whose simultaneous solution is the solution to the problem. Consider the following examples.

EXAMPLE 7 ▶ Investments

An investor has $300,000 to invest, part at 12% and the remainder in a less risky investment at 7%. If her investment goal is to have an annual income of $27,000, how much should she put in each investment?

SOLUTION

If we denote the amount invested at 12% as x and the amount invested at 7% as y, the sum of the investments is $x + y$, so we have the equation

$$x + y = 300,000$$

The annual income from the 12% investment is $0.12x$, and the annual income from the 7% investment is $0.07y$. Thus, the desired annual income from the two investments is

$$0.12x + 0.07y = 27,000$$

We can write the given information as a system of equations:

$$\begin{cases} x + y = 300,000 \\ 0.12x + 0.07y = 27,000 \end{cases}$$

To solve this system, we multiply the first equation by -0.12 and add the two equations. This results in an equation with one variable:

$$\begin{cases} -0.12x - 0.12y = -36,000 \\ 0.12x + 0.07y = 27,000 \end{cases}$$

$$-0.05y = -9000$$

$$y = 180,000$$

Substituting 180,000 for y in the first original equation and solving for x gives $x = 120,000$. Thus, $120,000 should be invested at 12%, and $180,000 should be invested at 7%.

To check this solution, we see that the total investment is $120,000 + $180,000, which equals $300,000. The interest earned at 12% is $120,000(0.12) = $14,400, and the interest earned at 7% is $180,000(0.07) = $12,600. The total interest is $14,400 + $12,600, which equals $27,000. This agrees with the given information.

EXAMPLE 8 ▶ Medication

A nurse has two solutions that contain different concentrations of a certain medication. One is a 12% concentration, and the other is an 8% concentration. How many cubic centimeters (cc) of each should she mix together to obtain 20 cc of a 9% solution?

SOLUTION

We begin by denoting the amount of the first solution by x and the amount of the second solution by y. The total amount of solution is the sum of x and y, so

$$x + y = 20$$

The total medication in the combined solution is 9% of 20 cc, or $0.09(20) = 1.8$ cc, and the mixture is obtained by adding $0.12x$ and $0.08y$, so

$$0.12x + 0.08y = 1.8$$

We can use substitution to solve the system

$$\begin{cases} x + y = 20 \\ 0.12x + 0.08y = 1.8 \end{cases}$$

Substituting $20 - x$ for y in $0.12x + 0.08y = 1.8$ gives $0.12x + 0.08(20 - x) = 1.8$, and solving this equation gives

$$0.12x + 0.08(20 - x) = 1.8$$
$$0.12x + 1.6 - 0.08x = 1.8$$
$$0.04x = 0.2$$
$$x = 5$$

Thus, combining 5 cc of the first solution with $20 - 5 = 15$ cc of the second solution gives 20 cc of the 9% solution.

Dependent and Inconsistent Systems

The system of linear equations discussed in Example 2 has a unique solution, shown as the point of intersection of the graphs. It is possible that two equations in a system of linear equations in two variables describe the same line. When this happens, the equations are equivalent, and the values that satisfy one equation are also solutions to the other equation and to the system. Such a system is a **dependent system**. If a system contains two equations whose graphs are parallel lines, they have no point in common, and thus the system has no solution. Such a system of equations is **inconsistent**. Figure 2.32(a)–(c) represents these three situations: systems that have a unique solution, many solutions (dependent system), and no solution (inconsistent system), respectively. Note that the slopes of the lines are equal in Figure 2.32(b) and in Figure 2.32(c).

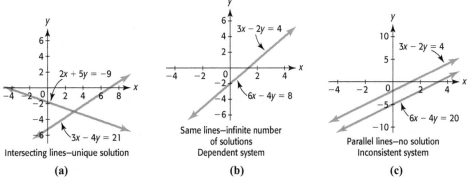

Intersecting lines—unique solution

(a)

Same lines—infinite number of solutions
Dependent system

(b)

Parallel lines—no solution
Inconsistent system

(c)

Figure 2.32

EXAMPLE 9 ▶   Systems with Nonunique Solutions

Use the elimination method to solve each of the following systems, if possible. Verify the solution graphically.

a. $\begin{cases} 2x - 3y = 4 \\ 6x - 9y = 12 \end{cases}$ **b.** $\begin{cases} 2x - 3y = 4 \\ 6x - 9y = 36 \end{cases}$

SOLUTION

a. To solve $\begin{cases} 2x - 3y = 4 \\ 6x - 9y = 12 \end{cases}$, we multiply the first equation by -3 and add the equations, getting

$$\begin{cases} -6x + 9y = -12 \\ 6x - 9y = 12 \end{cases}$$
$$0 = 0$$

This indicates that the graphs of the equations intersect when $0 = 0$, *which is always true*. Thus, any values of x and y that satisfy one of these equations also satisfy the other, and there are *infinitely many* solutions. Figure 2.33(a) shows that the graphs of the equations lie on the same line. Notice that the second equation is a multiple of the first, so the equations are equivalent. This system is *dependent*.

The infinitely many solutions all satisfy both of the two equations. That is, they are values of x and y that satisfy

$$2x - 3y = 4, \quad \text{or} \quad y = \frac{2}{3}x - \frac{4}{3}$$

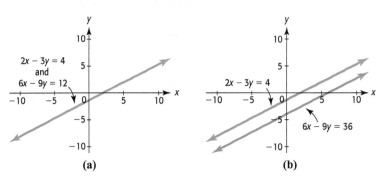

(a)   (b)

Figure 2.33

b. To solve $\begin{cases} 2x - 3y = 4 \\ 6x - 9y = 36 \end{cases}$, we multiply the first equation by -3 and add the equations, getting

$$\begin{cases} -6x + 9y = -12 \\ 6x - 9y = 36 \end{cases}$$
$$0 = 24$$

This indicates that the equations intersect when $0 = 24$, *which is never true*. Thus, no values of x and y satisfy both of the equations. Figure 2.33(b) shows that the graphs of the equations are parallel. This system is *inconsistent*.

EXAMPLE 10 ▶   Investment

Members of an investment club have set a goal of earning 15% on the money they invest in stocks. They are considering buying two stocks, for which the cost per share and the projected growth per share (both in dollars) are summarized in Table 2.13.

Table 2.13

	Utility	Technology
Cost/share	$30	$45
Growth/share	$4.50	$6.75

a. If they have $180,000 to invest, how many shares of each stock should they buy to meet their goal?

b. If they buy 1800 shares of the utility stock, how many shares of the technology stock should they buy to meet their goal?

SOLUTION

a. The money available to invest in stocks is $180,000, so if x is the number of utility shares and y is the number of technology shares purchased, we have

$$30x + 45y = 180,000$$

A 15% return on their investment would be $0.15(180,000) = 27,000$ dollars, so we have

$$4.50x + 6.75y = 27,000$$

To find x and y, we solve the system

$$\begin{cases} 30x + 45y = 180,000 \\ 4.50x + 6.75y = 27,000 \end{cases}$$

Multiplying 4.5 times both sides of the first equation and -30 times both sides of the second equation gives

$$\begin{cases} 135x + 202.5y = 810,000 \\ -135x - 202.5y = 810,000 \end{cases}$$

Adding the equations gives $0 = 0$, so the system is dependent, with many solutions.
The number of shares of each stock that can be purchased satisfies both of the two original equations. In particular, it satisfies $30x + 45y = 180,000$, so

$$y = \frac{180,000 - 30x}{45}, \quad \text{or} \quad y = \frac{12,000 - 2x}{3}$$

with x between 0 and 6000 shares and y between 0 and 4000 shares (because neither x nor y can be negative).

b. Substituting 1800 for x in the equation gives $y = 2800$, so if they buy 1800 shares of the utility stock, they should buy 2800 shares of the technology stock to meet their goal.

Skills CHECK 2.3

Determine if each ordered pair is a solution of the system of equations given.

1. $\begin{cases} 2x + 3y = -1 \\ x - 4y = -6 \end{cases}$

   a. $(2, 1)$ b. $(-2, 1)$

2. $\begin{cases} 4x - 2y = 7 \\ -2x + 2y = -4 \end{cases}$

   a. $\left(\dfrac{3}{2}, -\dfrac{1}{2}\right)$ b. $\left(\dfrac{1}{2}, -\dfrac{3}{2}\right)$

3. What are the coordinates of the point of intersection of $y = 3x - 2$ and $y = 3 - 2x$?

4. Give the coordinates of the point of intersection of $3x + 2y = 5$ and $5x - 3y = 21$.

In Exercises 5–8, solve the systems of equations graphically.

5. $\begin{cases} y = 3x - 12 \\ y = 4x + 2 \end{cases}$ 6. $\begin{cases} 2x - 4y = 6 \\ 3x + 5y = 20 \end{cases}$

7. $\begin{cases} 4x - 3y = -4 \\ 2x - 5y = -4 \end{cases}$ 8. $\begin{cases} 5x - 6y = 22 \\ 4x - 4y = 16 \end{cases}$

9. Does the system $\begin{cases} 2x + 5y = 6 \\ x + 2.5y = 3 \end{cases}$ have a unique solution, no solution, or many solutions? What does this mean graphically?

10. Does the system $\begin{cases} 6x + 4y = 3 \\ 3x + 2y = 3 \end{cases}$ have a unique solution, no solution, or many solutions? What does this mean graphically?

In Exercises 11–14, solve the systems of equations by substitution.

11. $\begin{cases} x = 5y + 12 \\ 3x + 4y = -2 \end{cases}$ 12. $\begin{cases} 2x - 3y = 2 \\ y = 5x - 18 \end{cases}$

13. $\begin{cases} 2x - 3y = 5 \\ 5x + 4y = 1 \end{cases}$ 14. $\begin{cases} 4x - 5y = -17 \\ 3x + 2y = -7 \end{cases}$

In Exercises 15–24, solve the systems of equations by elimination, if a solution exists.

15. $\begin{cases} x + 3y = 5 \\ 2x + 4y = 8 \end{cases}$ 16. $\begin{cases} 4x - 3y = -13 \\ 5x + 6y = 13 \end{cases}$

17. $\begin{cases} 5x = 8 - 3y \\ 2x + 4y = 8 \end{cases}$ 18. $\begin{cases} 3y = 5 - 3x \\ 2x + 4y = 8 \end{cases}$

19. $\begin{cases} 0.3x + 0.4y = 2.4 \\ 5x - 3y = 11 \end{cases}$ 20. $\begin{cases} 8x - 4y = 0 \\ 0.5x + 0.3y = 2.2 \end{cases}$

21. $\begin{cases} 3x + 6y = 12 \\ 4y - 8 = -2x \end{cases}$ 22. $\begin{cases} 6y - 12 = 4x \\ 10x - 15y = -30 \end{cases}$

23. $\begin{cases} 6x - 9y = 12 \\ 3x - 4.5y = -6 \end{cases}$ 24. $\begin{cases} 4x - 8y = 5 \\ 6x - 12y = 10 \end{cases}$

In Exercises 25–34, solve the systems of equations by any convenient method, if a solution exists.

25. $\begin{cases} y = 3x - 2 \\ y = 5x - 6 \end{cases}$ 26. $\begin{cases} y = 8x - 6 \\ y = 14x - 12 \end{cases}$

27. $\begin{cases} 4x + 6y = 4 \\ x = 4y + 8 \end{cases}$ 28. $\begin{cases} y = 4x - 5 \\ 3x - 4y = 7 \end{cases}$

29. $\begin{cases} 2x - 5y = 16 \\ 6x - 8y = 34 \end{cases}$ 30. $\begin{cases} 4x - y = 4 \\ 6x + 3y = 15 \end{cases}$

31. $\begin{cases} 3x = 7y - 1 \\ 4x = 11 - 3y \end{cases}$ 32. $\begin{cases} 5x = 12 + 3y \\ -5y = 8 - 3x \end{cases}$

33. $\begin{cases} 4x - 3y = 9 \\ 8x - 6y = 16 \end{cases}$ 34. $\begin{cases} 5x - 4y = 8 \\ -15x + 12y = -12 \end{cases}$

EXERCISES 2.3

35. *Break-Even* A manufacturer of kitchen sinks has total revenue given by $R = 76.50x$ and has total cost given by $C = 2970 + 27x$, where x is the number of sinks produced and sold. Use graphical methods to find the number of units that gives break-even for the product.

36. *Break-Even* A jewelry maker has total revenue for her bracelets given by $R = 89.75x$ and incurs a total cost of $C = 23.50x + 1192.50$, where x is the number of bracelets produced and sold. Use graphical methods to find the number of units that gives break-even for the product.

37. *Break-Even* A manufacturer of reading lamps has total revenue given by $R = 15.80x$ and total cost given by $C = 8593.20 + 3.20x$, where x is the number of units produced and sold. Use a nongraphical method to find the number of units that gives break-even for this product.

38. *Break-Even* A manufacturer of automobile air conditioners has total revenue given by $R = 136.50x$ and total cost given by $C = 9661.60 + 43.60x$, where x is the number of units produced and sold. Use a nongraphical method to find the number of units that gives break-even for this product.

39. *Market Equilibrium* The demand for a brand of clock radio is given by $p + 2q = 320$, and the supply for these radios is given by $p - 8q = 20$, where p is the price and q is the number of clock radios. Solve the system containing these two equations to find (a) the price at which the quantity demanded equals the quantity supplied and (b) the equilibrium quantity.

40. *Supply and Demand* A certain product has supply and demand functions given by $p = 5q + 20$ and $p = 128 - 4q$, respectively.

    a. If the price p is \$60, how many units q are supplied and how many are demanded?

    b. What price gives market equilibrium, and how many units are demanded and supplied at this price?

41. *Concerta and Ritalin* Concerta and Ritalin are two different brands of a drug used to treat ADHD.

    a. Use the fact that market share for Concerta was $y = 2.4\%$ for August 25, 2000 ($x = 0$), and was 10% eleven weeks later to write a linear function representing its market share as a function of time.

    b. Use the fact that market share for Ritalin was $y = 7.7\%$ for August 25, 2000 ($x = 0$), and was 6.9% eleven weeks later to write a linear function representing its market share as a function of time.

    c. Find the number of weeks past the release date of Concerta (August 25) that the weekly market share of Concerta reached that of Ritalin.
(Source: *Newsweek*, December 4, 2000)

42. *Market Equilibrium* Wholesalers' willingness to sell laser printers is given by the supply function $p = 50.50 + 0.80q$, and retailers' willingness to buy the printers is given by $p = 400 - 0.70q$, where p is the price per printer in dollars and q is the number of printers. What price will give market equilibrium for the printers?

43. *Military* The number of active-duty U.S. Navy personnel (in thousands) is given by $y = -5.686x + 676.173$, and the number of active-duty U.S. Air Force personnel is given by $y = -11.997x + 847.529$, where x is the number of years after 1960.

    a. Use graphical methods to find the year in which the number of Navy personnel reached the number of Air Force personnel.

    b. How many were in each service when the numbers of personnel were equal?
(Source: *World Almanac*)

44. *College Enrollment* Suppose the percent of males who enrolled in college within 12 months of high school graduation is given by $y = -0.126x + 55.72$ and the percent of females who enrolled in college within 12 months of high school graduation is given by $y = 0.73x + 39.7$, where x is the number of years after 1960. Use graphical methods to find the year these models indicate that the percent of females equaled the percent of males.
(Source: *Statistical Abstract of the United States*)

45. *U.S. Population* Using data and projections from 1980 through 2050, the percent of Hispanics in the U.S. civilian noninstitutional population is given by $y = 0.224x + 9.0$ and the percent of blacks is given by $y = 0.057x + 12.3$, where x is the number of years after 1990. During what year did the percent of Hispanics equal the percent of blacks in the United States?

46. *Earnings and Race* The median annual earnings for blacks (B) as a function of the median annual earnings for whites (W), both in thousands of dollars, can be modeled by $B = 0.6234W + 0.3785$ using one set of data and by $B = 1.05W - 18.691$ using more recent data. Use graphical or numerical methods to

find what annual earnings by whites will result in both models giving the same median annual earnings for blacks.
(Source: *Statistical Abstract of the United States*)

47. *Revenue* The sum of the 2011 revenue and twice the 2008 revenue for Mama Joan's International, Inc., is $2144.9 million. The difference between the 2011 and 2008 revenues is $135.5 million. If Mama Joan's revenue between 2008 and 2011 is an increasing linear function, find the 2008 and 2011 revenues.

48. *Stock Prices* The sum of the high and low prices of a share of stock in Johns, Inc., in 2012 is $83.50, and the difference between these two prices in 2012 is $21.88. Find the high and low prices.

49. *Pricing* A concert promoter needs to make $84,000 from the sale of 2400 tickets. The promoter charges $30 for some tickets and $45 for the others.

   a. If there are x of the $30 tickets sold and y of the $45 tickets sold, write an equation that states that the total number of tickets sold is 2400.

   b. How much money is received from the sale of x tickets for $30 each?

   c. How much money is received from the sale of y tickets for $45 each?

   d. Write an equation that states that the total amount received from the sale is $84,000.

   e. Solve the equations simultaneously to find how many tickets of each type must be sold to yield the $84,000.

50. *Rental Income* A woman has $500,000 invested in two rental properties. One yields an annual return of 10% of her investment, and the other returns 12% per year on her investment. Her total annual return from the two investments is $53,000. Let x represent the amount of the 10% investment and y represent the amount of the 12% investment.

   a. Write an equation that states that the sum of the investments is $500,000.

   b. What is the annual return on the 10% investment?

   c. What is the annual return on the 12% investment?

   d. Write an equation that states that the sum of the annual returns is $53,000.

   e. Solve these two equations simultaneously to find how much is invested in each property.

51. *Investment* One safe investment pays 8% per year, and a more risky investment pays 12% per year.

   a. How much must be invested in each account if an investor of $100,000 would like a return of $9000 per year?

   b. Why might the investor use two accounts rather than put all the money in the 12% investment?

52. *Investment* A woman invests $52,000 in two different mutual funds, one that averages 10% per year and another that averages 14% per year. If her average annual return on the two mutual funds is $5720, how much did she invest in each fund?

53. *Investment* Jake has $250,000 to invest. He chooses one money market fund that pays 6.6% and a mutual fund that has more risk but has averaged 8.6% per year. If his goal is to average 7% per year with minimal risk, how much should he invest in each fund?

54. *Investment* Sue chooses one money market fund that pays 6.2% and a mutual fund that has more risk but has averaged 9.2% per year. If she has $300,000 to invest and her goal is to average 7.6% per year with minimal risk, how much should she invest in each fund?

55. *Medication* A pharmacist wants to mix two solutions to obtain 100 cc of a solution that has an 8% concentration of a certain medicine. If one solution has a 10% concentration of the medicine and the second has a 5% concentration, how much of each of these solutions should she mix?

56. *Medication* A pharmacist wants to mix two solutions to obtain 200 cc of a solution that has a 12% concentration of a certain medicine. If one solution has a 16% concentration of the medicine and the second has a 6% concentration, how much of each solution should she mix?

57. *Nutrition* A glass of skim milk supplies 0.1 mg of iron and 8.5 g of protein. A quarter pound of lean meat provides 3.4 mg of iron and 22 g of protein. A person on a special diet is to have 7.1 mg of iron and 69.5 g of protein. How many glasses of skim milk and how many quarter-pound servings of meat will provide this?

58. *Nutrition* Each ounce of substance A supplies 6% of a nutrient a patient needs, and each ounce of substance B supplies 10% of the required nutrient. If the total number of ounces given to the patient was 14 and 100% of the nutrient was supplied, how many ounces of each substance was given?

59. *Alcohol Use* According to the National Household Survey on Drug Abuse by the U.S. Department of Health and Human Services, the pattern of higher rates of current alcohol use, binge alcohol use, and

heavy alcohol use among full-time college students than among others aged 18 to 22 has remained consistent since 2002. (See the figure.)

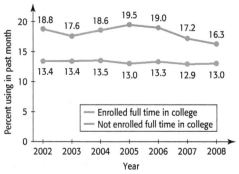

(Source: National Survey on Drug Abuse, U.S. Department of Health and Human Services)

Using data from 2002 to 2008, the equations representing the percents of young adults aged 18 to 22 who used alcohol are

Enrolled in college: $y = -0.282x + 19.553$

Not enrolled: $y = -0.086x + 13.643$

where x represents the number of years after 2000. Solve this system, if possible, to determine when the percent for those enrolled in college will equal that for those not enrolled. What will the percent be?

60. *Medication* A nurse has two solutions that contain different concentrations of a certain medication. One is a 30% concentration, and the other is a 15% concentration. How many cubic centimeters (cc) of each should she mix to obtain 45 cc of a 20% solution?

61. *Supply and Demand* The table below gives the quantity of graphing calculators demanded and the quantity supplied for selected prices.

a. Find the linear equation that gives the price as a function of the quantity demanded.

b. Find the linear equation that gives the price as a function of the quantity supplied.

c. Use these equations to find the market equilibrium price.

Price ($)	Quantity Demanded (thousands)	Quantity Supplied (thousands)
50	210	0
60	190	40
70	170	80
80	150	120
100	110	200

62. *Market Analysis* The supply function and the demand function for a product are linear and are determined by the table that follows. Create the supply and demand functions and find the price that gives market equilibrium.

Supply Price	Function Quantity	Demand Price	Function Quantity
200	400	400	400
400	800	200	800
600	1200	0	1200

63. *Asparagus Cultivation* The most successful asparagus cultivation has been in special micro-climates in the "rain shadow" of the Andes mountains, particularly in Peru. Consequently, the United States obtains most of its asparagus in the off-seasons from Peru. Asparagus cultivation in hectares for Peru and Mexico can be modeled by

Peru: $y = 0.682x + 11.727$

Mexico: $y = 0.127x + 11.509$

where x is the number of years after 1990.

a. Solve this system of equations for x.

b. If these models continue to be accurate, will Mexico's cultivation equal Peru's cultivation after 1990?

(Source: Sustainablog.org)

64. *Medication* Suppose combining x cubic centimeters (cc) of a 20% concentration of a medication and y cc of a 5% concentration of the medication gives $(x + y)$ cc of a 15.5% concentration. If 7 cc of the 20% concentration are added, by how much must the amount of 5% concentration be increased to keep the same concentration?

65. *Social Agency* A social agency provides emergency food and shelter to two groups of clients. The first group has x clients who need an average of $300 for emergencies, and the second group has y clients who need an average of $200 for emergencies. The agency has $100,000 to spend for these two groups.

a. Write an equation that describes the maximum number of clients who can be served with the $100,000.

b. If the first group has twice as many clients as the second group, how many clients are in each group if all the money is spent?

66. *Market Equilibrium* A retail chain will buy 800 televisions if the price is $350 each and 1200 if the price is $300. A wholesaler will supply 700 of these televisions at $280 each and 1400 at $385 each. Assuming that the supply and demand functions are linear, find the market equilibrium point and explain what it means.

67. *Market Equilibrium* A retail chain will buy 900 cordless phones if the price is $10 each and 400 if the price is $60. A wholesaler will supply 700 phones at $30 each and 1400 at $50 each. Assuming that the supply and demand functions are linear, find the market equilibrium point and explain what it means.

2.4   Solutions of Linear Inequalities

KEY OBJECTIVES

- Solve linear inequalities algebraically
- Solve linear inequalities graphically with the intersection and x-intercept methods
- Solve double inequalities algebraically and graphically

SECTION PREVIEW   Profit

For an electronic reading device, the respective weekly revenue and weekly cost are given by

$$R(x) = 400x \quad \text{and} \quad C(x) = 200x + 16{,}000$$

where x is the number of units produced and sold. For what levels of production will a profit result?

Profit will occur when revenue is greater than cost. So we find the level of production and sale x that gives a profit by solving the **linear inequality**

$$R(x) > C(x), \quad \text{or} \quad 400x > 200x + 16{,}000$$

(See Example 2.) In this section, we will solve linear inequalities of this type algebraically and graphically. ■

Algebraic Solution of Linear Inequalities

An **inequality** is a statement that one quantity or expression is greater than, less than, greater than or equal to, or less than or equal to another.

Linear Inequality

A linear inequality (or first-degree inequality) in the variable x is an inequality that can be written in the form $ax + b > 0$, where $a \neq 0$. (The inequality symbol can be $>$, $\geq$, $<$, or $\leq$.)

The inequality $4x + 3 < 7x - 6$ is a linear inequality (or first-degree inequality) because the highest power of the variable (x) is 1. The values of x that satisfy the inequality form the solution set for the inequality. For example, 5 is in the solution set of this inequality because substituting 5 into the inequality gives

$$4 \cdot 5 + 3 < 7 \cdot 5 - 6, \quad \text{or} \quad 23 < 29$$

which is a true statement. On the other hand, 2 is not in the solution set because

$$4 \cdot 2 + 3 \not< 7 \cdot 2 - 6$$

Solving an inequality means finding its solution set. The solution to an inequality can be written as an inequality or in interval notation. The solution can also be represented by a graph on a real number line.

Two inequalities are *equivalent* if they have the same solution set.

We use the properties of inequalities discussed in the Algebra Toolbox to solve an inequality. In general, the steps used to solve a linear inequality are the same as those used to solve a linear equation, except that the inequality symbol is reversed if both sides are multiplied (or divided) by a negative number.

Steps for Solving a Linear Inequality Algebraically

1. If a linear inequality contains fractions with constant denominators, multiply both sides of the inequality by a positive number that will remove all denominators in the inequality. If there are two or more fractions, use the least common denominator (LCD) of the fractions.

2. Remove any parentheses by multiplication.

3. Perform any additions or subtractions to get all terms containing the variable on one side and all other terms on the other side of the inequality. Combine like terms.

4. Divide both sides of the inequality by the coefficient of the variable. *Reverse the inequality symbol if this number is negative.*

5. Check the solution by substitution or with a graphing utility. If a real-world solution is desired, check the algebraic solution for reasonableness in the real-world situation.

EXAMPLE 1 ▶ **Solution of a Linear Inequality**

Solve the inequality $3x - \dfrac{1}{3} \le -4 + x$.

SOLUTION

To solve the inequality

$$3x - \frac{1}{3} \le -4 + x$$

first multiply both sides by 3:

$$3\left(3x - \frac{1}{3}\right) \le 3(-4 + x)$$

Removing parentheses gives

$$9x - 1 \le -12 + 3x$$

Performing additions and subtractions to both sides to get the variables on one side and the constants on the other side gives

$$6x \le -11$$

Dividing both sides by the coefficient of the variable gives

$$x \le -\frac{11}{6}$$

The solution set contains all real numbers less than or equal to $-\frac{11}{6}$. The graph of the solution set $\left(-\infty, -\frac{11}{6}\right]$ is shown in Figure 2.34.

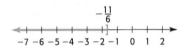

Figure 2.34

EXAMPLE 2 ▶   Profit

For an electronic reading device, the weekly revenue and weekly cost (in dollars) are given by

$$R(x) = 400x \quad \text{and} \quad C(x) = 200x + 16{,}000$$

respectively, where x is the number of units produced and sold. For what levels of production will a profit result?

SOLUTION

Profit will occur when revenue is greater than cost. So we find the level of production and sale that gives a profit by solving the linear inequality $R(x) > C(x)$, or $400x > 200x + 16{,}000$.

$$400x > 200x + 16{,}000$$
$$200x > 16{,}000 \qquad \text{Subtract } 200x \text{ from both sides.}$$
$$x > 80 \qquad \text{Divide both sides by 200.}$$

Thus, a profit occurs if more than 80 units are produced and sold.

EXAMPLE 3 ▶   Body Temperature

A child's health is at risk when his or her body temperature is 103°F or higher. What Celsius temperature reading would indicate that a child's health was at risk?

SOLUTION

A child's health is at risk if $F \ge 103$, and $F = \frac{9}{5}C + 32$, where F is the temperature in degrees Fahrenheit and C is the temperature in degrees Celsius. Substituting $\frac{9}{5}C + 32$ for F, we have

$$\frac{9}{5}C + 32 \ge 103$$

Now we solve the inequality for *C:*

$$\frac{9}{5}C + 32 \geq 103$$

$$9C + 160 \geq 515 \qquad \text{Multiply both sides by 5 to clear fractions.}$$

$$9C \geq 355 \qquad \text{Subtract 160 from both sides.}$$

$$C \geq 39.\overline{4} \qquad \text{Divide both sides by 9.}$$

Thus, a child's health is at risk if his or her Celsius temperature is approximately 39.4° or higher.

Graphical Solution of Linear Inequalities

In Section 2.1, we used graphical methods to solve linear equations. In a similar manner, graphical methods can be used to solve linear inequalities. We will illustrate both the intersection of graphs method and the *x*-intercept method.*

Intersection Method

To solve an inequality by the intersection method, we use the following steps.

Steps for Solving a Linear Inequality with the Intersection Method

1. Set the left side of the inequality equal to y_1, set the right side equal to y_2, and graph the equations using your graphing utility.

2. Choose a viewing window that contains the point of intersection and find the point of intersection, with *x*-coordinate *a*. This is the value of *x* where $y_1 = y_2$.

3. The values of *x* that satisfy the inequality represented by $y_1 < y_2$ are those values for which the graph of y_1 is below the graph of y_2. The values of *x* that satisfy the inequality represented by $y_1 > y_2$ are those values of *x* for which the graph of y_1 is above the graph of y_2.

To solve the inequality

$$5x + 2 < 2x + 6$$

by using the intersection method, let

$$y_1 = 5x + 2 \quad \text{and} \quad y_2 = 2x + 6$$

Entering y_1 and y_2 and graphing the equations using a graphing utility (Figure 2.35) shows that the point of intersection occurs at $x = \frac{4}{3}$. Some graphing utilities show this answer in the form $x = 1.3333333$. The exact *x*-value $\left(x = \frac{4}{3}\right)$ can also be found by solving the equation $5x + 2 = 2x + 6$ algebraically. Figure 2.35 shows that the graph

---

*See Appendix A, page 624.

of y_1 is below the graph of y_2 when x is less than $\dfrac{4}{3}$. Thus, the solution to the inequality $5x + 2 < 2x + 6$ is $x < \dfrac{4}{3}$, which can be written in interval notation as $\left(-\infty, \dfrac{4}{3} \right)$.

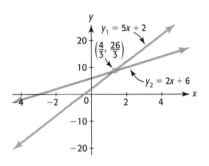

Figure 2.35

EXAMPLE 4 ▶   **Intersection Method of Solution**

Solve $\dfrac{5x + 2}{5} \geq \dfrac{4x - 7}{8}$ using the intersection of graphs method.

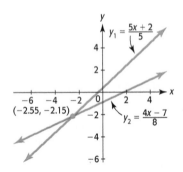

Figure 2.36

SOLUTION

Enter the left side of the inequality as $y_1 = (5x + 2)/5$, enter the right side of the inequality as $y_2 = (4x - 7)/8$, graph these lines, and find their point of intersection. As seen in Figure 2.36, the two lines intersect at the point where $x = -2.55$ and $y = -2.15$.

The solution to the inequality is the x-interval for which the graph of y_1 is above the graph of y_2, or the x-value for which the graph of y_1 intersects the graph of y_2. Figure 2.36 indicates that this is the interval to the right of and including the input value of the point of intersection of the two lines. Thus, the solution is $x \geq -2.55$, or $[-2.55, \infty)$.

---

x-Intercept Method

To use the x-intercept method to solve a linear inequality, we use the following steps.

Solving Linear Inequalities with the x-Intercept Method

1. Rewrite the inequality with all nonzero terms on one side of the inequality and combine like terms, getting $f(x) > 0, f(x) < 0, f(x) \leq 0,$ or $f(x) \geq 0$.

2. Graph the nonzero side of this inequality. (Any window in which the x-intercept can be clearly seen is appropriate.)

3. Find the x-intercept of the graph to find the solution to the equation $f(x) = 0$. (The exact solution can be found algebraically.)

4. Use the graph to determine where the inequality is satisfied.

To use the *x*-intercept method to solve the inequality $5x + 2 < 2x + 6$, we rewrite the inequality with all nonzero terms on one side of the inequality and combine like terms:

$$5x + 2 < 2x + 6$$

$$3x - 4 < 0 \qquad \text{Subtract 2x and 6 from both sides of the inequality.}$$

Graphing the nonzero side of this inequality as the linear function $f(x) = 3x - 4$ gives the graph in Figure 2.37. Finding the *x*-intercept of the graph (Figure 2.37) gives the solution to the equation $3x - 4 = 0$. The *x*-intercept (and zero of the function) is

$$x = 1.3333 \cdots = \frac{4}{3}.$$

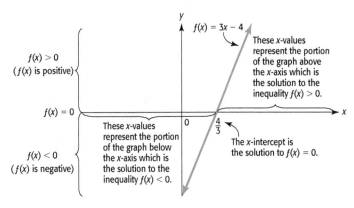

Figure 2.37

We now want to find where $f(x) = 3x - 4$ is less than 0. Notice that the portion of the graph *below* the *x*-axis gives $3x - 4 < 0$. Thus, the solution to $3x - 4 < 0$, and thus to $5x + 2 < 2x + 6$, is $x < \frac{4}{3}$, or $\left(-\infty, \frac{4}{3}\right)$.

EXAMPLE 5 ▶   **Apparent Temperature**

During a recent summer, Dallas, Texas, endured 29 consecutive days when the temperature was at least 100°F. On many of these days, the combination of heat and humidity made it feel even hotter than it was. When the temperature is 100°F, the apparent temperature *A* (or heat index) depends on the humidity *h* (expressed as a decimal) according to

$$A = 90.2 + 41.3h$$

For what humidity levels is the apparent temperature at least 110°F? (Source: W. Bosch and C. Cobb, "Temperature-Humidity Indices," *UMAP Journal*)

SOLUTION

If the apparent temperature is at least 110°F, the inequality to be solved is

$$A \geq 110, \quad \text{or} \quad 90.2 + 41.3h \geq 110$$

Rewriting this inequality with 0 on the right side gives $41.3h - 19.8 \geq 0$. Entering $y_1 = 41.3x - 19.8$ and graphing with a graphing utility gives the graph in Figure 2.38. The x-intercept of the graph is (approximately) 0.479.

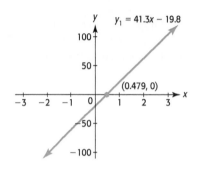

Figure 2.38

The x-interval where the graph is on or above the x-axis is the solution that we seek, so the solution to the inequality is $[0.479, \infty)$. However, humidity is limited to 100%, so the solution is $0.479 \leq h \leq 1.00$, and we say that the apparent temperature is at least 110° when the humidity is between 47.9% and 100%, inclusive.

Double Inequalities

The inequality $0.479 \leq h \leq 1.00$ in Example 5 is a **double inequality**. A double inequality represents two inequalities connected by the word *and* or *or*. The inequality $0.479 \leq h \leq 1.00$ is a compact way of saying $0.479 \leq h$ and $h \leq 1.00$. Double inequalities can be solved algebraically or graphically, as illustrated in the following example. Note that any arithmetic operation is performed to *all three* parts of a double inequality.

EXAMPLE 6 ▶   ## Course Grades

A student has taken four tests and has earned grades of 90%, 88%, 93%, and 85%. If all five tests count the same, what grade must the student earn on the final test so that his course average is a B (that is, so his average is at least 80% and less than 90%)?

ALGEBRAIC SOLUTION

To receive a B, the final test score, represented by x, must satisfy

$$80 \leq \frac{90 + 88 + 93 + 85 + x}{5} < 90$$

Solving this inequality gives

$$80 \leq \frac{356 + x}{5} < 90$$

$400 \leq 356 + x < 450$ Multiply all three parts by 5.

$44 \leq x < 94$ Subtract 356 from all three parts.

Thus, he will receive a grade of B if his final test score is at least 44 but less than 94.

GRAPHICAL SOLUTION

To solve this inequality graphically, we assign the left side of the inequality to y_1, the middle to y_2, and the right side to y_3, and we graph these equations to obtain the graph in Figure 2.39.

$$y_1 = 80$$

$$y_2 = (356 + x)/5$$

$$y_3 = 90$$

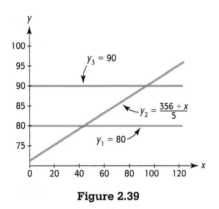

Figure 2.39

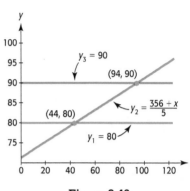

Figure 2.40

We seek the values of x where the graph of y_2 is above or on the graph of y_1 and below the graph of y_3. The left endpoint of this x-interval occurs at the point of intersection of y_2 and y_1, and the right endpoint of the interval occurs at the intersection of y_2 and y_3. These two points can be found using the intersection method. Figure 2.40 shows the points of intersection of these graphs.

The x-values of the points of intersection are 44 and 94, so the solution to $80 \leq \dfrac{356 + x}{5} < 90$ is $44 \leq x < 94$, which agrees with our algebraic solution.

EXAMPLE 7 ▶ Expected Prison Sentences

The mean (expected) time y served in prison for a serious crime can be approximated by a function of the mean sentence length x, with $y = 0.55x - 2.886$, where x and y are measured in months. According to this model, to how many months should a judge sentence a convicted criminal so that the criminal will serve between 37 and 78 months? (Source: National Center for Policy Analysis)

SOLUTION

We seek values of x that give y-values between 37 and 78, so we solve the inequality $37 \leq 0.55x - 2.886 \leq 78$ for x:

$$37 \leq 0.55x - 2.886 \leq 78$$

$$37 + 2.886 \leq 0.55x \leq 78 + 2.886$$

$$39.886 \leq 0.55x \leq 80.886$$

$$72.52 \leq x \leq 147.07$$

Thus, the judge could impose a sentence of 73 to 147 months if she wants the criminal to actually serve between 37 and 78 months.

Skills CHECK 2.4

In Exercises 1–12, solve the inequalities both algebraically and graphically. Draw a number line graph of each solution.

1. $3x - 7 \leq 5 - x$

2. $2x + 6 < 4x + 5$

3. $4(3x - 2) \leq 5x - 9$

4. $5(2x - 3) > 4x + 6$

5. $4x + 1 < -\dfrac{3}{5}x + 5$

6. $4x - \dfrac{1}{2} \leq -2 + \dfrac{x}{3}$

7. $\dfrac{x - 5}{2} < \dfrac{18}{5}$

8. $\dfrac{x - 3}{4} < \dfrac{16}{3}$

9. $\dfrac{3(x - 6)}{2} \geq \dfrac{2x}{5} - 12$

10. $\dfrac{2(x - 4)}{3} \geq \dfrac{3x}{5} - 8$

11. $2.2x - 2.6 \geq 6 - 0.8x$

12. $3.5x - 6.2 \leq 8 - 0.5x$

In Exercises 13 and 14, solve graphically by the intersection method. Give the solution in interval notation.

13. $7x + 3 < 2x - 7$

14. $3x + 4 \leq 6x - 5$

In Exercises 15 and 16, solve graphically by the x-intercept method. Give the solution in interval notation.

15. $5(2x + 4) \geq 6(x - 2)$

16. $-3(x - 4) < 2(3x - 1)$

17. The graphs of two linear functions f and g are shown in the following figure. (Domains are all real numbers.)

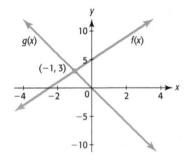

a. Solve the equation $f(x) = g(x)$.

b. Solve the inequality $f(x) < g(x)$.

18. The graphs of three linear functions f, g, and h are shown in the following figure.

a. Solve the equation $f(x) = g(x)$.

b. Solve the inequality $h(x) \leq g(x)$.

c. Solve the inequality $f(x) \leq g(x) \leq h(x)$.

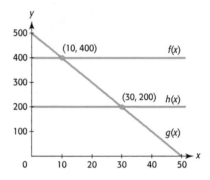

In Exercises 19–28, solve the double inequalities.

19. $17 \leq 3x - 5 < 31$

20. $120 < 20x - 40 \leq 160$

21. $2x + 1 \geq 6$ and $2x + 1 \leq 21$

22. $16x - 8 > 12$ and $16x - 8 < 32$

23. $3x + 1 < -7$ and $2x - 5 > 6$

24. $6x - 2 \leq -5$ or $3x + 4 > 9$

25. $\dfrac{3}{4}x - 2 \geq 6 - 2x$ or $\dfrac{2}{3}x - 1 \geq 2x - 2$

26. $\dfrac{1}{2}x - 3 < 5x$ or $\dfrac{2}{5}x - 5 > 6x$

27. $37.002 \leq 0.554x - 2.886 \leq 77.998$

28. $70 \leq \dfrac{60 + 88 + 73 + 65 + x}{5} < 80$

EXERCISES 2.4

29. *Depreciation* Suppose a business purchases equipment for $12,000 and depreciates it over 5 years with the straight-line method until it reaches its salvage value of $2000 (see the figure below). Assuming that the depreciation can be for any part of a year, do the following:

a. Write an equation that represents the depreciated value V as a function of the years t.

b. Write an inequality that indicates that the depreciated value V of the equipment is less than $8000.

c. Write an inequality that describes the time t during which the depreciated value is at least half of the original value.

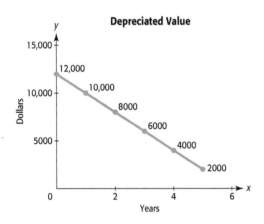

Depreciated Value

30. *Blood Alcohol Percent* The blood alcohol percent p of a 220-pound male is a function of the number of 12-oz beers consumed, and the percent at which a person is considered legally intoxicated (and guilty of DUI if driving) is 0.1% or higher (see the following figure).

a. Use an inequality to indicate the percent of alcohol in the blood when a person is considered legally intoxicated.

b. If x is the number of beers consumed by a 220-pound male, write an inequality that gives the number of beers that will cause him to be legally intoxicated.

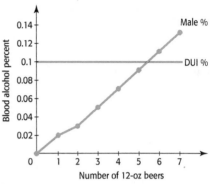

(Source: Pennsylvania Liquor Control Board)

31. *Freezing* The equation $F = \dfrac{9}{5}C + 32$ gives the relationship between temperatures measured in degrees Celsius and degrees Fahrenheit. We know that a temperature at or below 32°F is "freezing." Use an inequality to represent the corresponding "freezing" Celsius temperature.

32. *Boiling* The equation $C = \dfrac{5}{9}(F - 32)$ gives the relationship between temperatures measured in degrees Celsius and degrees Fahrenheit. We know that a temperature at or above 100°C is "boiling." Use an inequality to represent the corresponding "boiling" Fahrenheit temperature.

33. *Job Selection* Deb Cook is given the choice of two positions, one paying $3100 per month and the other paying $2000 per month plus a 5% commission on all sales made during the month. What amount must she sell in a month for the second position to be more profitable?

34. *Stock Market* Susan Mason purchased 1000 shares of stock for $22 per share, and 3 months later the value had dropped by 20%. What is the minimum percent increase required for her to make a profit?

35. *Grades* If Stan Cook has a course average score between 80 and 89, he will earn a grade of B in his algebra course. Suppose that he has four exam scores of 78, 69, 92, and 81 and that his teacher said the final exam score has twice the weight of each of the other exams. What range of scores on the final exam will result in Stan earning a grade of B?

36. *Grades* If John Deal has a course average score between 70 and 79, he will earn a grade of C in his algebra course. Suppose that he has three exam scores of 78, 62, and 82 and that his teacher said the final exam score has twice the weight of the other exams. What range of scores on the final exam will result in John earning a grade of C?

37. *Cigarette Use* For the period 1997–2009, the percent y of students in grade 12 who used cigarettes can be modeled by $2.1x + y = 82.1$, where x is the number of years after 1990.

   a. Solve the equation for x to represent the number of years after 1990 as a function of the percent.

   b. Use the equation from part (a) to determine the range of percent of cigarette use for the years 2000 to 2009.
   (Source: MonitoringtheFuture.org)

38. *SAT Scores* The College Board began reporting SAT scores with a new scale in 1996, with the new scale score y defined as a function of the old scale score x by the equation $y = 0.97x + 128.3829$. Suppose a college requires a new scale score greater than or equal to 1000 to admit a student. To determine what old score values would be equivalent to the new scores that would result in admission to this college, do the following:

   a. Write an inequality to represent the problem, and solve it algebraically.

   b. Solve the inequality from part (a) graphically to verify your result.

39. *Doctorates* For the period 2005–2009, the number of new doctorates in mathematics employed in academic positions can be modeled by $y = 28.5x + 50.5$, where x is the number of years after 2000.

   a. If the model is accurate, algebraically determine the year in which the number of doctorates employed was 250.

   b. Use a graph to verify your answer to part (a).

   c. Use your graph to find when the number of doctorates employed was below 250.
   (Source: www.ams.org)

40. *Internet Access* The percent of households in the United States with Internet access is given by $y = 5.8x + 7.13$, where x is the number of years after

1995. In what years does this model call for the percent to be greater than 88?
(Source: U.S. Census Bureau)

41. *HID Headlights* The new high-intensity discharge (HID) headlights containing xenon gas have an expected life of 1500 hours. Because a complete system costs $1000, it is hoped that these lights will last for the life of the car. Suppose that the actual life of the lights could be 10% longer or shorter than the advertised expected life. Write an inequality that gives the range of life of these new lights.
(Source: *Automobile*, July 2000)

42. *Prison Sentences* The mean time y spent in prison for a crime can be found from the mean sentence length x, using the equation $y = 0.554x - 2.886$, where x and y are measured in months. To how many months should a judge sentence a convicted criminal if she wants the criminal to actually serve between 4 and 6 years?
(Source: Index of Leading Cultural Indicators)

43. *Marriage Rate* According to data from the *National Vital Statistics Report 2010*, the marriage rate (marriages per 1000) can be described by $y = -0.146x + 11.074$, where x is the number of years after 1980. For what years does this model indicate that the marriage rate was above 9 marriages per 1000? Was below 8 marriages per 1000?

44. *Earnings and Minorities* The relation between the median annual salaries of blacks and whites can be modeled by the function $B = 1.05W - 18.691$, where B and W represent the median annual salaries (in thousands of dollars) for blacks and whites, respectively. What is the median salary range for whites that corresponds to a salary range of at least $100,000 for blacks?
(Source: *Statistical Abstract of the United States*)

45. *Home Appraisal* A home purchased in 1996 for $190,000 was appraised at $270,000 in 2000. Assuming the rate of increase in the value of the home is constant, do the following:

   a. Write an equation for the value of the home as a function of the number of years, x, after 1996.

   b. Assuming that the equation in part (a) remained accurate, write an inequality that gives the range of years (until the end of 2010) when the value of the home was greater than $400,000.

   c. Does it seem reasonable that this model remained accurate until the end of 2010?

46. *Car Sales Profit* A car dealer purchases 12 new cars for $32,500 each and sells 11 of them at a profit of 5.5%. For how much must he sell the remaining car to average a profit of at least 6% on the 12 cars?

47. *Electrical Components Profit* A company's daily profit from the production and sale of electrical components can be described by the equation $P(x) = 6.45x - 2000$ dollars, where x is the number of units produced and sold. What level of production and sales will give a daily profit of more than $10,900?

48. *Profit* The yearly profit from the production and sale of Plumber's Helpers is $P(x) = -40,255 + 9.80x$ dollars, where x is the number of Plumber's Helpers produced and sold. What level of production and sales gives a yearly profit of more than $84,355?

49. *Break-Even* A large hardware store's monthly profit from the sale of PVC pipe can be described by the equation $P(x) = 6.45x - 9675$ dollars, where x is the number of feet of PVC pipe sold. What level of monthly sales is necessary to avoid a loss?

50. *Break-Even* The yearly profit from the production and sale of Plumber's Helpers is $P(x) = -40,255 + 9.80x$ dollars, where x is the number of Plumber's Helpers produced and sold. What level of production and sales will result in a loss?

51. *Break-Even* A company produces a logic board for computers. The annual fixed cost for the board is $345,000, and the variable cost is $125 per board. If the logic board sells for $489, write an inequality that gives the number of logic boards that will give a profit for the product.

52. *Temperature* The temperature T (in degrees Fahrenheit) inside a concert hall m minutes after a 40-minute power outage during a summer rock concert is given by $T = 0.43m + 76.8$. Write and solve an inequality that describes when the temperature in the hall is not more than 85°F.

53. *Hispanic Population* Using data and projections from 1990 through 2050, the percent of Hispanics in the U.S. population is given by $H(x) = 0.224x + 9.0$, where x is the number of years after 1990. Find the years when the Hispanic population is projected to be at least 14.6% of the U.S. population.
(Source: U.S. Census Bureau)

54. *Reading Tests* The average reading score of 17-year-olds on the National Assessment of Progress tests is given by $y = 0.155x + 244.37$ points, where x is the number of years after 1970. Assuming that this model was valid, write and solve an inequality that describes when the average 17-year-old's reading score on this test was between but not including 245 and 248. (Your answer should be interpreted discretely.)
(Source: U.S. Department of Education)

55. *Cigarette Use* The percent p of high school seniors who used cigarettes can be modeled by

$$p = 82.074 - 2.088x$$

where x is the number of years after 1990.

   a. What percent does this model estimate for the year 2008?

   b. Test integer values of x with the TABLE feature of your graphing utility to find the values of x for which $p \geq 57.018$.

   c. For what years does this model say the percent is at least 57.018%?

56. *Black Population* Using data and projections from 1990 through 2050, the percent of the U.S. population that is black can be modeled by $B(x) = 0.057x + 12.3$, where x is the number of years after 1990. When does this model call for the percent of blacks to be at least 13.44%?
(Source: U.S. Census Bureau)

chapter 2 ▸ SUMMARY

In this chapter, we studied the solution of linear equations and systems of linear equations. We used graphing utilities to solve linear equations. We solved business and economics problems involving linear functions, solved application problems, solved linear inequalities, and used graphing utilities to model linear functions.

Key Concepts and Formulas

2.1 Algebraic and Graphical Solution of Linear Equations

Algebraic solution of linear equations	If a linear equation contains fractions, multiply both sides of the equation by a number that will remove all denominators in the equation. Next remove any parentheses or other symbols of grouping, and then perform any additions or subtractions to get all terms containing the variable on one side and all other terms on the other side of the equation. Combine like terms. Divide both sides of the equation by the coefficient of the variable. Check the solution by substitution in the original equation.
Solving real-world application problems	To solve an application problem that is set in a real-world context, use the same solution methods. However, remember to include units of measure with your answer and check that your answer makes sense in the problem situation.
Zero of a function	Any number a for which $f(a) = 0$ is called a zero of the function $f(x)$.
Solutions, zeros, and x-intercepts	If a is an x-intercept of the graph of a function f, then a is a real zero of the function f, and a is a real solution to the equation $f(x) = 0$.
Graphical solution of linear equations	
• x-intercept method	Rewrite the equation with 0 on one side, enter the nonzero side into the equation editor of a graphing calculator, and find the x-intercept of the graph. This is the solution to the equation.
• Intersection method	Enter the left side of the equation into y_1, enter the right side of the equation into y_2, and find the point of intersection. The x-coordinate of the point of intersection is the solution to the equation.
Literal equations; solving an equation for a specified linear variable	To solve an equation for one variable if two or more variables are in the equation and if that variable is to the first power in the equation, solve for that variable by treating the other variables as constants and using the same steps as are used to solve a linear equation in one variable.
Direct variation	Two variables x and y are directly proportional to each other if their quotient is a constant.

2.2 Fitting Lines to Data Points; Modeling Linear Functions

Fitting lines to data points	When real-world information is collected as numerical information called *data*, technology can be used to determine the pattern exhibited by the data (provided that a recognizable pattern exists). Such patterns can often be described by mathematical functions.
Constant first differences	If the first differences of data outputs are constant (for equally spaced inputs), a linear model can be found that fits the data exactly. If the first differences are "nearly constant," a linear model can be found that is an approximate fit for the data.

Linear regression	We can determine the equation of the line that is the best fit for a set of points by using a procedure called linear regression (or the least squares method), which defines the best-fit line as the line for which the sum of the squares of the vertical distances from the data points to the line is a minimum.
Modeling data	We can model a set of data by entering the data into a graphing utility, obtaining a scatter plot, and using the graphing utility to obtain the linear equation that is the best fit for the data. The equation and/or numerical results should be reported in a way that makes sense in the context of the problem, with the appropriate units and with the variables identified.
Discrete versus continuous	We use the term *discrete* to describe data or a function that is presented in the form of a table or a scatter plot. We use the term *continuous* to describe a function or graph when the inputs can be any real number or any real number between two specified values.
Applying models	Using a model to find an output for an input between two given data points is called *interpolation*. When a model is used to predict an output for an input outside the given data points, the process is called *extrapolation*.
Goodness of fit	The goodness of fit of a linear model can be observed from a graph of the model and the data points and/or measured with the correlation coefficient.

2.3 Systems of Linear Equations in Two Variables

System of equations	A system of linear equations is a set of equations in two or more variables. A solution of the system must satisfy every equation in the system.
Solving a system of linear equations in two variables	
• Graphing	Graph the equations and find their point of intersection.
• Substitution	Solve one of the equations for one variable and substitute that expression into the other equation, thus giving an equation in one variable.
• Elimination	Rewrite one or both equations in a form that allows us to eliminate one of the variables by adding or subtracting the equations.
Break-even analysis	A company is said to break even from the production and sale of a product if the total revenue equals the total cost—that is, if the profit for that product is zero.
Market equilibrium	*Market equilibrium* is said to occur when the quantity of a commodity demanded is equal to the quantity supplied. The price at this point is called the *equilibrium price*, and the quantity at this point is called the *equilibrium quantity*.
Dependent and inconsistent systems	
• Unique solution	Graphs are intersecting lines.
• No solution	Graphs are parallel lines; the system is *inconsistent*.
• Many solutions	Graphs are the same line; the system is *dependent*.
Modeling systems of equations	Solution of real problems sometimes requires us to create two or more equations whose simultaneous solution is the solution to the problem.

2.4 Solutions of Linear Inequalities

Linear inequality	A linear inequality (or first-degree inequality) is an inequality that can be written in the form $ax + b > 0$, where $a \neq 0$. (The inequality symbol can be $>$, $\geq$, $<$, or $\leq$.)
Algebraically solving linear inequalities	The steps used to solve a linear inequality are the same as those used to solve a linear equation, except that the inequality symbol is reversed if both sides are multiplied (or divided) by a negative number.
Graphical solution of linear inequalities	
• **Intersection method**	Set the left side of the inequality equal to y_1 and set the right side equal to y_2, graph the equations using a graphing utility, and find the x-coordinate of the point of intersection. The values of x that satisfy the inequality represented by $y_1 < y_2$ are those values for which the graph of y_1 is below the graph of y_2.
• **x-intercept method**	To use the x-intercept method to solve an inequality, rewrite the inequality with all nonzero terms on one side and zero on the other side of the inequality and combine like terms. Graph the nonzero side of this inequality and find the x-intercept of the graph. If the inequality to be solved is $f(x) > 0$, the solution will be the interval of x-values representing the portion of the graph above the x-axis. If the inequality to be solved is $f(x) < 0$, the solution will be the interval of x-values representing the portion of the graph below the x-axis.
Double inequalities	A double inequality represents two inequalities connected by the word *and* or *or*. Double inequalities can be solved algebraically or graphically. Any operation performed on a double inequality must be performed on *all three* parts.

chapter 2 ▶ SKILLS CHECK

In Exercises 1–6, solve the equation for x algebraically and graphically.

1. $3x + 22 = 8x - 12$

2. $2(x - 7) = 5(x + 3) - x$

3. $\dfrac{3(x - 2)}{5} - x = 8 - \dfrac{x}{3}$

4. $\dfrac{6x + 5}{2} = \dfrac{5(2 - x)}{3}$

5. $\dfrac{3x}{4} - \dfrac{1}{3} = 1 - \dfrac{2}{3}\left(x - \dfrac{1}{6}\right)$

6. $3.259x - 198.8546 = -3.8(8.625x + 4.917)$

7. For the function $f(x) = 7x - 105$, (a) find the zero of the function, (b) find the x-intercept of the graph of the function, and (c) solve the equation $f(x) = 0$.

8. Solve $P(a - y) = 1 + \dfrac{m}{3}$ for y.

9. Solve $4x - 3y = 6$ for y and graph it on a graphing utility with a standard window.

Use the table of data below in Exercises 10–13.

x	1	3	6	8	10
y	−9	−1	5	12	18

10. Create a scatter plot of the data.

11. Find the linear function that is the best fit for the data in the table.

12. Use a graphing utility to graph the function found in Exercise 11 on the same set of axes as the scatter plot in Exercise 10, with $x_{\min} = 0$, $x_{\max} = 15$, $y_{\min} = -12$, and $y_{\max} = 20$.

13. Do the data points in the table fit exactly on the graph of the function from Exercise 12?

Solve the systems of linear equations in Exercises 14–19, if possible.

14. $\begin{cases} 3x + 2y = 0 \\ 2x - y = 7 \end{cases}$ **15.** $\begin{cases} 3x + 2y = -3 \\ 2x - 3y = 3 \end{cases}$

16. $\begin{cases} -4x + 2y = -14 \\ 2x - y = 7 \end{cases}$ **17.** $\begin{cases} -6x + 4y = 10 \\ 3x - 2y = 5 \end{cases}$

18. $\begin{cases} 2x + 3y = 9 \\ -x - y = -2 \end{cases}$ **19.** $\begin{cases} 2x + y = -3 \\ 4x - 2y = 10 \end{cases}$

In Exercises 20–22, solve the inequalities both algebraically and graphically.

20. $3x + 8 < 4 - 2x$

21. $3x - \dfrac{1}{2} \le \dfrac{x}{5} + 2$

22. $18 \le 2x + 6 < 42$

chapter 2   REVIEW

When money is borrowed to purchase an automobile, the amount borrowed A determines the monthly payment P. In particular, if a dealership offers a 5-year loan at 2.9% interest, then the amount borrowed for the car determines the payment according to the following table. Use the table to define the function $P = f(A)$ in Exercises 23–24.

Amount Borrowed ($)	Monthly Payment ($)
10,000	179.25
15,000	268.87
20,000	358.49
25,000	448.11
30,000	537.73

(Source: Sky Financial)

23. *Car Loans*

  a. Are the first differences of the outputs in the table constant?

  b. Is there a line on which these data points fit exactly?

  c. Write the equation $P = f(A)$ of the line that fits the data points in the table, with coefficients rounded to three decimal places.

24. *Car Loans*

  a. Use the rounded linear model found in part (c) of Exercise 23 to find $P = f(28{,}000)$ and explain what it means.

  b. Can the function f be used to find the monthly payment for any dollar amount A of a loan if the interest rate and length of loan are unchanged?

  c. Determine the amount of a loan that will keep the payment less than or equal to $500, using the unrounded model.

25. *Teacher Salaries* The average salary of a classroom teacher in the United States is given by the function $f(t) = 982.06t + 32{,}903.77$, where t is the number of years after 1990. In what year was the average teacher salary $40,760.25, according to this model? (Source: www.ors2.state.sc.us/abstract)

26. *Fuel* The table below shows data for the number of gallons of gas purchased each day of a certain week by the 250 taxis owned by the Inner City Transportation Taxi Company. Write the equation that models the data.

Days from First Day	0	1	2	3	4	5	6
Gas Used (gal)	4500	4500	4500	4500	4500	4500	4500

27. *Asparagus Production* The total production of asparagus, in metric tons, in Switzerland for the years 1993–2004 is given in the table below.

Year	Production
1993	200
1994	200
1995	200
1996	200
1997	200
1998	200
1999	200
2000	200
2001	200
2002	200
2003	200
2004	200

(Source: U.S. Department of Agriculture)

a. Write the equation of a function that describes the asparagus production in Switzerland using an input equal to the number of years after 1990.

b. What type of function is this?

28. *Job Selection* A job candidate is given the choice of two positions, one paying $2100 per month and one paying $1000 per month plus a 5% commission on all sales made during the month.

a. How much (in dollars) must the employee sell in a month for the second position to pay as much as the first?

b. To be sure that the second position will pay more than the first, how much (in dollars) must the employee sell each month?

29. *Marketing* A car dealer purchased 12 automobiles for $24,000 each. If she sells 8 of them at an average profit of 12%, for how much must she sell the remaining 4 to obtain an average profit of 10% on all 12?

30. *Investment* A retired couple has $420,000 to invest. They chose one relatively safe investment fund that has an annual yield of 6% and another riskier investment that has a 10% annual yield. How much should they invest in each fund to earn $30,000 per year?

31. *Writing Scores* The average writing scores of 11th-graders on the National Assessment of Educational Progress tests have changed over the years since 1984, with the average score given by $y = -0.629x + 293.871$, where x is the number of years after 1980. For what year does this model give an average score of 285?
(Source: U.S. Department of Education)

32. *Profit* A company has revenue given by $R(x) = 500x$ dollars and total costs given by $C(x) = 48,000 + 100x$ dollars, where x is the number of units produced and sold. How many units will give a profit?

33. *Profit* A company has revenue given by $R(x) = 564x$ dollars and total cost given by $C(x) = 40,000 + 64x$ dollars, where x is the number of units produced and sold. The profit can be found by forming the function $P(x) = R(x) - C(x)$.

a. Write the profit function.

b. For what values of x is $P(x) > 0$?

c. For how many units is there a profit?

34. *Depreciation* A business property can be depreciated for tax purposes by using the formula $y + 15,000x = 300,000$, where y is the value of the property x years after it was purchased.

a. For what x-values is the property value below $150,000?

b. After how many years is the property value below $150,000?

35. *Marginal Profit* A company has determined that its profit for a product can be described by a linear function. The profit from the production and sale of 150 units is $455, and the profit from 250 units is $895.

a. Write the equation of the profit function for this product.

b. How many units must be produced and sold to make a profit on this product?

36. *Life Expectancy*

a. Find a linear function $y = f(x)$ that models the data shown in the figure that follows, with x equal to the number of years after 1950 and y equal to the number of years the average 65-year-old woman is estimated to live beyond age 65.

b. Graph the data and the model on the same set of axes.

c. Use the model to estimate $f(99)$ and explain what it means.

d. Determine the time period (in years) for which the average 65-year-old woman can expect to live more than 84 years.

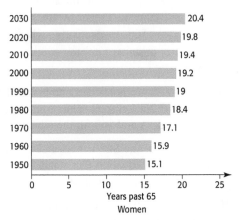

(Source: National Center for Health Statistics)

37. *Life Expectancy*

a. Find a linear function $y = g(x)$ that models the data shown in the figure, with x equal to the number of years after 1950 and y equal to the number of years the average 65-year-old man is estimated to live beyond age 65.

b. Graph the data and the model on the same set of axes.

c. Use the model to estimate $g(130)$ and explain what it means.

d. In what year would the average 65-year-old man expect to live to age 90?

e. Determine the time period for which the average man could expect to live less than 81 years.

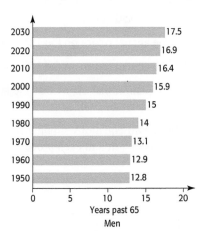

Source: National Center for Health Statistics

38. *Population Growth* The resident population of Florida (in thousands) is given in the following table. Let x equal the number of years after 2000 and y equal the number of thousands of residents.

a. Graph the data points to determine if a linear equation is a reasonable model for the data.

b. If it is reasonable, find the linear model that is the best fit for the data, with x equal to the number of years after 2000.

c. What does this unrounded model predict that the population will be in 2020?

Year	2002	2003	2004	2005
Population (thousands)	16,713	17,019	17,375	17,784

Year	2006	2007	2008	2009
Population (thousands)	18,089	18,278	18,424	18,538

(Source: U.S. Census Bureau)

39. *Education Spending* Personal expenditures for higher education rose dramatically from 1990 to 2008. The chart shows the personal expenditures for higher education in billions of dollars for selected years from 1990 to 2008.

a. Graph the data points or find the first differences to determine if a linear equation is a reasonable model for these data.

b. If it is reasonable, find the linear model that is the best fit for the data, with x equal to the number of years after 1990.

c. Use the unrounded model to predict spending in 2012.

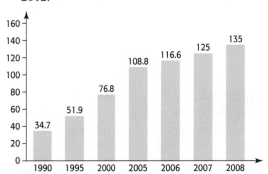

(Source: Bureau of Economic Analysis, U.S. Department of Commerce)

40. *Earnings per Share* The table on the next page gives the earnings ($ thousands) per share (EPS) for ACS stock for the years 2005–2010. Let x equal the number of years after 2000 and y equal the EPS, in thousands of dollars.

a. Graph the data points to determine if a linear equation is a reasonable model for the data.

b. If it is reasonable, find the linear model that is the best fit for the data.

c. Graph the data points and the function on the same axes and discuss the goodness of fit.

Year	2005	2006	2007	2008	2009	2010
EPS ($ thousands)	0.85	1.05	1.29	1.46	1.68	1.90

41. *Marriage Rate* According to data from the *National Vital Statistics Report*, the marriage rate (the number of marriages per 1000 people) can be described by $y = -0.146x + 11.074$, where x is the number of years after 1980. For what years does this model indicate that the rate will be above 6 per 1000? Will be below 7 per 1000?

42. *Marijuana Use* The percent p of high school seniors who use marijuana daily is given by the equation $30p - 19x = 1$, where x is the number of years after 1990. If this model is accurate, what is the range of years for which the percent of use is 3.2%–7%? (Source: Index of Leading Cultural Indicators)

43. *Prison Sentences* The mean time y in prison for a crime can be found as a function of the mean sentence length x, using $y = 0.554x - 2.886$, where x and y are in months. If a judge sentences a convicted criminal to serve between 3 and 5 years, how many months would we expect the criminal to serve? (Source: Index of Leading Cultural Indicators)

44. *Investment* A retired couple has $240,000 to invest. They chose one relatively safe investment fund that has an annual yield of 8% and another riskier investment that has a 12% annual yield. How much should they invest in each fund to earn $23,200 per year?

45. *Break-Even* A computer manufacturer has a new product with daily total revenue given by $R = 565x$ and daily total cost given by $C = 6000 + 325x$. How many units per day must be produced and sold to give break-even for the product?

46. *Medication* Medication A is given six times per day, and medication B is given twice per day. For a certain patient, the total intake of the two medications is limited to 25.2 mg per day. If the ratio of the dosage of medication A to the dosage of medication B is 2 to 3, how many milligrams are in each dosage?

47. *Market Equilibrium* The demand for a certain brand of women's shoes is given by $3q + p = 340$, and the supply of these shoes is given by $p - 4q = -220$, where p is the price in dollars and q is the number of pairs at price p. Solve the system containing these two equations to find the equilibrium price and the equilibrium quantity.

48. *Market Analysis* Suppose that, for a certain product, the supply and demand functions are $p = \dfrac{q}{10} + 8$ and $10p + q = 1500$, respectively, where p is in dollars and q is in units. Find the equilibrium price and quantity.

49. *Pricing* A concert promoter needs to make $120,000 from the sale of 2600 tickets. The promoter charges $40 for some tickets and $60 for the others.

a. If there are x of the $40 tickets and y of the $60 tickets, write an equation that states that the total number of the tickets sold is 2600.

b. How much money is made from the sale of x tickets for $40 each?

c. How much money is made from the sale of y tickets for $60 each?

d. Write an equation that states that the total amount made from the sale is $120,000.

e. Solve the equations simultaneously to find how many tickets of each type must be sold to yield the $120,000.

50. *Rental Income* A woman has $500,000 invested in two rental properties. One yields an annual return of 12% of her investment, and the other returns 15% per year on her investment. Her total annual return from the two investments is $64,500. If x represents the 12% investment and y represents the 15% investment, answer the following:

a. Write an equation that states that the sum of the investments is $500,000.

b. What is the annual return on the 12% investment?

c. What is the annual return on the 15% investment?

d. Write an equation that states that the sum of the annual returns is $64,500.

e. Solve these two equations simultaneously to find how much is invested in each property.

Group Activities
▶ EXTENDED APPLICATIONS

1. Taxes

The table below gives the income tax due, $f(x)$, on each given taxable income, x.

1. What are the domain and the range of the function in the table?
2. Create a scatter plot of the data.
3. Do the points appear to lie on a line?
4. Do the inputs change by the same amount? Do the outputs change by the same amount?
5. Is the rate of change in tax per \$1 of income constant? What is the rate of change?
6. Will a linear function fit the data points exactly?
7. Write a linear function $y = g(x)$ that fits the data points.
8. Verify that the linear model fits the data points by evaluating the linear function at $x = 63,900$ and $x = 64,100$ and comparing the resulting y-values with the income tax due for these taxable incomes.
9. Is the model a discrete or continuous function?
10. Can the model be used to find the tax due on any taxable income between \$63,700 and \$64,300?
11. What is the tax due on a taxable income of \$64,150, according to this model?

U.S. Federal Taxes

Taxable Income ($)	Income Tax Due ($)
63,700	8779
63,800	8804
63,900	8829
64,000	8854
64,100	8879
64,200	8904
64,300	8929

(Source: U.S. Internal Revenue Service)

2. Research

Linear functions can be used to model many types of real data. Graphs displaying linear growth are frequently found in periodicals such as *Newsweek* and *Time*, in newspapers such as *USA Today* and the *Wall Street Journal*, and on numerous websites on the Internet. Tables of data can also be found in these sources, especially on federal and state government websites, such as www.census.gov. (This website is the source of the Florida resident population data used in the Chapter Review, for example.)

Your mission is to find a company sales record, a stock price, a biological growth pattern, or a sociological trend over a period of years (using at least four points) that is linear or "nearly linear," and then to create a linear function that is a model for the data.

A linear model will be a good fit for the data if

- the data are presented as a graph that is linear or "nearly linear."
- the data are presented in a table and the plot of the data points lies near some line.
- the data are presented in a table and the first differences of the outputs are nearly constant for equally spaced inputs.

After you have created the model, you should test the goodness of fit of the model to the data and discuss uses that you could make of the model.

Your completed project should include

a. A complete citation of the source of the data you are using.
b. An original copy or photocopy of the data being used.
c. A scatter plot of the data.
d. The equation that you have created.
e. A graph containing the scatter plot and the modeled equation.
f. A statement about how the model could be used to make estimations or predictions about the trend you are observing.

Some helpful hints:

1. If you decide to use a relation determined by a graph that you have found, read the graph very carefully to determine the data points or contact the source of the data to get the data from which the graph was drawn.
2. Align the independent variable by letting x represent the number of years after some convenient year and then enter the data into a graphing utility and create a scatter plot.
3. Use your graphing utility to create the equation of the function that is the best fit for the data. Graph this equation and the data points on the same axes to see if the equation is reasonable.

3

Quadratic, Piecewise-Defined, and Power Functions

Revenue and profit from the sale of products frequently cannot be modeled by linear functions because they increase at rates that are not constant. In this chapter, we use nonlinear functions, including quadratic and power functions, to model numerous applications in business, economics, and the life and social sciences.

objectives

Graph quadratic functions; find vertices of parabolas; identify increasing and decreasing functions

Solve equations by factoring; solve equations graphically; combine graphs and factoring methods; solve equations with the square root method and by completing the square; solve equations using the quadratic formula; find complex solutions; find discriminants

Graph and apply power, root, reciprocal, piecewise-defined, and absolute value functions; apply direct variation as a power; apply inverse variation

Model with quadratic functions; compare linear and quadratic models; model with power functions; compare power and quadratic models

applications

Maximum revenue from sales, foreign-born population, height of a ball, minimizing cost

Gasoline prices, height of a ball, profit, marijuana use

Residential power costs, postage, wind chill factor, service calls, wingspan, production, illumination

Mobile Internet advertising, Starbucks stores, diabetes, cohabitating households, purchasing power

Algebra TOOLBOX

KEY OBJECTIVES

- Simplify expressions involving integer and rational exponents
- Find the absolute values of numbers
- Simplify expressions involving radicals
- Convert rational exponents to radicals and vice versa
- Find the product of monomials and binomials
- Factor polynomial expressions completely
- Identify numbers as real, imaginary, or pure imaginary
- Determine values that make complex numbers equal

In this Toolbox, we discuss absolute value, integer and rational exponents, and radicals. We also discuss the multiplication of monomials and binomials, factoring, and complex numbers.

Integer Exponents

In this chapter and future ones, we will discuss functions and equations containing integer powers of variables. For example, we will discuss the function $y = x^{-1} = \dfrac{1}{x}$.

> If a is a real number and n is a positive integer, then a^n represents a as a factor n times in a product
>
> $$a^n = \underbrace{a \cdot a \cdot a \ldots \cdot a}_{n \text{ times}}$$
>
> In a^n, a is called the base and n is called the exponent.

In particular, $a^2 = a \cdot a$ and $a^1 = a$. Note that, for positive integers m and n,

$$a^m \cdot a^n = \underbrace{a \cdot a \cdot a \ldots \cdot a}_{m \text{ times}} \underbrace{a \cdot a \ldots \cdot a}_{n \text{ times}} = \underbrace{a \cdot a \cdot a \ldots \cdot a}_{m+n \text{ times}} = a^{m+n}$$

and that, for $m \geq n$,

$$\frac{a^m}{a^n} = \frac{\overbrace{a \cdot a \cdot a \ldots \cdot a}^{m \text{ times}}}{\underbrace{a \cdot a \ldots \cdot a}_{n \text{ times}}} = \overbrace{a \cdot a \cdot a \ldots \cdot a}^{m-n \text{ times}} = a^{m-n} \quad \text{if } a \neq 0$$

These two important properties of exponents can be extended to all integers.

Properties of Exponents

For real numbers a and b and integers m and n,

1. $a^m \cdot a^n = a^{m+n}$ (Product Property)

2. $\dfrac{a^m}{a^n} = a^{m-n}, a \neq 0$ (Quotient Property)

We define an expression raised to zero and to a negative power as follows.

Zero and Negative Exponents

For $a \neq 0, b \neq 0$,

1. $a^0 = 1$ 2. $a^{-1} = \dfrac{1}{a}$

3. $a^{-n} = \dfrac{1}{a^n}$ 4. $\left(\dfrac{a}{b}\right)^{-n} = \left(\dfrac{b}{a}\right)^n$

EXAMPLE 1 ▶ Zero and Negative Exponents

Simplify the following expressions by removing all zero and negative exponents, for nonzero a, b, and c.

a. $(4c)^0$ **b.** $4c^0$ **c.** $(5b)^{-1}$ **d.** $5b^{-1}$ **e.** $\left(\dfrac{a}{b}\right)^{-3}$ **f.** $6a^{-3}$

SOLUTION

a. $(4c)^0 = 1$ **b.** $4c^0 = 4(1) = 4$ **c.** $(5b)^{-1} = \dfrac{1}{(5b)} = \dfrac{1}{5b}$

d. $5b^{-1} = 5 \cdot \dfrac{1}{b} = \dfrac{5}{b}$ **e.** $\left(\dfrac{a}{b}\right)^{-3} = \left(\dfrac{b}{a}\right)^3 = \dfrac{b^3}{a^3}$ **f.** $6a^{-3} = 6 \cdot \dfrac{1}{a^3} = \dfrac{6}{a^3}$

Absolute Value

The distance the number a is from 0 on a number line is the **absolute value** of a, denoted by $|a|$. The absolute value of any nonzero number is positive, and the absolute value of 0 is 0. For example, the distance from 5 to 0 is 5, so $|5| = 5$, and the distance from -8 to 0 is 8, so $|-8| = 8$. Note that if a is a nonnegative number, then $|a| = a$, but if a is negative, then $|a|$ is the positive number $-a$. Formally, we say

$$|a| = \begin{cases} a & \text{if } a \geq 0 \\ -a & \text{if } a < 0 \end{cases}$$

For example, $|5| = 5$ and $|-5| = -(-5) = 5$.

Rational Exponents and Radicals

In this chapter we will study functions involving a variable raised to a rational power (called **power functions**), and we will solve equations involving rational exponents and radicals. This may involve converting expressions involving radicals to expressions involving rational exponents, or vice versa.

Exponential expressions are defined for rational numbers in terms of radicals. Note that, for $a \geq 0$ and $b \geq 0$,

$$\sqrt{a} = b \text{ only if } a = b^2$$

Thus,

$$(\sqrt{a})^2 = b^2 = a, \text{ so } (\sqrt{a})^2 = a$$

We define $a^{1/2} = \sqrt{a}$, so $(a^{1/2})^2 = a$ for $a \geq 0$.

The following definitions show the connection between rational exponents and radicals.

Rational Exponents

1. If a is a real number, variable, or algebraic expression and n is a positive integer $n \geq 2$, then

$$a^{1/n} = \sqrt[n]{a}$$

provided that $\sqrt[n]{a}$ exists.

2. If a is a real number and if m and n are integers containing no common factor with $n \geq 2$, then

$$a^{m/n} = \sqrt[n]{a^m} = \left(\sqrt[n]{a}\right)^m$$

provided that $\sqrt[n]{a}$ exists.

EXAMPLE 2 ▶ Write the following expressions with exponents rather than radicals.

a. $\sqrt[3]{x^2}$ **b.** $\sqrt[4]{x^3}$ **c.** $\sqrt{(3xy)^5}$ **d.** $3\sqrt{(xy)^5}$

SOLUTION

a. $\sqrt[3]{x^2} = x^{2/3}$ **b.** $\sqrt[4]{x^3} = x^{3/4}$ **c.** $\sqrt{(3xy)^5} = (3xy)^{5/2}$ **d.** $3(xy)^{5/2}$

EXAMPLE 3 ▶ Write the following in radical form.

a. $y^{1/2}$ **b.** $(3x)^{3/7}$ **c.** $12x^{3/5}$

SOLUTION

a. $y^{1/2} = \sqrt{y}$ **b.** $(3x)^{3/7} = \sqrt[7]{(3x)^3} = \sqrt[7]{27x^3}$ **c.** $12x^{3/5} = 12\sqrt[5]{x^3}$

Multiplication of Monomials and Binomials

Polynomials with one term are called monomials, those with two terms are called binomials, and those with three terms are called trinomials. In this chapter, we will factor monomials from polynomials and we will factor trinomials into two binomials. To better see how this factoring is accomplished, we will review multiplying by monomials and binomials.

We multiply two monomials by multiplying the coefficients and adding the exponents of the respective variables that are in both monomials. For example,

$$(3x^3y^2)(4x^2y) = 3 \cdot 4 \cdot x^3 \cdot x^2 \cdot y^2 \cdot y = 3 \cdot 4x^{3+2}y^{2+1} = 12x^5y^3$$

We can multiply more than two monomials in the same manner.

EXAMPLE 4 ▶ Find the product:

$$(-2x^4z)(4x^2y^3)(yz^5)$$

SOLUTION

$$(-2x^4z)(4x^2y^3)(yz^5) = -2 \cdot 4x^{4+2}y^{3+1}z^{1+5} = -8x^6y^4z^6$$

We can use the **distributive property**,

$$a(b + c) = ab + ac$$

to multiply a monomial times a polynomial. For example,

$$x(3x + y) = x \cdot 3x + x \cdot y = 3x^2 + xy$$

We can extend the property $a(b + c) = ab + ac$ to multiply a monomial times any polynomial. For example,

$$3x(2x + xy + 6) = 3x \cdot 2x + 3x \cdot xy + 3x \cdot 6 = 6x^2 + 3x^2y + 18x$$

The product of two binomials can be found by using the distributive property as follows:

$$(a + b)(c + d) = a(c + d) + b(c + d) = ac + ad + bc + bd$$

Note that this product can be remembered as the sum of the products of the First, Outer, Inner, and Last terms of the binomials, and we use the word **FOIL** to denote this method.

EXAMPLE 5 ▶ Find the following products.

a. $(x - 4)(x - 5)$ **b.** $(2x - 3)(3x + 2)$ **c.** $(3x - 5y)(3x + 5y)$

SOLUTION

a. $(x - 4)(x - 5) = x \cdot x + x(-5) + (-4)x + (-4)(-5)$
$$= x^2 - 5x - 4x + 20 = x^2 - 9x + 20$$

b. $(2x - 3)(3x + 2) = (2x)(3x) + (2x)2 + (-3)(3x) + (-3)2$
$$= 6x^2 + 4x - 9x - 6 = 6x^2 - 5x - 6$$

c. $(3x - 5y)(3x + 5y) = (3x)(3x) + (3x)(5y) + (-5y)(3x) + (-5y)(5y)$
$$= 9x^2 + 15xy - 15xy - 25y^2 = 9x^2 - 25y^2$$

Certain products and powers involving binomials occur frequently, so the following special products should be remembered.

Special Binomial Products

1. $(x + a)(x - a) = x^2 - a^2$ (difference of two squares)
2. $(x + a)^2 = x^2 + 2ax + a^2$ (perfect square trinomial)
3. $(x - a)^2 = x^2 - 2ax + a^2$ (perfect square trinomial)

EXAMPLE 6 ▶ Find the following products by using the special binomial products formulas.

a. $(5x + 1)^2$ **b.** $(2x - 5)(2x + 5)$ **c.** $(3x - 4)^2$

SOLUTION

a. $(5x + 1)^2 = (5x)^2 + 2(5x)(1) + 1^2 = 25x^2 + 10x + 1$

b. $(2x - 5)(2x + 5) = (2x)^2 - 5^2 = 4x^2 - 25$

c. $(3x - 4)^2 = (3x)^2 - 2(3x)(4) + 4^2 = 9x^2 - 24x + 16$

Factoring

Factoring is the process of writing a number or an algebraic expression as the product of two or more numbers or expressions. For example, the distributive property justifies factoring of monomials from polynomials, as in

$$5x^2 - 10x = 5x(x - 2)$$

Factoring out the **greatest common factor** (gcf) from a polynomial is the first step in factoring.

EXAMPLE 7 ▶ Factor out the greatest common factor.

a. $4x^2y^3 - 18xy^4$ **b.** $3x(a - b) - 2y(a - b)$

SOLUTION

a. The gcf of 4 and 18 is 2. The gcf of x^2 and x is the lower power of x, which is x. The gcf of y^3 and y^4 is the lower power of y, which is y^3. Thus, the gcf of $4x^2y^3$ and $18xy^4$ is $2xy^3$. Factoring out the gcf gives

$$4x^2y^3 - 18xy^4 = 2xy^3(2x - 9y)$$

b. The gcf of $3x(a - b)$ and $2y(a - b)$ is the binomial $a - b$. Factoring out $a - b$ from each term gives

$$3x(a - b) - 2y(a - b) = (a - b)(3x - 2y)$$

By recognizing that a polynomial has the form of one of the special products given on the previous page, we can factor that polynomial.

EXAMPLE 8 ▶ Use knowledge of binomial products to factor the following algebraic expressions.

a. $9x^2 - 25$ **b.** $4x^2 - 12x + 9$

SOLUTION

a. Both terms are squares, so the polynomial can be recognized as the **difference of two squares**. It will then factor as the product of the sum and the difference of the square roots of the terms (see Special Binomial Products, Formula 1).

$$9x^2 - 25 = (3x + 5)(3x - 5)$$

b. Recognizing that the second-degree term and the constant term are squares leads us to investigate whether $12x$ is twice the product of the square roots of these two terms (see Special Binomial Products, Formula 3). The answer is yes, so the polynomial is a **perfect square**, and it can be factored as follows:

$$4x^2 - 12x + 9 = (2x - 3)^2$$

This can be verified by expanding $(2x - 3)^2$.

The first step in factoring is to look for common factors. All applicable factoring techniques should be applied to factor a polynomial completely.

EXAMPLE 9 ▶   Factor the following polynomials completely.

a. $3x^2 - 33x + 72$ b. $6x^2 - x - 1$

SOLUTION

a. The number 3 can be factored from all three terms, giving

$$3x^2 - 33x + 72 = 3(x^2 - 11x + 24)$$

If the trinomial can be factored into the product of two binomials, the first term of each binomial must be x, and we seek two numbers whose product is 24 and whose sum is -11. Because -3 and -8 satisfy these requirements, we get

$$3(x^2 - 11x + 24) = 3(x - 3)(x - 8)$$

b. The four possible factorizations of $6x^2 - x - 1$ that give $6x^2$ as the product of the first terms and -1 as the product of the last terms follow:

$$(6x - 1)(x + 1) \qquad (6x + 1)(x - 1)$$
$$(2x - 1)(3x + 1) \qquad (2x + 1)(3x - 1)$$

The factorization that gives a product with middle term $-x$ is the correct factorization.

$$(2x - 1)(3x + 1) = 6x^2 - x - 1$$

Some polynomials, such as $6x^2 + 9x - 8x - 12$, can be factored by **grouping**. To do this, we factor out common factors from pairs of terms and then factor out a common binomial expression if it exists. For example,

$$6x^2 + 9x - 8x - 12 = 3x(2x + 3) - 4(2x + 3) = (2x + 3)(3x - 4)$$

When a second-degree trinomial can be factored but there are many possible factors to test to find the correct one, an alternative method of factoring can be used. The steps used to factor a trinomial using factoring by grouping techniques follow.

Factoring a Trinomial into the Product of Two Binomials Using Grouping

Steps	Example
To factor a quadratic trinomial in the variable x:	Factor $5x - 6 + 6x^2$:
1. Arrange the trinomial with the powers of x in descending order.	1. $6x^2 + 5x - 6$
2. Form the product of the second-degree term and the constant term (first and third terms).	2. $6x^2(-6) = -36x^2$
3. Determine if there are two factors of the product in step 2 that will sum to the middle (first-degree) term. (If there are no such factors, the trinomial will not factor into two binomials.)	3. $-36x^2 = (-4x)(9x)$ and $-4x + 9x = 5x$
4. Rewrite the middle term from step 1 as a sum of the two factors from step 3.	4. $6x^2 + 5x - 6 = 6x^2 - 4x + 9x - 6$
5. Factor the four-term polynomial from step 4 by grouping.	5. $6x^2 - 4x + 9x - 6 = 2x(3x - 2) + 3(3x - 2) = (3x - 2)(2x + 3)$

EXAMPLE 10 ▶ Factor $10x^2 + 23x - 5$, using grouping.

SOLUTION

To use this method to factor $10x^2 + 23x - 5$, we

1. Note that the trinomial has powers of x in descending order.

$$10x^2 + 23x - 5$$

2. Multiply the second-degree and constant terms.

$$10x^2(-5) = -50x^2$$

3. Factor $-50x^2$ so the sum of the factors is $23x$.

$$-50x^2 = 25x(-2x), \quad 25x + (-2x) = 23x$$

4. Rewrite the middle term of the expression as a sum of the two factors from step 3.

$$10x^2 + 23x - 5 = 10x^2 + 25x - 2x - 5$$

5. Factor by grouping.

$$10x^2 + 25x - 2x - 5$$
$$= 5x(2x + 5) - (2x + 5)$$
$$= (2x + 5)(5x - 1)$$

Complex Numbers

In this chapter, some equations do not have real number solutions, but do have solutions that are complex numbers.

The numbers discussed up to this point are real numbers (either rational numbers such as 2, -3, $\frac{5}{8}$, and $-\frac{2}{3}$ or irrational numbers such as $\sqrt{3}$, $\sqrt[3]{6}$, and π). But some equations do not have real solutions. For example, if

$$x^2 + 1 = 0, \quad \text{then} \quad x^2 = -1$$

but there is no real number that, when squared, will equal -1. However, we can denote one solution to this equation as the **imaginary unit i**, defined by

$$i = \sqrt{-1}$$

Note that $i^2 = \sqrt{-1} \cdot \sqrt{-1} = -1$.

The set of **complex numbers** is formed by adding real numbers and multiples of i.

Complex Number

The number $a + bi$, in which a and b are real numbers, is said to be a **complex number in standard form**. The a is the real part of the number, and bi is the imaginary part. If $b = 0$, the number $a + bi = a$ is a real number, and if $b \neq 0$, the number $a + bi$ is an **imaginary number**. If $a = 0$, bi is a **pure imaginary number**.

The complex number system includes the real numbers as well as the imaginary numbers (Figure 3.1). Examples of imaginary numbers are $3 + 2i$, $5 - 4i$, and $\sqrt{3} - \frac{1}{2}i$; examples of pure imaginary numbers are $-i$, $2i$, $12i$, $-4i$, $i\sqrt{3}$, and πi.

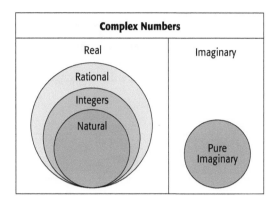

Figure 3.1

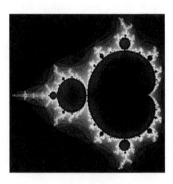

Figure 3.2

The complex number system is an extension of the real numbers that includes imaginary numbers. The term *imaginary number* seems to imply that the numbers do not exist, but in fact they do have important theoretical and technical applications. Complex numbers are used in the design of electrical circuits and airplanes, and they were used in the development of quantum physics. For example, one way of accounting for the amount as well as the phase of the current or voltage in an alternating current electrical circuit involves complex numbers. Special sets of complex numbers can be graphed on the **complex coordinate system** to create pictures called **fractal images**. (Figure 3.2 shows a fractal image called the Mandelbrot set, which can be generated with the complex number *i*.)

EXAMPLE 11 ▶ ## Simplifying Complex Numbers

Simplify the following numbers by writing them in the form a, bi, or $a + bi$.

a. $6 + \sqrt{-8}$ **b.** $\dfrac{4 - \sqrt{-6}}{2}$

SOLUTION

a. $6 + \sqrt{-8} = 6 + i\sqrt{8} = 6 + 2i\sqrt{2}$
    because $\sqrt{8} = \sqrt{4 \cdot 2} = \sqrt{4} \cdot \sqrt{2} = 2\sqrt{2}$

b. $\dfrac{4 - \sqrt{-6}}{2} = \dfrac{4 - i\sqrt{6}}{2} = \dfrac{4}{2} - \dfrac{i\sqrt{6}}{2} = 2 - \dfrac{i\sqrt{6}}{2}$

EXAMPLE 12 ▶ ## Complex Numbers

Identify each number as one or more of the following: real, imaginary, pure imaginary.

a. $6 + 4i$ **b.** $3i - \sqrt{4}$ **c.** $3 - 2i^2$ **d.** $4 - \sqrt{-16}$ **e.** $\sqrt{-3}$

SOLUTION

a. Imaginary because it contains *i*.

b. Imaginary because it contains *i*.

c. Real because $3 - 2i^2 = 3 - 2(-1) = 3 + 2 = 5$.

d. Imaginary because it contains $\sqrt{-16} = 4i$, which gives $4 - 4i$.

e. Imaginary because it contains $\sqrt{-3} = i\sqrt{3}$, and pure imaginary because the real part is 0.

Toolbox EXERCISES

In Exercises 1–6, use the rules of exponents to simplify the following expressions and remove all zero and negative exponents. Assume that all variables are nonzero.

1. $\left(\dfrac{2}{3}\right)^{-2}$

2. $\left(\dfrac{3}{2}\right)^{-3}$

3. $10^{-2} \cdot 10^0$

4. $8^{-2} \cdot 8^0$

5. $(2^{-1})^3$

6. $(4^{-2})^2$

Find the absolute values in Exercises 7 and 8.

7. $|-6|$

8. $|7 - 11|$

9. Write each of the following expressions in simplified exponential form.

    a. $\sqrt{x^3}$ **b.** $\sqrt[4]{x^3}$ **c.** $\sqrt[5]{x^3}$

    d. $\sqrt[6]{27y^9}$ **e.** $27\sqrt[6]{y^9}$

10. Write each of the following in radical form.

    a. $a^{3/4}$ **b.** $-15x^{5/8}$

    c. $(-15x)^{5/8}$

In Exercises 11–15, find the products.

11. $(4x^2y^3)(-3a^2x^3)$

12. $2xy^3(2x^2y + 4xz - 3z^2)$

13. $(x - 7)(2x + 3)$ **14.** $(k - 3)^2$

15. $(4x - 7y)(4x + 7y)$

In Exercises 16–26, factor each of the polynomials completely.

16. $3x^2 - 12x$ **17.** $12x^5 - 24x^3$

18. $9x^2 - 25m^2$ **19.** $x^2 - 8x + 15$

20. $x^2 - 2x - 35$ **21.** $3x^2 - 5x - 2$

22. $8x^2 - 22x + 5$

23. $6n^2 + 18 + 39n$

24. $3y^4 + 9y^2 - 12y^2 - 36$

25. $18p^2 + 12p - 3p - 2$

26. $5x^2 - 10xy - 3x + 6y$

In Exercises 27 and 28, identify each number as one or more of the following: real, imaginary, pure imaginary.

27. a. $2 - i\sqrt{2}$ **b.** $5i$ **c.** $4 + 0i$

    d. $2 - 5i^2$

28. a. $3 + i\sqrt{5}$ **b.** $3 + 0i$ **c.** $8i$

    d. $2i^2 - i$

In Exercises 29–31, find values for a and b that make the statement true.

29. $a + bi = 4$ **30.** $a + 3i = 15 - bi$

31. $a + bi = 2 + 4i$

3.1 Quadratic Functions; Parabolas

KEY OBJECTIVES

- Determine if a function is quadratic
- Determine if the graph of a quadratic function is a parabola that opens up or down
- Determine if the vertex of the graph of a quadratic function is a maximum or a minimum
- Determine if a function increases or decreases over a given interval
- Find the vertex of the graph of a quadratic function
- Graph a quadratic function
- Write the equation of a quadratic function given information about its graph
- Find the vertex form of the equation of a quadratic function

SECTION PREVIEW Revenue

When products are sold with variable discounts or with prices affected by supply and demand, revenue functions for these products may be nonlinear. Suppose the monthly revenue from the sale of Carlson 42-inch 3D televisions is given by the function

$$R(x) = -0.1x^2 + 600x \text{ dollars}$$

where x is the number of TVs sold. In this case, the revenue of this product is represented by a **second-degree function**, or **quadratic function**. A quadratic function is a function that can be written in the form

$$f(x) = ax^2 + bx + c$$

where a, b, and c are real numbers with $a \neq 0$.

The graph of the quadratic function $f(x) = ax^2 + bx + c$ is a **parabola** with a turning point called the **vertex**. Figure 3.3 shows the graph of the function $R(x) = -0.1x^2 + 600x$, which is a parabola that opens downward, and the vertex occurs where the function has its maximum value. Notice also that the graph of this revenue function is symmetric about a vertical line through the vertex. This vertical line is called the **axis of symmetry**.

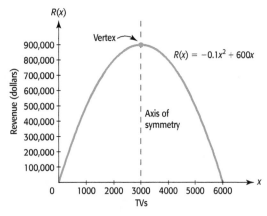

Figure 3.3

Knowing where the maximum value of $R(x)$ occurs can show the company how many units must be sold to obtain the largest revenue. Knowing the maximum output of $R(x)$ helps the company plan its sales campaign. (See Example 2.) In this section, we graph and apply quadratic functions. ■

Parabolas

The graph of every quadratic function has the distinctive shape known as a parabola. The graph of a quadratic function is determined by the location of the vertex and whether the parabola opens upward or downward.

Consider the basic quadratic function $y = x^2$. Each output y is obtained by squaring an input x (Table 3.1). The graph of $y = x^2$, shown in Figure 3.4, is a parabola that opens upward with the vertex (turning point) at the origin, $(0, 0)$.

Table 3.1

x	y
−4	16
−3	9
−2	4
−1	1
0	0
1	1
2	4
3	9
4	16

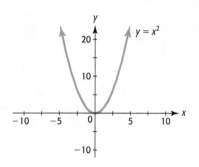

Figure 3.4

Observe that for $x > 0$, the graph of $y = x^2$ rises as it moves from left to right (that is, as the x-values increase), so the function $y = x^2$ is **increasing** for $x > 0$. For values of $x < 0$, the graph falls as it moves from left to right (as the x-values increase), so the function $y = x^2$ is **decreasing** for $x < 0$.

Increasing and Decreasing Functions

A function f is **increasing** on an interval if, for any x_1 and x_2 in the interval, when $x_2 > x_1$, it is true that $f(x_2) > f(x_1)$.

A function f is **decreasing** on an interval if, for any x_1 and x_2 in the interval, when $x_2 > x_1$, it is true that $f(x_2) < f(x_1)$.

The quadratic function $y = -x^2$ has the form $y = ax^2$ with $a < 0$, and its graph is a parabola that opens downward with vertex at $(0, 0)$ (Figure 3.5(a)). The function $y = \dfrac{1}{2}x^2$ has the form $y = ax^2$ with $a > 0$, and its graph is a parabola opening upward (Figure 3.5(b)).

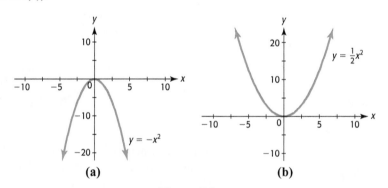

(a)   (b)

Figure 3.5

In general, the graph of a quadratic function of the form $y = ax^2$ is a parabola that opens upward (is **concave up**) if a is positive and opens downward (is **concave down**) if a is negative.* The vertex, which is the point where the parabola turns, is a **minimum point** if a is positive and is a **maximum point** if a is negative. The vertical line through the vertex is called the **axis of symmetry** because this line divides the graph into two halves that are reflections of each other (Figure 3.6(a)).

---

* A parabola that is concave up will appear to "hold water," and a parabola that is concave down will appear to "shed water."

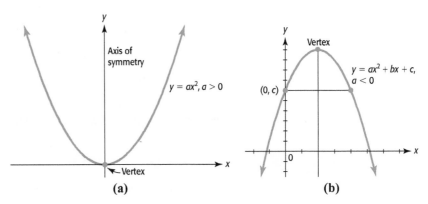

Figure 3.6

We can find the x-coordinate of the vertex of the graph of $y = ax^2 + bx + c$ by using the fact that the axis of symmetry of a parabola passes through the vertex. As Figure 3.6(b) shows, the y-intercept of the graph of $y = ax^2 + bx + c$ is $(0, c)$, and there is another point on the graph with y-coordinate c.

The x-coordinates of the points on this graph with y-coordinate c satisfy

$$c = ax^2 + bx + c$$

Solving this equation gives

$$0 = ax^2 + bx$$
$$0 = x(ax + b)$$
$$x = 0 \quad \text{or} \quad x = \frac{-b}{a}$$

The x-coordinate of the vertex is on the axis of symmetry, which is halfway from $x = 0$ to $x = \frac{-b}{a}$, so it is at $x = \frac{-b}{2a}$. The y-coordinate of the vertex can be found by evaluating the function at the x-coordinate of the vertex.

Graph of a Quadratic Function

The graph of the function

$$f(x) = ax^2 + bx + c$$

is a parabola that opens upward, and the vertex is a minimum, if $a > 0$. The parabola opens downward, and the vertex is a maximum, if $a < 0$.

The larger the value of $|a|$, the more narrow the parabola will be. Its vertex is at the point $\left(\dfrac{-b}{2a}, f\left(\dfrac{-b}{2a} \right) \right)$ (Figure 3.7).

The axis of symmetry of the parabola has equation $x = \dfrac{-b}{2a}$.

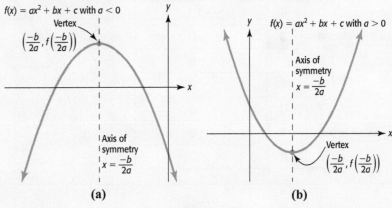

Figure 3.7

Observe that the graph of $y = x^2$ (Figure 3.4) is narrower than the graph of $y = \dfrac{1}{2}x^2$ (Figure 3.5(b)), and that 1 (the coefficient of x^2 in $y = x^2$) is larger than $\dfrac{1}{2}$ $\left(\text{the coefficient of } x^2 \text{ in } y = \dfrac{1}{2}x^2\right)$.

If we know the location of the vertex and the direction in which the parabola opens, we can make a good sketch of the graph by plotting just a few more points.

EXAMPLE 1 ▶ Graphing a Quadratic Function

Find the vertex and graph the quadratic function $f(x) = -2x^2 - 4x + 6$.

SOLUTION

Note that a, the coefficient of x^2, is -2, so the parabola opens downward. The x-coordinate of the vertex is $\dfrac{-b}{2a} = \dfrac{-(-4)}{2(-2)} = -1$, and the y-coordinate of the vertex is $f(-1) = -2(-1)^2 - 4(-1) + 6 = 8$. Thus, the vertex is $(-1, 8)$. The x-intercepts can be found by setting $f(x) = 0$ and solving for x:

$$-2x^2 - 4x + 6 = 0$$
$$-2(x^2 + 2x - 3) = 0$$
$$-2(x + 3)(x - 1) = 0$$
$$x = -3 \quad \text{or} \quad x = 1$$

The y-intercept is easily found by computing $f(0)$:

$$f(0) = -2(0)^2 - 4(0) + 6 = 6$$

The axis of symmetry is the vertical line $x = -1$. The graph is shown in Figure 3.8.

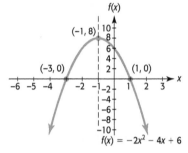

Figure 3.8

Spreadsheet ▶ SOLUTION To graph the function $f(x) = -2x^2 - 4x + 6$ using Excel, create a table (Table 3.2) containing values for x and $f(x)$ and highlight the two columns containing these values. Select XY (Scatter) chart type with the smooth curve option* to get Figure 3.9.

Table 3.2

	A	B
1	x	f(x) = −2x² − 4x + 6
2	−5	−24
3	−4	−10
4	−3	0
5	−2	6
6	−1	8
7	0	6
8	1	0
9	2	−10

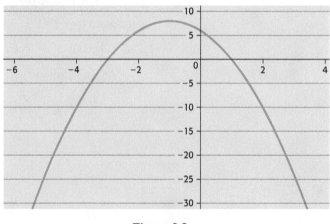

Figure 3.9

---

* For more details, see the Excel Guide in Appendix B, page 642.

EXAMPLE 2 ▶   Maximizing Revenue

Suppose the monthly revenue from the sale of Carlson 42-inch 3D televisions is given by the function

$$R(x) = -0.1x^2 + 600x \text{ dollars}$$

where x is the number of televisions sold.

a. Find the vertex and the axis of symmetry of the graph of this function.

b. Determine if the vertex represents a maximum or minimum point.

c. Interpret the vertex in the context of the application.

d. Graph the function.

e. For what x-values is the function increasing? decreasing? What does this mean in the context of the application?

SOLUTION

a. The function is a quadratic function with $a = -0.1$, $b = 600$, and $c = 0$. The x-coordinate of the vertex is $\dfrac{-b}{2a} = \dfrac{-600}{2(-0.1)} = 3000$, and the axis of symmetry is the line $x = 3000$. The y-coordinate of the vertex is

$$R(3000) = -0.1(3000)^2 + 600(3000) = 900,000$$

so the vertex is (3000, 900,000).

b. Because $a < 0$, the parabola opens downward, so the vertex is a maximum point.

c. The x-coordinate of the vertex gives the number of televisions that must be sold to maximize revenue, so selling 3000 sets will result in the maximum revenue. The y-coordinate of the vertex gives the maximum revenue, $900,000.

d. Table 3.3 lists some values that satisfy $R(x) = -0.1x^2 + 600x$. The graph is shown in Figure 3.10.

Table 3.3

x	R(x)
0	0
1000	500,000
2000	800,000
3000	900,000
4000	800,000
5000	500,000
6000	0

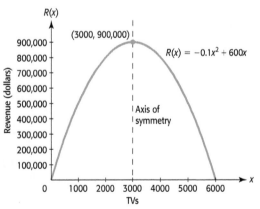

Figure 3.10

e. The function is increasing on the interval $(-\infty, 3000)$ and decreasing on $(3000, \infty)$. However, in the context of the application, negative inputs and outputs do not make sense, so we may say the revenue increases on (0, 3000) and decreases on (3000, 6000).

Even when using a graphing utility to graph a quadratic function, it is important to recognize that the graph is a parabola and to locate the vertex. Determining which way the parabola opens and the x-coordinate of the vertex is very useful in setting the viewing window so that a complete graph (that includes the vertex and the intercepts) is shown.

EXAMPLE 3 ▶ **Foreign-Born Population**

Using data from 1900 through 2008, the percent of the U.S. population that was foreign born can be modeled by the equation

$$y = 0.0034x^2 - 0.439x + 20.185$$

where x is the number of years after 1900.

a. During what year does the model indicate that the percent of foreign-born population was a minimum?

b. What is the minimum percent?

SOLUTION

a. This equation is in the form $f(x) = ax^2 + bx + c$, so $a = 0.0034$. Because $a > 0$, the parabola opens upward, the vertex is a minimum, and the x-coordinate of the vertex is where the minimum percent occurs.

$$x = \frac{-b}{2a} = \frac{-(-0.439)}{2(0.0034)} = 64.6$$

So the percent of the U.S. population that was foreign born was a minimum in the 65th year after 1900, or in 1965.

b. The minimum for the model is found by evaluating the function at $x = 64.6$. This value, 6.0, can be found with TRACE or TABLE or with direct evaluation (Figure 3.11). So the minimum percent is 6%.

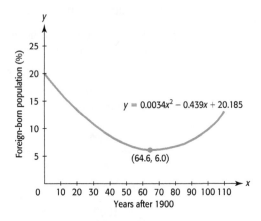

Figure 3.11

Suppose an object is shot or thrown into the air and then falls. If air resistance is ignored, the height in feet of the object after t seconds can be modeled by

$$S(t) = -16t^2 + v_0 t + h_0$$

where -16 ft/sec^2 is the acceleration due to gravity, v_0 ft/sec is the initial velocity (at $t = 0$ sec), and h_0 is the initial height in feet (at $t = 0$).

EXAMPLE 4 ▶ **Height of a Ball**

A ball is thrown upward at 64 feet per second from the top of an 80-foot-high building.

a. Write the quadratic function that models the height (in feet) of the ball as a function of the time t (in seconds).

b. Find the t-coordinate and S-coordinate of the vertex of the graph of this quadratic function.

c. Graph the model.

d. Explain the meaning of the coordinates of the vertex for this model.

SOLUTION

a. The model has the form $S(t) = -16t^2 + v_0 t + h_0$, where $v_0 = 64$ and $h_0 = 80$. Thus, the model is

$$S(t) = -16t^2 + 64t + 80 \text{ (feet)}$$

b. The height S is a function of the time t, and the t-coordinate of the vertex is

$$t = \frac{-b}{2a} = \frac{-64}{2(-16)} = 2$$

The S-coordinate of the vertex is the value of S at $t = 2$, so $S = -16(2)^2 + 64(2) + 80 = 144$ is the S-coordinate of the vertex. The vertex is $(2, 144)$.

c. The function is quadratic and the coefficient in the second-degree term is negative, so the graph is a parabola that opens down with vertex $(2, 144)$. To graph the function, we choose a window that includes the vertex $(2, 144)$ near the center top of the screen. Using the window $[0, 6]$ by $[-20, 150]$ gives the graph shown in Figure 3.12(a). Using 2ND CALC maximum* verifies that the vertex is $(2, 144)$ (see Figure 3.12(b)).

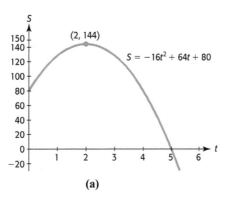
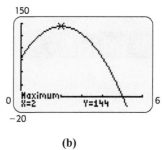

(a)   (b)

Figure 3.12

d. The graph is a parabola that opens down, so the vertex is the highest point on the graph and the function has its maximum there. The t-coordinate of the vertex, 2, is the time (in seconds) at which the ball reaches its maximum height, and the S-coordinate, 144, is the maximum height (in feet) that the ball reaches.

Vertex Form of a Quadratic Function

When a quadratic function is written in the form $f(x) = ax^2 + bx + c$, we can calculate the coordinates of the vertex. But if a quadratic function is written in the form

$$y = a(x - h)^2 + k$$

the vertex of the parabola is at (h, k) (Figure 3.13(a)). For example, the graph of $y = (x - 2)^2 + 3$ is a parabola opening upward with vertex $(2, 3)$ (Figure 3.13(b)).

---

* For more details, see Appendix A, page 621.

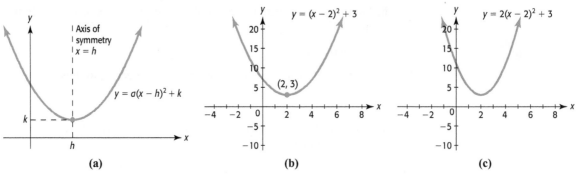

Figure 3.13

Graph of a Quadratic Function

In general, the graph of the function

$$y = a(x - h)^2 + k$$

is a parabola with its vertex at the point (h, k).

The parabola opens upward if $a > 0$, and the vertex is a minimum.
The parabola opens downward if $a < 0$, and the vertex is a maximum.
The axis of symmetry of the parabola has equation $x = h$.
The a is the same as the leading coefficient in $y = ax^2 + bx + c$, so the larger the value of $|a|$, the narrower the parabola will be.

Note that the graph of $y = 2(x - 2)^2 + 3$ in Figure 3.13(c) is narrower than the graph of $y = (x - 2)^2 + 3$ in Figure 3.13(b).

EXAMPLE 5 ▶ **Minimizing Cost**

The cost for producing x Champions golf hats is given by the function

$$C(x) = 0.2(x - 40)^2 + 200 \text{ dollars}$$

a. Find the vertex of this function.

b. Is the vertex a maximum or minimum? Interpret the vertex in the context of the application.

c. Graph the function using a window that includes the vertex.

d. Describe what happens to the function between $x = 0$ and the x-coordinate of the vertex. What does this mean in the context of the application?

SOLUTION

a. This function is in the form $y = a(x - h)^2 + k$ with $h = 40$ and $k = 200$. Thus, the vertex of $C(x) = 0.2(x - 40)^2 + 200$ is (40, 200).

b. Because $a = 0.2$, which is positive, the vertex is a minimum. This means that the cost of producing golf hats is at a minimum of $200 when 40 hats are produced.

c. We know that the vertex of the graph of this function is (40, 200) and that it is a minimum, so we can choose a window with the $x = 40$ near the center of the screen and $y = 200$ near the bottom of the screen. The graph using the window [0, 100] by [−50, 1000] is shown in Figure 3.14.

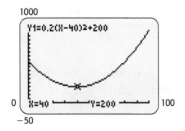

Figure 3.14

d. For x-values between 0 and 40, the graph decreases. Thus, the cost of producing golf hats is decreasing until 40 hats are produced; after 40 hats are produced, the cost begins to increase.

We can use the vertex form $y = a(x - h)^2 + k$ to write the equation of a quadratic function if we know the vertex and a point on its graph.

EXAMPLE 6 ▶ **Profit**

Right Sports Management had its monthly maximum profit, $450,000, when it produced and sold 5500 Waist Trimmers. Its fixed cost is $155,000. If the profit can be modeled by a quadratic function of x, the number of Waist Trimmers produced and sold each month, find this quadratic function $P(x)$.

SOLUTION

When 0 units are produced, the cost is $155,000 and the revenue is $0. Thus, the profit is $-\$155,000$ when 0 units are produced, and the y-intercept of the graph of the function is $(0, -155,000)$. The vertex of the graph of the quadratic function is $(5500, 450,000)$. Using these points gives

$$P(x) = a(x - 5500)^2 + 450,000$$

and

$$-155,000 = a(0 - 5500)^2 + 450,000$$

which gives

$$a = -0.02$$

Thus, the quadratic function that models the profit is $P(x) = -0.02(x - 5500)^2 + 450,000$, or $P(x) = -0.02x^2 + 220x - 155,000$, where $P(x)$ is in dollars and x is the number of units produced and sold.

EXAMPLE 7 ▶ **Equation of a Quadratic Function**

If the points in the table lie on a parabola, write the equation whose graph is the parabola.

x	−1	0	1	2
y	13	−2	−7	−2

SOLUTION

The x-values are a uniform distance apart. Because the points $(0, -2)$ and $(2, -2)$ both have a y-coordinate of -2, the symmetry of a parabola indicates that the vertex will be halfway between $x = 0$ and $x = 2$. Thus, the vertex of this parabola is at $(1, -7)$. (See Figure 3.15.)

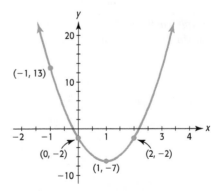

Figure 3.15

The equation of the function is

$$y = a(x - 1)^2 - 7$$

The point $(2, -2)$ or any other point in the table besides $(1, -7)$ can be used to find a.

$$-2 = a(2 - 1)^2 - 7, \quad \text{or} \quad a = 5$$

Thus, the equation is

$$y = 5(x - 1)^2 - 7, \quad \text{or} \quad y = 5x^2 - 10x - 2$$

EXAMPLE 8 ▶ Vertex Form of a Quadratic Function

Write the vertex form of the equation of the quadratic function from the general form $y = 2x^2 - 8x + 5$ by first finding the vertex and a point on the parabola.

SOLUTION

The vertex is at $x = \dfrac{-b}{2a} = \dfrac{-(-8)}{2(2)} = 2$, and $y = 2(2^2) - 8(2) + 5 = -3$. We know a is 2, because a is the same in both forms. Thus,

$$y = 2(x - 2)^2 - 3$$

is the vertex form of the equation.

Skills CHECK 3.1

In Exercises 1–6, (a) determine if the function is quadratic. If it is, (b) determine if the graph is concave up or concave down. (c) Determine if the vertex of the graph is a maximum point or a minimum point.

1. $y = 2x^2 - 8x + 6$ **2.** $y = 4x - 3$

3. $y = 2x^3 - 3x^2$ **4.** $f(x) = x^2 + 4x + 4$

5. $g(x) = -5x^2 - 6x + 8$

6. $h(x) = -2x^2 - 4x + 6$

In Exercises 7–14, (a) graph each quadratic function on $[-10, 10]$ by $[-10, 10]$. (b) Does this window give a complete graph?

7. $y = 2x^2 - 8x + 6$ **8.** $f(x) = x^2 + 4x + 4$

9. $g(x) = -5x^2 - 6x + 8$

10. $h(x) = -2x^2 - 4x + 6$

11. $y = x^2 + 8x + 19$

12. $y = x^2 - 4x + 5$

13. $y = 0.01x^2 - 8x$

14. $y = 0.1x^2 + 8x + 2$

15. Write the equation of the quadratic function whose graph is shown.

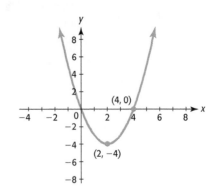

16. Write the equation of the quadratic function whose graph is shown.

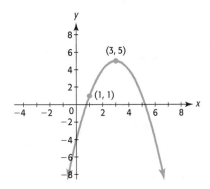

17. The two graphs shown have equations of the form $y = a(x - 2)^2 + 1$. Is the value of a larger for y_1 or y_2?

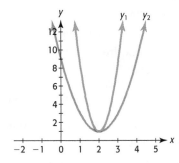

18. The two graphs shown have equations of the form $y = -a(x - 3)^2 + 5$. Is the value of $|a|$ larger for y_1 or y_2?

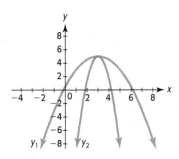

19. If the points in the table lie on a parabola, write the equation whose graph is the parabola.

x	−1	1	3	5
y	−7	13	−7	−67

20. If the points in the table lie on a parabola, write the equation whose graph is the parabola.

x	−6	−5	−4	−3	−2	−1
y	33	12	−3	−12	−15	−12

In Exercises 21–30, (a) give the coordinates of the vertex of the graph of each function. (b) Graph each function on a window that includes the vertex.

21. $y = (x - 1)^2 + 3$ **22.** $y = (x + 10)^2 - 6$

23. $y = (x + 8)^2 + 8$ **24.** $y = (x - 12)^2 + 1$

25. $f(x) = 2(x - 4)^2 - 6$

26. $f(x) = -0.5(x - 2)^2 + 1$

27. $y = 12x - 3x^2$ **28.** $y = 3x + 18x^2$

29. $y = 3x^2 + 18x - 3$ **30.** $y = 5x^2 + 75x + 8$

For Exercises 31–34, (a) find the x-coordinate of the vertex of the graph. (b) Set the viewing window so that the x-coordinate of the vertex is near the center of the window and the vertex is visible, and then graph the given equation. (c) State the coordinates of the vertex.

31. $y = 2x^2 - 40x + 10$

32. $y = -3x^2 - 66x + 12$

33. $y = -0.2x^2 - 32x + 2$

34. $y = 0.3x^2 + 12x - 8$

In Exercises 35–40, sketch complete graphs of the functions.

35. $y = x^2 + 24x + 144$

36. $y = x^2 - 36x + 324$

37. $y = -x^2 - 100x + 1600$

38. $y = -x^2 - 80x - 2000$

39. $y = 2x^2 + 10x - 600$

40. $y = 2x^2 - 75x - 450$

Use the graph of each function in Exercises 41–46 to estimate the x-intercepts.

41. $y = 2x^2 - 8x + 6$ **42.** $f(x) = x^2 + 4x + 4$

43. $y = x^2 - x - 110$ **44.** $y = x^2 + 9x - 36$

45. $g(x) = -5x^2 - 6x + 8$

46. $h(x) = -2x^2 - 4x + 6$

47. *Profit* The daily profit for a product is given by $P = 32x - 0.01x^2 - 1000$, where x is the number of units produced and sold.

   a. Graph this function for x between 0 and 3200.

   b. Describe what happens to the profit for this product when the number of units produced is between 1 and 1600.

   c. What happens to the profit after 1600 units are produced?

48. *Profit* The daily profit for a product is given by $P = 420x - 0.1x^2 - 4100$ dollars, where x is the number of units produced and sold.

   a. Graph this function for x between 0 and 4200.

   b. Is the graph of the function concave up or down?

49. *Juvenile Arrests* The number of juvenile arrests for property crimes is given by $y = 2.252x^2 - 30.227x + 524.216$, where x is the number of years after 2000 and y is the number of arrests in thousands.

   a. Graph this function for the years 2000–2010.

   b. What does this model estimate the number of arrests for property crimes to be in 2015?

   (Source: Office of Juvenile Justice and Delinquency Prevention)

50. *World Population* A low-projection scenario of world population for the years 1995–2150 by the United Nations is given by the function $y = -0.36x^2 + 38.52x + 5822.86$, where x is the number of years after 1990 and the world population is measured in millions of people.

   a. Graph this function for $x = 0$ to $x = 120$.

   b. What would the world population have been in 2010 if the projections made using this model had been accurate?

   (Source: *World Population Prospects,* United Nations)

51. *Tourism Spending* The equation

$$y = 1.69x^2 - 0.92x + 324.10,$$

with x equal to the number of years after 1998, models the global spending (in billions of dollars) on travel and tourism from 1998 to 2009.

   a. Graph this function for $x = 0$ to $x = 20$.

   b. If $x = 0$ in 1998, find the spending projected by this model for 2015.

   c. Is the value in part (b) an interpolation or an extrapolation?

   (Source: *Statistical Abstract of the United States*)

52. *Flight of a Ball* If a ball is thrown upward at 96 feet per second from the top of a building that is 100 feet high, the height of the ball can be modeled by $S(t) = 100 + 96t - 16t^2$ feet, where t is the number of seconds after the ball is thrown.

   a. Describe the graph of the model.

   b. Find the t-coordinate and S-coordinate of the vertex of the graph of this quadratic function.

   c. Explain the meaning of the coordinates of the vertex for this model.

53. *Flight of a Ball* If a ball is thrown upward at 39.2 meters per second from the top of a building that is 30 meters high, the height of the ball can be modeled by $S(t) = 30 + 39.2t - 9.8t^2$ meters, where t is the number of seconds after the ball is thrown.

   a. Find the t-coordinate and S-coordinate of the vertex of the graph of this quadratic function.

   b. Explain the meaning of the coordinates of the vertex for this function.

   c. Over what time interval is the function increasing? What does this mean in relation to the ball?

54. *Photosynthesis* The rate of photosynthesis R for a certain plant depends on the intensity of light x, in lumens, according to $R(x) = 270x - 90x^2$.

   a. Sketch the graph of this function on a meaningful window.

   b. Determine the intensity x that gives the maximum rate of photosynthesis.

55. *Union Membership* U.S. union membership, in thousands, is given by $f(x) = -5.864x^2 + 947.522x - 19,022.113$, where x is the number of years after 1900.

   a. Graph this function for $x = 40$ to $x = 110$.

   b. Does the model indicate that union membership increased or decreased for the years 1940 to 1980?

   c. Until 1980, the data include dues-paying members of traditional trade unions, regardless of employment status. After that, the data include employed only. How is this reflected in the graph of the function?

   (Source: Bureau of Labor Statistics, U.S. Department of Labor)

56. *Workers and Output* The weekly output of graphing calculators is $Q(x) = 200x + 6x^2$. Graph this function for values of x and Q that make sense in this application, if x is the number of weeks, $x \leq 10$.

57. *Profit* The profit for a certain brand of MP3 player can be described by the function $P(x) = 40x - 3000 - 0.01x^2$ dollars, where x is the number of MP3 players produced and sold.

   a. To maximize profit, how many MP3 players must be produced and sold?

   b. What is the maximum possible profit?

58. *Profit* The profit for Easy-Cut lawnmowers can be described by the function $P(x) = 840x - 75.6 - 0.4x^2$ dollars, where x is the number of mowers produced and sold.

   a. To maximize profit, how many mowers must be produced and sold?

   b. What is the maximum possible profit?

59. *Revenue* The annual total revenue for Pilot V5 pens is given by $R(x) = 1500x - 0.02x^2$ dollars, where x is the number of pens sold.

   a. To maximize the annual revenue, how many pens must be sold?

   b. What is the maximum possible annual revenue?

60. *Revenue* The monthly total revenue for satellite radios is given by $R(x) = 300x - 0.01x^2$ dollars, where x is the number of radios sold.

   a. To maximize the monthly revenue, how many radios must be sold?

   b. What is the maximum possible monthly revenue?

61. *Area* If 200 feet of fence are used to enclose a rectangular pen, the resulting area of the pen is $A = x(100 - x)$, where x is the width of the pen.

   a. Is A a quadratic function of x?

   b. What is the maximum possible area of the pen?

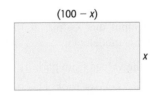

(100 − x)

x

62. *Area* If 25,000 feet of fence are used to enclose a rectangular field, the resulting area of the field is $A = (12,500 - x)x$, where x is the width of the pen. What is the maximum possible area of the pen?

63. *Marijuana Use* The percent of U.S. high school seniors who used marijuana during 1990–2006 can be

modeled by $y = -0.0584x^2 + 1.096x + 24.3657$ percent, where x is the number of years after 1990.

   a. What is the vertex of the graph of this model?

   b. In what year does the model estimate that the use reached a maximum?

   c. What is the maximum percent of usage, according to the model?
   (Source: monitoringthefuture.org)

64. *U.S. Visitors* The number of international visitors to the United States, in millions, can be modeled by $y = 0.592x^2 - 3.277x + 48.493$, where x is the number of years after 2000.

   a. Does the model estimate that a maximum or minimum number of visitors will occur during this period? How can you tell without graphing the equation?

   b. Find the input and output at the vertex.

   c. Interpret the results of part (b).
   (Source: *Statistical Abstract of the United States*)

65. *Poverty* The number of millions of people in the United States who lived below the poverty level for the years 1993 to 2009 can be modeled by $y = 0.114x^2 - 2.322x + 45.445$, where x is the number of years after 1990.

   a. Is the vertex of the graph of this function a maximum or minimum?

   b. Find the coordinates of the vertex and interpret them in the context of the problem.

   c. Use the vertex to set the window and graph the model.
   (Source: U.S. Census Bureau, U.S. Department of Commerce)

66. *Abortions* Using data from 1975 to 2008, the number of abortions in the United States per 1000 women aged 15–44 can be modeled by the function $y = -0.026x^2 + 0.951x + 18.161$, where x is the number of years after 1970 and y is measured in thousands.

   a. In what year does this model indicate the number was a maximum?

   b. What was the maximum number of abortions in this year if this model is accurate?

   c. For what values of x can we be sure that this model no longer applies?
   (Source: Alan Guttmacher Institute)

67. *Wind and Pollution* The amount of particulate pollution p in the air depends on the wind speed s, among other things, with the relationship between p and s approximated by $p = 25 - 0.01s^2$, where p is in ounces per cubic yard and s is in miles per hour.

a. Sketch the graph of this model with s on the horizontal axis and with nonnegative values of s and p.

b. Is the function increasing or decreasing on this domain?

c. What is the p-intercept of the graph?

d. What does the p-intercept mean in the context of this application?

68. *Drug Sensitivity* The sensitivity S to a drug is related to the dosage size x by $S = 1000x - x^2$.

a. Sketch the graph of this model using a domain and range with nonnegative x and S.

b. Is the function increasing or decreasing for x between 0 and 500?

c. What is the positive x-intercept of the graph?

d. Why is this x-intercept important in the context of this application?

69. *Falling Object* A tennis ball is thrown downward into a swimming pool from the top of a tall hotel. The height of the ball from the pool is given by $D(t) = -16t^2 - 4t + 210$ feet, where t is the time, in seconds, after the ball is thrown. Graphically find the t-intercepts for this function. Interpret the value(s) that make sense in this problem context.

70. *Break-Even* The profit for a product is given by $P = 1600 - 100x + x^2$, where x is the number of units produced and sold. Graphically find the x-intercepts of the graph of this function to find how many units will give break-even (that is, return a profit of zero).

71. *Flight of a Ball* A softball is hit with upward velocity 32 feet per second when $t = 0$, from a height of 3 feet.

a. Find the function that models the height of the ball as a function of time.

b. Find the maximum height of the ball.

72. *Flight of a Ball* A baseball is hit with upward velocity 48 feet per second when $t = 0$, from a height of 4 feet.

a. Find the function that models the height of the ball as a function of time.

b. Find the maximum height of the ball and in how many seconds the ball will reach that height.

73. *Apartment Rental* The owner of an apartment building can rent all 100 apartments if he charges $1200 per apartment per month, but the number of apartments rented is reduced by 2 for every $40 increase in the monthly rent.

a. Construct a table that gives the revenue if the rent charged is $1240, $1280, and $1320.

b. Does $R(x) = (1200 + 40x)(100 - 2x)$ model the revenue from these apartments if x represents the number of $40 increases?

c. What monthly rent gives the maximum revenue for the apartments?

74. *Rink Rental* The owner of a skating rink rents the rink for parties at $720 if 60 or fewer skaters attend, so the cost is $12 per person if 60 attend. For each 6 skaters above 60, she reduces the price per skater by $0.50.

a. Construct a table that gives the revenue if the number attending is 66, 72, and 78.

b. Does the function $R(x) = (60 + 6x)(12 - 0.5x)$ model the revenue from the party if x represents the number of increases of 6 people each?

c. How many people should attend for the rink's revenue to be a maximum?

75. *World Population* A low-projection scenario for world population for 1995–2150 by the United Nations is given by the function $y = -0.36x^2 + 38.52x + 5822.86$, where x is the number of years after 1990 and the world population is measured in millions of people.

a. Find the input and output at the vertex of the graph of this model.

b. Interpret the values from part (a).

c. For what years after 1995 does this model predict that the population will increase?
(Source: *World Population Prospects*, United Nations)

KEY OBJECTIVES

- Solve quadratic equations using factoring

- Solve quadratic equations graphically using the *x*-intercept method and the intersection method

- Solve quadratic equations by combining graphical and factoring methods

- Solve quadratic equations using the square root method

- Solve quadratic equations by completing the square

- Solve quadratic equations using the quadratic formula

- Solve quadratic equations having complex solutions

SECTION PREVIEW Gasoline Prices

The U.S. Energy Information Administration expected retail prices for regular-grade automotive gasoline to average $3.15 per gallon in 2011, 37 cents per gallon higher than the 2010 average, and $3.30 per gallon in 2012, with prices forecast to average about 5 cents per gallon higher in each year during the peak driving season (April through September). The real price of a gallon of gasoline (price adjusted for inflation) from 1990 and projected to 2012 can be described by the function

$$G(x) = 0.006x^2 - 0.054x + 1.780$$

where *x* is the number of years after 1990. The price of gasoline has a large impact on consumer spending, so viewing and understanding the pattern indicated by this model can help consumers make decisions about taking action to reduce gasoline use, such as switching to alternative fuel vehicles, using public transit, or reducing commuting time. (Source: U.S. Energy Information Administration)

To find the year in which gas prices are expected to reach $5.00 per gallon, we solve the quadratic equation

$$0.006x^2 - 0.054x + 1.780 = 5, \quad \text{or} \quad 0.006x^2 - 0.054x - 3.220 = 0$$

The values of *x* that satisfy this equation are the solutions of the equation. They are also zeros of the function

$$y = 0.006x^2 - 0.054x - 3.220$$

and they are the *x*-intercepts of the graph of this function. (See Example 9.) In this section, we learn how to solve quadratic equations by factoring methods, graphical methods, the square root method, completing the square, and the quadratic formula. ▪

Factoring Methods

An equation that can be written in the form $ax^2 + bx + c = 0$, with $a \neq 0$, is called a **quadratic equation**. Solutions to some quadratic equations can be found exactly by factoring; other quadratic equations require different types of solution methods to find or to approximate solutions.

Solution by factoring is based on the following property of real numbers.

Zero Product Property

If the product of two real numbers is 0, then at least one of them must be 0. That is, for real numbers *a* and *b*, if the product $ab = 0$, then either $a = 0$ or $b = 0$ or both *a* and *b* are equal to 0.

To use this property to solve a quadratic equation by factoring, we must first make sure that the equation is written in a form with zero on one side. If the resulting nonzero expression is factorable, we factor it and use the zero product property to convert the equation into two linear equations that are easily solved.* Before applying this technique to real-world applications, we consider the following example.

---

* For a review of factoring methods, see the Algebra Toolbox.

EXAMPLE 1 ▶   ## Solving a Quadratic Equation by Factoring

Solve the equation $3x^2 + 7x = 6$.

SOLUTION

We first subtract 6 from both sides of the equation to rewrite the equation with 0 on one side:

$$3x^2 + 7x - 6 = 0$$

To begin factoring the trinomial $3x^2 + 7x - 6$, we seek factors of $3x^2$ (that is, $3x$ and x) as the first terms of two binomials and factors of -6 as the last terms of the binomials. The factorization whose inner and outer products combine to $7x$ is $(3x - 2)(x + 3)$, so we have

$$(3x - 2)(x + 3) = 0$$

Using the zero product property gives

$$3x - 2 = 0, \quad \text{or} \quad x + 3 = 0$$

Solving these linear equations gives the two solutions to the original equation:

$$x = \frac{2}{3} \quad \text{or} \quad x = -3$$

EXAMPLE 2 ▶   ## Height of a Ball

The height above ground of a ball thrown upward at 64 feet per second from the top of an 80-foot-high building is modeled by $S(t) = 80 + 64t - 16t^2$ feet, where t is the number of seconds after the ball is thrown. How long will the ball be in the air?

SOLUTION

The ball will be in the air from $t = 0$ (with the height $S = 80$) until it reaches the ground ($S = 0$). Thus, we can find the time in the air by solving

$$0 = -16t^2 + 64t + 80$$

Because 16 is a factor of each of the terms, we can get a simpler but equivalent equation by dividing both sides of the equation by -16:

$$0 = t^2 - 4t - 5$$

This equation can be solved easily by factoring the right side:

$$0 = (t - 5)(t + 1)$$
$$0 = t - 5 \quad \text{or} \quad 0 = t + 1$$
$$t = 5 \quad \text{or} \quad t = -1$$

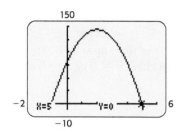

Figure 3.16

The time in the air starts at $t = 0$, so $t = -1$ has no meaning in this application. S also equals 0 at $t = 5$, which means that the ball is on the ground 5 seconds after it was thrown; that is, the ball is in the air 5 seconds. Figure 3.16 shows a graph of the function, which confirms that the height of the ball is 0 at $t = 5$.

Graphical Methods

In cases where factoring $f(x)$ to solve $f(x) = 0$ is difficult or impossible, graphing $y = f(x)$ can be helpful in finding the solution. Recall that if a is a real number, the following three statements are equivalent:

- a is a real solution to the equation $f(x) = 0$.
- a is a real zero of the function f.
- a is an x-intercept of the graph of $y = f(x)$.

It is important to remember that the above three statements are equivalent because connecting these concepts allows us to use different methods for solving equations. Sometimes graphical methods are the easiest way to find or approximate solutions to real data problems. If the x-intercepts of the graph of $y = f(x)$ are easily found, then graphical methods may be helpful in finding the solutions. Note that if the graph of $y = f(x)$ does not cross or touch the x-axis, there are no real solutions to the equation $f(x) = 0$.

We can also find solutions or decimal approximations of solutions to quadratic equations by using the intersection method with a graphing utility.

EXAMPLE 3 ▶ Profit

Consider the daily profit from the production and sale of x units of a product, given by

$$P(x) = -0.01x^2 + 20x - 500 \text{ dollars}$$

a. Use a graph to find the levels of production and sales that give a daily profit of $1400.

b. Is it possible for the profit to be greater than $1400?

SOLUTION

a. To find the level of production and sales, x, that gives a daily profit of 1400 dollars, we solve

$$1400 = -0.01x^2 + 20x - 500$$

To solve this equation by the intersection method, we graph

$$y_1 = -0.01x^2 + 20x - 500 \quad \text{and} \quad y_2 = 1400$$

To find the appropriate window for this graph, we note that the graph of the function $y_1 = -0.01x^2 + 20x - 500$ is a parabola with the vertex at

$$x = \frac{-b}{2a} = \frac{-20}{2(-0.01)} = 1000 \quad \text{and} \quad y = P(1000) = 9500$$

We use a viewing window containing this point, with $x = 1000$ near the center, to graph the function (Figure 3.17). Using the intersection method, we see that $(100, 1400)$ and $(1900, 1400)$ are the points of intersection of the graphs of the two functions. Thus, $x = 100$ and $x = 1900$ are solutions to the equation $1400 = -0.01x^2 + 20x - 500$, and the profit is $1400 when $x = 100$ units or $x = 1900$ units of the product are produced and sold.

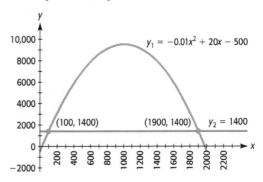

Figure 3.17

b. We can see from the graph in Figure 3.17 that the profit is more than $1400 for many values of x. Because the graph of this profit function is a parabola that opens down, the maximum profit occurs at the vertex of the graph. As we found in part (a), the vertex occurs at $x = 1000$, and the maximum possible profit is $P(1000) = \$9500$, which is more than $1400.

Combining Graphs and Factoring

If the factors of a quadratic function are not easily found, its graph may be helpful in finding the factors. Note once more the important correspondences that exist among solutions, zeros, and x-intercepts: the x-intercepts of the graph of $y = P(x)$ are the real solutions of the equation $0 = P(x)$ and the real zeros of $P(x)$. In addition, the factors of $P(x)$ are related to the zeros of $P(x)$. The relationship among the factors, solutions, and zeros is true for any polynomial function f and can be generalized by the following theorem.

Factor Theorem

The polynomial function f has a factor $(x - a)$ if and only if $f(a) = 0$. Thus, $(x - a)$ is a factor of $f(x)$ if and only if $x = a$ is a solution to $f(x) = 0$.

This means that we can verify our factorization of f and the real solutions to $0 = f(x)$ by graphing $y = f(x)$ and observing where the graph crosses the x-axis. We can also sometimes use our observation of the graph to assist us in the factorization of a quadratic function:

> *If one solution can be found exactly from the graph, it can be used to find one of the factors of the function. The second factor can then be found easily, leading to the second solution.*

EXAMPLE 4 ▶ ## Graphing and Factoring Methods Combined

Solve $0 = 3x^2 - x - 10$ by using the following steps.

a. Graphically find one of the x-intercepts of $y = 3x^2 - x - 10$.

b. Algebraically verify that the zero found in part (a) is an exact solution to $0 = 3x^2 - x - 10$.

c. Use the method of factoring to find the other solution to $0 = 3x^2 - x - 10$.

SOLUTION

a. The vertex of the graph of $y = 3x^2 - x - 10$ is at $x = \dfrac{1}{6}$ and $y \approx -10.08$, and because the parabola opens up, we set a viewing window that includes this point near the bottom center of the window. Graphing the function and using the x-intercept method, we find that the graph crosses the x-axis at $x = 2$ (Figure 3.18). This means that 2 is a zero of $f(x) = 3x^2 - x - 10$.

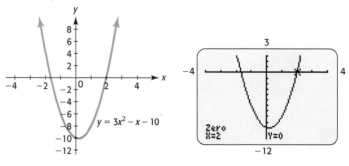

Figure 3.18

b. Because $x = 2$ was obtained graphically, it may be an approximation of the exact solution. To verify that it is exact, we show that $x = 2$ makes the equation $0 = 3x^2 - x - 10$ a true statement:

$$3(2)^2 - 2 - 10 = 12 - 2 - 10 = 0$$

c. Because $x = 2$ is a solution of $0 = 3x^2 - x - 10$, one factor of $3x^2 - x - 10$ is $(x - 2)$. Thus, we use $(x - 2)$ as one factor and seek a second binomial factor so that the product of the two binomials is $3x^2 - x - 10$:

$$3x^2 - x - 10 = 0$$
$$(x - 2)(\qquad) = 0$$
$$(x - 2)(3x + 5) = 0$$

The remaining factor is $3x + 5$, which we set equal to 0 to obtain the other solution:

$$3x = -5$$
$$x = -\frac{5}{3}$$

Thus, the two solutions to $0 = 3x^2 - x - 10$ are $x = 2$ and $x = -\frac{5}{3}$.

Graphical and Numerical Methods

When approximate solutions to quadratic equations are sufficient, graphical and/or numerical solution methods can be used. These methods of solving are illustrated in Example 5.

EXAMPLE 5 ▶ **Marijuana Use**

The percent p of high school seniors who have tried marijuana can be considered as a function of the time t according to the model

$$p = -0.1967t^2 + 4.0630t + 27.7455$$

where t is the number of years after 1990.

a. Find the year(s) after 1995 during which the percent is predicted to be 40, using a graphical method.

b. Verify the solution(s) numerically. (Source: National Institute on Drug Abuse)

SOLUTION

a. Note that the output p is measured in percent, so $p = 40$. Thus, to find where the percent is 40, we solve

$$40 = -0.1967t^2 + 4.0630t + 27.7455$$

We solve this equation with the intersection method. Figure 3.19(a) shows the graphs of $y_1 = -0.1967x^2 + 4.0630x + 27.7455$ and $y_2 = 40$, and Figure 3.19(b) and Figure 3.19(c) show two points where the graphs intersect.

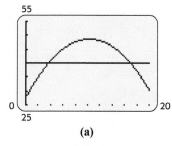

(a)

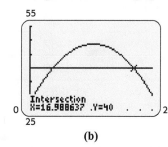

(b)

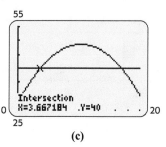

(c)

Figure 3.19

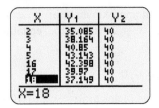

Figure 3.20

One point indicates that the percent is 40 when $t = 17$, so the model predicts that 40% of high school seniors will have tried marijuana 17 years after 1990, in 2007. Another point has $t = 3.7$, which indicates that the percent will be 40 four years after 1990, in 1994. But this is not after 1995, so it is not a solution to the problem.

b. Figure 3.20 shows a table of inputs near 4 and near 17 and the corresponding outputs. The table shows that the percent is 39.97 in the 17th year after 1990, in 2007.

Spreadsheet ▸ SOLUTION   The intersection of the graphs of $y = -0.1967x^2 + 4.0630x + 27.7455$ and $y = 40$ can be found using the Goal Seek command in Excel (Figure 3.21). The x-coordinates of the points of intersection are shown in cells A2 and A17 of Table 3.4 to be approximately 3.7 and 17.* These solutions agree with those found with a calculator in Figure 3.19.

Table 3.4

	A	B	C	D
1	x	y = 40	P(x) = −.1967x² + 4.063x + 27.7455	
2	3.667184	40	39.99995613	4.39E-05
3	1	40	31.6118	
4	2	40	35.0847	
5	3	40	38.1642	
6	4	40	40.8503	
7	5	40	43.143	
8	6	40	45.0423	
9	7	40	46.5482	
10	8	40	47.6607	
11	9	40	48.3798	
12	10	40	48.7055	
13	11	40	48.6378	
14	12	40	48.1767	
15	13	40	47.3222	
16	14	40	46.0743	
17	16.98863	40	40.00001588	−1.6E-05
18	16	40	42.3983	
19	17	40	39.9702	
20	18	40	37.1487	
21	19	40	33.9338	
22	20	40	30.3255	

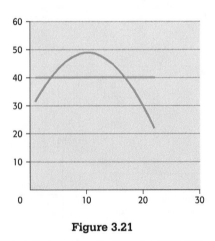

Figure 3.21

---

* For more information, see the Excel Guide in Appendix B, page 647.

The Square Root Method

We have solved quadratic equations by factoring and by graphical methods. Another method can be used to solve quadratic equations that are written in a particular form. In general, we can find the solutions of quadratic equations of the form $x^2 = C$, where C is a constant, by taking the square root of both sides. For example, to solve $x^2 = 25$, we take the square root of both sides of the equation, getting $x = \pm 5$. Note that there are two solutions because $5^2 = 25$ and $(-5)^2 = 25$.

Square Root Method

The solutions of the quadratic equation $x^2 = C$ are $x = \pm \sqrt{C}$. Note that, when we take the square root of both sides, we use a $\pm$ symbol because there are both a positive and a negative value that, when squared, give C.

Note that this method can also be used to solve equations of the form $(ax + b)^2 = C$.

EXAMPLE 6 ▶ **Square Root Solution Method**

Solve the following equations by using the square root method.

a. $2x^2 - 16 = 0$

b. $(x - 6)^2 = 18$

SOLUTION

a. This equation can be written in the form $x^2 = C$, so the square root method can be used. We want to rewrite the equation so that the x^2-term is isolated and its coefficient is 1 and then take the square root of both sides.

$$2x^2 - 16 = 0$$
$$2x^2 = 16$$
$$x^2 = 8$$
$$x = \pm\sqrt{8} = \pm2\sqrt{2}$$

The exact solutions to $2x^2 - 16 = 0$ are $x = 2\sqrt{2}$ and $x = -2\sqrt{2}$.

b. The left side of this equation is a square, so we can take the square root of both sides to find x.

$$(x - 6)^2 = 18$$
$$x - 6 = \pm\sqrt{18}$$
$$x = 6 \pm \sqrt{18}$$
$$= 6 \pm \sqrt{9}\sqrt{2}$$
$$= 6 \pm 3\sqrt{2}$$

Completing the Square

When we convert one side of a quadratic equation to a perfect binomial square and use the square root method to solve the equation, the method used is called **completing the square**.

EXAMPLE 7 ▶ **Completing the Square**

Solve the equation $x^2 - 12x + 7 = 0$ by completing the square.

SOLUTION

Because the left side of this equation is not a perfect square, we rewrite it in a form where we can more easily get a perfect square. We subtract 7 from both sides, getting

$$x^2 - 12x = -7$$

We complete the square on the left side of this equation by adding the appropriate number to both sides of the equation to make the left side a perfect square trinomial. Because $(x + a)^2 = x^2 + 2ax + a^2$, the constant term of a perfect square trinomial will be the *square of half the coefficient of x*. Using this rule with the equation $x^2 - 12x = -7$, we would need to take half the coefficient of x and add the square of this number to get a perfect square trinomial. Half of -12 is -6, so adding $(-6)^2 = 36$ to $x^2 - 12x$ gives the perfect square $x^2 - 12x + 36$. We also have to add 36 to the other side of the equation (to preserve the equality), giving

$$x^2 - 12x + 36 = -7 + 36$$

Factoring this perfect square trinomial gives

$$(x - 6)^2 = 29$$

We can now solve the equation with the square root method.

$$(x - 6)^2 = 29$$
$$x - 6 = \pm\sqrt{29}$$
$$x = 6 \pm \sqrt{29}$$

The Quadratic Formula

We can generalize the method of completing the square to derive a general formula that gives the solution to any quadratic equation. We use this method to find the general solution to $ax^2 + bx + c = 0, a \neq 0$.

$$ax^2 + bx + c = 0 \qquad \text{Standard form}$$
$$ax^2 + bx = -c \qquad \text{Subtract } c \text{ from both sides.}$$
$$x^2 + \frac{b}{a}x = -\frac{c}{a} \qquad \text{Divide both sides by } a.$$

We would like to make the left side of the last equation a perfect square trinomial. Half the coefficient of x is $\frac{b}{2a}$, and squaring this gives $\frac{b^2}{4a^2}$. Hence, adding $\frac{b^2}{4a^2}$ to both sides of the equation gives a perfect square trinomial on the left side, and we can continue with the solution.

$$x^2 + \frac{b}{a}x + \frac{b^2}{4a^2} = \frac{b^2}{4a^2} - \frac{c}{a}$$ Add $\dfrac{b^2}{4a^2}$ to both sides of the equation.

$$\left(x + \frac{b}{2a}\right)^2 = \frac{b^2 - 4ac}{4a^2}$$ Factor the left side and combine the fractions on the right side.

$$x + \frac{b}{2a} = \pm \sqrt{\frac{b^2 - 4ac}{4a^2}}$$ Take the square root of both sides.

$$x = -\frac{b}{2a} \pm \frac{\sqrt{b^2 - 4ac}}{2|a|}$$ Subtract $\dfrac{b}{2a}$ from both sides and simplify.

$$x = -\frac{b}{2a} \pm \frac{\sqrt{b^2 - 4ac}}{2a}$$ $\pm 2|a| = \pm 2a$

$$x = \frac{-b \pm \sqrt{b^2 - 4ac}}{2a}$$ Combine the fractions.

The formula we have developed is called the **quadratic formula**.

Quadratic Formula

The solutions of the quadratic equation $ax^2 + bx + c = 0$ are given by the formula

$$x = \frac{-b \pm \sqrt{b^2 - 4ac}}{2a}$$

Note that a is the coefficient of x^2, b is the coefficient of x, and c is the constant term.

Because of the $\pm$ sign, the solutions can be written as

$$x = \frac{-b + \sqrt{b^2 - 4ac}}{2a} \quad \text{and} \quad x = \frac{-b - \sqrt{b^2 - 4ac}}{2a}$$

We can use the quadratic formula to solve all quadratic equations exactly, but it is especially useful for finding exact solutions to those equations for which factorization is difficult or impossible. For example, the solutions to $40 = -0.1967t^2 + 4.0630t + 27.7455$ were found approximately by graphical methods in Example 5. If we need to find the exact solutions, we could use the quadratic formula.

EXAMPLE 8 ▶   Solving Using the Quadratic Formula

Solve $6 - 3x^2 + 4x = 0$ using the quadratic formula.

SOLUTION

The equation $6 - 3x^2 + 4x = 0$ can be rewritten as $-3x^2 + 4x + 6 = 0$, so $a = -3$, $b = 4$, and $c = 6$. The two solutions to this equation are

$$x = \frac{-4 \pm \sqrt{4^2 - 4(-3)(6)}}{2(-3)} = \frac{-4 \pm \sqrt{88}}{-6} = \frac{-4 \pm 2\sqrt{22}}{-6} = \frac{2 \pm \sqrt{22}}{3}$$

Thus, the exact solutions are the irrational numbers $x = \dfrac{2 + \sqrt{22}}{3}$ and $x = \dfrac{2 - \sqrt{22}}{3}$.

Three-place decimal approximations for these solutions are $x \approx 2.230$ and $x \approx -0.897$.

Decimal approximations of irrational solutions found with the quadratic formula will often suffice as answers to an applied problem. The quadratic formula is especially useful when the coefficients of a quadratic equation are decimal values that make factorization impractical. This occurs in many applied problems, such as the one discussed at the beginning of this section. If the graph of the quadratic function $y = f(x)$ does not intersect the x-axis at "nice" values of x, the solutions may be irrational numbers, and using the quadratic formula allows us to find these solutions exactly.

EXAMPLE 9 ▶ Gasoline Prices

The real price of a gallon of gasoline (price adjusted for inflation) from 1990 to 2012 can be described by the function

$$G(x) = 0.006x^2 - 0.054x + 1.780$$

where x is the number of years after 1990. If the model remains valid, in what year after 2000 will the price of a gallon of gasoline be $5.00?
(Source: U.S. Energy Information Administration)

SOLUTION

To answer this question, we solve (using an algebraic, numerical, or graphical method) the equation

$$0.006x^2 - 0.054x + 1.780 = 5$$

We choose to use the quadratic formula (even though an approximate answer is all that is needed). We first write $0.006x^2 - 0.054x + 1.780 = 5$ with 0 on one side:

$$0.006x^2 - 0.054x + 1.780 - 5 = 0$$
$$0.006x^2 - 0.054x - 3.22 = 0$$

This gives $a = 0.006$, $b = -0.054$, and $c = -3.22$. Substituting in the quadratic formula gives

$$x = \frac{-(-0.054) \pm \sqrt{(-0.054)^2 - 4(0.006)(-3.22)}}{2(0.006)}$$

$$= \frac{0.054 \pm \sqrt{0.080196}}{0.012}$$

$$x = 28.1 \quad \text{or} \quad x = -19.1$$

Thus, the price of a gallon of gasoline will reach $5.00 in the 29th year after 1990, or in 2019. (Note that a negative solution for x will give a year before 2000.) Figure 3.22(a) shows the outputs of

$$Y_1 = 0.006x^2 - 0.054x + 1.78$$

at integer values near the solution above, and Figure 3.22(b) shows a graph of the intersection of $Y_1 = 0.006x^2 - 0.054x + 1.78$ and $Y_2 = 5$, which confirms our conclusions numerically and graphically.

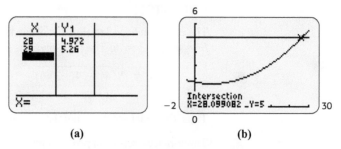

(a)   (b)

Figure 3.22

The Discriminant

We can also determine the type of solutions a quadratic equation has by looking at the expression $b^2 - 4ac$, which is called the **discriminant** of the quadratic equation $ax^2 + bx + c = 0$. The discriminant is the expression inside the radical in the quadratic formula $x = \dfrac{-b \pm \sqrt{b^2 - 4ac}}{2a}$, so it determines if the quantity inside the radical is positive, zero, or negative.

- If $b^2 - 4ac > 0$, there are two different real solutions.
- If $b^2 - 4ac = 0$, there is one real solution.
- If $b^2 - 4ac < 0$, there is no real solution.

For example, the equation $3x^2 + 4x + 2 = 0$ has no real solution because $4^2 - 4(3)(2) = -8 < 0$, and the equation $x^2 + 4x + 2 = 0$ has two different real solutions because $4^2 - 4(1)(2) = 8 > 0$.

Aids for Solving Quadratic Equations

The x-intercepts of the graph of a quadratic function $y = f(x)$ can be used to determine how to solve the quadratic equation $f(x) = 0$. Table 3.5 summarizes these ideas with suggested methods for solving and the graphical representations of the solutions.

Table 3.5 Connections Between Graphs of Quadratic Functions and Solution Methods

Graph			
Type of x-Intercepts	Graph crosses x-axis twice.	Graph touches but does not cross x-axis.	Graph does not cross x-axis.
Type of Solutions	Equation has two real solutions.	Equation has one real solution.	Equation has no real solutions.
Suggested Solution Methods*	Use factoring, graphing, or the quadratic formula.	Use factoring, graphing, or the square root method.	Use the quadratic formula to verify that solutions are not real.

Equations with Complex Solutions

Recall that the solutions of the quadratic equation $x^2 = C$ are $x = \pm\sqrt{C}$, so the solutions of the equation $x^2 = -a$ for $a > 0$ are

$$x = \pm\sqrt{-a} = \pm\sqrt{-1}\sqrt{a} = \pm i\sqrt{a}$$

EXAMPLE 10 ▶ **Solution Using the Square Root Method**

Solve the equations.

a. $x^2 = -9$ **b.** $3x^2 + 24 = 0$

---

* Solution methods other than the suggested methods may also be successful.

SOLUTION

a. Taking the square root of both sides of the equation gives the solution of $x^2 = -9$:

$$x = \pm\sqrt{-9} = \pm\sqrt{-1}\sqrt{9} = \pm 3i$$

b. We solve $3x^2 + 24 = 0$ using the square root method, as follows:

$$3x^2 = -24$$
$$x^2 = -8$$
$$x = \pm\sqrt{-8} = \pm\sqrt{-1}\sqrt{4 \cdot 2} = \pm 2i\sqrt{2}$$

We used the square root method to solve the equations in Example 10 because neither equation contained a first-degree term (that is, a term containing x to the first power). We can also find complex solutions by using the quadratic formula.* Recall that the solutions of the quadratic equation $ax^2 + bx + c = 0$ are given by the formula

$$x = \frac{-b \pm \sqrt{b^2 - 4ac}}{2a}$$

EXAMPLE 11 ▶ Complex Solutions of Quadratic Equations

Solve the equations.

a. $x^2 - 3x + 5 = 0$ **b.** $3x^2 + 4x = -3$

SOLUTION

a. Using the quadratic formula, with $a = 1$, $b = -3$, and $c = 5$, gives

$$x = \frac{-(-3) \pm \sqrt{(-3)^2 - 4(1)(5)}}{2(1)} = \frac{3 \pm \sqrt{-11}}{2} = \frac{3 \pm i\sqrt{11}}{2}$$

Note that the solutions can also be written in the form $\dfrac{3}{2} \pm \dfrac{\sqrt{11}}{2}i$.

Thus, the solutions are the complex numbers $\dfrac{3}{2} + \dfrac{\sqrt{11}}{2}i$ and $\dfrac{3}{2} - \dfrac{\sqrt{11}}{2}i$.

b. Writing $3x^2 + 4x = -3$ in the form $3x^2 + 4x + 3 = 0$ gives $a = 3$, $b = 4$, and $c = 3$, so the solutions are

$$x = \frac{-4 \pm \sqrt{(4)^2 - 4(3)(3)}}{2(3)} = \frac{-4 \pm \sqrt{-20}}{6} = \frac{-4 \pm \sqrt{-1}\sqrt{4}\sqrt{5}}{6}$$

$$= \frac{-4 \pm 2i\sqrt{5}}{6} = \frac{2(-2 \pm i\sqrt{5})}{2 \cdot 3} = \frac{-2 \pm i\sqrt{5}}{3}$$

Thus,

$$x = -\frac{2}{3} + \frac{\sqrt{5}}{3}i \quad \text{and} \quad x = -\frac{2}{3} - \frac{\sqrt{5}}{3}i$$

---

* The solutions could also be found by completing the square. Recall that the quadratic formula was developed by completing the square on the quadratic equation $ax^2 + bx + c = 0$.

Skills CHECK 3.2

In Exercises 1–10, use factoring to solve the equations.

1. $x^2 - 3x - 10 = 0$ **2.** $x^2 - 9x + 18 = 0$

3. $x^2 - 11x + 24 = 0$ **4.** $x^2 + 3x - 10 = 0$

5. $2x^2 + 2x - 12 = 0$ **6.** $2s^2 + s - 6 = 0$

7. $0 = 2t^2 - 11t + 12$ **8.** $6x^2 - 13x + 6 = 0$

9. $6x^2 + 10x = 4$ **10.** $10x^2 + 11x = 6$

Use a graphing utility to find or to approximate the x-intercepts of the graph of each function in Exercises 11–16.

11. $y = x^2 - 3x - 10$ **12.** $y = x^2 + 4x - 32$

13. $y = 3x^2 - 8x + 4$ **14.** $y = 2x^2 + 8x - 10$

15. $y = 2x^2 + 7x - 4$ **16.** $y = 5x^2 - 17x + 6$

Use a graphing utility as an aid in factoring to solve the equations in Exercises 17–22.

17. $2w^2 - 5w - 3 = 0$ **18.** $3x^2 - 4x - 4 = 0$

19. $x^2 - 40x + 256 = 0$ **20.** $x^2 - 32x + 112 = 0$

21. $2s^2 - 70s = 1500$ **22.** $3s^2 - 130s = -1000$

In Exercises 23–26, use the square root method to solve the quadratic equations.

23. $4x^2 - 9 = 0$ **24.** $x^2 - 20 = 0$

25. $x^2 - 32 = 0$ **26.** $5x^2 - 25 = 0$

In Exercises 27–30, complete the square to solve the quadratic equations.

27. $x^2 - 4x - 9 = 0$ **28.** $x^2 - 6x + 1 = 0$

29. $x^2 - 3x + 2 = 0$ **30.** $2x^2 - 9x + 8 = 0$

In Exercises 31–34, use the quadratic formula to solve the equations.

31. $x^2 - 5x + 2 = 0$ **32.** $3x^2 - 6x - 12 = 0$

33. $5x + 3x^2 = 8$ **34.** $3x^2 - 30x - 180 = 0$

In Exercises 35–40, use a graphing utility to find or approximate solutions of the equations.

35. $2x^2 + 2x - 12 = 0$ **36.** $2x^2 + x - 6 = 0$

37. $0 = 6x^2 + 5x - 6$ **38.** $10x^2 = 22x - 4$

39. $4x + 2 = 6x^2 + 3x$ **40.** $(x - 3)(x + 2) = -4$

In Exercises 41–46, find the exact solutions to $f(x) = 0$ in the complex numbers and confirm that the solutions are not real by showing that the graph of $y = f(x)$ does not cross the x-axis.

41. $x^2 + 25 = 0$ **42.** $2x^2 + 40 = 0$

43. $(x - 1)^2 = -4$ **44.** $(2x + 1)^2 + 7 = 0$

45. $x^2 + 4x + 8 = 0$ **46.** $x^2 - 5x + 7 = 0$

In Exercises 47 and 48, you are given the graphs of several functions of the form $f(x) = ax^2 + bx + c$ for different values of a, b, and c. For each function,

a. *Determine if the discriminant is positive, negative, or zero.*

b. *Determine if there are 0, 1, or 2 real solutions to $f(x) = 0$.*

c. *Solve the equation $f(x) = 0$.*

47.

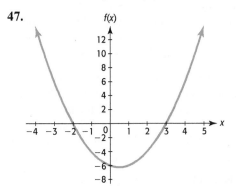

48.

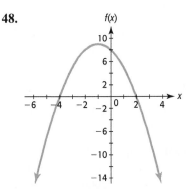

EXERCISES 3.2

In Exercises 49–60, solve analytically and then check graphically.

49. *Flight of a Ball* If a ball is thrown upward at 96 feet per second from the top of a building that is 100 feet high, the height of the ball can be modeled by $S(t) = 100 + 96t - 16t^2$ feet, where t is the number of seconds after the ball is thrown. How long after the ball is thrown is the height 228 feet?

50. *Falling Object* A tennis ball is thrown into a swimming pool from the top of a tall hotel. The height of the ball above the pool is modeled by $D(t) = -16t^2 - 4t + 200$ feet, where t is the time, in seconds, after the ball is thrown. How long after the ball is thrown is it 44 feet above the pool?

51. *Break-Even* The profit for an electronic reader is given by $P(x) = -12x^2 + 1320x - 21,600$, where x is the number of readers produced and sold. How many readers give break-even (that is, give zero profit) for this product?

52. *Break-Even* The profit for Coffee Exchange coffee beans is given by $P(x) = -15x^2 + 180x - 405$ thousand dollars, where x is the number of tons of coffee beans produced and sold. How many tons give break-even (that is, give zero profit) for this product?

53. *Break-Even* The total revenue function for French door refrigerators is given by $R = 550x$ dollars, and the total cost function for this same product is given by $C = 10,000 + 30x + x^2$, where C is measured in dollars. For both functions, the input x is the number of refrigerators produced and sold.

   a. Form the profit function for the refrigerators from the two given functions.

   b. What is the profit when 18 refrigerators are produced and sold?

   c. What is the profit when 32 refrigerators are produced and sold?

   d. How many refrigerators must be sold to break even on this product?

54. *Break-Even* The total revenue function for a home theater system is given by $R = 266x$, and the total cost function for the system is $C = 2000 + 46x + 2x^2$, where R and C are each measured in dollars and x is the number of units produced and sold.

   a. Form the profit function for this product from the two given functions.

   b. What is the profit when 55 systems are produced and sold?

   c. How many systems must be sold to break even on this product?

55. *Wind and Pollution* The amount of particulate pollution p from a power plant in the air above the plant depends on the wind speed s, among other things, with the relationship between p and s approximated by $p = 25 - 0.01s^2$, with s in miles per hour.

   a. Find the value(s) of s that will make $p = 0$.

   b. What does $p = 0$ mean in this application?

   c. What solution to $0 = 25 - 0.01s^2$ makes sense in the context of this application?

56. *Velocity of Blood* Because of friction from the walls of an artery, the velocity of a blood corpuscle in an artery is greatest at the center of the artery and decreases as the distance r from the center increases. The velocity of the blood in the artery can be modeled by the function

$$v = k(R^2 - r^2)$$

where R is the radius of the artery and k is a constant that is determined by the pressure, viscosity of the blood, and the length of the artery. In the case where $k = 2$ and $R = 0.1$ centimeter, the velocity is $v = 2(0.01 - r^2)$ centimeters per second (cm/sec).

   a. What distance r would give a velocity of 0.02 cm/sec?

   b. What distance r would give a velocity of 0.015 cm/sec?

   c. What distance r would give a velocity of 0 cm/sec? Where is the blood corpuscle?

57. *Drug Sensitivity* The sensitivity S to a drug is related to the dosage size x by $S = 100x - x^2$, where x is the dosage size in milliliters.

   a. What dosage(s) would give zero sensitivity?

   b. Explain what your answer in part (a) might mean.

58. *Body-Heat Loss* The model for body-heat loss depends on the coefficient of convection K, which depends on wind speed s according to the equation $K^2 = 16s + 4$, where s is in miles per hour. Find the positive coefficient of convection when the wind speed is

   a. 20 mph.

b. 60 mph.

c. What is the change in K for a change in speed from 20 mph to 60 mph?

59. *Market Equilibrium* Suppose that the demand for artificial Christmas trees is given by the function

$$p = 109.70 - 0.10q$$

and that the supply of these trees is given by

$$p = 0.01q^2 + 5.91$$

where p is the price of a tree in dollars and q is the quantity of trees that are demanded/supplied in hundreds. Find the price that gives the market equilibrium price and the number of trees that will be sold/bought at this price.

60. *Market Equilibrium* The demand for diamond-studded watches is given by $p = 7000 - 2x$ dollars, and the supply of watches is given by $p = 0.01x^2 + 2x + 1000$ dollars, where x is the number of watches demanded and supplied when the price per watch is p dollars. Find the equilibrium quantity and the equilibrium price.

61. *Foreign-Born Population* Suppose the percent of the U.S. population that is foreign born can be modeled by the equation $y = 0.003x^2 - 0.438x + 20.18$, where x is the number of years after 1900.

a. Verify graphically that one solution to $12.62 = 0.003x^2 - 0.438x + 20.18$ is $x = 20$.

b. Use this information and factoring to find the year or years when the percent of the U.S. population that is foreign born is 12.62.
(Source: U.S. Census Bureau)

62. *World Population* A low-projection scenario for world population for the years 1995–2150 by the United Nations is given by the function $y = -0.36x^2 + 38.52x + 5822.86$, where x is the number of years after 1990 and the world population is measured in millions of people. In what year will the world population reach 6581 million?

63. *U.S. Energy Consumption* Energy consumption in the United States in quadrillion BTUs can be modeled by $C(x) = -0.013x^2 + 1.281x + 67.147$, where x is the number of years after 1970.

a. One solution to the equation $87.567 = -0.013x^2 + 1.281x + 67.147$ is $x = 20$. What does this mean?

b. Graphically verify that $x = 20$ is a solution to $87.567 = -0.013x^2 + 1.281x + 67.147$.

c. To find when after 2020 U.S. energy consumption will be 87.567 quadrillion BTUs according to the model, do we need to find the second solution to this equation? Why or why not?
(Source: U.S. Energy Information Administration)

64. *Tourism Spending* The global spending on travel and tourism (in billions of dollars) can be described by the equation $y = 1.69x^2 - 0.92x + 324.10$, where x equals the number of years after 1990. Graphically find the year in which spending is projected to reach $1817.5 billion.
(Source: *Statistical Abstract of the United States*)

65. *High School Smokers* The percent of high school students who smoked cigarettes on 1 or more of the 30 days preceding the Youth Risk Behavior Survey can be modeled by $y = -0.061x^2 + 0.275x + 33.698$, where x is the number of years after 1990.

a. How would you set the viewing window to show a graph of the model for the years 1990 to 2015? Graph the function.

b. Graphically estimate this model's prediction for when the percent of high school students who smoke would be 12.
(Source: Centers for Disease Control)

66. *High School Smokers* The percent of high school students who smoked cigarettes on 1 or more of the 30 days preceding the Youth Risk Behavior Survey can be modeled by $y = -0.061x^2 + 0.275x + 33.698$, where x is the number of years after 1990. In what year does this model predict the value of y to be negative, making it certain that the model is no longer valid?
(Source: Centers for Disease Control)

67. *Federal Funds* The amount of federal funds, in billions, spent on child nutrition programs can be modeled by $N = 0.645x^2 - 0.165x + 10.298$, where x is the number of years after 2000.

a. What was the increase in federal funds spent on child nutrition programs between 2000 and 2010?

b. Use graphical methods to find in what year after 2000 the model predicts the amount of funds spent to be $134.408 billion.

c. Use the model to estimate the funds that will be spent in 2020. Is this interpolation or extrapolation? What assumptions are you making when you make this estimation?
(Source: *Statistical Abstract of the United States*)

68. *International Travel to the United States* The number of millions of visitors to the United States can be modeled by $V(x) = 0.592x^2 - 3.277x + 48.493$, where x is the number of years after 2000.

    a. Use the model to find the increase in the number of visitors to the United States from 2000 to 2008.

    b. When would the number of visitors exceed 84,000,000, according to the model?
(Source: Office of Travel and Tourism Industries, U.S. Department of Commerce)

69. *Gold Prices* The price of an ounce of gold in U.S. dollars for the years 1997–2011 can be modeled by the function $G(x) = 11.532x^2 - 259.978x + 1666.555$, where x is the number of years after 1990.

    a. Graph this function for values of x representing 1997–2011.

    b. According to the model, what will the price of gold be in 2020?

    c. Use graphical or numerical methods to estimate when the price of gold will reach \$2702.80 per ounce.
(Source: goldprice.org)

70. *Quitting Smoking* The percent of people over 19 years of age who have ever smoked and quit is given by the equation $y = 0.010x^2 + 0.344x + 28.74$, where x is the number of years since 1960.

    a. Graph this function for values of x representing 1960–2010.

    b. Assuming that the pattern indicated by this model continues through 2015, what would be the percent in 2015?

    c. When does this model indicate that the percent reached 40%?
(Source: *Substance Abuse*, Princeton, N.J.)

71. *Retail Sales* November and December retail sales, excluding autos, for the years 2001–2010 can be modeled by the function $S(x) = -1.751x^2 + 38.167x + 388.997$ billion dollars, where x is the number of years after 2000.

    a. Graph the function for values of x representing 2001–2010.

    b. During what years does the model estimate the sales to be \$550 billion?

    c. The recession in 2008 caused retail sales to drop. Does the model agree with the facts; that is, does it indicate that a maximum occurred in 2008?
(Source: U.S. Census Bureau)

72. *Cell Phones* Using the CTIA Wireless Survey for 1985–2009, the number of U.S. cell phone subscribers (in millions) can be modeled by

$$y = 0.632x^2 - 2.651x + 1.209$$

where x is the number of years after 1985.

    a. Graphically find when the number of U.S. subscribers was 301,617,000.

    b. When does the model estimate that the number of U.S. subscribers would reach 359,515,000?

    c. What does the answer to (b) tell about this model?
(Source: U.S. Census Bureau)

73. *World Population* One projection of the world population by the United Nations (a low-projection scenario) is given in the table below. The data can be modeled by $y = -0.36x^2 + 38.52x + 5822.86$ million people, where x is the number of years after 1990. In what year after 1990 does this model predict the world population will first reach 6,702,000,000?

Year	Projected Population (millions)
1995	5666
2000	6028
2025	7275
2050	7343
2075	6402
2100	5153
2125	4074
2150	3236

(Source: *World Population Prospects,* United Nations)

3.3

Piecewise-Defined Functions and Power Functions

KEY OBJECTIVES

- Evaluate and graph piecewise-defined functions
- Graph the absolute function
- Solve absolute value equations
- Evaluate and graph power functions
- Graph root functions and the reciprocal function
- Solve problems involving direct variation as the nth power
- Solve problems involving inverse variation

SECTION PREVIEW Residential Power Costs

The data in Table 3.6 give the rates that Georgia Power Company charges its residential customers for electricity during the months of October through May, excluding fuel adjustment costs and taxes. The monthly charges can be modeled by a **piecewise-defined function**, as we will see in Example 3. Piecewise-defined functions are used in applications in the life, social, and physical sciences, as well as in business, when there is not a single function that accurately represents the application.

Power functions are another type of function that is a basic building block for many mathematical applications. In this section, we evaluate, graph, and apply piecewise-defined and power functions.

Table 3.6

Monthly Kilowatt-hours (kWh)	Monthly Charge
0 to 650	$7.50 plus $0.045991 per kWh
More than 650, up to 1000	$37.40 plus $0.03946 per kWh above 650
More than 1000	$51.21 plus $0.038737 per kWh above 1000

Piecewise-Defined Functions

It is possible that a set of data cannot be modeled with a single equation. We can use a piecewise-defined function when there is not a single function that accurately represents the situation. A piecewise-defined function is so named because it is defined with different pieces for different parts of its domain rather than one equation.

EXAMPLE 1 ▶ Postage

The postage paid for a first-class letter "jumps" by 20 cents for each ounce after the first ounce, but does not increase until the weight increases by 1 ounce. Table 3.7 gives the 2011 postage for letters up to 3.5 ounces. Write a piecewise-defined function that models the price of postage and graph the function.

Table 3.7

Weight x (oz)	Postage y (cents)
$0 < x \leq 1$	44
$1 < x \leq 2$	64
$2 < x \leq 3$	84
$3 < x \leq 3.5$	104

SOLUTION

The function that models the price P of postage for x ounces, where x is between 0 and 3.5, is

$$P(x) = \begin{cases} 44 & \text{if } 0 < x \leq 1 \\ 64 & \text{if } 1 < x \leq 2 \\ 84 & \text{if } 2 < x \leq 3 \\ 104 & \text{if } 3 < x \leq 3.5 \end{cases}$$

The graph of this function (Figure 3.23) shows that each output is a constant with discontinuous "steps" at $x = 1, 2,$ and 3. This is a special piecewise-defined function, called a **step function**. Other piecewise-defined functions can be defined by polynomial functions over limited domains.

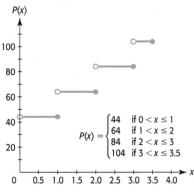

Figure 3.23

EXAMPLE 2 ▶   ## Piecewise-Defined Function

Graph the function

$$f(x) = \begin{cases} 5x + 2 & \text{if } 0 \le x < 3 \\ x^3 & \text{if } 3 \le x \le 5 \end{cases}$$

SOLUTION

Table 3.8(a)

x	0	1	2	2.99
f(x)	2	7	12	16.95

Table 3.8(b)

x	3	4	5
f(x)	27	64	125

To graph this function, we can construct a table of values for each of the pieces. Table 3.8(a) gives outputs of the function for some sample inputs x on the interval $[0, 3)$; on this interval, $f(x)$ is defined by $f(x) = 5x + 2$. Table 3.8(b) gives outputs for some sample inputs on the interval $[3, 5]$; on this interval, $f(x)$ is defined by $f(x) = x^3$.

Plotting the points from Table 3.8(a) and connecting them with a smooth curve gives the graph of $y = f(x)$ on the x-interval $[0, 3)$ (Figure 3.24). The open circle on this piece of the graph indicates that this piece of the function is not defined for $x = 3$. Plotting the points from Table 3.8(b) and connecting them with a smooth curve gives the graph of $y = f(x)$ on the x-interval $[3, 5]$ (Figure 3.24). The closed circles on this piece of the graph indicate that this piece of the function is defined for $x = 3$ and $x = 5$. The graph of $y = f(x)$, shown in Figure 3.24, consists of these two pieces.

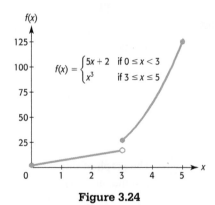

Figure 3.24

EXAMPLE 3 ▶ Residential Power Costs

Excluding fuel adjustment costs and taxes, the rates Georgia Power Company charges its residential customers for electricity during the months of October through May are shown in Table 3.9.

Table 3.9

Monthly Kilowatt-hours (kWh)	Monthly Charge
0 to 650	$7.50 plus $0.045991 per kWh
More than 650, up to 1000	$37.40 plus $0.03946 per kWh above 650
More than 1000	$51.21 plus $0.038737 per kWh above 1000

a. Write the piecewise-defined function C that gives the monthly charge for residential electricity, with input x equal to the monthly number of kilowatt-hours.

b. Find $C(950)$ and explain what it means.

c. Find the charge for using 1560 kWh in a month.

SOLUTION

a. The monthly charge is given by the function

$$C(x) = \begin{cases} 7.50 + 0.045991x & \text{if } 0 \le x \le 650 \\ 37.40 + 0.03946(x - 650) & \text{if } 650 < x \le 1000 \\ 51.21 + 0.038737(x - 1000) & \text{if } x > 1000 \end{cases}$$

where $C(x)$ is the charge in dollars for x kWh of electricity.

b. To evaluate $C(950)$, we must determine which "piece" defines the function when $x = 950$. Because 950 is between 650 and 1000, we use the "middle piece" of the function:

$$C(950) = 37.40 + 0.03946(950 - 650) = 49.238$$

Companies regularly round charges *up* to the next cent if any part of a cent is due. This means that if 950 kWh are used in a month, the bill is $49.24.

c. To find the charge for 1560 kWh, we evaluate $C(1560)$, using the "bottom piece" of the function because $1560 > 1000$. Evaluating gives

$$C(1560) = 51.21 + 0.038737(1560 - 1000) = 72.903$$

so the charge for the month is $72.91.

EXAMPLE 4 ▶ Wind Chill Factor

Wind chill factors are used to measure the effect of the combination of temperature and wind speed on human comfort, by providing equivalent air temperatures with no wind blowing. One formula that gives the wind chill factor for a 30°F temperature and a wind with velocity V in miles per hour is

$$W = \begin{cases} 30 & \text{if } 0 \le V < 4 \\ 1.259V - 18.611\sqrt{V} + 62.255 & \text{if } 4 \le V \le 55.9 \\ -6.5 & \text{if } V > 55.9 \end{cases}$$

a. Find the wind chill factor for the 30°F temperature if the wind is 40 mph.

b. Find the wind chill factor for the 30°F temperature if the wind is 65 mph.

c. Graph this function for $0 \leq V \leq 70$.

d. What are the domain and range of the function graphed in (c)?
(Source: The National Weather Service)

SOLUTION

a. Because $V = 40$ is in the interval $4 \leq V \leq 55.9$, the wind chill factor is

$$1.259(40) - 18.611\sqrt{40} + 62.255 \approx -5.091 \approx -5$$

This means that, if the wind is 40 mph, a temperature of 30°F would actually feel like −5°F.

b. Because $V = 65$ is in the interval $V > 55.9$, the wind chill factor is −6.5. This means that, if the wind is 65 mph, a temperature of 30°F would actually feel like −6.5°F.

c. The graph is shown in Figure 3.25.

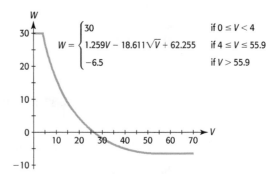

Figure 3.25

d. The domain is specified to be $0 \leq V \leq 70$, and the range, which is the set of outputs for W, is $-6.5 \leq W \leq 30$. That is, for wind speeds between 0 and 70 mph, the wind chill factor is no lower than −6.5°F and no higher than 30°F.

EXAMPLE 5 ▶   Service Calls

The cost of weekend service calls by Airtech Services is shown by the graph in Figure 3.26. Write a piecewise-defined function that represents the cost for service during the first five hours, as defined by the graph.

SOLUTION

The graph indicates that there are three pieces to the function, and each piece is a constant function. For t-values between 0 and 1, $C = 330$; for t-values between 1 and 2, $C = 550$; and for t-values between 2 and 5, $C = 770$. The open and closed circles on the graph tell us where the endpoints of each interval of the domain are defined. Thus, the cost function, with C in dollars and t in hours, is

$$C = \begin{cases} 330 & \text{if } 0 < t \leq 1 \\ 550 & \text{if } 1 < t \leq 2 \\ 770 & \text{if } 2 < t \leq 5 \end{cases}$$

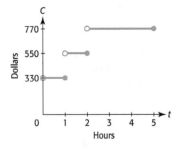

Figure 3.26

Absolute Value Function

We can construct a new function by combining two functions into a special piecewise-defined function. For example, we can write a function with the form

$$f(x) = \begin{cases} x & \text{if } x \geq 0 \\ -x & \text{if } x < 0 \end{cases}$$

This function is called the **absolute value function**, which is denoted by $f(x) = |x|$ and is derived from the definition of the absolute value of a number. Recall that the definition of the absolute value of a number is

$$|x| = \begin{cases} x & \text{if } x \geq 0 \\ -x & \text{if } x < 0 \end{cases}$$

To graph $f(x) = |x|$, we graph the portion of the line $y_1 = x$ for $x \geq 0$ (Figure 3.27(a)) and the portion of the line $y_2 = -x$ for $x < 0$ (Figure 3.27(b)). When these pieces are joined on the same graph (Figure 3.27(c)), we have the graph of $y = |x|$.

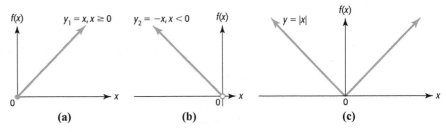

Figure 3.27

Solving Absolute Value Equations

We now consider the solution of equations that contain absolute value symbols, called **absolute value equations**. Consider the equation $|x| = 5$. We know that $|5| = 5$ and $|-5| = 5$. Therefore, the solution to $|x| = 5$ is $x = 5$ or $x = -5$. Also, if $|x| = 0$, then x must be 0. Finally, because the absolute value of a number is never negative, we cannot solve $|x| = a$ if a is negative. We generalize this as follows.

Absolute Value Equation

If $|x| = a$ and $a > 0$, then $x = a$ or $x = -a$.
There is no solution to $|x| = a$ if $a < 0$; $|x| = 0$ has solution $x = 0$.

If one side of an equation is a function contained in an absolute value and the other side is a nonnegative constant, we can solve the equation by using the method above.

EXAMPLE 6 ▶   ## Absolute Value Equations

Solve the following equations.

a. $|x - 3| = 9$ **b.** $|2x - 4| = 8$ **c.** $|x^2 - 5x| = 6$

SOLUTION

a. If $|x - 3| = 9$, then $x - 3 = 9$ or $x - 3 = -9$. Thus, the solution is $x = 12$ or $x = -6$.

b. If $|2x - 4| = 8$, then

$$2x - 4 = 8 \quad \text{or} \quad 2x - 4 = -8$$

$2x = 12$	$2x = -4$
$x = 6$	$x = -2$

Thus, the solution is $x = 6$ or $x = -2$.

c. If $|x^2 - 5x| = 6$, then

$$x^2 - 5x = 6 \quad \text{or} \quad x^2 - 5x = -6$$

$x^2 - 5x - 6 = 0$	$x^2 - 5x + 6 = 0$
$(x - 6)(x + 1) = 0$	$(x - 3)(x - 2) = 0$
$x = 6$ or $x = -1$	$x = 3$ or $x = 2$

Thus, four values of x satisfy the equation. We can check these solutions by graphing or by substitution. Using the intersection method, we graph $y_1 = |x^2 - 5x|$ and $y_2 = 6$ and find the points of intersection. The solutions to the equation, $x = 6, -1, 3$, and 2 found above, are also the x-values of the points of intersection (Figure 3.28).

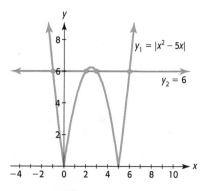

Figure 3.28

Power Functions

We define power functions as follows.

Power Functions

A **power function** is a function of the form $y = ax^b$, where a and b are real numbers, $b \neq 0$.

We know that the area of a rectangle is $A = lw$ square units, where l and w are the length and the width, respectively, of the rectangle. Because $l = w$ when the rectangle is a square, the area of a square can be found with the formula $A = x^2$ if each side is x units long.

This is the function $A = x^2$ for $x \geq 0$. A related function is $y = x^2$, defined on the set of all real numbers; this is a power function with power 2. This function is also a basic quadratic function, called the **squaring function**, and its complete graph is a parabola (see Figure 3.29(a)).

If we fill a larger cube with cubes that are 1 unit on each edge and have a volume of 1 cubic unit, the number of these small cubes that fit gives the volume of the larger cube. It is easily seen that if the length of an edge of a cube is x units, its volume is x^3 cubic units for any nonnegative value of x. The related function $y = x^3$ defined for all real numbers x is called the **cubing function**. A complete graph of this function is shown in Figure 3.29(b).

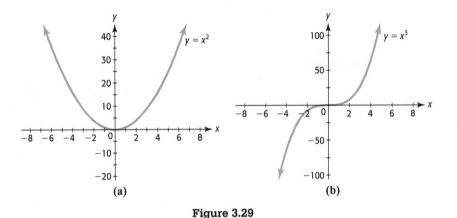

Figure 3.29

The functions $y = x$, $y = x^2$, and $y = x^3$ are examples of **power functions** with positive integer powers. Additional examples of power functions include functions with noninteger powers of x. For example, the graphs of the functions $y = x^{2/3}$ and $y = x^{3/2}$ are shown in Figures 3.30(a) and 3.30(b).

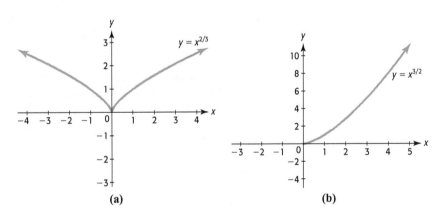

Figure 3.30

Note that $y = x^{2/3}$ is a power function with power less than 1, and its graph increases in the first quadrant but not as fast as $y = x$ does for values of x greater than 1. We say that the graph is *concave down* in the first quadrant. Also, $y = x^{3/2}$ is a power function with power greater than 1, and its graph increases in the first quadrant faster than $y = x$ does for values of x greater than 1. We say that the graph is *concave up* in the first quadrant.

In general, if $a > 0$ and $x > 0$, the graph of $y = ax^b$ is concave up if $b > 1$ and concave down if $0 < b < 1$. Figures 3.31(a) and 3.31(b) show the first-quadrant portion of typical graphs of $y = ax^b$ for $a > 0$ and different values of b.

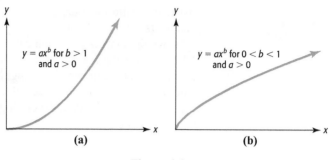

Figure 3.31

EXAMPLE 7 ▶

Wingspan of Birds

At 17 feet, the wingspan of a prehistoric bony-toothed bird may exceed that of any other bird that ever existed. The animal weighed about 64 pounds and soared the Chilean skies 5–10 million years ago. The wingspan of a bird can be estimated using the model

$$L = 2.43W^{0.3326}$$

where L is the wingspan in feet and W is the weight in pounds.
(Source: Discovery News)

a. Graph this function in the context of the application.

b. Use the function to compute the wingspan of a mute swan that weighs 40 pounds.

c. Use the function to approximate the weight of an albatross whose wingspan is 11 feet 4 inches.

SOLUTION

a. It is appropriate to graph this model only in the first quadrant because both the weight and the wingspan must be nonnegative. The graph is shown in Figure 3.32(a).

b. To compute the wingspan of a mute swan that weighs 40 pounds, we evaluate the function with an input of 40, giving an output of $L = 2.43(40)^{0.3326} = 8.288$. Thus, the wingspan of the mute swan is approximately 8.3 feet.

c. If the wingspan of an albatross is 11 feet 4 inches, we graphically solve the equation $11.333 = 2.43W^{0.3326}$. Using the intersection method (Figure 3.32(b)) gives the weight of the albatross to be approximately 102.48 pounds.

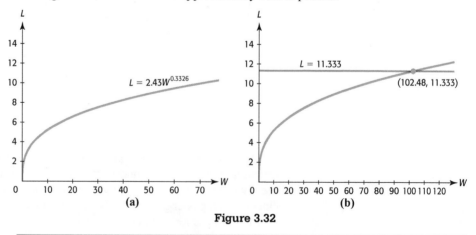

Figure 3.32

Note that the power in $L = 2.43W^{0.3326}$ is 0.3326, which can be written in the form $\dfrac{3326}{10,000}$ and is therefore a rational number.

Root Functions

Functions with rational powers can also be written with radicals. For example, $y = x^{1/3}$ can be written in the form $y = \sqrt[3]{x}$, and $y = x^{1/2}$ can be written in the form $y = \sqrt{x}$. Functions like $y = \sqrt[3]{x}$ and $y = \sqrt{x}$ are special power functions called **root functions**.

Root Functions

A root function is a function of the form $y = ax^{1/n}$, or $y = a\sqrt[n]{x}$, where n is an integer, $n \geq 2$.

The graphs of $y = \sqrt{x}$ and $y = \sqrt[3]{x}$ are shown in parts (a) and (b) of Figure 3.33. Note that the domain of $y = \sqrt{x}$ is $x \geq 0$ because $\sqrt{x}$ is undefined for negative values of x. Note also that these root functions can be written as power functions.

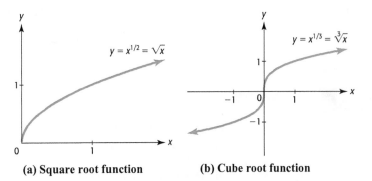

(a) Square root function (b) Cube root function

Figure 3.33

All root functions of the form $f(x) = \sqrt[n]{x}$, with n an even positive integer, have the same shape as shown in Figure 3.33(a); the domain and the range are both $[0, \infty)$. Root functions with n odd have the same shape as shown in Figure 3.33(b); the domain and range are both $(-\infty, \infty)$.

Reciprocal Function

One special power function is the function

$$y = x^{-1} = \frac{1}{x}$$

This function is usually called the **reciprocal function**. Its graph, shown in Figure 3.34, is called a rectangular hyperbola. Note that x cannot equal 0, so 0 is not in the domain of this function. As the graph of $y = \dfrac{1}{x}$ shows, as x gets close to 0, $|y|$ gets large and the graph approaches but does not touch the y-axis. We say that the y-axis is a **vertical asymptote** of this graph.

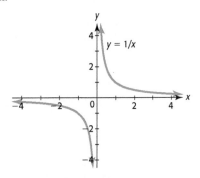

Figure 3.34

Note also that the values of $y = \dfrac{1}{x}$ get very small as $|x|$ gets large, and the graph approaches but does not touch the x-axis. We say that the x-axis is a **horizontal asymptote** for this graph.

Variation

In Section 2.1, we discussed direct variation. If a quantity y varies directly as a power of x, we say that y is directly proportional to the nth power of x.

Direct Variation as the nth Power

A quantity y varies directly as the nth power ($n > 0$) of x if there is a constant k such that

$$y = kx^n$$

The number k is called the constant of variation or the constant of proportionality.

EXAMPLE 8 ▶ **Production**

In the production of an item, the number of units of one raw material required varies as the cube of the number of units of a second raw material that is required. Suppose 500 units of the first and 5 units of the second raw material are required to produce 100 units of the item. How many units of the first raw material are required if the number of units produced requires 10 units of the second raw material?

SOLUTION

If x is the number of units of the second raw material required and y is the number of units of the first raw material required, then y varies as the cube of x, or

$$y = kx^3$$

Because $y = 500$ when $x = 5$, we have

$$500 = k \cdot 5^3, \quad \text{or} \quad k = 4$$

Then $y = 4x^3$ and $y = 4(10^3) = 4000$. Thus, 4000 units of the first raw material are required if the number of units produced requires 10 units of the second raw material.

When two variables x and y are **inversely proportional** (or vary inversely), an increase in one variable results in a decrease in the other.

Inverse Variation

A quantity y is inversely proportional to x (or y varies inversely as x) if there exists a nonzero number k such that

$$y = \frac{k}{x}, \quad \text{or} \quad y = kx^{-1}$$

Also, y is inversely proportional to the nth power of x if there exists a nonzero number k such that

$$y = \frac{k}{x^n}, \quad \text{or} \quad y = kx^{-n}$$

We can also say that y varies inversely as the nth power of x.

EXAMPLE 9 ▶ **Illumination**

The illumination produced by a light varies inversely as the square of the distance from the source of the light. If the illumination 30 feet from a light source is 60 candela, what is the illumination 20 feet from the source?

SOLUTION

If L represents the illumination and d represents the distance, the relation is

$$L = \frac{k}{d^2}$$

Substituting for L and d and solving for k gives

$$60 = \frac{k}{30^2} \Rightarrow k = 54{,}000$$

Thus, the relation is $L = \dfrac{54{,}000}{d^2}$, and when $d = 20$ feet,

$$L = \frac{54{,}000}{20^2} = 135 \text{ candela}$$

So the illumination 20 feet from the source of the light is 135 candela.

Skills CHECK 3.3

In Exercises 1–10, sketch the graph of each function using a window that gives a complete graph.

1. $y = x^3$

2. $y = x^4$

3. $y = x^{1/3}$

4. $y = x^{4/3}$

5. $y = \sqrt{x} + 2$

6. $y = \sqrt[3]{x} - 2$

7. $y = \dfrac{1}{x} - 3$

8. $y = 4 - \dfrac{1}{x}$

9. $y = \begin{cases} -1 & \text{if } x < 0 \\ 1 & \text{if } x \geq 0 \end{cases}$

10. $y = \begin{cases} 2 & \text{if } x \geq 2 \\ 6 & \text{if } x < 2 \end{cases}$

11. a. Graph the function $f(x) = \begin{cases} 5 & \text{if } 0 \leq x < 2 \\ 10 & \text{if } 2 \leq x < 4 \\ 15 & \text{if } 4 \leq x < 6 \\ 20 & \text{if } 6 \leq x < 8 \end{cases}$

b. What type of function is this?

12. a. Graph the function

$$f(x) = \begin{cases} 100 & \text{if } 0 \leq x < 20 \\ 200 & \text{if } 20 \leq x < 40 \\ 300 & \text{if } 40 \leq x < 60 \\ 400 & \text{if } 60 \leq x < 80 \end{cases}$$

b. What type of function is this?

13. a. Graph $f(x) = \begin{cases} 4x - 3 & \text{if } x \leq 3 \\ x^2 & \text{if } x > 3 \end{cases}$

b. Find $f(2)$ and $f(4)$.

c. State the domain of the function.

14. a. Graph $f(x) = \begin{cases} 3 - x & \text{if } x \leq 2 \\ x^2 & \text{if } x > 2 \end{cases}$

b. Find $f(2)$ and $f(3)$.

c. State the domain of the function.

15. One type of step function is the **greatest integer function**, denoted int(x) or $[x]$, where $[x] = $ the greatest integer that is less than or equal to x.

a. If $f(x) = [x]$, find $f(0.2)$, $f(3.8)$, $f(-2.6)$, and $f(5)$.

b. Graph $f(x) = [x]$ for domain $[-5, 5]$.

16. To graph $f(x) = [x]$ on a graphing calculator, use the function int(x), found in the Catalog. Graph this function for domain $[-5, 5]$ with your calculator in Dot mode.

17. a. Graph $f(x) = |x|$.

b. Find $f(-2)$ and $f(5)$.

c. State the domain of the function.

18. a. Graph $f(x) = |x - 4|$.

   b. Find $f(-2)$ and $f(5)$.

   c. State the domain of the function.

For each of the functions in Exercises 19–22, find the value of (a) $f(-1)$ and (b) $f(3)$, if possible.

19. $y = \begin{cases} 5 & \text{if } x \le 1 \\ 6 & \text{if } x > 1 \end{cases}$ **20.** $y = \begin{cases} -2 & \text{if } x < -1 \\ 4 & \text{if } x \ge -1 \end{cases}$

21. $y = \begin{cases} x^2 - 1 & \text{if } x \le 0 \\ x^3 + 2 & \text{if } x > 0 \end{cases}$

22. $y = \begin{cases} 3x + 1 & \text{if } x < 3 \\ x^2 & \text{if } x \ge 3 \end{cases}$

23. Determine if the function $y = 4x^3$ is increasing or decreasing for

   a. $x < 0$.   **b.** $x > 0$.

24. Determine if the function $y = -3x^4$ is increasing or decreasing for

   a. $x < 0$.   **b.** $x > 0$.

For each of the functions in Exercises 25–28, determine if the function is concave up or concave down in the first quadrant.

25. $y = x^{1/2}$ **26.** $y = x^{3/2}$

27. $y = x^{1.4}$ **28.** $y = x^{0.6}$

29. Graph $f(x) = \begin{cases} x & \text{if } x \ge 0 \\ -x & \text{if } x < 0 \end{cases}$

30. Graph $f(x) = \begin{cases} x - 4 & \text{if } x \ge 4 \\ 4 - x & \text{if } x < 4 \end{cases}$

31. Graph $f(x) = \begin{cases} x & \text{if } x < 0 \\ -x & \text{if } x \ge 0 \end{cases}$

32. Compare the graph in Exercise 29 with the graph in Exercise 17(a).

33. Compare the graph in Exercise 30 with the graph in Exercise 18(a).

In Exercises 34–38, solve the equations and check graphically.

34. $|2x - 5| = 3$ **35.** $\left|x - \dfrac{1}{2}\right| = 3$

36. $|x| = x^2 + 4x$ **37.** $|3x - 1| = 4x$

38. $|x - 5| = x^2 - 5x$

39. Suppose that S varies directly as the 2/3 power of T, and that $S = 64$ when $T = 64$. Find S when $T = 8$.

40. Suppose that y varies directly as the square root of x, and that $y = 16$ when $x = 4$. Find x when $y = 24$.

41. If y varies inversely as the 4th power of x and $y = 5$ when $x = -1$, what is y when $x = 0.5$?

42. If S varies inversely as the square root of T and $S = 4$ when $T = 4$, what is S when $T = 16$?

EXERCISES 3.3

43. *Electric Charges* For the nonextreme weather months, Palmetto Electric charges $7.10 plus 6.747 cents per kilowatt-hour (kWh) for customers using up to 1200 kWh, and charges $88.06 plus 5.788 cents per kWh above 1200 for customers using more than 1200 kWh.

   a. Write the function that gives the monthly charge in dollars as a function of the kilowatt-hours used.

   b. What is the monthly charge if 960 kWh are used?

   c. What is the monthly charge if 1580 kWh are used?

44. *Postal Rates* The table that follows gives the local postal rates as a function of the weight for media mail.

Write a step function that gives the postage P as a function of the weight in pounds x for $0 < x \le 5$.

Weight (lb)	Postal Rate ($)
$0 < x \le 1$	2.38
$1 < x \le 2$	2.77
$2 < x \le 3$	3.16
$3 < x \le 4$	3.55
$4 < x \le 5$	3.94

(Source: USPS)

45. *First-Class Postage* The postage charged for first-class mail is a function of its weight. The U.S. Postal Service uses the following table to describe the rates for 2011.

Weight Increment x (oz)	First-Class Postage P(x)
First ounce or fraction of an ounce	44¢
Each additional ounce or fraction	20¢

(Source: pe.usps.gov/text)

a. Convert this table to a piecewise-defined function that represents first-class postage for letters weighing up to 4 ounces, using x as the weight in ounces and P as the postage in cents.

b. Find $P(1.2)$ and explain what it means.

c. Give the domain of P as it is defined above.

d. Find $P(2)$ and $P(2.01)$.

e. Find the postage for a 2-ounce letter and for a 2.01-ounce letter.

46. *Income Tax* The 2010 U.S. federal income tax owed by a married couple filing jointly can be found from the following table.

Filing Status: Married Filing Jointly

If Taxable Income Is between	Tax Due Is	of the Amount over
$0–$16,750	10%	$0
$16,750–$68,000	$1675 + 15%	$16,750
$68,000–$137,300	$9362.50 + 25%	$68,000
$137,300–$209,250	$26,687.50 + 28%	$137,300
$209,250–$373,650	$46,833.50 + 33%	$209,250
$373,650–up	$101,085.50 + 35%	$373,650

(Source: Internal Revenue Service, 2010, Form 1040 Instructions)

a. Write the piecewise-defined function T with input x that models the federal tax dollars owed as a function of x, the taxable income dollars earned, with $0 < x \le 137,300$.

b. Use the function to find $T(42,000)$.

c. Find the tax owed on a taxable income of $65,000.

d. A friend tells Jack Waddell not to earn anything over $68,000 because it would raise his tax rate to 25% on all of his taxable income. Test this statement by finding the tax on $68,000 and $68,000 + $1. What do you conclude?

47. *Banks* The number of banks in the United States for the years 1935 through 2009 is given by

$$f(x) = \begin{cases} 84.3x + 12,365 & \text{if } x < 80 \\ -376.1x + 48,681 & \text{if } x \ge 80 \end{cases}$$

where x is the number of years after 1900.

a. What does this model give as the number of banks in 1970? 1990? 2010?

b. Graph the function for $50 \le x \le 112$.

48. *Wind Chill* The formula that gives the wind chill factor for a 60°F temperature and a wind with velocity V in miles per hour is

$$W = \begin{cases} 60 & \text{if } 0 \le V < 4 \\ 0.644V - 9.518\sqrt{V} + 76.495 & \text{if } 4 \le V \le 55.9 \\ 41 & \text{if } V > 55.9 \end{cases}$$

a. Find the wind chill factor for the 60° temperature if the wind is 20 mph.

b. Find the wind chill factor for the 60° temperature if the wind is 65 mph.

c. Graph the function for $0 \le V \le 80$.

d. What are the domain and range of the function graphed in part (c)?

49. *Female Physicians* Representation of females in medicine continues to show steady increases. The number of female physicians can be modeled by $F(x) = 0.623x^{1.552}$, where x is the number of years after 1960 and $F(x)$ is the number of female physicians in thousands.

a. What type of function is this?

b. What is $F(35)$? What does this mean?

c. How many female physicians will there be in 2020, according to the model?

(Source: American Medical Association)

50. *Taxi Miles* The Inner City Taxi Company estimated, on the basis of collected data, that the number of taxi miles driven each day can be modeled by the function $Q = 489L^{0.6}$, when they employ L drivers per day.

a. Graph this function for $0 \le L \le 35$.

b. How many taxi miles are driven each day if there are 32 drivers employed?

c. Does this model indicate that the number of taxi miles increases or decreases as the number of drivers increases? Is this reasonable?

51. *College Enrollment* The total percent of individuals aged 16 to 24 enrolled in college as of October of each year who completed high school during the preceding 12 months is $f(x) = 61.925x^{0.041}$, where x is the number of years after 1999.

a. Use this model to estimate the total percent in 2000 and 2008.

b. Graph this function for the years 2000 to 2015.

c. According to the model, will the percent enrolled in college ever reach 100%?
(Source: National Center for Education Statistics)

52. *U.S. Population* The U.S. population can be modeled by the function $y = 165.6x^{1.345}$, where y is in thousands and x is the number of years after 1800.

a. What was the population in 1960, according to this model?

b. Is the graph of this function concave up or concave down?

c. Use numerical or graphical methods to find when the model estimates the population to be 93,330,000.

53. *Personal Expenditures* Personal consumption expenditures for durable goods in the United States, in billions of dollars, can be modeled by the function $P(x) = 306.472x^{0.464}$, where x is the number of years after 1990.

a. Is this function increasing or decreasing for the years 1990–2010?

b. Is this function concave up or concave down during this period of time?

c. Use numerical or graphical methods to find when the model predicts that personal consumption expenditures will reach $1532.35 billion.
(Source: Bureau of Economic Analysis, U.S. Department of Commerce)

54. *Production Output* The monthly output of a product (in units) is given by $P = 1200x^{5/2}$, where x is the capital investment in thousands of dollars.

a. Graph this function for x from 0 to 10 and P from 0 to 200,000.

b. Is the graph concave up or concave down?

55. *Diabetes* The projected percent of the U.S. adult population with diabetes (diagnosed and undiagnosed) can be modeled by $y = 4.97x^{0.495}$, where x is the number of years after 2000.

a. Does this model indicate that the percent of the U.S. adult population with diabetes is projected to increase or decrease?

b. What percent is projected for 2022?

c. In what year does this model project the percent to be 17?

56. *Harvesting* A farmer's main cash crop is tomatoes, and the tomato harvest begins in the month of May. The number of bushels of tomatoes harvested on the xth day of May is given by the equation $B(x) = 6(x + 1)^{3/2}$. How many bushels did the farmer harvest on May 8?

57. *Purchasing Power* The purchasing power of a 1983 dollar is given by the function $y = 34.394x^{-1.1088}$, where x is the number of years after 1960.

a. Graph the function for $10 \leq x \leq 60$.

b. What is the projected purchasing power of a 1983 dollar in 2020?

58. *Trust in the Government* The percent of people who say they trust the government in Washington always or most of the time is given by $y = 154.131x^{-0.492}$, with x equal to the number of years after 1960.

a. Graph the function.

b. Does this model indicate that trust in the government is increasing or decreasing?

c. Use the graph to estimate the percent of people who trusted the government in 1998.
(Source: Pew Research Center)

59. *Body Weight* The weight of a body varies inversely as the square of its distance from the center of Earth. If the radius of Earth is 4000 miles, how much would a 180-pound man weigh 1000 miles above the surface of Earth?

60. *Concentration of Body Substances* The concentration C of a substance in the body depends on the quantity of substance Q and the volume V through which it is distributed. For a static substance, the concentration is given by

$$C = \frac{Q}{V}$$

a. For $Q = 1000$ milliliters (mL) of a substance, graph the concentration as a function of the volume on the interval from $V = 1000$ mL to $V = 5000$ mL.

b. For a fixed quantity of a substance, does the concentration of the substance in the body increase or decrease as the volume through which it is distributed increases?

61. *Investing* If money is invested for 3 years with interest compounded annually, the future value of the investment varies directly as the cube of $1 + r$, where r is the annual interest rate. If the future value of the investment is $6298.56 when the interest rate is 8%, what rate gives a future value of $5955.08?

62. *Investing* If money is invested for 4 years with interest compounded annually, the future value of the investment varies directly as the fourth power of $1 + r$, where r is the annual interest rate. If the future value of the investment is $17,569.20 when the interest rate is 10%, what rate gives a future value of $24,883.20?

63. *Investing* The present value that will give a future value S in 3 years with interest compounded annually

varies inversely as the cube of $1 + r$, where r is the annual interest rate. If the present value of $8396.19 gives a future value of $10,000, what would be the present value of $16,500?

64. *Investing* The present value that will give a future value S in 4 years with interest compounded annually varies inversely as the fourth power of $1 + r$, where r is the annual interest rate. If the present value of $1525.79 gives a future value of $2000, what would be the present value of $8000?

3.4 | Quadratic and Power Models

KEY OBJECTIVES

- Find the exact quadratic function that fits three points on a parabola

- Model data approximately using quadratic functions

- Model data using power functions

- Use first and second differences and visual comparison to determine if a linear or quadratic function is the better fit to a set of data

- Determine whether a quadratic or power function gives the better fit to a given set of data

SECTION PREVIEW Mobile Internet Advertising

Table 3.10 shows the dollar value or projected dollar value of the U.S. mobile Internet advertising market for the years from 2006 to 2012. The scatter plot of the data, in Figure 3.35, shows that a curve would fit the data better than a line, so a linear function is not a good model for the data. (See Example 3.) Graphing calculators and Excel permit us to model nonlinear data with other types of functions by using steps similar to those that we used to model data with linear functions. In this section, we model data with quadratic and power functions.

Table 3.10

Year	Market (billions of $)
2006	0.045
2007	0.114
2008	0.273
2009	0.409
2010	0.545
2011	0.791
2012	1.023

(Source: *Wall Street Journal*, August 16, 2010)

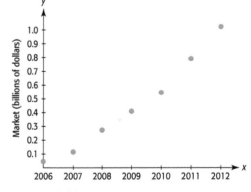

Figure 3.35

Modeling a Quadratic Function from Three Points on Its Graph

If we know three (or more) points that fit exactly on a parabola, we can find the quadratic function whose graph is the parabola.

EXAMPLE 1 ▶   Writing the Equation of a Quadratic Function

A parabola passes through the points $(0, 5)$, $(4, 13)$, and $(-2, 25)$. Write the equation of the quadratic function whose graph is this parabola.

SOLUTION

The general form of a quadratic function is $y = ax^2 + bx + c$. Because $(0, 5)$ is on the graph of the parabola, we can substitute 0 for x and 5 for y in $y = ax^2 + bx + c$, which gives $5 = a(0)^2 + b(0) + c$, so $c = 5$. Substituting the values of x and y for the other two points on the parabola and $c = 5$ in $y = ax^2 + bx + c$ gives

$$\begin{cases} 13 = a \cdot 4^2 + b \cdot 4 + 5 \\ 25 = a \cdot (-2)^2 + b \cdot (-2) + 5 \end{cases} \quad \text{or} \quad \begin{cases} 8 = 16a + 4b \\ 20 = 4a - 2b \end{cases}$$

Multiplying both sides of the bottom equation by 2 and adding the equations gives

$$\begin{cases} 8 = 16a + 4b \\ 40 = 8a - 4b \end{cases}$$
$$\overline{48 = 24a \quad \Rightarrow a = 2}$$

Substituting $a = 2$ in $8 = 16a + 4b$ gives $8 = 32 + 4b$, so $b = -6$. Thus, $a = 2$, $b = -6$, and $c = 2$, and the equation of this parabola is

$$y = 2x^2 - 6x + 5$$

EXAMPLE 2 ▶   Equation of a Quadratic Function

Find the equation of the quadratic function whose graph is a parabola containing the points $(-1, 9)$, $(2, 6)$, and $(3, 13)$.

SOLUTION

Using the three points $(-1, 9)$, $(2, 6)$, and $(3, 13)$, we substitute the values for x and y in the general equation $y = ax^2 + bx + c$, getting three equations.

$$\begin{cases} 9 = a(-1)^2 + b(-1) + c \\ 6 = a(2)^2 + b(2) + c \\ 13 = a(3)^2 + b(3) + c \end{cases} \quad \text{or} \quad \begin{cases} 9 = a - b + c \\ 6 = 4a + 2b + c \\ 13 = 9a + 3b + c \end{cases}$$

We use these three equations to solve for a, b, and c, using techniques similar to those used to solve two equations in two variables.*

Subtracting the third equation, $13 = 9a + 3b + c$, from each of the first and second equations gives a system of two equations in two variables.

$$\begin{cases} -4 = -8a - 4b \\ -7 = -5a - b \end{cases}$$

Multiplying the second equation by -4 and adding the equations gives

$$\begin{cases} -4 = -8a - 4b \\ 28 = 20a + 4b \end{cases}$$
$$\overline{24 = 12a \quad \Rightarrow a = 2}$$

---

* We will discuss solution of systems of three equations in three variables further in Chapter 7.

Substituting $a = 2$ in $-4 = -8a - 4b$ gives

$$-4 = -16 - 4b, \quad \text{or} \quad b = -3$$

and substituting $a = 2$ and $b = -3$ in the original first equation, $9 = a - b + c$, gives

$$9 = 2 - (-3) + c, \quad \text{or} \quad c = 4$$

Thus, $a = 2$, $b = -3$, and $c = 4$, and the quadratic function whose graph contains the points is

$$y = 2x^2 - 3x + 4$$

Modeling with Quadratic Functions

If the graph of a set of data has a pattern that approximates the shape of a parabola or part of a parabola, a quadratic function may be appropriate to model the data. In Example 3, we will use technology to model the dollar value of the U.S. mobile advertising market as a function of the number of years after 2000.

EXAMPLE 3 ▶ Mobile Internet Advertising

Table 3.11 shows the dollar value or projected dollar value of the U.S. mobile Internet advertising market for the years from 2006 to 2012.

Table 3.11

Year	Advertising Market (billions of $)	Year	Advertising Market (billions of $)
2006	0.045	2010	0.545
2007	0.114	2011	0.791
2008	0.273	2012	1.023
2009	0.409		

(Source: *Wall Street Journal*, August 16, 2010)

a. Create a scatter plot of the data, with x equal to the number of years after 2000.

b. Find the quadratic function that is the best fit for the data, with x equal to the number of years after 2000.

c. Graph the aligned data and the model on the same axes. Does this model seem like a reasonable fit?

d. Use the model to find the year in which the mobile Internet advertising market is projected to reach $3 billion.

SOLUTION

a. The scatter plot of the data is shown in Figure 3.36(a). The shape looks as though it could be part of a parabola, so it is reasonable to find a quadratic function using the data points.

b. Enter the aligned inputs 6 through 12 in one list of a graphing calculator and the corresponding outputs in a second list. Using quadratic regression (select STAT, CALC, 5: QuadReg) gives a quadratic function that models the data. The function, with the coefficients rounded to three decimal places, is $y = 0.0149x^2 - 0.105x + 0.137$, with $x = 0$ for 2000.

c. The scatter plot of the aligned data and the graph of the function are shown in Figure 3.36(b). The model is a good fit for the data.

d. Intersecting the (unrounded) model with $y = 3$ gives $x = 17.84$, so the mobile Internet advertising market should reach $3 billion during 2018 (Figure 3.36(c)).

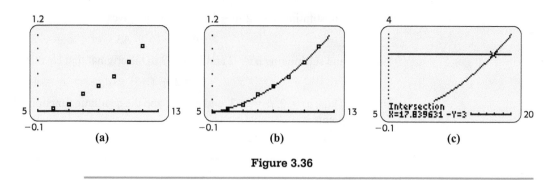

(a)   **(b)**   **(c)**

Figure 3.36

EXAMPLE 4 ▶   Starbucks Stores

Table 3.12 gives the number of Starbucks stores in the United States for the years 1992 through 2009.

Table 3.12

Year	Starbucks Stores	Year	Starbucks Stores
1992	113	2001	2925
1993	163	2002	3756
1994	264	2003	4453
1995	430	2004	5452
1996	663	2005	7353
1997	974	2006	8896
1998	1321	2007	10,684
1999	1657	2008	11,567
2000	2119	2009	11,128

a. Create a scatter plot of the data points, with x equal to the number of years after 1990.

b. Create a quadratic function that models the data, using the number of years after 1990 as the input x.

c. Graph the aligned data and the quadratic function on the same axes. Does this model seem like a reasonable fit?

d. Use the model to estimate the number of stores in 2008. Is the estimate close to the actual number?

e. Use the model to estimate the number of stores in 2012. Discuss the reliability of this estimate.

SOLUTION

a. Figure 3.37(a) shows a scatter plot of the data. The shape looks as though it could be part of a parabola, so it is reasonable to find a quadratic function using the data points.

b. Enter the aligned inputs 2 through 19 in one list of a graphing calculator and the corresponding outputs in a second list. Using quadratic regression on a graphing calculator gives a quadratic function that models the data. The function, with the coefficients rounded to three decimal places, is

$$y = 50.076x^2 - 327.938x + 681.179$$

where x is the number of years after 1990. Note that we will use the unrounded model in graphing and performing calculations.

c. The scatter plot of the aligned data and the graph of the function are shown in Figure 3.37(b). The function appears to be an excellent fit to the data.

d. Evaluating the function at $x = 18$ gives the number of U.S. stores in 2008 as 11,003 (Figure 3.37(c)). This is a reasonable estimate of the actual number of stores, which is 11,567.

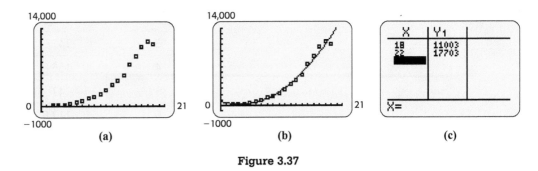

(a)   (b)   (c)

Figure 3.37

e. Figure 3.37(c) also gives the prediction for 2012, which is 17,703. The number of stores fell in 2009, and it seems unlikely it will reach 17,703 in 2012.

Spreadsheet ▸ SOLUTION We can use graphing utilities, software programs, and spreadsheets to find the quadratic function that is the best fit for data. Table 3.13 shows a partial Excel spreadsheet for the aligned data of Example 4. Selecting the cells containing the data, getting the scatter plot of the data, selecting Add Trendline, and picking Polynomial with order 2 gives the equation of the quadratic function that is the best fit for the data, along with the scatter plot and the graph of the best-fitting parabola (see Figure 3.38).*

Table 3.13

	A	B
	Year x	Starbucks Stores y
1		
2	2	113
3	3	163
4	4	264
5	5	430
6	6	663
7	7	974
8	8	1321

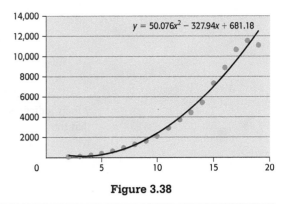

$$y = 50.076x^2 - 327.94x + 681.18$$

Figure 3.38

Comparison of Linear and Quadratic Models

Recall that when the changes in inputs are constant and the (first) differences of the outputs are constant or nearly constant, a linear model will give a good fit for the data. In a similar manner, we can compare the differences of the first differences, which are

* See Appendix B, page 643.

called the **second differences**. If the second differences are constant for equally spaced inputs, the data can be modeled exactly by a quadratic function.

Consider the data in Table 3.14, which gives the measured height y of a toy rocket x seconds after it has been shot into the air from the ground.

Table 3.14   Height of a Rocket

Time x (seconds)	Height (meters)
1	68.6
2	117.6
3	147
4	156.8
5	147
6	117.6
7	68.6

Table 3.15 gives the first differences and second differences for the equally spaced inputs for the rocket height data in Table 3.14. Excel is especially useful for finding first and second differences.

Table 3.15

	A	B	C	D
1	Time	Height	1st Differences	2nd Differences
2	1	68.6		
3	2	117.6	49	
4	3	147	29.4	−19.6
5	4	156.8	9.8	−19.6
6	5	147	−9.8	−19.6
7	6	117.6	−29.4	−19.6
8	7	68.6	−49	−19.6

The first differences are not constant, but each of the second differences is −19.6, which indicates that the data can be fit exactly by a quadratic function. We can find that the quadratic model for the height of the toy rocket is $y = 78.4x - 9.8x^2$ meters, where x is the time in seconds.

Modeling with Power Functions

We can model some experimental data by observing patterns. For example, by observing how the area of each of the (square) faces of a cube is found and that a cube has six faces, we can deduce that the surface area of a cube that is x units on each edge is

$$S = 6x^2 \text{ square units}$$

We can also measure and record the surface areas for cubes of different sizes to investigate the relationship between the edge length and the surface area for cubes. Table 3.16 contains selected measures of edges and the resulting surface areas.

Table 3.16

Edge Length x (units)	Surface Area of Cube (square units)
1	6
2	24
3	54
4	96
5	150

We can enter the lengths from the table as the independent (x) variable and the corresponding surface areas as the dependent (y) variable in a graphing utility, and then have the utility create the **power function** that is the best model for the data (Figure 3.39). This model also has the equation $y = 6x^2$ square units.

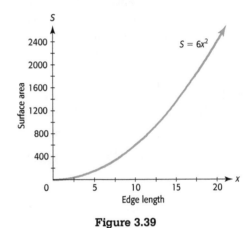

Figure 3.39

EXAMPLE 5 ▶   **Diabetes**

Figure 3.40 and Table 3.17 show that the percent of the U.S. adult population with diabetes (diagnosed and undiagnosed) is projected to grow rapidly in the future.

a. Find the power model that fits the data, with x equal to the number of years after 2000.

b. Use the model to predict the percent of U.S. adults with diabetes in 2015.

c. In what year does this model predict the percent to be 29.6?

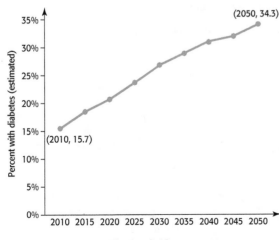

Figure 3.40

Table 3.17

Year	Percent	Year	Percent	Year	Percent
2010	15.7	2025	24.2	2040	31.4
2015	18.9	2030	27.2	2045	32.1
2020	21.1	2035	29.0	2050	34.3

(Source: Centers for Disease Control and Prevention)

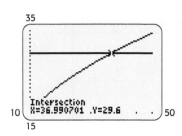

Figure 3.41

SOLUTION

a. The power model that fits the data is $f(x) = 4.947x^{0.495}$, with x equal to the number of years after 2000.

b. Evaluating the unrounded model at 15 gives the percent of U.S. adults with diabetes in 2015 to be

$$f(15) = 18.9\%$$

c. By intersecting the graphs of the unrounded model and $y = 29.6$ (Figure 3.41), we find the percent to be 29.6 when $x = 36.99$, or during 2037.

EXAMPLE 6 ▶ Cohabiting Households

The data in Table 3.18 give the number of cohabiting (without marriage) households (in thousands) for selected years from 1960 to 2008. The scatter plot of the data, with x representing the number of years after 1950 and y representing thousands of households, is shown in Figure 3.42.

Table 3.18 Cohabiting Households

Year	Cohabiting Households (thousands)	Year	Cohabiting Households (thousands)
1960	439	1995	3668
1970	523	2002	4898
1980	1589	2004	5080
1985	1983	2007	6209
1990	2856	2008	6214

(Source: Index of Leading Cultural Indicators)

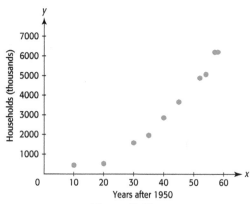

Figure 3.42

a. Find the power function that models the data.

b. Graph the data and the model on the same axes.

SOLUTION

a. Using power regression in a graphing utility gives the power function that models the data. The power function that is the best fit is

$$y = 6.057x^{1.675}$$

where y is in thousands and x is the number of years after 1950.

b. The graphs of the data and the power function that models it are shown in Figure 3.43.

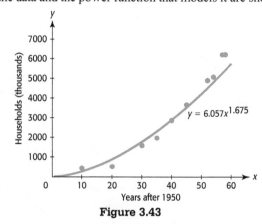

Figure 3.43

Power models can also be found with Excel. The Excel spreadsheet in Table 3.19 shows the number of cohabiting households (in thousands) for 1960–2008, listed as years after 1950. Selecting the cells containing the data, getting the scatter plot of the data, selecting Add Trendline, and picking Power gives the equation of the power function that is the best fit for the data, along with the scatter plot and the graph of the best-fitting power function. The equation and graph of the function are shown in Figure 3.44.

Table 3.19

	A	B
1	Years after 1950	Households (thousands)
2	10	439
3	20	523
4	30	1589
5	35	1983
6	40	2856
7	45	3668
8	52	4898
9	54	5080
10	57	6209
11	58	6210

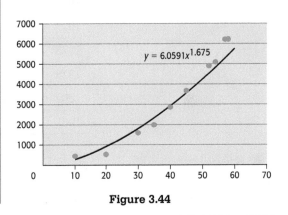

$$y = 6.0591x^{1.675}$$

Figure 3.44

Comparison of Power and Quadratic Models

We found a power function that is a good fit for the data in Table 3.19, but a linear function or a quadratic function may also be a good fit for the data. A quadratic function may fit data points even if there is no obvious turning point in the graph of the data points. If the data points appear to rise (or fall) more rapidly than a line, then a quadratic model or a power model may fit the data well. In some cases, it may be necessary to find both models to determine which is the better fit for the data.

EXAMPLE 7 ▶   Purchasing Power

Inflation causes a decrease in the value of money used to purchase goods and services. Table 3.20 gives the purchasing power of a 1983 dollar based on consumer prices for 1968–2010. Note that a 2006 dollar will purchase half as much as a 1983 dollar; that is, a given good or service costs twice as much in 2006.

a. Find a quadratic function that models the data if x is the number of years after 1960 and y is the purchasing power.

b. Find a power function that models the data if x is the number of years after 1960 and y is the purchasing power.

c. Graph each model with the aligned data points and discuss which model is the better fit.

d. Use the power model to predict the purchasing power of a 1983 dollar in 2055.

e. Find when the purchasing power is $0.40, according to the power model.

Table 3.20

Year	Purchasing Power	Year	Purchasing Power	Year	Purchasing Power
1968	2.873	1982	1.035	1996	0.637
1969	2.726	1983	1.003	1997	0.623
1970	2.574	1984	0.961	1998	0.613
1971	2.466	1985	0.928	1999	0.600
1972	2.391	1986	0.913	2000	0.581
1973	2.251	1987	0.88	2001	0.565
1974	2.029	1988	0.846	2002	0.556
1975	1.859	1989	0.807	2003	0.543
1976	1.757	1990	0.766	2004	0.529
1977	1.649	1991	0.734	2005	0.512
1978	1.532	1992	0.713	2006	0.500
1979	1.38	1993	0.692	2007	0.482
1980	1.215	1994	0.675	2008	0.464
1981	1.098	1995	0.656	2009	0.463
				2010	0.456

(Source: Bureau of Labor Statistics)

SOLUTION

a. The quadratic function that is the best fit is $y = 0.00211x^2 - 0.173x + 4.030$.

b. The power function that is the best fit is $y = 34.394x^{-1.109}$.

c. The graphs are shown in Figure 3.45. The graph of the power model in part (a) appears to be a better fit than the graph of the quadratic model in part (b). The quadratic model begins to rise after 2001, while the actual purchasing power continues to fall.

d. The value of y is 0.221 when x is 95, so the purchasing power of a 1983 dollar is predicted to be 22 cents in 2055.

e. The value of x that gives 0.40 for y is 55.55, so the purchasing power of a 1983 dollar is predicted to be 40 cents in 2016.

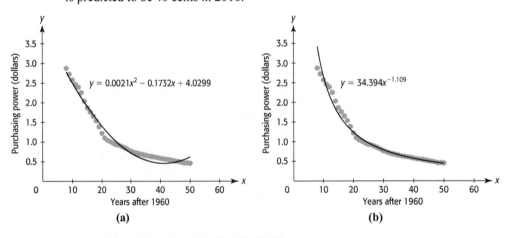

(a)   (b)

Figure 3.45

Skills CHECK 3.4

In Exercises 1–6, write the equation of the quadratic function whose graph is a parabola containing the given points.

1. $(0, 1)$, $(3, 10)$, and $(−2, 15)$

2. $(0, −3)$, $(4, 37)$, and $(−3, 30)$

3. $(6, 30)$, $(0, −3)$, and $(−3, 7.5)$

4. $(6, −22)$, $(−3, 23)$, and $(0, 2)$

5. $(0, 6)$, $\left(2, \frac{22}{3}\right)$, and $\left(−9, \frac{99}{2}\right)$

6. $(0, 7)$, $(2, 8.5)$, and $(−3, 12.25)$

7. A ball is thrown upward from the top of a 48-foot-high building. The ball is 64 feet above ground level after 1 second, and it reaches ground level in 3 seconds. The height above ground is a quadratic function of the time after the ball is thrown. Write the equation of this function.

8. A ball is dropped from the top of a 256-foot-high building. The ball is 192 feet above ground after 2 seconds, and it reaches ground level in 4 seconds. The height above ground is a quadratic function of the time after the ball is thrown. Write the equation of this function.

9. Find the equation of the quadratic function whose graph is a parabola containing the points $(−1, 6)$, $(2, 3)$, and $(3, 10)$.

10. Find the equation of the quadratic function whose graph is a parabola containing the points $(−2, −4)$, $(3, 1)$, and $(2, 4)$.

11. Find the quadratic function that models the data in the table below.

x	−2	−1	0	1	2	3	4
y	16	5	0	1	8	21	40

x	5	6	7	8	9	10
y	65	96	133	176	225	280

12. The following table has the inputs, x, and the outputs for three functions, f, g, and h. Use second differences to determine which function is exactly quadratic, which is approximately quadratic, and which is not quadratic.

x	f(x)	g(x)	h(x)
0	0	2	0
2	399	0.8	110
4	1601	1.2	300
6	3600	3.2	195
8	6402	6.8	230
10	9998	12	290

13. As you can verify, the following data points give constant second differences, but the points do not fit on the graph of a quadratic function. How can this be?

x	1	2	4	8	16	32	64
y	1	6	15	28	45	66	91

14. **a.** Make a scatter plot of the data in the table below.

    b. Does it appear that a quadratic model or a power model is the better fit for the data?

x	y
1	4
2	9
3	11
4	21
5	32
6	45

15. Find the quadratic function that is the best fit for $f(x)$ defined by the table in Exercise 12.

16. Find the quadratic function that is the best fit for $g(x)$ defined by the table in Exercise 12.

17. **a.** Find a power function that models the data in the table in Exercise 14.

    b. Find a linear function that models the data.

    c. Visually determine which function is the better fit for the data.

18. **a.** Make a scatter plot of the data in the table below.

    b. Does it appear that a linear model or a power model is the better fit for the data?

x	y
3	4.7
5	8.6
7	13
9	17

19. **a.** Find a power function that models the data in the table in Exercise 18.

    b. Find a linear function that models the data.

    c. Visually determine if each model is a good fit.

20. Find the quadratic function that models the data in the tables that follow.

x	−2	−1	0	1	2	3
y	15	5	2	1	3	10

x	4	5	6	7	8
y	20	35	55	75	176

21. Find the power function that models the data in the table below.

x	1	2	3	4	5	6	7	8
y	3	4.5	5.8	7	8	9	10	10.5

22. Is a power function or a quadratic function the better model for the data below?

x	0.5	1	2	3	4	5	6
y	1	7	17	32	49	70	90

EXERCISES 3.4

Calculate numerical results with the unrounded models, unless otherwise instructed. Report models to 3 decimal places unless otherwise stated.

23. *Income* The median annual income of males in the United States for selected years between 1960 and 2009 is shown in the table below.

Year	Median Annual Income ($)	Year	Median Annual Income ($)
1960	4080	2003	29,931
1970	6670	2004	30,513
1980	12,530	2005	31,275
1990	20,293	2007	33,196
2000	28,343	2008	33,161
2001	29,101	2009	32,184
2002	29,238		

(Source: U.S. Census Bureau)

    a. Find the quadratic function that models the median income as a function of x, the number of years after 1960.

    b. Use the model from part (a) to estimate the median annual income of males in 1997 and in 2014.

    c. Do you feel that the estimate for 2014 is reliable? Explain.

24. *National Health Care* The table shows national expenditures for health care in the United States for selected years, with projections to 2015.

    a. Use a scatter plot with x as the number of years after 1950 and y as the total expenditures for health care (in billions) to identify what type (or types) of function would make a good model for these data.

    b. Find a power model and a quadratic model for the data.

    c. Which model from (b) more accurately estimates the 2010 expenditures for national health care?

    d. Use the better model from (c) to estimate the 2020 expenditures for national health care.

Year	National Expenditures for Health Care (in billions)
1960	$28
1970	75
1980	255
1990	717
1995	1020
2000	1359
2005	2016
2010	2879
2015	4032

(Source: U.S. Centers for Medicare and Medicaid Services)

25. *Age and Income* The 2008 median annual income of householders for certain average ages is given in the table.

   a. Find the quadratic function that models the annual income as a function of the average age.

   b. Does the model appear to be a good fit?

   c. Find and interpret the vertex of the graph of the *rounded* function.

Average Age (yr)	Median Income ($)	Average Age (yr)	Median Income ($)
19.5	32,270	49.5	64,349
29.5	51,400	59.5	57,265
39.5	62,954	69.5	29,744

(Source: *Statistical Abstract of the United States*)

26. *Unemployment* The percent of unemployment in the United States for the years 2004–2010 is given by the data in the table below.

   a. Create a scatter plot for the data, with x equal to the number of years after 2004.

   b. Does it appear that a quadratic model will fit the data? If so, find the best-fitting quadratic model.

   c. Does the y-intercept of the function in part (b) have meaning in the context of this problem? If so, interpret the value.

Year	Percent Unemployment	Year	Percent Unemployment
2004	5.4	2008	7.3
2005	4.9	2009	9.9
2006	4.4	2010	9.4
2007	5.0		

(Source: Bureau of Labor Statistics)

27. *Wind Chill* The table gives the wind chill temperature when the outside temperature is 20°F.

Wind (mph)	Wind Chill (°F)	Wind (mph)	Wind Chill (°F)
5	13	35	0
10	9	40	−1
15	6	45	−2
20	4	50	−3
25	3	55	−3
30	1	60	−4

(Source: National Weather Service)

   a. Use x as the wind speed and create a quadratic model for the data.

   b. At what wind speed does the model estimate that the wind chill temperature will be −3°F?

   c. Do you think the model found in part (a) is valid for $x > 60$? Explain.

28. *Foreign-Born Population* The following table gives the percent of the U.S. population that is foreign born.

   a. Create a scatter plot for the data, with x equal to the number of years after 1900 and y equal to the percent.

   b. Does it appear that the data could be modeled with a quadratic function?

   c. Find the best-fitting quadratic function for the data. Report your answer to 4 decimal places.

   d. Use the function to estimate the percent in 2015.

Year	Foreign Born (%)	Year	Foreign Born (%)
1900	13.6	1970	4.7
1910	14.7	1980	6.2
1920	13.2	1990	8.0
1930	11.6	2000	10.4
1940	8.8	2005	11.7
1950	6.9	2007	12.6
1960	5.4	2008	12.2

(Source: U.S. Census Bureau)

29. *Super Bowl Ads* A one-minute ad during Super Bowl VII in 1973 cost $200,000. The price tag for a 30-second ad during the 2011 Super Bowl was $3 million. The table below gives the cost for a 30-second ad during Super Bowls for selected years from the first Super Bowl in 1967 to 2011.

Year	Cost ($)	Year	Cost ($)
1967	42,500	2003	2,100,000
1973	100,000	2004	2,250,000
1990	700,000	2005	2,400,000
1992	800,000	2006	2,500,000
1994	900,000	2007	2,385,000
1996	1,100,000	2008	2,700,000
1998	1,300,000	2009	3,000,000
2000	2,100,000	2010	2,974,000
2001	2,050,000	2011	3,000,000
2002	1,900,000		

(Source: Nielsen Monitor-Plus)

a. Create a scatter plot of the data, with x equal to the number of years after 1960 and y equal to the number of millions of dollars.

b. Does it appear that the data could be modeled by a quadratic function?

c. Find the best-fitting quadratic function for the data.

d. According to the model, when will the cost for a 30-second ad be $4,000,000?

30. *U.S. Population* The table gives the U.S. population, in millions, for selected years, with projections to 2050.

a. Create a scatter plot for the data in the table, with x equal to the number of years after 1960.

b. Use the scatter plot to determine the type of function that can be used to model the data, and create a function that best fits the data, with x equal to the number of years after 1960.

Year	U.S. Population (millions)	Year	U.S. Population (millions)
1960	180.671	1995	263.044
1965	194.303	1998	270.561
1970	205.052	2000	281.422
1975	215.973	2003	294.043
1980	227.726	2025	358.030
1985	238.466	2050	408.695
1990	249.948		

(Source: U.S. Census Bureau)

31. *Violent Crime* The following table gives the rate of violent crimes (per 100,000 residents) in the United States, as a function of the year x.

a. Find the power model that is the best fit for the data, with x equal to the number of years after 1999.

b. What rate of violent crimes did the model predict for 2012?

Year	Violent Crimes per 100,000	Year	Violent Crimes per 100,000
2000	506.5	2004	463.2
2001	504.5	2005	469.2
2002	494.4	2006	473.6
2003	475.8	2007	466.9

(Source: U.S. Census Bureau)

32. *Volume* The measured volume of a pyramid with each edge of the base equal to x units and with its altitude (height) equal to x units is given in the table below.

a. Determine if the second differences of the outputs are constant.

b. If the answer is yes, find the quadratic model that is the best fit for the data. Otherwise, find the power function that is the best fit.

Edge Length x (units)	Volume of Pyramid (cubic units)
1	1/3
2	8/3
3	9
4	64/3
5	125/3
6	72

33. *Cell Phones* The table gives the number, in millions, of U.S. cellular telephone subscribers.

a. Create a scatter plot for the data, with x equal to the number of years after 1985. Does it appear that the data could be modeled with a quadratic function?

b. Find the quadratic function that is the best fit for these data, with x equal to the number of years after 1985 and y equal to the number of subscribers in millions.

c. Use the model to estimate the number in 2011.

d. What part of the U.S. population does this estimate equal if the U.S. population is 310 million?

Year	Subscribers (millions)	Year	Subscribers (millions)
1985	0.340	1998	69.209
1986	0.682	1999	86.047
1987	1.231	2000	109.478
1988	2.069	2001	128.375
1989	3.509	2002	140.767
1990	5.283	2003	158.722
1991	7.557	2004	182.140
1992	11.033	2005	207.896
1993	16.009	2006	233.041
1994	24.134	2007	255.396
1995	33.786	2008	270.334
1996	44.043	2009	285.694
1997	55.312		

(Source: Semiannual CTIA Wireless Industry Survey)

34. *World Population* One projection of the world population by the United Nations for selected years (a
low-projection scenario) is given in the table below.

Year	Projected Population (millions)	Year	Projected Population (millions)
1995	5666	2075	6402
2000	6028	2100	5153
2025	7275	2125	4074
2050	7343	2150	3236

(Source: *World Population Prospects*, United Nations)

a. Find a quadratic function that fits the data, using
the number of years after 1990 as the input.

b. Find the positive x-intercept of this graph, to the
nearest year.

c. When can we be certain that this model no longer
applies?

35. *Personal Savings* The following table gives Americans'
personal savings rate for selected years from 1980 to
2009.

a. Find a quadratic function that models the personal
savings rate as a function of the number of years
after 1960.

b. Find and interpret the vertex of the graph of the
rounded model from part (a).

c. When after 1990 will the personal savings rate
reach 6%?

Year	Personal Savings Rate (%)
1980	9.4
1990	6.0
2000	3.4
2002	3.26
2004	3.23
2006	3.39
2008	3.76
2009	4.03

(Source: Bureau of Economic Analysis)

36. *Mortgages* The balance owed y on a $50,000 mort-
gage after x monthly payments is shown in the table
that follows. Graph the data points with each of the

equations below to determine which is the better
model for the data, if x is the number of months that
payments have been made.

a. $y = 338{,}111.278x^{-0.676}$

b. $y = 4700\sqrt{110 - x}$

Monthly Payments	Balance Owed ($)
12	47,243
24	44,136
48	36,693
72	27,241
96	15,239
108	8074

37. *International Visitors* The number of international
visitors to the United States for selected years 1986–
2010 is given in the table below.

Year	U.S. Visitors (millions)	Year	U.S. Visitors (millions)
1986	26	2000	50.9
1987	29.5	2001	44.9
1988	34.1	2002	41.9
1989	36.6	2003	41.2
1990	39.5	2004	46.1
1991	43	2005	49.4
1992	47.3	2006	51.0
1994	45.5	2007	56.0
1995	44	2008	57.9
1996	46.3	2010	55
1997	48.9		

(Source: Office of Travel and Tourism, U.S. Department of Commerce)

a. Using an input equal to the number of years after
1980, graph the aligned data points and both of
the given equations to determine which is the better model for the aligned data.

i. $y = 17\sqrt[3]{x}$

ii. $y = -0.031x^2 + 2x + 21$

b. If you had to pick one of these models to predict
the number of international visitors in the year
2020, which model would be the more reasonable
choice?

38. *Consumer Price Index* The effect of inflation is described by the Consumer Price Index, which tells how much it costs to buy an item in a given year if it cost $1 in an earlier year. The following table gives the prices of all goods and services for urban households, based on a price of $1 for them in 1913.

a. Find the quadratic function that is the best fit for the data, with x equal to the number of years after 1900.

b. The model gives the minimum index in what year?

c. What event in history explains why the index is lower after 1920?

Year	Amount It Took to Equal $1 in 1913	Year	Amount It Took to Equal $1 in 1913
1920	2.02	1970	3.92
1925	1.77	1975	5.43
1930	1.69	1980	8.32
1935	1.38	1985	10.87
1940	1.41	1990	13.20
1945	1.82	1995	15.39
1950	2.43	2000	17.39
1955	2.71	2005	19.73
1960	2.99	2006	20.18
1965	3.18		

39. *Modeling Personal Income* Total personal income in the United States (in billions of dollars) for selected years from 1960 to 2009 is given in the following table.

Year	Personal Income (billions of $)	Year	Personal Income (billions of $)
1960	411.3	2003	9378.1
1970	838.6	2004	9937.2
1980	2301.5	2005	10,485.9
1990	4846.7	2006	11,268.1
1995	6200.9	2007	11,912.3
2000	8559.7	2008	12,391.1
2001	8883.3	2009	12,174.9
2002	9060.1		

(Source: Bureau of Economic Analysis, U.S. Department of Commerce)

a. These data can be modeled by a power function. Write the equation of this function, with x as the number of years after 1950.

b. What does this model predict for total U.S. personal income in 2012?

c. Find the quadratic function that is the best fit for the data, with x as the number of years after 1950.

d. Which model is the better fit for the data?

40. *Poverty* The table shows the number of millions of people in the United States who lived below the poverty level for selected years.

a. Find a quadratic model that approximately fits the data, using x as the number of years after 1990.

b. Use the model to predict the number of people living below the poverty level in 2013.

Year	People Living Below the Poverty Level (millions)	Year	People Living Below the Poverty Level (millions)
1993	39.3	2002	34.6
1994	38.1	2003	35.9
1995	36.4	2004	37.0
1996	36.5	2005	37.0
1997	35.6	2006	36.5
1998	34.5	2007	37.2
1999	32.3	2008	39.8
2000	31.1	2009	43.6

(Source: U.S. Bureau of the Census, U.S. Department of Commerce)

41. *Income Tax per Capita* The U.S. federal income tax per civilian is given in the table.

Year	Tax per Capita ($)	Year	Tax per Capita ($)
1960	252.31	1999	3607.05
1965	280.06	2000	4047.59
1970	513.40	2001	4151.49
1975	731.56	2002	3622.61
1980	1274.47	2003	3416.05
1985	1679.20	2004	3395.55
1990	2178.49	2005	3762.28
1995	2550.81	2006	4159.93
1997	3039.94	2007	4552.09

(Source: U.S. Census Bureau; IRS)

a. Use x as the number of years after 1960 and write the equation of the function that is the best fit for the data.

b. In what year does the model predict the tax per capita will reach $7500?

42. *Medicare Trust Fund Balance* The year 1994 marked the 30th anniversary of Medicare. A 1994 pamphlet by Representative Lindsey O. Graham gave projections for the Medicare Trust Fund balance, shown in the table below.

a. Representative Graham stated, "The fact of the matter is that Medicare is going broke." Use the data in the table to find when he predicted that this would happen.

b. Find a quadratic model to fit the data, with x equal to the number of years after 1990. When did this model predict that the Medicare Trust Fund balance would be 0?

c. The pamphlet reported the Medicare trustees as saying "the present financing schedule for the hospital insurance program is sufficient to ensure the payment of benefits only over the next 7 years." Does the function in part (b) confirm or refute the value 7 in this statement?

d. Find and interpret the vertex of the quadratic function in part (b).

Year	Medicare Trust Fund Balance (billions of $)	Year	Medicare Trust Fund Balance (billions of $)
1993	128	1999	98
1994	133	2000	72
1995	136	2001	37
1996	135	2002	−7
1997	129	2003	−61
1998	117	2004	−126

43. *Travel and Tourism Spending* Global spending on travel and tourism (in billions of dollars) for the years 1991–2009 is given in the table.

a. Write the equation of a power function that models the data, letting your input represent the number of years after 1990.

b. Find the best-fitting quadratic model for the data, with $x = 0$ in 1990.

c. Compare the two models by graphing each model on the same axes as the data points. Which model appears to be the best fit?

Year	Spending (billions of $)	Year	Spending (billions of $)
1991	278	2001	472
1992	317	2002	487
1993	323	2003	533
1994	356	2004	634
1995	413	2005	679
1996	439	2006	744
1997	443	2007	859
1998	445	2008	942
1999	455	2009	852
2000	483		

(Source: *World Almanac*)

44. *Insurance Rates* The following table gives the monthly insurance rates for a $100,000 life insurance policy for smokers 35–50 years of age.

a. Create a scatter plot for the data.

b. Does it appear that a quadratic function can be used to model the data? If so, find the best-fitting quadratic model.

c. Find the power model that is the best fit for the data.

d. Compare the two models by graphing each model on the same axes with the data points. Which model appears to be the better fit?

Age (yr)	Monthly Insurance Rate ($)	Age (yr)	Monthly Insurance Rate ($)
35	17.32	43	23.71
36	17.67	44	25.11
37	18.02	45	26.60
38	18.46	46	28.00
39	19.07	47	29.40
40	19.95	48	30.80
41	21.00	49	32.55
42	22.22	50	34.47

(Source: American General Life Insurance Company)

45. *U.S. Gross Domestic Product* The table gives the U.S. gross domestic product (in billions of dollars) for selected years from 1940 through 2005.

a. Find the best-fitting quadratic model for the data, with x equal to number of years after 1900.

b. Find the power model that is the best fit for the data, with x equal to number of years after 1900.

c. Compare the two models by graphing each model on the same axes with the data points. Which model appears to be the better fit?

Year	Domestic Gross Product	Year	Domestic Gross Product
1940	837	1985	4207
1945	1559	1990	4853
1950	1328	1995	5439
1955	1700	2000	9817
1960	1934	2001	10,128
1965	2373	2002	10,470
1970	2847	2003	10,976
1975	3173	2004	11,713
1980	3746	2005	12,456

(Source: U.S. Bureau of Economic Analysis)

46. *Auto Noise* The noise level of a Volvo S60 increases as the speed of the car increases. The table gives the noise, in decibels (db), at different speeds.

Speed (mph)	Noise Level (db)
10	42
30	57
50	64
70	66
100	71

(Source: *Road & Track*, 2011)

a. Fit a power function model to the data.

b. Graph the data points and model on the same axes.

c. Use the result from part (a) to estimate the noise level at 80 mph.

47. *Banks* The table gives the number of banks in the United States for selected years from 1935 to 2009.

a. Create a scatter plot of the data, with x equal to the number of years after 1900.

b. Find a quadratic function that models the data.

c. Use the model to estimate the number of banks in 2007.

d. In what years does the model estimate that the number of banks in the U.S. equals 5818?

Year	Number of Banks
1935	15,295
1940	15,772
1950	16,500
1960	17,549
1970	18,205
1980	18,763
1990	15,158
2000	9905
2009	8185

48. *Box-Office Revenues* The data in the table below give the box-office revenues, in billions of dollars, for movies released in selected years between 1980 and 2009.

Year	Revenue (billions of $)	Year	Revenue (billions of $)
1980	2.75	2000	7.47
1985	3.75	2005	8.83
1990	5.02	2009	10.54
1995	5.27		

(Source: Motion Picture Association of America)

a. Find the power model that best fits the revenues as a function of the number of years after 1970. Round to four decimal places.

b. What does the unrounded model estimate as the revenue in 2015?

c. Discuss the reliability of this estimation.

chapter 3 ▶ SUMMARY

In this chapter, we discussed in depth quadratic functions, including realistic applications that involve the vertex and x-intercepts of parabolas and solving quadratic equations. We then studied piecewise-defined functions, power functions, and other nonlinear functions. Real-world data are provided throughout the chapter, and we fit power and quadratic functions to some of these data.

Key Concepts and Formulas

3.1 Quadratic Functions; Parabolas

Quadratic function	Also called a *second-degree polynomial function*, this function can be written in the form $f(x) = ax^2 + bx + c$, where $a \neq 0$.
Parabola	A parabola is the graph of a quadratic function.
Vertex	The turning point on a parabola is the vertex.
Maximum point	If a quadratic function has $a < 0$, the vertex of the parabola is the maximum point, and the parabola opens downward.
Minimum point	If a quadratic function has $a > 0$, the vertex of the parabola is the minimum point, and the parabola opens upward.
Axis of symmetry	The axis of symmetry is the vertical line through the vertex of the parabola.
Forms of quadratic functions	$y = a(x - h)^2 + k$ — Vertex at (h, k); Axis of symmetry is the line $x = h$; Parabola opens up if a is positive and down if a is negative. $y = ax^2 + bx + c$ — x-coordinate of vertex at $x = -\dfrac{b}{2a}$; Axis of symmetry is the line $x = -\dfrac{b}{2a}$; Parabola opens up if a is positive and down if a is negative.

3.2 Solving Quadratic Equations

Solving quadratic equations	An equation that can be written in the form $ax^2 + bx + c = 0$, $a \neq 0$, is called a *quadratic equation*.
Zero product property	For real numbers a and b, if the product $ab = 0$, then at least one of the numbers a or b must equal 0.
Solving by factoring	To solve a quadratic equation by factoring, write the equation in a form with 0 on one side. Then factor the nonzero side of the equation, if possible, and use the zero product property to convert the equation into two linear equations that are easily solved.
Solving and checking graphically	Solutions or decimal approximations of solutions to quadratic equations can be found by using TRACE, ZERO, or INTERSECT with a graphing utility.
Factor Theorem	The factorization of a polynomial function $f(x)$ and the real solutions to $f(x) = 0$ can be verified by graphing $y = f(x)$ and observing where the graph crosses the x-axis. If $x = a$ is a solution to $f(x) = 0$, then $(x - a)$ is a factor of f.
Solving using the square root method	When a quadratic equation has the simplified form $x^2 = C$, the solutions are $x = \pm\sqrt{C}$. This method can also be used to solve equations of the form $(ax + b)^2 = C$.
Completing the square	A quadratic equation can be solved by converting one side to a perfect binomial square and taking the square root of both sides.
Solving using the quadratic formula	The solutions of the quadratic equation $ax^2 + bx + c = 0$, with $a \neq 0$, are given by the formula $$x = \frac{-b \pm \sqrt{b^2 - 4ac}}{2a}$$
Solutions, zeros, x-intercepts, and factors	If a is a real number, the following three statements are equivalent: • a is a real solution to the equation $f(x) = 0$. • a is a real zero of the function $f(x)$. • a is an x-intercept of the graph of $y = f(x)$.

3.3 Piecewise-Defined Functions and Power Functions

Piecewise-defined functions	This is a function that is defined in pieces for different intervals of the domain. Piecewise-defined functions can be graphed by graphing their pieces on the same axes. A familiar example of a piecewise-defined function is the *absolute value function*.
Absolute value function	$$\lvert x \rvert = \begin{cases} x & \text{if } x \geq 0 \\ -x & \text{if } x < 0 \end{cases}$$
Absolute value equation	The solution of the absolute value equation $\lvert x \rvert = a$ is $x = a$ or $x = -a$ when $a \geq 0$. There is no solution to $\lvert x \rvert = a$ if $a < 0$.
Power functions	A power function is a function of the form $y = ax^b$, where a and b are real numbers, $b \neq 0$.
Cubing function	A special power function, $y = x^3$, is called the cubing function.
Root functions	A root function is a function of the form $y = ax^{1/n}$, or $y = a\sqrt[n]{x}$, where n is an integer, $n \geq 2$.

Reciprocal function	This function is formed by the quotient of one and the identity function, $f(x) = \dfrac{1}{x}$, and can be written $f(x) = x^{-1}$.
Direct variation as the nth power	If a quantity y varies directly as the nth power of x, then y is directly proportional to the nth power of x and $$y = kx^n$$ The number k is called the constant of variation or the constant of proportionality.
Inverse variation	A quantity y is inversely proportional to the nth power of x if there exists a nonzero number k such that $y = \dfrac{k}{x^n}$.

3.4 Quadratic and Power Models

Modeling a quadratic function from three points on its graph	If we know three (or more) points that fit exactly on a parabola, we can find the quadratic function whose graph is the parabola.
Quadratic modeling	The use of graphing utilities permits us to fit quadratic functions to data by using technology.
Second differences	If the second differences of data are constant for equally spaced inputs, a quadratic function is an exact fit for the data.
Power modeling	The use of graphing utilities permits us to fit power functions to nonlinear data by using technology.

chapter 3 ▶ SKILLS CHECK

In Exercises 1–8, (a) give the coordinates of the vertex of the graph of each quadratic function and (b) graph each function in a window that includes the vertex and all intercepts.

1. $y = (x - 5)^2 + 3$
2. $y = (x + 7)^2 - 2$
3. $y = 3x^2 - 6x - 24$
4. $y = 2x^2 + 8x - 10$
5. $y = -x^2 + 30x - 145$
6. $y = -2x^2 + 120x - 2200$
7. $y = x^2 - 0.1x - 59.998$
8. $y = x^2 + 0.4x - 99.96$

In Exercises 9 and 10, use factoring to solve the equations.

9. $x^2 - 5x + 4 = 0$
10. $6x^2 + x - 2 = 0$

Use a graphing utility as an aid in factoring to solve the equations in Exercises 11 and 12.

11. $5x^2 - x - 4 = 0$
12. $3x^2 + 4x - 4 = 0$

In Exercises 13 and 14, use the quadratic formula to solve the equations.

13. $x^2 - 4x + 3 = 0$
14. $4x^2 + 4x - 3 = 0$

15. **a.** Use graphical and algebraic methods to find the x-intercepts of the graph of $f(x) = 3x^2 - 6x - 24$.

    b. Find the solutions to $f(x) = 0$ if $f(x) = 3x^2 - 6x - 24$.

16. **a.** Use graphical and algebraic methods to find the x-intercepts of the graph of $f(x) = 2x^2 + 8x - 10$.

    b. Find the solutions to $f(x) = 0$ if $f(x) = 2x^2 + 8x - 10$.

In Exercises 17 and 18, use the square root method to solve the equations.

17. $5x^2 - 20 = 0$ **18.** $(x - 4)^2 = 25$

In Exercises 19–22, find the exact solutions to the equations in the complex numbers.

19. $z^2 - 4z + 6 = 0$ **20.** $w^2 - 4w + 5 = 0$

21. $4x^2 - 5x + 3 = 0$ **22.** $4x^2 + 2x + 1 = 0$

In Exercises 23–30, graph each function.

23. $f(x) = \begin{cases} 3x - 2 & \text{if } x < -1 \\ 4 - x^2 & \text{if } x \geq -1 \end{cases}$

24. $f(x) = \begin{cases} 4 - x & \text{if } x \leq 3 \\ x^2 - 5 & \text{if } x > 3 \end{cases}$

25. $f(x) = 2x^3$ **26.** $f(x) = x^{3/2}$

27. $f(x) = \sqrt{x - 4}$ **28.** $f(x) = \dfrac{1}{x} - 2$

29. $y = x^{4/5}$ **30.** $y = \sqrt[3]{x + 2}$

31. Determine if the function $y = -3x^2$ is increasing or decreasing

    a. For $x < 0$.   **b.** For $x > 0$.

32. For each of the functions, determine if the function is concave up or concave down.

    a. $y = x^{5/4}$ **b.** $y = x^{4/5}$ for $x > 0$

33. Solve $|3x - 6| = 24$.

34. Solve $|2x + 3| = 13$.

35. Find the equation of a quadratic function whose graph is a parabola passing through the points $(0, -2)$, $(-2, 12)$, and $(3, 7)$.

36. Find the equation of a quadratic function whose graph is a parabola passing through the points $(-2, -9)$, $(2, 7)$, and $(4, -9)$.

37. Find a power function that models the data below.

x	1	2	3	4	5	6
y	4	9	11	21	32	45

38. Find a quadratic function that models the data below.

x	1	3	4	6	8
y	2	8	15	37	63

39. Suppose that q varies directly as the 3/2 power of p and that $q = 16$ when $p = 4$. Find q when $p = 16$.

40. If $f(x) = \begin{cases} 3x - 2 & \text{if } -8 \leq x < 0 \\ x^2 - 4 & \text{if } 0 \leq x < 3, \\ -5 & \text{if } x \geq 3 \end{cases}$ find

    $f(-8), f(0),$ and $f(4)$.

chapter 3 ▸ REVIEW

41. *Maximizing Profit* The monthly profit from producing and selling x units of a product is given by the function $P(x) = -0.01x^2 + 62x - 12{,}000$.

    a. Producing and selling how many units will result in the maximum profit for this product?

    b. What is the maximum possible profit for the product?

42. *Profit* The revenue from sales of x units of a product is given by $R(x) = 200x - 0.01x^2$, and the cost of producing and selling the product can be described by $C(x) = 38x + 0.01x^2 + 16{,}000$.

    a. Producing and selling how many units will give maximum profit?

    b. What is the maximum possible profit from producing and selling the product?

43. *Height of a Ball* If a ball is thrown into the air at 64 feet per second from a height of 192 feet, its height (in feet) is given by $S(t) = 192 + 64t - 16t^2$, where t is in seconds.

    a. In how many seconds will the ball reach its maximum height?

    b. What is the maximum possible height for the ball?

44. *Height of a Ball* If a ball is thrown into the air at 29.4 meters per second from a height of 60 meters, its height (in meters) is given by $S(t) = 60 + 29.4t - 9.8t^2$, where t is in seconds.

    a. In how many seconds will the ball reach its maximum height?

    b. What is the maximum possible height for the ball?

45. *Visas* The number of skilled workers' visas issued in the United States can be modeled by $y = -1.48x^2 + 38.901x - 118.429$, where x is the number of years after 1990 and y is the number of visas in thousands.

  a. Use numerical methods to determine the year when the number of visas issued was a maximum.

  b. What was the maximum number of visas issued?

  c. During what year after 2000 does the model indicate that the number of visas issued would be 100 thousand?
  (Source: Department of Homeland Security)

46. *Break-Even* The profit for a product is given by $P = -3600 + 150x - x^2$, where x is the number of units produced and sold. How many units will give break-even (that is, return a profit of 0)?

47. *Falling Ball* If a ball is dropped from the top of a 400-foot-high building, its height S in feet is given by $S(t) = 400 - 16t^2$, where t is in seconds. In how many seconds will it hit the ground?

48. *Profit* The profit from producing and selling x units of a product is given by the function $P(x) = -0.3x^2 + 1230x - 120,000$ dollars. Producing and selling how many units will result in a profit of $324,000 for this product?

49. *Millionaire's Tax Rate* The effective tax rate for a head of household earning the equivalent of $1 million in non-investment income in 2010 dollars can be modeled by the function

$$T(x) = \begin{cases} 0.08x^2 - 2.64x + 22.35 & \text{if } 15 \le x \le 45 \\ -0.525x + 89.82 & \text{if } 45 < x \le 110 \end{cases}$$

where x is the number of years after 1900.

  a. Graph the function for $15 \le x \le 110$.

  b. According to the model, what was the tax rate for a millionaire head of household in 1990?

  c. In 2010, with President Bush's tax cuts in effect, the tax rate was 32.4%. Does the model agree with this rate?
  (Source: The Tax Foundation)

50. *Diabetes* As the following table shows, projections indicate that the percent of U.S. adults with diabetes could dramatically increase.

Year	Percent	Year	Percent	Year	Percent
2010	15.7	2025	24.2	2040	31.4
2015	18.9	2030	27.2	2045	32.1
2020	21.1	2035	29.0	2050	34.3

  a. Find a quadratic model that fits the data in the table, with $x = 0$ in 2000.

  b. Use the model to predict the percent of U.S. adults with diabetes in 2022.

  c. In what year does this model predict the percent of U.S. adults with diabetes will be 30.2%?

51. *ATV Deaths* The number of ATV-related deaths for people of all ages from 1999 to 2009 can be modeled by the function

$$f(x) = \begin{cases} 3.607x^2 - 16.607x + 254.000 & \text{if } 9 \le x \le 15 \\ -43.25x^2 + 1361.65x - 9881.75 & \text{if } 15 < x \le 19 \end{cases}$$

where x is the number of years after 1990.

  a. Graph the function for $9 \le x \le 19$.

  b. According to the model, how many deaths were there in 2003?

  c. According to the model, how many deaths were there in 2009?

  d. Use the model to predict when, after 2000, the number of deaths would be 200.
  (Source: U.S. Consumer Product Safety Commission)

52. *Home Range* The home range of an animal is the region to which the animal confines its movements. The area, in hectares, of the home range of a meat-eating mammal can be modeled by the function $H(x) = 0.11x^{1.36}$, where x is the mass of the animal in grams. What would the home range be for a bobcat weighing 1.6 kg?
  (Source: The American Naturalist)

53. *Internet Usage* Worldwide Internet usage from 1997 through 2007 is shown in the table below.

  a. Find the quadratic function that models the data, with x equal to the number of years after 1990.

  b. Graph the data and the function that models the data.

  c. Does this model appear to be a good fit for the data?

  d. Use the result of part (a) to estimate when Internet usage would be 2 billion.

Year	Internet Usage (millions)	Year	Internet Usage (millions)
1997	75	2001	340
1998	105	2002	580
1999	175	2005	1018
2000	255	2007	1215

(Source: Infoplane.com)

54. *Personal Income* The income received by people from all sources minus their personal contributions for Social Security insurance is called *personal income*. The table lists the personal income, in billions of dollars, of people living in the United States for the indicated years.

Year	Personal Income (billions of $)
1990	4847
2000	8559
2004	9937
2005	10,486
2006	11,268
2007	11,894
2008	12,239
2009	12,026

(Source: U.S. Census Bureau)

a. Using an input equal to the number of years after 1980, find a quadratic function that models these data.

b. Use your unrounded model to estimate when after 1980 the personal income was $5500 billion.

c. When does the model estimate that the personal income will be double its 2009 value?

55. *Resident Population* The U.S. resident population from 15 to 19 years of age is given for selected years in the table below. Write the quadratic function (to three decimal places) that models the 15- to 19-year-old population as a function of years after 1980. Include a description of this population and the unit of measure.

Year	Age 15–19 Population (thousands)	Year	Age 15–19 Population (thousands)
1980	21,168	2000	20,262
1985	18,727	2002	20,366
1990	17,890	2004	20,724
1995	18,152	2005	21,039
1997	19,068		

(Source: U.S. Census Bureau, *Current Population Reports*)

56. *Insurance Premiums* The following table gives the monthly premiums required for a $250,000 term life insurance policy on a 35-year-old female nonsmoker for different guaranteed term periods.

a. Find a quadratic function that models the monthly premium as a function of the length of term for a 35-year-old female nonsmoking policyholder. Report your answer to five decimal places.

b. Assuming that the domain of the function contains integer values between 10 years and 30 years, what term in years could a 35-year-old nonsmoking female purchase for $130 a month?

Term Period (years)	Monthly Premium for 35-Year-Old Female ($)
10	103
15	125
20	145
25	183
30	205

(Source: Quotesmith.com)

57. *Health Services Employment* The numbers of people, in thousands, employed in psychiatric and substance abuse hospitals for selected years from 1990 to 2009 are given in the table.

Year	People Employed (thousands)
1990	113
2000	86
2004	92
2005	93
2006	98
2007	99
2008	102
2009	105

(Source: U.S. Bureau of Labor Statistics)

a. Fit a linear model to the 1990 and 2000 data, with x equal to the number of years after 1990.

b. Fit a power model to the 2004–2009 data, with x equal to the number of years after 1990.

c. Combine the results of parts (a) and (b) to form a piecewise function that models the number employed for the years 1990–2009.

d. Use the piecewise function to answer the following:

i. Find and interpret the output for $x = 1995$.

ii. When was the number employed 90,000, according to the model?

iii. If the model remains accurate, when will there be 120 thousand people employed in psychiatric and substance abuse hospitals?

58. *Hawaii Population* The data in the table below give the population of Hawaii from 1990 through 2010.

Year	Population (thousands)	Year	Population (thousands)
1990	1108	1996	1183
1991	1131	1997	1187
1992	1150	2000	1212
1993	1160	2006	1285
1994	1173	2009	1295
1995	1179	2010	1341

(Source: *Statistical Abstract of the United States*)

a. Align the data with $x =$ the number of years after 1980 and find a power function to fit the aligned data.

b. If the pattern indicated by the model remains valid, estimate in what year Hawaii's population will rise to 1.5 million people.

59. *Pet Industry* The table shows U.S. pet industry expenditures for selected years from 1994 to 2010.

Year	Expenditure (billions of $)
1994	17
1996	21
1998	23
2001	28.5
2002	29.5
2003	32.4
2004	34.4
2005	36.3
2006	38.5
2007	41.2
2008	43.2
2009	45.5
2010	47.7

a. Find a linear model for the data, with x equal to the number of years after 1990, and discuss the fit.

b. Find a quadratic model for the data, with x equal to the number of years after 1990, and discuss the fit.

c. Use both models to predict the amount of money Americans will spend on their pets in 2015.

60. *Prescription Drug Sales* Retail prescription drug sales in the United States for selected years from 1995 to 2009 are given in the table. Align the data, with x equal to the number of years after 1990.

Year	Sales (billions of $)
1995	72.2
2000	145.6
2001	161.3
2002	182.7
2003	204.2
2004	216.7
2005	226.1
2006	243.2
2007	249.2
2008	253.6
2009	268.9

(Source: U.S. Census Bureau)

a. Find a quadratic model for the data.

b. Find a power model for the data.

c. Which model appears more appropriate for predicting the amount of sales after 2009? Why?

Group Activities
▶ EXTENDED APPLICATIONS

1. Modeling

Graphs displaying linear and nonlinear growth are frequently found in periodicals such as *Newsweek* and *Time*, in newspapers such as *USA Today* and the *Wall Street Journal*, and on websites on the Internet. Tables of data can also be found in these sources, especially on federal and state government websites such as www.census.gov.

Your mission is to find a company sales, stock price, biological growth, or sociological trend over a period of years that is nonlinear, to determine which type of nonlinear function is the best fit for the data, and to find a nonlinear function that is a model for the data.

A quadratic model will be a good fit for the data under the following conditions:

- If the data are presented as a graph that resembles a parabola or part of a parabola.
- If the data are presented in a table and a plot of the data points shows that they lie near some parabola or part of a parabola.
- If the data are presented in a table and the second differences of the outputs are nearly constant for equally spaced inputs.

A power model may be a good fit if the shape of the graph or of the plot of the data points resembles part of a parabola but a parabola is not a good fit for the data.

After creating the model, you should test the goodness of fit of the model to the data and discuss uses that you could make of the model.

Your completed project should include

a. A complete citation of the source of the data that you are using.
b. An original copy or photocopy of the data being used.
c. A scatter plot of the data.
d. The equation that you have created.
e. A graph containing the scatter plot and the modeled equation.
f. A statement about how the model could be used to make estimations or predictions about the trend you are observing.

Some helpful hints:

1. If you decide to use a relation determined by a graph that you have found, read the graph very carefully to determine the data points or contact the source of the data to get the data from which the graph was drawn.
2. Align the independent variable by letting x represent the number of years after some convenient year and then enter the data into a graphing utility and create a scatter plot.
3. Use your graphing utility to create the equation of the function that is the best fit for the data. Graph this equation and the data points on the same axes to see if the equation is reasonable.

2. Gender and Earnings

Median Earnings of Full-Time, Year-Round Workers by Gender

Year	Males		Females		Female-to-Male Earnings Ratio
	Number (millions)	**Earnings (thousands of $)**	**Number (millions)**	**Earnings (thousands of $)**	
1970	36.132	40.656	15.476	24.137	
1975	37.267	42.493	17.452	24.994	
1980	41.881	43.360	22.859	26.085	
1985	44.943	43.236	27.383	27.920	
1990	49.171	41.391	31.682	29.643	
1995	52.667	41.375	35.482	29.554	
2000	59.602	43.615	41.719	32.153	
2005	61.500	42.743	43.351	32.903	
2006	63.055	42.261	44.663	32.515	

(Source: U.S. Census Bureau)

The table gives the numbers of male and female employees in the United States and the median earnings (in thousands of 2006 dollars) of full-time, year-round male and female workers for selected years from 1970 to 2006. To investigate the equity of employment for females:

1. Find the quadratic function $y = f(x)$ that models the number of employed males as a function of the number of years after 1970.
2. Find the linear function $y = g(x)$ that models the number of employed females as a function of the number of years after 1970.
3. Graph $y = f(x)$ and $y = g(x)$ to determine if the number of female workers will equal the number of male workers. If so, estimate the year when the number of female workers will equal the number of male workers.

4. Calculate the female-to-male earnings ratio for each of the given years, and enter the ratios in the table.
5. Use each of the given years and the corresponding female-to-male earnings ratio to create a quadratic function that models the female-to-male earnings ratio as a function of number of years after 1970.
6. Write an equation to find an estimate of the year when median female earnings reaches median male earnings, and solve the equation algebraically.
7. Confirm your solution graphically.
8. One possible reason that it might take a long time for median female earnings to reach median male earnings is gender discrimination. What other reasons are possible?

4

Additional Topics with Functions

A profit function can be formed by algebraically combining cost and revenue functions. An average cost function can be constructed by finding the quotient of two functions. Other new functions can be created using function composition and inverse functions. In this chapter, we study additional topics with functions to create functions and solve applied problems.

Algebra TOOLBOX

KEY OBJECTIVES

- Identify functions
- Find domains and ranges
- Determine if a function is increasing or decreasing over given intervals
- Determine over what intervals a function is increasing or decreasing

In this Toolbox, we create a library of functions studied so far, which sets the stage for the study of properties and operations with functions.

Linear Functions

$y = mx + b$

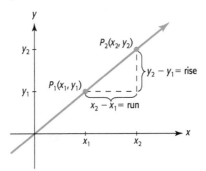

- m is the slope of the line, b is the y-intercept:

$$m = \frac{y_2 - y_1}{x_2 - x_1}$$

- The function is increasing if $m > 0$.
- The function is decreasing if $m < 0$.

Special Linear Functions

IDENTITY FUNCTION	CONSTANT FUNCTION

IDENTITY FUNCTION

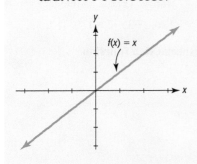

Domain: $(-\infty, \infty)$

Range: $(-\infty, \infty)$

- Increases on its domain

CONSTANT FUNCTION

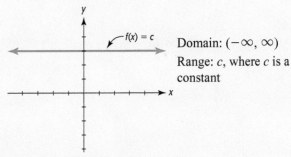

Domain: $(-\infty, \infty)$

Range: c, where c is a constant

- Graph is a horizontal line.

Quadratic Functions

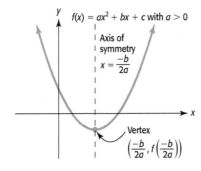

- Vertex is at the point $\left(\dfrac{-b}{2a}, f\left(\dfrac{-b}{2a} \right) \right)$.

Piecewise-Defined Functions

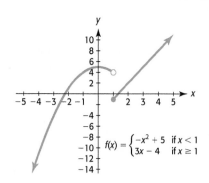

$$f(x) = \begin{cases} -x^2 + 5 & \text{if } x < 1 \\ 3x - 4 & \text{if } x \geq 1 \end{cases}$$

Special Piecewise-Defined Functions

ABSOLUTE VALUE FUNCTION

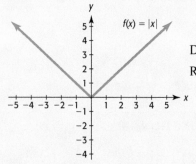

$f(x) = |x|$

Domain: $(-\infty, \infty)$

Range: $[0, \infty)$

- Decreases on $(-\infty, 0)$, increases on $(0, \infty)$

GREATEST INTEGER FUNCTION

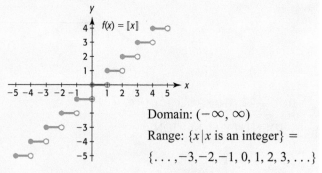

$f(x) = [\![x]\!]$

Domain: $(-\infty, \infty)$

Range: $\{x \mid x \text{ is an integer}\} =$
$\{\dots, -3, -2, -1, 0, 1, 2, 3, \dots\}$

- $f(x) = [\![x]\!]$ is a special **step function**.

Power Functions

POWER FUNCTION, n EVEN

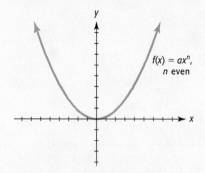

$f(x) = ax^n$, n even

- Decreases on $(-\infty, 0)$, increases on $(0, \infty)$ for $a > 0$
- Increases on $(-\infty, 0)$, decreases on $(0, \infty)$ for $a < 0$

POWER FUNCTION, n ODD

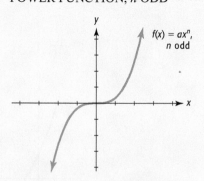

$f(x) = ax^n$, n odd

- Increases on $(-\infty, \infty)$ for $a > 0$
- Decreases on $(-\infty, \infty)$ for $a < 0$

Special Power Functions

ROOT FUNCTION, *n* EVEN

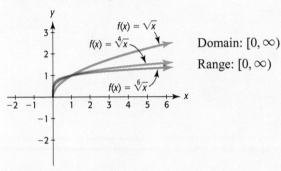

Domain: $[0, \infty)$

Range: $[0, \infty)$

• Increases on its domain

ROOT FUNCTION, *n* ODD

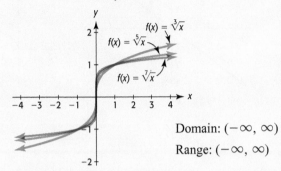

Domain: $(-\infty, \infty)$

Range: $(-\infty, \infty)$

• Increases on its domain

RECIPROCAL FUNCTION

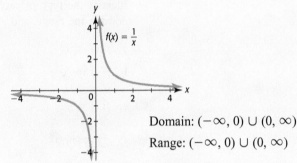

Domain: $(-\infty, 0) \cup (0, \infty)$

Range: $(-\infty, 0) \cup (0, \infty)$

• Decreases on $(-\infty, 0)$ and $(0, \infty)$

EXAMPLE 1 ▶   **Special Functions**

Identify the type of each of the following functions from its equation. Then find the domain and range, and graph the function.

a. $f(x) = x$

b. $f(x) = \sqrt{x}$

c. $f(x) = 3$

SOLUTION

a. This is the identity function, a special linear function. The domain and range are both the set of all real numbers, $(-\infty, \infty)$. See Figure 4.1(a).

b. This is a special power function, $f(x) = x^{1/2}$, called the square root function. Only nonnegative values can be used for x, so the domain is $[0, \infty)$. The outputs will also be nonnegative values, so the range is $[0, \infty)$. See Figure 4.1(b).

c. This is a constant function, a special linear function. The domain is all real numbers, $(-\infty, \infty)$, and the range consists of the single value 3. See Figure 4.1(c).

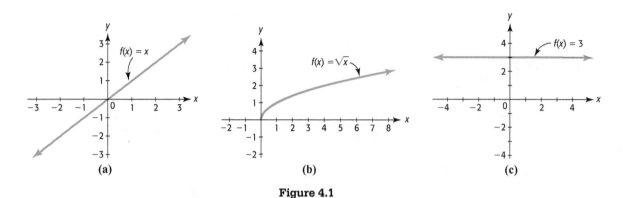

Figure 4.1

EXAMPLE 2 ▶ **Special Functions**

Identify the type of each of the following functions from its equation. Then find the domain and range, and graph the function.

a. $f(x) = x^3$

b. $f(x) = \sqrt[3]{x}$

c. $f(x) = \dfrac{5}{x^2}$

SOLUTION

a. This is a special power function, with power 3. It is called the cubing function, and both the domain and the range are the set of all real numbers, $(-\infty, \infty)$. See Figure 4.2(a).

b. This is a special power function, with power $1/3$. It is called the cube root function. Any real number input results in a real number output, so both the domain and the range are the set of all real numbers, $(-\infty, \infty)$. See Figure 4.2(b).

c. This function can be written as $f(x) = 5x^{-2}$, so it is a special power function, with power -2. If 0 is substituted for x, the result is undefined, so only nonzero real numbers can be used as inputs and only positive real numbers result as outputs. Thus, the domain is $(-\infty, 0) \cup (0, \infty)$, and the range is $(0, \infty)$. We will see in Chapter 6 that $f(x) = \dfrac{5}{x^2}$ is also called a **rational function**. See Figure 4.2(c).

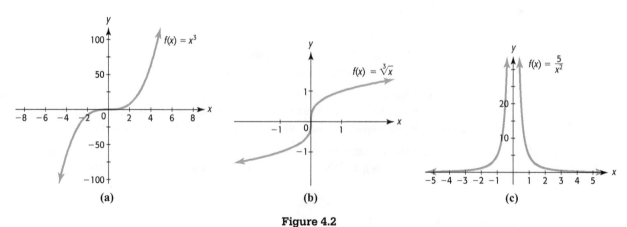

Figure 4.2

EXAMPLE 3 ▶   Functions

Identify the type of each of the following functions from its equation. Then find the domain and range, and graph the function.

a. $g(x) = 2x^2 - 4x + 3$ **b.** $f(x) = 3x - 18$ **c.** $f(x) = \begin{cases} 3x - 4 \text{ if } x < 2 \\ 8 - 2x \text{ if } x \geq 2 \end{cases}$

SOLUTION

a. This is a quadratic function. Any real number can be used as an input, and the resulting outputs are all real numbers greater than or equal to 1. Thus, the domain is $(-\infty, \infty)$, and the range is $[1, \infty)$. Its graph is a parabola that opens up. See Figure 4.3(a).

b. This is a linear function. Its graph is a line that rises as x increases, because its slope is positive. The domain and range are both $(-\infty, \infty)$. See Figure 4.3(b).

c. This is a piecewise-defined function, where each piece is a linear function. The domain is $(-\infty, \infty)$, and the range is $(-\infty, 4]$. See Figure 4.3(c).

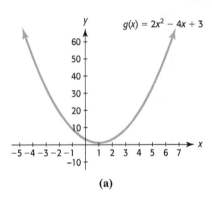

(a)

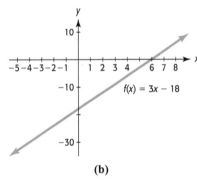

(b)

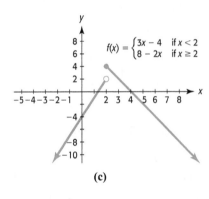

(c)

Figure 4.3

EXAMPLE 4 ▶   Functions

Determine if each of the graphs represents a function.

a.

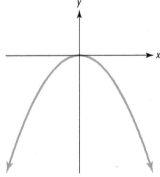

b.

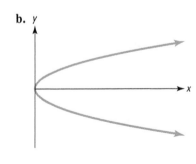

SOLUTION

a. The vertical line test confirms that this graph represents a function. It is a parabola opening downward, so it is the graph of a quadratic function.

b. By the vertical line test, we see that this is not the graph of a function. This graph has the shape of a parabola and is called a **horizontal parabola.**

Toolbox EXERCISES

1. The domain of the reciprocal function is _____, and its range is _____.

2. The domain of the constant function $g(x) = k$ is _____, and its range is _____.

3. The reciprocal function decreases on _____.

4. The absolute value function increases on the interval _____ and decreases on _____.

5. The range of the squaring function is _____.

6. The domain of the squaring function is _____.

In Exercises 7–12, determine if the function is increasing or decreasing on the given interval.

7. $g(x) = \sqrt[5]{x}; (-\infty, \infty)$ 8. $h(x) = \sqrt[4]{x}; [0, \infty)$

9. $f(x) = 5 - 0.8x; (-\infty, \infty)$

10. $f(x) = \dfrac{-1}{x}; (-\infty, 0)$

11. $g(x) = -2x^{2/3}; (0, \infty)$

12. $h(x) = 3x^{1/6}; [0, \infty)$

Identify the type of each of the following functions from its equation. Then graph the function.

13. $f(x) = x^3$ 14. $f(x) = \sqrt{x}$

15. $f(x) = 8\sqrt[6]{x}$ 16. $f(x) = \dfrac{-4}{x^3}$

Which of the following graphs represents a function?

17.

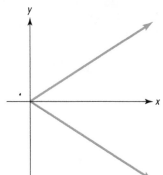

18.

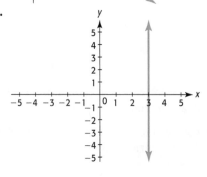

19.

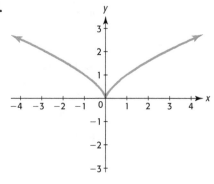

20.

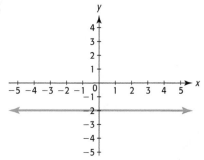

21.

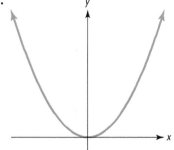

22.

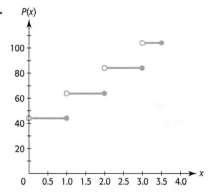

Transformations of Graphs and Symmetry

KEY OBJECTIVES

- Find equations and graphs of functions whose graphs have been vertically shifted, horizontally shifted, stretched, compressed, and reflected

- Determine if a graph is symmetric about the y-axis

- Determine if a graph is symmetric about the origin

- Determine if a graph is symmetric about the x-axis

- Determine if a function is even, odd, or neither

SECTION PREVIEW Diabetes

The projected percent of the U.S. adult population with diabetes (diagnosed and undiagnosed) for 2010 through 2050 can be modeled by the power function $f(x) = 4.947x^{0.495}$, with x equal to the number of years after 2000. To find the model that gives this projection with x equal to the number of years after 2010, we can rewrite the equation in a "shifted" form. (See Example 5.)

In this section, we discuss shifting, stretching, compressing, and reflecting the graph of a function. This is useful in obtaining graphs of additional functions and also in finding windows in which to graph them. We also discuss symmetry of graphs. ∎

Shifts of Graphs of Functions

Consider the quadratic function in the form $y = a(x - h)^2 + k$. Substituting h for x in this equation gives $y = k$, so the graph of this function contains the point (h, k). Because the graph is symmetric about the vertical line through the vertex, there is a second point on the graph with y-coordinate k, *unless* the point (h, k) is the vertex of the parabola (Figure 4.4). To see if the second point exists or if (h, k) is the vertex, we solve

$$k = a(x - h)^2 + k$$

This gives

$$0 = a(x - h)^2$$
$$0 = (x - h)^2$$
$$0 = x - h$$
$$x = h$$

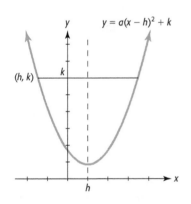

Figure 4.4

Thus, only one value of x corresponds to $y = k$, so the point (h, k) is the vertex of any parabola with an equation of the form $y = a(x - h)^2 + k$.* We can say that the vertex of $y = a(x - h)^2 + k$ has been **shifted** from $(0, 0)$ to the point (h, k), and this entire graph is in fact the graph of $y = x^2$ shifted h units horizontally and k units vertically. We will see that graphs of other functions can be shifted similarly.

In Section 3.3, we studied absolute value functions. In the following example, we investigate shifts of the absolute value function.

EXAMPLE 1 ▶ Shifts of Functions

Graph the functions $f(x) = |x|$ and $g(x) = |x| + 3$ on the same axes. What relationship do you notice between the two graphs?

SOLUTION

Table 4.1 shows inputs for x and outputs for both $f(x) = |x|$ and $g(x) = |x| + 3$. Observe that for a given input, each output value of $g(x) = |x| + 3$ is 3 more than the corresponding output value for $f(x) = |x|$. Thus, the y-coordinate of each point on the graph of $g(x) = |x| + 3$ is 3 more than the y-coordinate of the point on the graph of $f(x) = |x|$ with the same x-coordinate. We say that the graph of $g(x) = |x| + 3$ is the graph of $f(x) = |x|$ shifted up 3 units (Figure 4.5).

---

*This confirms use of the vertex form of a quadratic function in Section 3.1.

Table 4.1

x	$f(x) = |x|$	$g(x) = |x| + 3$
−3	3	6
−2	2	5
−1	1	4
0	0	3
1	1	4
2	2	5
3	3	6
4	4	7

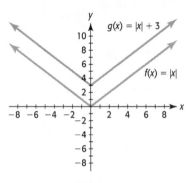

Figure 4.5

In Example 1, the equation of the shifted graph is $g(x) = |x| + 3 = f(x) + 3$. In general, we have the following.

Vertical Shifts of Graphs

If k is a positive real number:

The graph of $g(x) = f(x) + k$ can be obtained by shifting the graph of $f(x)$ upward k units.

The graph of $g(x) = f(x) - k$ can be obtained by shifting the graph of $f(x)$ downward k units.

EXAMPLE 2 ▶ ## Shifts of Functions

Graph the functions $f(x) = |x|$ and $g(x) = |x - 5|$ on the same axes. What relationship do you notice between the two graphs?

SOLUTION

Table 4.2 shows inputs for x and outputs for both $f(x) = |x|$ and $g(x) = |x - 5|$. Observe that the output values for both functions are equal *if* the input value for $g(x) = |x - 5|$ is 5 more than the corresponding input value for $f(x) = |x|$. Thus, the x-coordinate of each point on the graph of $g(x) = |x - 5|$ is 5 more than the x-coordinate of the point on the graph of $f(x) = |x|$ with the same y-coordinate. We say that the graph of $g(x) = |x - 5|$ is the graph of $f(x) = |x|$ shifted 5 units to the right (Figure 4.6).

Table 4.2

x	$f(x) = |x|$	x	$g(x) = |x - 5|$
−3	3	2	3
−2	2	3	2
−1	1	4	1
0	0	5	0
1	1	6	1
2	2	7	2
3	3	8	3
4	4	9	4

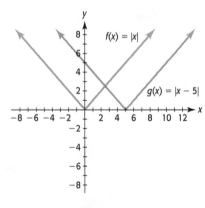

Figure 4.6

In Example 2, the equation of the shifted graph is $g(x) = |x - 5| = f(x - 5)$. In general, we have the following.

Horizontal Shifts of Graphs

If h is a positive real number:

The graph of $g(x) = f(x - h)$ can be obtained by shifting the graph of $f(x)$ to the right h units.
The graph of $g(x) = f(x + h)$ can be obtained by shifting the graph of $f(x)$ to the left h units.

EXAMPLE 3 ▶ Shifts of Functions

Graph the functions $f(x) = \sqrt{x}$ and $g(x) = \sqrt{x + 3} - 4$ on the same axes. What relationship do you notice between the two graphs?

SOLUTION

The graphs of $f(x) = \sqrt{x}$ and $g(x) = \sqrt{x + 3} - 4$ are shown in Figure 4.7. We see that the graph of $g(x) = \sqrt{x + 3} - 4$ can be obtained by shifting the graph of $f(x) = \sqrt{x}$ to the left 3 units and down 4 units. To verify this, consider a few points from each graph in Table 4.3.

Table 4.3

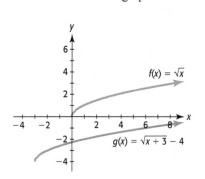

Figure 4.7

$f(x) = \sqrt{x}$		$g(x) = \sqrt{x + 3} - 4$	
x	$f(x) = \sqrt{x}$	x	$g(x) = \sqrt{x + 3} - 4$
0	0	-3	-4
1	1	-2	-3
2	$\sqrt{2}$	-1	$\sqrt{2} - 4$
3	$\sqrt{3}$	0	$\sqrt{3} - 4$
4	2	1	-2

EXAMPLE 4 ▶ Profit

Right Sports Management sells elliptical machines, and the monthly profit from these machines can be modeled by

$$P(x) = -(x - 55)^2 + 4500$$

where x is the number of elliptical machines produced and sold.

a. Use the graph of $y = -x^2$ and the appropriate transformation to set the window and graph the profit function.

b. How many elliptical machines must be produced and sold to yield the maximum monthly profit? What is the maximum monthly profit?

SOLUTION

a. Shifting the graph of $y = -x^2$ (Figure 4.8(a)) 55 units to the right and up 4500 units gives the graph of $P(x) = -(x - 55)^2 + 4500$. Note that we graph the function only in the first quadrant, because only nonnegative values make sense for this application.

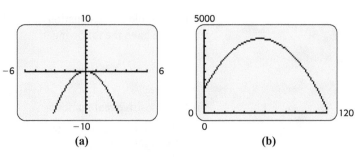

Figure 4.8

b. The maximum value of the function occurs at the vertex, and the vertex of the graph is (55, 4500). Thus, maximum monthly profit of $4500 is made when 55 elliptical machines are produced and sold.

EXAMPLE 5 ▶   Diabetes

The projected percent of the U.S. adult population with diabetes (diagnosed and undiagnosed) for 2010 through 2050 can be modeled by the power function $f(x) = 4.947x^{0.495}$, with x equal to the number of years after 2000. Find the power function that models this projection with x equal to the number of years after 2010.

SOLUTION

Because y represents the percent in both models, there is no vertical shift from one function to the other. To find the power model that will give the percent where x is the number of years after 2010 rather than the number of years after 2000, we rewrite the equation in a new form where $x = 0$ represents the same input as $x = 10$ does in the original model. This can be accomplished by replacing x in the original model with $x + 10$, giving the new model

$$f(x) = 4.947(x + 10)^{0.495}$$

where x is the number of years after 2010.

To see that these two models are equivalent, observe the inputs and outputs for selected years in Table 4.4 and the graphs in Figure 4.9.

Table 4.4

Original Model		New Model	
Years from 2000	Percent	Years from 2010	Percent
10	15.46	0	15.46
20	21.80	10	21.80
30	26.64	20	26.64
50	34.30	40	34.30

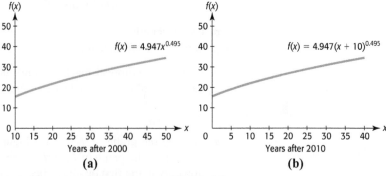

Figure 4.9

Stretching and Compressing Graphs

EXAMPLE 6 ▶ ### Stimulus-Response

One of the early results in psychology relates the magnitude of a stimulus x to the magnitude of the response y with the model $y = kx^2$, where k is an experimental constant. Compare the graphs of $y = kx^2$ for $k = 1, 2,$ and $\frac{1}{2}$.

SOLUTION

Table 4.5 shows the values of y for $y = f(x) = x^2, y = 2f(x) = 2x^2,$ and $y = \frac{1}{2}f(x) = \frac{1}{2}x^2$ for selected values of x. Observe that for these x-values, the y-values for $y = 2x^2$ are 2 times the y-values for $y = x^2$, and the points on the graph of $y = 2x^2$ have y-values that are 2 times the y-values on the graph of $y = x^2$ for equal x-values. We say that the graph of $y = 2x^2$ is a **vertical stretch** of $y = x^2$ by a factor of 2. (Compare Figure 4.10(a) and Figure 4.10 (b).)

Observe also that for these x-values, the y-values for $y = \frac{1}{2}x^2$ are $\frac{1}{2}$ of the y-values for $y = x^2$, and the points on the graph of $y = \frac{1}{2}x^2$ have y-values that are $\frac{1}{2}$ of the y-values on the graph of $y = x^2$ for equal x-values. We say that the graph of $y = \frac{1}{2}x^2$ is a **vertical compression** of $y = x^2$ using a factor of $\frac{1}{2}$ (Figure 4.10(c)).

Table 4.5

x	0	1	2
f(x)	0	1	4

x	0	1	2
2f(x)	0	2	8

x	0	1	2
$\frac{1}{2}f(x)$	0	$\frac{1}{2}$	2

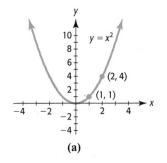

(a)

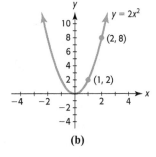

(b)

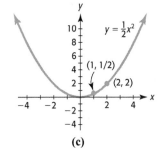

(c)

Figure 4.10

Stretching and Compressing Graphs

The graph of $y = af(x)$ is obtained by vertically stretching the graph of $f(x)$ using a factor of $|a|$ if $|a| > 1$ and vertically compressing the graph of $f(x)$ using a factor of $|a|$ if $0 < |a| < 1$.

Reflections of Graphs

In Figure 4.11(a), we see that the graph of $y = -x^2$ is a parabola that *opens down*. It can be obtained by reflecting the graph of $f(x) = x^2$, which *opens up*, across the x-axis. We can compare the y-coordinates of the graphs of these two functions by looking at

Table 4.6 and Figure 4.11(b). Notice that for a given value of x, the y-coordinates of $y = x^2$ and $y = -x^2$ are negatives of each other.

Table 4.6

x	$y = x^2$	$y = -x^2$
-2	4	-4
-1	1	-1
0	0	0
1	1	-1
2	4	-4

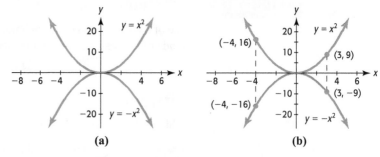

(a)   (b)

Figure 4.11

We can also see that the graph of $y = (-x)^3$ is a reflection of the graph of $y = x^3$ across the y-axis by looking at Table 4.7 and Figure 4.12. Notice that for a given value of y, the x-coordinates of $y = x^3$ and $y = (-x)^3$ are negatives of each other.

Table 4.7

x	$y = x^3$	x	$y = (-x)^3$
2	8	-2	8
1	1	-1	1
0	0	0	0
-1	-1	1	-1
-2	-8	2	-8

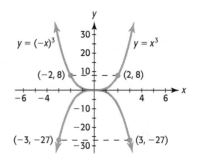

Figure 4.12

In general, we have the following:

Reflections of Graphs Across the Coordinate Axes

1. The graph of $y = -f(x)$ can be obtained by reflecting the graph of $y = f(x)$ across the x-axis.

2. The graph of $y = f(-x)$ can be obtained by reflecting the graph of $y = f(x)$ across the y-axis.

A summary of the transformations of a graph follows.

Graph Transformations

For a given function $y = f(x)$,

Vertical Shift	$y = f(x) + k$	Graph is shifted k units up if $k > 0$ and k units down if $k < 0$.
Horizontal Shift	$y = f(x - h)$	Graph is shifted h units right if $h > 0$ and h units left if $h < 0$.

(continued)

Stretch/Compress	$y = a\,f(x)$	Graph is vertically stretched using a factor of $\lvert a \rvert$ if $\lvert a \rvert > 1$. Graph is compressed using a factor of $\lvert a \rvert$ if $\lvert a \rvert < 1$.
Reflection	$y = -f(x)$	Graph is reflected across the x-axis.
	$y = f(-x)$	Graph is reflected across the y-axis.

EXAMPLE 7 ▶ Pollution

Suppose that for a certain city the cost C of obtaining drinking water that contains $p\%$ impurities (by volume) is given by

$$C = \frac{120{,}000}{p} - 1200$$

a. Determine the domain of this function and graph the function without concern for the context of the problem.

b. Use knowledge of the context of the problem to graph the function in a window that applies to the application.

c. What is the cost of obtaining drinking water that contains 5% impurities?

SOLUTION

a. All values of p except $p = 0$ result in real values for the function. Thus, the domain of C is all real numbers except 0. The graph of this function is a transformation of the graph of $C = \dfrac{1}{p}$, stretched by a factor of 120,000, then shifted downward 1200 units. The viewing window should have its center near $p = 0$ (horizontally) and $C = -1200$ (vertically). We increase the vertical view to allow for the large stretching factor, using the viewing window $[-100, 100]$ by $[-20{,}000, 20{,}000]$. The graph, shown in Figure 4.13(a), has the same shape as the graph of $y = \dfrac{1}{x}$.

b. Because p represents the percent of impurities, the viewing window is set for values of p from 0 to 100. (Recall from part (a) that p cannot be 0.) The cost of reducing the impurities cannot be negative, so the vertical view is set from 0 to 20,000. The graph is shown in Figure 4.13(b).

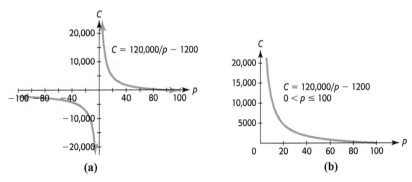

(a)   **(b)**

Figure 4.13

c. To determine the cost of obtaining drinking water that contains 5% impurities, we substitute 5 for p in $C = \dfrac{120{,}000}{p} - 1200$, giving

$$C = \frac{120{,}000}{5} - 1200 = 22{,}800$$

Thus, the cost is $22,800 to obtain drinking water that contains 5% impurities.

EXAMPLE 8 ▶ **Velocity of Blood**

Because of friction from the walls of an artery, the velocity of blood is greatest at the center of the artery and decreases as the distance r from the center increases. The velocity of the blood in an artery can be modeled by the function

$$v = k(R^2 - r^2)$$

where R is the radius of the artery and k is a constant that is determined by the pressure, the viscosity of the blood, and the length of the artery. In the case where $k = 2$ and $R = 0.1$ centimeter, the velocity is

$$v = 2(0.01 - r^2) \text{ centimeter per second}$$

a. Graph this function and the functions $v = r^2$ and $v = -2r^2$ in the viewing window $[-0.1, 0.1]$ by $[-0.05, 0.05]$.

b. How is the function $v = 2(0.01 - r^2)$ related to the second-degree power function $v = r^2$?

SOLUTION

a. The graph of the function $v = 2(0.01 - r^2)$ is shown in Figure 4.14(a), the graph of $v = r^2$ is shown in Figure 4.14(b), and the graph of $v = -2r^2$ is shown in Figure 4.14(c).

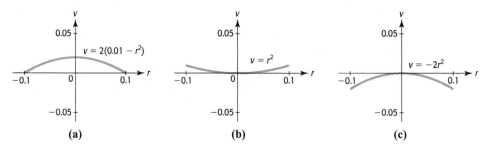

Figure 4.14

b. The function $v = 2(0.01 - r^2)$ can also be written in the form $v = -2r^2 + 0.02$, so it should be no surprise that the graph is a reflected, stretched, and shifted form of the graph of $v = r^2$. If the graph of $v = r^2$ is reflected about the r-axis, has each r^2-value doubled, and then is shifted upward 0.02 unit, the graph of $v = -2r^2 + 0.02 = 2(0.01 - r^2)$ results.

Symmetry; Even and Odd Functions

In Chapter 3, we observed that any parabola that is a graph of a quadratic function is symmetric about a vertical line called the axis of symmetry. This means that the two halves of the parabola are reflections of each other. In particular, the graph of $y = x^2$ in

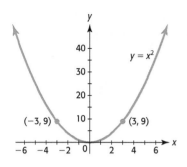

Figure 4.15

Figure 4.15 is symmetric (that is, is a reflection of itself) about the y-axis. For example, the point $(3, 9)$ is on the graph, and its mirror image across the y-axis is the point $(-3, 9)$. This suggests an algebraic way to determine whether the graph of an equation is symmetric with respect to the y-axis—that is, replace x with $-x$ and simplify. If the resulting equation is equivalent to the original equation, then the graph is symmetric with respect to the y-axis.

Symmetry with Respect to the y-axis

The graph of $y = f(x)$ is symmetric with respect to the y-axis if, for every point (x, y) on the graph, the point $(-x, y)$ is also on the graph—that is,

$$f(-x) = f(x)$$

for all x in the domain of f. Such a function is called an **even function**.

All power functions with even integer exponents will have graphs that are symmetric about the y-axis. This is why all functions satisfying this condition are called even functions, although functions other than power functions exist that are even functions.

A graph of $y = x^3$ is shown in Figure 4.16(a). Notice in Figure 4.16(b) that if we draw a line through the origin and any point on the graph, it will intersect the graph at a second point, and the distances from the two points to the origin will be equal.

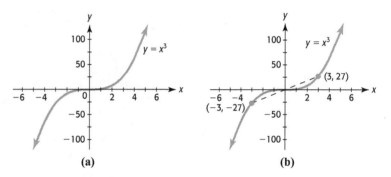

(a)   **(b)**

Figure 4.16

For example, the point $(3, 27)$ is on the graph of $y = x^3$, and the line through $(3, 27)$ and the origin intersects the graph at the point $(-3, -27)$, which is the same distance from the origin as $(3, 27)$. In general, the graph of a function is called symmetric with respect to the origin if for every point (x, y) on the graph, the point $(-x, -y)$ is also on the graph.

To determine algebraically whether the graph of an equation is symmetric about the origin, replace x with $-x$, replace y with $-y$, and simplify. If the resulting equation is equivalent to the original equation, then the graph is symmetric with respect to the origin.

Symmetry with Respect to the Origin

The graph of $y = f(x)$ is symmetric with respect to the origin if, for every point (x, y) on the graph, the point $(-x, -y)$ is also on the graph—that is,

$$f(-x) = -f(x)$$

for all x in the domain of f. Such a function is called an **odd function**.

The graphs of some equations can be symmetric with respect to the x-axis (see Figure 4.17). Such graphs do not represent y as a function of x (because they do not pass the vertical line test).

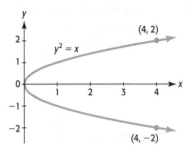

Figure 4.17

For example, we can graph the equation $x^2 + y^2 = 9$ by hand or with technology after solving it for y. Solving gives

$$y^2 = 9 - x^2$$
$$y = \pm \sqrt{9 - x^2}$$

Entering values for x in the equation gives points on the **circle,** and entering the two functions in a calculator, as in Figure 4.18(a), gives the graph. Note that the calculator graph will not look like a circle unless the calculator window is square. Figure 4.18(b) gives the graph with the standard window, Figure 4.18(c) gives the graph after Zoom Square, and Figure 4.18(d) gives the graph with Zoom Decimal.* In general, if the equation contains one and only one even power of y, its graph will be symmetric about the x-axis.

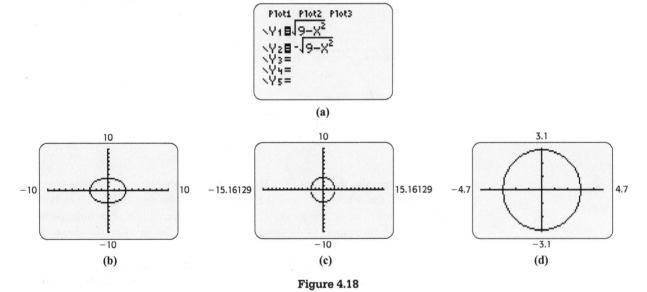

Figure 4.18

To determine algebraically whether the graph of an equation is symmetric about the x-axis, replace y with $-y$ and simplify. If the resulting equation is equivalent to the original equation, then the graph is symmetric with respect to the x-axis.

Symmetry with Respect to the x-axis

The graph of an equation is symmetric with respect to the x-axis if, for every point (x, y) on the graph, the point $(x, -y)$ is also on the graph.

*See Appendix A, page 618.

EXAMPLE 9 ▶ Symmetry

Determine algebraically whether the graph of each equation is symmetric with respect to the x-axis, y-axis, or origin. Confirm your conclusion graphically.

a. $y = \dfrac{2x^2}{x^2 + 1}$ **b.** $y = x^3 - 3x$ **c.** $x^2 + y^2 = 16$

SOLUTION

a. To test the graph of $y = \dfrac{2x^2}{x^2 + 1}$ for symmetry with respect to the x-axis, we replace y with $-y$. Because the result,

$$-y = \frac{2x^2}{x^2 + 1}$$

is not equivalent to the original equation, the graph is not symmetric with respect to the x-axis. This equation gives y as a function of x.

To test for y-axis symmetry, we replace x with $-x$:

$$y = \frac{2(-x)^2}{(-x)^2 + 1} = \frac{2x^2}{x^2 + 1}$$

which is equivalent to the original equation. Thus, the graph is symmetric with respect to the y-axis, and y is an even function of x.

To test for symmetry with respect to the origin, we replace x with $-x$ and y with $-y$:

$$-y = \frac{2(-x)^2}{(-x)^2 + 1} = \frac{2x^2}{x^2 + 1}$$

which is not equivalent to the original equation. Thus, the graph is not symmetric with respect to the origin.

Figure 4.19 confirms the fact that the graph of $y = \dfrac{2x^2}{x^2 + 1}$ is symmetric with respect to the y-axis.

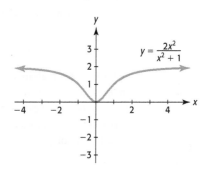

Figure 4.19

b. To test the graph of $y = x^3 - 3x$ for symmetry with respect to the x-axis, we replace y with $-y$. Because the result,

$$-y = x^3 - 3x$$

is not equivalent to the original equation, the graph is not symmetric with respect to the x-axis.

To test for y-axis symmetry, we replace x with $-x$:

$$y = (-x)^3 - 3(-x) = -x^3 + 3x$$

which is not equivalent to the original equation. Thus, the graph is not symmetric with respect to the y-axis.

To test for symmetry with respect to the origin, we replace x with $-x$ and y with $-y$:

$$-y = (-x)^3 - 3(-x) = -x^3 + 3x, \text{ or } y = x^3 - 3x$$

which is equivalent to the original equation. Thus, the graph is symmetric with respect to the origin, and y is an odd function of x.

Figure 4.20 confirms this.

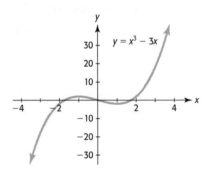

Figure 4.20

c. To test the graph of $x^2 + y^2 = 16$ for symmetry with respect to the x-axis, we replace y with $-y$:

$$x^2 + (-y)^2 = 16 \quad \text{or} \quad x^2 + y^2 = 16$$

This is equivalent to the original equation, so the graph is symmetric with respect to the x-axis. Thus, y is not a function of x in this equation.

To test for y-axis symmetry, we replace x with $-x$:

$$(-x)^2 + y^2 = 16 \quad \text{or} \quad x^2 + y^2 = 16$$

This is equivalent to the original equation, so the graph is symmetric with respect to the y-axis.

To test for symmetry with respect to the origin, we replace x with $-x$ and y with $-y$:

$$(-x)^2 + (-y)^2 = 16, \quad \text{or} \quad x^2 + y^2 = 16$$

which is equivalent to the original equation. Thus, the graph is symmetric with respect to the origin.

Solving for y gives the two functions graphed in Figure 4.21, and the graph confirms the fact that the graph of $x^2 + y^2 = 16$ is symmetric with respect to the x-axis, y-axis, and origin.

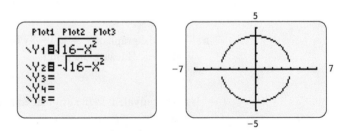

Figure 4.21

Skills CHECK 4.1

In Exercises 1–16, (a) sketch the graph of each pair of functions using a standard window, and (b) describe the transformations used to obtain the graph of the second function from the first function.

1. $y = x^3, y = x^3 + 5$ 2. $y = x^2, y = x^2 + 3$

3. $y = \sqrt{x}, y = \sqrt{x - 4}$

4. $y = x^2, y = (x + 2)^2$

5. $y = \sqrt[3]{x}, y = \sqrt[3]{x + 2} - 1$

6. $y = x^3, y = (x - 5)^3 - 3$

7. $y = |x|, y = |x - 2| + 1$

8. $y = |x|, y = |x + 3| - 4$

9. $y = x^2, y = -x^2 + 5$

10. $y = \sqrt{x}, y = -\sqrt{x - 2}$

11. $y = \dfrac{1}{x}, y = \dfrac{1}{x} - 3$

12. $y = \dfrac{1}{x}, y = \dfrac{2}{x - 1}$

13. $f(x) = x^2, g(x) = \dfrac{1}{3}x^2$

14. $f(x) = x^3, g(x) = 0.4x^3$

15. $f(x) = |x|, g(x) = 3|x|$

16. $f(x) = \sqrt{x}, g(x) = 4\sqrt{x}$

17. How is the graph of $y = (x - 2)^2 + 3$ transformed from the graph of $y = x^2$?

18. How is the graph of $y = (x + 4)^3 - 2$ transformed from the graph of $y = x^3$?

19. Suppose the graph of $y = x^{3/2}$ is shifted to the left 4 units. What is the equation that gives the new graph?

20. Suppose the graph of $y = x^{3/2}$ is shifted down 5 units and to the right 4 units. What is the equation that gives the new graph?

21. Suppose the graph of $y = x^{3/2}$ is stretched by a factor of 3 and then shifted up 5 units. What is the equation that gives the new graph?

22. Suppose the graph of $y = x^{2/3}$ is compressed by a factor of $\dfrac{1}{5}$ and then shifted right 6 units. What is the equation that gives the new graph?

In Exercises 23–26, write the equation of the function $g(x)$ that is transformed from the given function $f(x)$ and whose graph is shown.

23. $f(x) = x^2$

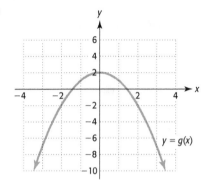

24. $f(x) = x^2$

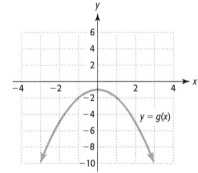

25. $f(x) = |x|$

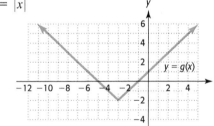

26. $f(x) = x^2$

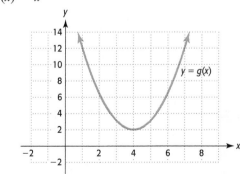

In Exercises 27–30, determine visually whether each of the graphs is symmetric about the x-axis, y-axis, or neither.

27.

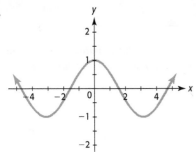

28.

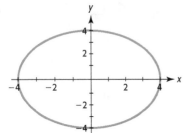

29.

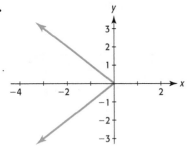

30.

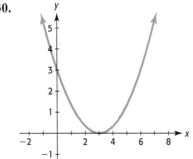

In Exercises 31 and 32, determine whether the graph of the equation in Y_1 is symmetric about the y-axis, based on the ordered pairs shown in each pair of tables.

31.

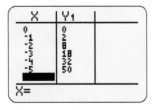

32.

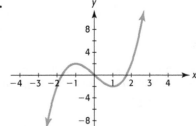

In Exercises 33–40, determine algebraically whether the graph of the given equation is symmetric with respect to the x-axis, the y-axis, and/or the origin. Confirm graphically.

33. $y = 2x^2 - 3$ **34.** $y = -x^2 + 4$

35. $y = x^3 - x$ **36.** $y = -x^3 + 5x$

37. $y = \dfrac{6}{x}$ **38.** $x = 3y^2$

39. $x^2 + y^2 = 25$ **40.** $x^2 - y^2 = 25$

In Exercises 41–46, determine whether the function is even, odd, or neither.

41. $f(x) = |x| - 5$ **42.** $f(x) = |x - 2|$

43. $g(x) = \sqrt{x^2 + 3}$ **44.** $f(x) = \dfrac{1}{2}x^3 - x$

45. $g(x) = \dfrac{5}{x}$ **46.** $g(x) = 4x + x^2$

Determine if each of the complete graphs in Exercises 47–48 represents functions that are even, odd, or neither.

47.

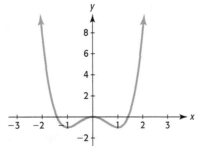

48.

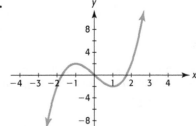

49. Graph $(y + 1)^2 = x + 4$.

50. About what line is the function $y = 3 \pm \sqrt{x - 2}$ symmetric?

EXERCISES 4.1

51. *Marijuana Use* The number of millions of people age 12 and older in the United States who used marijuana during the years 2003 to 2008 is described by the function $M(x) = -0.062(x - 4.8)^2 + 25.4$ for $3 \le x \le 8$, where x is the number of years after 2000.

a. The graph of this function is a shifted graph of which basic function?

b. Find and interpret $M(3)$.

c. Sketch a graph of $y = M(x)$ for $3 \le x \le 8$.
(Source: 2008 National Survey on Drug Use and Health, U.S. Department of Health and Human Services)

52. *Ballistics* Ballistic experts are able to identify the weapon that fired a certain bullet by studying the markings on the bullet. If a test is conducted by firing a bullet into a water tank, the distance that a bullet will travel is given by $s = 27 - (3 - 10t)^3$ inches for $0 \le t \le 0.3$, t in seconds.

a. The graph of this function is a shifted graph of which basic function?

b. Graph this function for $0 \le t \le 0.3$.

c. How far does the bullet travel during this time period?

53. *Supply and Demand* The price per unit of a product is $\$p$, and the number of units of the product is denoted by q. The supply function for a product is given by $p = \dfrac{180 + q}{6}$, and the demand for the product is given by $p = \dfrac{30{,}000}{q} - 20$.

a. Is the supply function a linear function or a shifted reciprocal function?

b. Is the demand function a shifted linear function or a shifted reciprocal function? Describe the transformations needed to obtain the specific function from the basic function.

54. *Supply and Demand* The supply function for a commodity is given by $p = 58 + \dfrac{q}{2}$, and the demand function for this commodity is given by $p = \dfrac{2555}{q + 5}$.

a. Is the supply function a linear function or a shifted reciprocal function?

b. Is the demand function a shifted linear function or a shifted reciprocal function? Describe the transformations needed to obtain the specific function from the basic function.

55. *Mob Behavior* In a study of lynchings between 1899 and 1946, psychologist Brian Mullin concluded that the size of the mob relative to the number of victims predicted the level of brutality. The formula he developed gives y, the other-total ratio that predicts the level of self-attentiveness of people in a crowd of size x with one victim. Mullin's formula is $y = \dfrac{1}{x + 1}$; the lower the value of y, the more likely an individual is to be influenced by "mob psychology."

a. This function is a shifted version of what basic function, and how is it shifted?

b. Graph this function for $x > 0$.

c. Will the amount of self-attentiveness of a person increase or decrease as the crowd size becomes larger?

56. *Pollution* The daily cost C (in dollars) of removing pollution from the smokestack of a coal-fired electric power plant is related to the percent of pollution p being removed according to the equation

$$C = \frac{10{,}500}{100 - p}$$

a. Describe the transformations needed to obtain this function from the function $C = \dfrac{1}{p}$.

b. Graph the function for $0 \le p < 100$.

c. What is the daily cost of removing 80% of the pollution?

57. *Population Growth* Suppose the population of a certain microorganism at time t (in minutes) is given by

$$P = -1000\left(\frac{1}{t + 10} - 1\right)$$

a. Describe the transformations needed to obtain this function from the function $f(t) = \dfrac{1}{t}$.

b. Graph this function for values of t representing 0 to 100 minutes.

58. *Mortgages* The balance owed y on a \$50,000 mortgage after x monthly payments is shown in the table that follows. The function that models the data is

$$y = 4700\sqrt{110 - x}$$

Monthly Payments	Balance Owed (\$)
12	47,243
24	44,136
48	36,693
72	27,241
96	15,239
108	8074

a. Is this a shifted root function?

b. What is the domain of the function in the context of this application?

c. Describe the transformations needed to obtain the graph from the graph of $y = \sqrt{x}$.

59. *Personal Expenditures* Personal consumption expenditures for durable goods in the United States, in billions of dollars, can be modeled by the function $P(x) = 306.472x^{0.464}$, where x is the number of years after 1990. Rewrite the function with x equal to the number of years after 1995. (*Hint:* Shift the x-value by the difference in years.)
(Source: Bureau of Economic Analysis, U.S. Department of Commerce)

60. *Unconventional Vehicle Sales* The number of E85 flex fuel vehicles, in millions, projected to be sold in the United States can be modeled by the function $F(x) = 0.084x^{0.675}$, where x is the number of years after 2000. Convert the function so that x equals the number of years after 1990.
(Source: www.eia.gov)

61. *Tobacco Sales* Sales of fine-cut cigarettes (in millions) in Canada can be described by $y = -13.898x^2 + 255.467x + 5425.618$, where x is the number of years after 1980. Convert the function so that x represents the number of years after 1990.

62. *Aircraft Accidents* The number of U.S. aircraft accidents can be modeled by $y = -17.560x^2 + 126.488x + 7986.786$, with $x = 0$ in 2000. Rewrite the model with x equal to the number of years after 1990.

63. *U.S. Cell Phone Subscribers* The following table shows the number of U.S. cellular telephone subscribers from 1985 to 2009. The number of U.S. cell phone subscribers, in millions, can be modeled by the function $S(t) = 0.00056t^4$, where t is the number of years after 1980.

Year	Subscribers (millions)	Year	Subscribers (millions)
1985	0.340	1998	69.209
1986	0.682	1999	86.047
1987	1.231	2000	109.478
1988	2.069	2001	128.375
1989	3.509	2002	140.767
1990	5.283	2003	158.722
1991	7.557	2004	182.140
1992	11.033	2005	207.896
1993	16.009	2006	233.041
1994	24.134	2007	255.396
1995	33.786	2008	270.334
1996	44.043	2009	285.694
1997	55.312		

(Source: Semiannual CTIA Wireless Survey)

a. Use the model to find the number of subscribers in 2005.

b. Write a new model $C(t)$ with t equal to the number of years after 1985.

c. With the model in part (b), what value of t should be used to determine the number of subscribers in 2005? Does the value of $C(t)$ agree with the answer to part (a)?

64. *Cost-Benefit* Suppose for a certain city the cost C of obtaining drinking water that contains $p\%$ impurities (by volume) is given by

$$C = \frac{120,000}{p} - 1200$$

a. What is the cost of drinking water that is 100% impure?

b. What is the cost of drinking water that is 50% impure?

c. What transformations of the graph of the reciprocal function give the graph of this function?

65. *U.S. Poverty Threshold* The table below shows the yearly income poverty thresholds for a single person for selected years from 1990 to 2009.

Year	Income ($)	Year	Income ($)
1990	6652	2002	9182
1995	7763	2003	9573
1998	8316	2004	9827
1999	8501	2005	10,160
2000	8794	2008	10,991
2001	9039	2009	10,956

(Source: U.S. Census Bureau)

The poverty threshold income for a single person can be modeled by the function $P(t) = 638.57t^{0.775}$, where t is the number of years after 1970.

a. Assuming that the model is accurate after 2009, find the poverty threshold income in 2013.

b. Write a new model $S(t)$ with t equal to the number of years after 1990.

c. Using the model in part (b), what value of t should be used to determine the poverty threshold income in 2013? Does the value of $S(t)$ agree with the answer to part (a)?

66. *Pollution* The daily cost C (in dollars) of removing pollution from the smokestack of a coal-fired electric power plant is related to the percent of pollution p being removed according to the equation $100C - Cp = 10,500$.

a. Solve this equation for C to write the daily cost as a function of p, the percent of pollution removed.

b. What is the daily cost of removing 50% of the pollution?

c. Why would this company probably resist removing 99% of the pollution?

4.2 Combining Functions; Composite Functions

KEY OBJECTIVES

- Find sums, differences, products, and quotients of two functions
- Form average cost functions
- Find the composition of two functions

SECTION PREVIEW **Profit**

If the daily total cost to produce x units of a product is

$$C(x) = 360 + 40x + 0.1x^2 \text{ thousand dollars}$$

and the daily revenue from the sale of x units of this product is

$$R(x) = 60x \text{ thousand dollars}$$

we can model the profit function as $P(x) = R(x) - C(x)$, giving

$$P(x) = 60x - (360 + 40x + 0.1x^2)$$

or

$$P(x) = -0.1x^2 + 20x - 360$$

In a manner similar to the one used to form this function, we can construct new functions by performing algebraic operations with two or more functions. For example, we can build an average cost function by finding the quotient of two functions. (See Example 4.) In addition to using arithmetic operations with functions, we can create new functions using function composition. ■

Operations with Functions

New functions that are the sum, difference, product, and quotient of two functions are defined as follows:

Operation	Formula	Example with $f(x) = \sqrt{x}$ and $g(x) = x^3$
Sum	$(f + g)(x) = f(x) + g(x)$	$(f + g)(x) = \sqrt{x} + x^3$
Difference	$(f - g)(x) = f(x) - g(x)$	$(f - g)(x) = \sqrt{x} - x^3$
Product	$(f \cdot g)(x) = f(x) \cdot g(x)$	$(f \cdot g)(x) = \sqrt{x} \cdot x^3 = x^3\sqrt{x}$
Quotient	$\left(\dfrac{f}{g}\right)(x) = \dfrac{f(x)}{g(x)} \quad (g(x) \neq 0)$	$\left(\dfrac{f}{g}\right)(x) = \dfrac{\sqrt{x}}{x^3} \quad (x \neq 0)$

The domain of the sum, difference, and product of f and g consists of all real numbers of the input variable for which f and g are defined. The domain of the quotient function consists of all real numbers for which f and g are defined and $g \neq 0$.

EXAMPLE 1 ▶   Operations with Functions

If $f(x) = x^3$ and $g(x) = x - 1$, find the following functions and give their domains.

a. $(f + g)(x)$ **b.** $(f - g)(x)$ **c.** $(f \cdot g)(x)$ **d.** $\left(\dfrac{f}{g}\right)(x)$

SOLUTION

a. $(f + g)(x) = f(x) + g(x) = x^3 + x - 1$; all real numbers

b. $(f - g)(x) = f(x) - g(x) = x^3 - (x - 1) = x^3 - x + 1$; all real numbers

c. $(f \cdot g)(x) = f(x) \cdot g(x) = x^3(x - 1) = x^4 - x^3$; all real numbers

d. $\left(\dfrac{f}{g}\right)(x) = \dfrac{f(x)}{g(x)} = \dfrac{x^3}{x - 1}$; all real numbers except 1

EXAMPLE 2 ▶   Revenue, Cost, and Profit

The demand for a certain electronic component is given by $p(x) = 1000 - 2x$. Producing and selling x components involves a monthly fixed cost of \$1999 and a production cost of \$4 for each component.

a. Write the equations that model total revenue and total cost as functions of the components produced and sold in a month.

b. Write the equation that models the profit as a function of the components produced and sold during a month.

c. Find the maximum possible monthly profit.

SOLUTION

a. The revenue for the components is given by the product of $p(x) = 1000 - 2x$ and x, the number of units sold.

$$R(x) = p(x) \cdot x = (1000 - 2x)x = 1000x - 2x^2 \text{ dollars}$$

The monthly total cost is the sum of the variable cost, $4x$, and the fixed cost, 1999.

$$C(x) = 4x + 1999 \text{ dollars}$$

b. The monthly profit for the production and sale of the electronic components is the difference between the revenue and cost functions.

$$P(x) = R(x) - C(x) = (1000x - 2x^2) - (4x + 1999) = -2x^2 + 996x - 1999$$

c. The maximum monthly profit occurs where $x = \dfrac{-996}{2(-2)} = 249$. The maximum profit is $P(249) = 122{,}003$ dollars.

EXAMPLE 3 ▶ Unconventional Vehicle Sales

In December 2010, the U.S. Energy Information Administration (EIA) presented results projecting rapid growth in sales of unconventional vehicles through 2035. The EIA expects unconventional vehicles—vehicles using diesel or alternative fuels and/ or hybrid electric systems—to account for over 40% of U.S. light-duty vehicles sold in 2035. Figure 4.22 shows the number of light-duty cars and trucks in each category sold, in millions, projected to 2035.

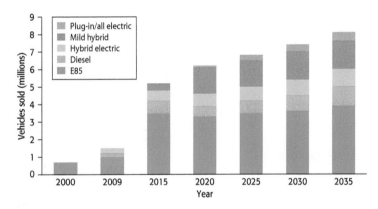

Figure 4.22

The functions that model the number of each type of vehicle sold, in millions, are

E85 flex fuel: $\qquad\qquad\qquad\qquad\quad F(x) = -0.0027x^2 + 0.193x + 0.439$

Diesel: $\qquad\qquad\qquad\qquad\qquad\quad D(x) = 0.028x + 0.077$

Hybrid electric: $\qquad\qquad\qquad\qquad H(x) = 0.025x + 0.154$

Mild hybrid: $\qquad\qquad\qquad\qquad\quad M(x) = -0.006x^2 + 0.350x - 3.380$

Plug-in electric and all electric: $\quad E(x) = -0.001x^2 + 0.081x - 1.115$

where x is the number of years after 2000.

a. Add functions F and D to obtain a function that gives the total number of flex fuel and diesel vehicles sold.

b. Add functions H, M, and E to obtain a function that gives the total number of hybrid and electric vehicles sold.

c. Graph the functions from parts (a) and (b) together on the same axes for the years 2000–2035.

d. Graph the functions from parts (a) and (b) together on the same axes through 2050 ($x = 50$). Will the number of hybrid and electric vehicles exceed the number of flex fuel and diesel vehicles before 2050, according to the models? Based on this graph, do the models appear to be good predictors of unconventional vehicle sales after 2035?

e. Add the functions from parts (a) and (b) to obtain a function that gives the total number of unconventional vehicles sold. According to the EIA projection, there will be 8.1 million unconventional vehicles sold in 2035. Does the model agree with this projection?

SOLUTION

a. Adding functions F and D gives

$$F(x) + D(x) = -0.0027x^2 + 0.193x + 0.439 + 0.028x + 0.077$$
$$= -0.0027x^2 + 0.221x + 0.516$$

b. Adding functions H, M, and E gives

$$H(x) + M(x) + E(x) = 0.025x + 0.154 - 0.006x^2 + 0.350x - 3.380$$
$$- 0.001x^2 + 0.081x - 1.115$$
$$= -0.007x^2 + 0.456x - 4.341$$

c. The graphs of the functions from parts (a) and (b) through $x = 35$ are shown in Figure 4.23.

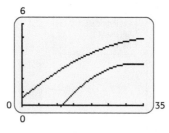

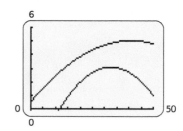

Figure 4.23   **Figure 4.24**

d. The graphs of the functions from parts (a) and (b) through $x = 50$ are shown in Figure 4.24. The graphs do not intersect in this time interval, so the number of hybrid and electric vehicles will not exceed the number of flex fuel and diesel vehicles before 2050. Based on these graphs, the models do not appear to be good predictors after 2035.

e. Adding the two functions from parts (a) and (b) gives

$$U(x) = -0.0027x^2 + 0.221x + 0.516 - 0.007x^2 + 0.456x - 4.341$$
$$= -0.0097x^2 + 0.677x - 3.825$$

Substituting $x = 35$ into function U gives

$$U(35) = -0.0097(35)^2 + 0.677(35) - 3.825 = 7.99$$

Thus, the model predicts $7.99 \approx 8.0$ million, which is close to the EIA projection.

Average Cost

A company's **average cost** per unit, when x units are produced, is the quotient of the function $C(x)$ (the total production cost) and the identity function $I(x) = x$ (the number of units produced). That is, the **average cost** function is

$$\overline{C}(x) = \frac{C(x)}{x}$$

EXAMPLE 4 ▶   Average Cost

Sunny's Greenhouse produces roses, and their total cost for the production of x hundred roses is

$$C(x) = 50x + 500$$

a. Form the average cost function.

b. For which input values is $\overline{C}$ defined? Give a real-world explanation of this answer.

c. Graph $\overline{C}$ for 0 to 5000 roses (50 units). What can you say about the average cost?

SOLUTION

a. The average cost function is the quotient of the cost function $C(x) = 50x + 500$ and the identity function $I(x) = x$.

$$\overline{C}(x) = \frac{50x + 500}{x}$$

b. The average cost function is defined for all real numbers such that $x > 0$ because producing negative units is not possible and the function is undefined for $x = 0$. (This is reasonable, because if nothing is produced, it does not make sense to discuss average cost per unit of product.)

c. The graph is shown in Figure 4.25. It can be seen from the graph that as the number of roses produced increases, the average cost decreases.

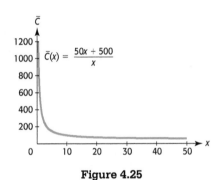

Figure 4.25

Composition of Functions

We have seen that we can convert a temperature of $x°$ Celsius to a Fahrenheit temperature by using the formula $F(x) = \dfrac{9}{5}x + 32$. There are also temperature scales in which $0°$ is absolute zero, which is set at $-273.15°C$. One of these, called the Kelvin temperature scale (K), uses the same degree size as Celsius until it reaches absolute zero, which is 0 K. To convert a Celsius temperature to a Kelvin temperature, we can add 273.15 to the Celsius temperature, and to convert x Kelvin to a Celsius temperature, we can use the formula

$$C(x) = x - 273.15$$

To convert from x Kelvin to a Fahrenheit temperature, we find $F(C(x))$ by substituting $C(x) = x - 273.15$ for x in $F(x) = \dfrac{9}{5}x + 32$, getting

$$F(C(x)) = F(x - 273.15) = \frac{9}{5}(x - 273.15) + 32 = \frac{9}{5}x - 459.67$$

That is, we can convert from the Kelvin scale to the Fahrenheit scale by using

$$F(K) = \frac{9}{5}K - 459.67$$

The process we have used to get a new function from these two functions is called **composition of functions**. The function $F(K(x))$ is called a **composite function**.

Composite Function

The composite function f of g is denoted by $f \circ g$ and defined by

$$(f \circ g)(x) = f(g(x))$$

The domain of $f \circ g$ is the subset of the domain of g for which $f \circ g$ is defined.
The composite function $g \circ f$ is defined by $(g \circ f)(x) = g(f(x))$.
The domain of $g \circ f$ is the subset of the domain of f for which $g \circ f$ is defined.

When computing the composite function $f \circ g$, keep in mind that the output of g becomes the input for f and that the rule for f is applied to this new input.

A composite function machine can be thought of as a machine within a machine. Figure 4.26 shows a composite "function machine" in which the input is denoted by x, the inside function is g, the outside function is f, the composite function rule is denoted by $f \circ g$, and the composite function output is symbolized by $f(g(x))$. Note that for most functions f and g, $f \circ g \neq g \circ f$.

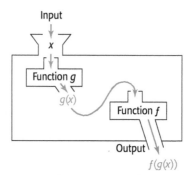

Figure 4.26

EXAMPLE 5 ▶ Function Composition and Orange Juice

There are two machines in a room; one of the machines squeezes oranges to make orange juice, and the other is a canning machine that puts the orange juice into cans. Thinking of these two processes as functions, we name the squeezing function g and the canning function f.

a. Describe the composite function $(f \circ g)(\text{orange}) = f(g(\text{orange}))$.

b. Describe the composite function $(g \circ f)(\text{orange}) = g(f(\text{orange}))$.

c. Which function, $f(g(\text{orange}))$, $g(f(\text{orange}))$, neither, or both, makes sense in context?

SOLUTION

a. The process by which the output $f(g(\text{orange}))$ is obtained is easiest to understand by thinking "from the inside out." Think of the input as an orange. The orange first goes into the *g* machine, so it is squeezed. The output of *g*, liquid orange juice, is then put into the *f* machine, which puts it into a can. The result is a can containing orange juice (Figure 4.27).

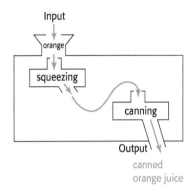

Figure 4.27

b. Again thinking from the inside out, the output *g*(*f*(orange)) is obtained by first putting an orange into the *f* machine, which puts it into a can. The output of *f*, the canned orange, is then put into the squeezing machine, *g*. The result is a compacted can, containing an orange (Figure 4.28).

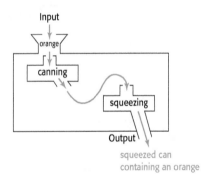

Figure 4.28

c. As can be seen from the results of parts (a) and (b), the order in which the functions appear in the composite function symbol does make a difference. The process that makes sense is the one in part (a), *f*(*g*(orange)).

EXAMPLE 6 ▶   **Finding Composite Function Outputs**

Find the following composite function outputs, using $f(x) = 2x - 5$, $g(x) = 6 - x^2$, and $h(x) = \dfrac{1}{x}$. Give the domain of each new function formed.

a. $(h \circ f)(x) = h(f(x))$ **b.** $(f \circ g)(x) = f(g(x))$ **c.** $(g \circ f)(x) = g(f(x))$

SOLUTION

a. First, the output of *f* becomes the input for *h*.

$$(h \circ f)(x) = h(\overbrace{f(x)}) = h(2x - 5) = \frac{1}{2x - 5}$$

Next, the rule for *h* is applied to this input.

The domain of this function is all $x \neq \dfrac{5}{2}$ because $\dfrac{1}{2x - 5}$ is undefined if $x = \dfrac{5}{2}$.

b. $(f \circ g)(x) = f(g(x)) = f(6 - x^2) = 2(6 - x^2) - 5 = 12 - 2x^2 - 5$
$$= -2x^2 + 7$$

The domain of this function is the set of all real numbers.

c. $(g \circ f)(x) = g(f(x)) = g(2x - 5) = 6 - (2x - 5)^2 = 6 - (4x^2 - 20x + 25)$
$$= 6 - 4x^2 + 20x - 25 = -4x^2 + 20x - 19$$

The domain of this function is the set of all real numbers.

Technology Note

If you have two functions input as Y_1 and Y_2, you can enter the sum, difference, product, quotient, or composition of the two functions in Y_3. These combinations of functions can then be graphed or evaluated for input values of x.

EXAMPLE 7 ▶ **Combinations with Functions**

If $f(x) = \sqrt{x - 5}$ and $g(x) = 2x^2 - 4$,

a. Graph $(f + g)(x)$ using the window $[-1, 10]$ by $[0, 200]$. What is the domain of $f + g$?

b. Graph $\left(\dfrac{f}{g}\right)(x)$ using the window $[5, 15]$ by $[0, 0.016]$. What is the domain of $\dfrac{f}{g}$?

c. Compute $(f \circ g)(-3)$ and $(g \circ f)(9)$.

SOLUTION

a. Functions f and g are entered as Y_1 and Y_2 in a graphing calculator and the sum of f and g is entered in Y_3, as shown in Figure 4.29(a). The graph of $(f + g)(x)$ is shown in Figure 4.29(b).

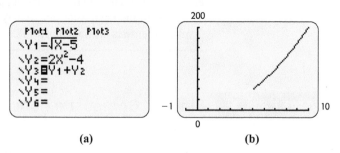

(a)   (b)

Figure 4.29

The domain of $(f + g)(x)$ is $x \geq 5$ because values of x less than 5 make $f(x) + g(x)$ undefined. This can also be seen in the graph in Figure 4.29(b).

b. The quotient of f and g is entered in Y_3, as shown in Figure 4.30(a). The graph of $\left(\dfrac{f}{g}\right)(x)$ is shown in Figure 4.30(b). The domain of $\left(\dfrac{f}{g}\right)(x)$ is $x \geq 5$ because values of x less than 5 make $\dfrac{f(x)}{g(x)}$ undefined. This can also be seen in the graph in Figure 4.30(b).

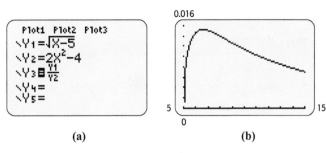

Figure 4.30

c. To compute $(f \circ g)(-3)$, we enter $Y_1(Y_2(-3))$, obtaining 3 (Figure 4.31(a)). To compute $(g \circ f)(9)$, we enter $Y_2(Y_1(9))$, obtaining 4 (Figure 4.31(b)).

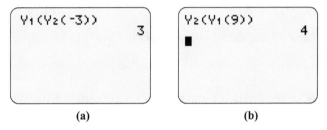

Figure 4.31

Skills CHECK 4.2

In Exercises 1–8, find the following: (a) $(f + g)(x)$, (b) $(f - g)(x)$, (c) $(f \cdot g)(x)$, (d) $\left(\dfrac{f}{g}\right)(x)$, and (e) the domain of $\dfrac{f}{g}$.

1. $f(x) = 3x - 5; g(x) = 4 - x$

2. $f(x) = 2x - 3; g(x) = 5 - x$

3. $f(x) = x^2 - 2x; g(x) = 1 + x$

4. $f(x) = 2x^2 - x; g(x) = 2x + 1$

5. $f(x) = \dfrac{1}{x}; g(x) = \dfrac{x + 1}{5}$

6. $f(x) = \dfrac{x - 2}{3}; g(x) = \dfrac{1}{x}$

7. $f(x) = \sqrt{x}; g(x) = 1 - x^2$

8. $f(x) = x^3; g(x) = \sqrt{x + 3}$

9. If $f(x) = x^2 - 5x$ and $g(x) = 6 - x^3$, evaluate

    a. $(f + g)(2)$ **b.** $(g - f)(-1)$

    c. $(f \cdot g)(-2)$ **d.** $\left(\dfrac{g}{f}\right)(3)$

10. If $f(x) = 4 - x^2$ and $g(x) = x^3 + x$, evaluate

    a. $(f + g)(1)$ **b.** $(f - g)(-2)$

    c. $(f \cdot g)(-3)$ **d.** $\left(\dfrac{g}{f}\right)(2)$

In Exercises 11–20, find (a) $(f \circ g)(x)$ and (b) $(g \circ f)(x)$.

11. $f(x) = 2x - 6; g(x) = 3x - 1$

12. $f(x) = 3x - 2; g(x) = 2x - 2$

13. $f(x) = x^2; g(x) = \dfrac{1}{x}$ **14.** $f(x) = x^3; g(x) = \dfrac{2}{x}$

15. $f(x) = \sqrt{x - 1}; g(x) = 2x - 7$

16. $f(x) = \sqrt{3 - x}; g(x) = x - 5$

17. $f(x) = |x - 3|; g(x) = 4x$

18. $f(x) = |4 - x|; g(x) = 2x + 1$

19. $f(x) = \dfrac{3x + 1}{2}; g(x) = \dfrac{2x - 1}{3}$

20. $f(x) = \sqrt[3]{x + 1}; g(x) = x^3 + 1$

In Exercises 21 and 22, use f(x) and g(x) to evaluate each expression.

21. $f(x) = 2x^2; g(x) = \dfrac{x - 5}{3}$

   a. $(f \circ g)(2)$ **b.** $(g \circ f)(-2)$

22. $f(x) = (x - 1)^2; g(x) = 3x - 1$

   a. $(f \circ g)(2)$ **b.** $(g \circ f)(-2)$

In Exercises 23 and 24, use the following graphs of f and g to evaluate the functions.

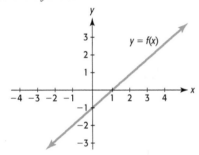

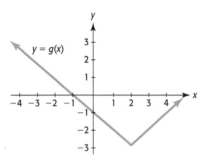

23. a. $(f + g)(2)$ **b.** $(f \circ g)(-1)$

   c. $\left(\dfrac{f}{g}\right)(4)$ **d.** $(f \circ g)(1)$

   e. $(g \circ f)(-2)$

24. a. $(g - f)(-2)$ **b.** $(f \circ g)(3)$

   c. $\left(\dfrac{f}{g}\right)(0)$ **d.** $(f \circ g)(-2)$

   e. $(g \circ f)(2)$

EXERCISES 4.2

25. *Profit* Suppose that the total weekly cost for the production and sale of x bicycles is $C(x) = 23x + 3420$ dollars and that the total revenue is given by $R(x) = 89x$ dollars, where x is the number of bicycles.

   a. Write the equation of the function that models the weekly profit from the production and sale of x bicycles.

   b. What is the profit on the production and sale of 150 bicycles?

26. *Profit* Suppose that the total weekly cost for the production and sale of television sets is $C(x) = 189x + 5460$ and that the total revenue is given by $R(x) = 988x$, where x is the number of televisions and $C(x)$ and $R(x)$ are in dollars.

   a. Write the equation of the function that models the weekly profit from the production and sale of x television sets.

   b. What is the profit on the production and sale of 80 television sets in a given week?

27. *Revenue and Cost* The total revenue function for LED TVs is given by $R = 1050x$ dollars, and the total cost function for the TVs is $C = 10,000 + 30x + x^2$ dollars, where x is the number of TVs that are produced and sold.

   a. Which function is quadratic, and which is linear?

   b. Form the profit function for the TVs from these two functions.

   c. Is the profit function a linear function, a quadratic function, or neither of these?

28. *Revenue and Cost* The total monthly revenue function for Easy-Ride golf carts is given by $R = 26,600x$ dollars, and the total monthly cost function for the carts is $C = 200,000 + 4600x + 2x^2$ dollars, where x is the number of golf carts that are produced and sold.

a. Which function is quadratic, and which is linear?

b. Form the profit function for the golf carts from these two functions.

c. Is the profit function a linear function, a quadratic function, or neither of these?

29. *Revenue and Cost* The total weekly revenue function for a certain digital camera is given by $R = 550x$ dollars, and the total weekly cost function for the cameras is $C = 10,000 + 30x + x^2$ dollars, where x is the number of cameras that are produced and sold.

a. Find the profit function.

b. Find the number of cameras that gives maximum profit.

c. Find the maximum possible profit.

30. *Revenue and Cost* The total monthly revenue function for camcorders is given by $R = 6600x$ dollars, and the total monthly cost function for the camcorders is $C = 2000 + 4800x + 2x^2$ dollars, where x is the number of camcorders that are produced and sold.

a. Find the profit function.

b. Find the number of camcorders that gives maximum profit.

c. Find the maximum possible profit.

31. *Average Cost* If the monthly total cost of producing 27-inch television sets is given by $C(x) = 50,000 + 105x$, where x is the number of sets produced per month, then the average cost per set is given by

$$\overline{C}(x) = \frac{50,000 + 105x}{x}$$

a. Explain how $C(x)$ and another function can be combined to obtain the average cost function.

b. What is the average cost per set if 3000 sets are produced?

32. *Cost-Benefit* Suppose that for a certain city the cost C of obtaining drinking water that contains $p\%$ impurities (by volume) is given by

$$C = \frac{120,000}{p} - 1200$$

a. This function can be considered as the difference of what two functions?

b. What is the cost of drinking water that is 80% impure?

33. *Printers* The weekly total cost function for producing a dot matrix printer is $C(x) = 3000 + 72x$, where x is the number of printers produced per week.

a. Form the weekly average cost function for this product.

b. Find the average cost for the production of 100 printers.

34. *Electronic Components* The monthly cost of producing x electronic components is $C(x) = 2.15x + 2350$.

a. Find the monthly average cost function.

b. Find the average cost for the production of 100 components.

35. *Football Tickets* At a certain school, the number of student tickets sold for a home football game can be modeled by $S(p) = 62p + 8500$, where p is the winning percent of the home team. The number of non-student tickets sold for these home games is given by $N(p) = 0.5p^2 + 16p + 4400$.

a. Write an equation for the total number of tickets sold for a home football game at this school as a function of the winning percent p.

b. What is the domain for the function in part (a) in this context?

c. Assuming that the football stadium is filled to capacity when the team wins 90% of its home games, what is the capacity of the school's stadium?

36. *T-Shirt Sales* Let $T(c)$ be the number of T-shirts that are sold when the shirts have c colors and $P(c)$ be the price, in dollars, of a T-shirt that has c colors. Write a sentence explaining the meaning of the function $(T \cdot P)(c)$.

37. *Harvesting* A farmer's main cash crop is tomatoes, and the tomato harvest begins in the month of May. The number of bushels of tomatoes harvested on the xth day of May is given by the equation $B(x) = 6(x + 1)^{3/2}$. The market price in dollars of 1 bushel of tomatoes on the xth day of May is given by the formula $P(x) = 8.5 - 0.12x$.

a. How many bushels did the farmer harvest on May 8?

b. What was the market price of tomatoes on May 8?

c. How much was the farmer's tomato harvest worth on May 8?

d. Write a model for the worth W of the tomato harvest on the xth day of May.

38. *Total Cost* If the fixed cost for producing a product is $3000 and the variable cost for the production is $3.30x^2$ dollars, where x is the number of units produced, form the total cost function $C(x)$.

39. *Profit* A manufacturer of satellite systems has monthly fixed costs of $32,000 and variable costs of $432 per system, and it sells the systems for $592 per unit.

   a. Write the function that models the profit P from the production and sale of x units of the system.

   b. What is the profit if 600 satellite systems are produced and sold in 1 month?

   c. At what rate does the profit grow as the number of units increases?

40. *Profit* A manufacturer of computers has monthly fixed costs of $87,500 and variable costs of $87 per computer, and it sells the computers for $295 per unit.

   a. Write the function that models the profit P from the production and sale of x computers.

   b. What is the profit if 700 computers are produced and sold in 1 month?

   c. What is the y-intercept of the graph of the profit function? What does it mean?

41. *Educational Attainment* The table below shows the percent of male and female individuals aged 16 to 24 enrolled in college as of October of each year who completed high school during the preceding 12 months.

Year	Male (%)	Female (%)
2000	59.9	66.2
2001	60.1	63.5
2002	62.1	68.4
2003	61.2	66.5
2004	61.4	71.5
2005	66.5	70.4
2006	65.8	66.1
2007	66.1	68.3
2008	65.9	71.6

(Source: National Center for Education Statistics)

   a. Will adding the percents for males and females create a new function that gives the total percent of all 16–24-year-old high school graduates enrolled in college? Why or why not?

   b. A function that approximately models the total percent of all 16–24-year-old high school graduates enrolled in college is $f(t) = 61.925t^{0.041}$, where t is the number of years after 1999. Use this model to estimate the total percent in 2002 and 2008.

   c. Add the percent of males and the percent of females for the 2002 data and the 2008 data in the table and divide by 2. Do these values conflict with your answers in part (b)?

42. *Population of Children* The following table gives the estimated population (in millions) of U.S. boys age 5 and under and U.S. girls age 5 and under in selected years.

	1995	2000	2005	2010
Boys	10.02	9.71	9.79	10.24
Girls	9.57	9.27	9.43	9.77

(Source: U.S. Department of Commerce)

A function that models the population (in millions) of U.S. boys age 5 and under t years after 1990 is $B(t) = 0.0076t^2 - 0.1752t + 10.705$, and a function that models the population (in millions) of U.S. girls age 5 and under t years after 1990 is $G(t) = 0.0064t^2 - 0.1448t + 10.12$.

   a. Find the equation of a function that models the estimated U.S. population (in millions) of children age 5 and under t years after 1990.

   b. Use the result of part (a) to estimate the U.S. population of children age 5 and under in 2003.

43. *Function Composition* Think of each of the following processes as a function designated by the indicated letter: f, placing in a styrofoam container; g, grinding. Describe each of the functions in parts (a) to (e), then answer part (f).

   a. $f(\text{meat})$ **b.** $g(\text{meat})$ **c.** $(g \circ g)(\text{meat})$

   d. $f(g(\text{meat}))$ **e.** $g(f(\text{meat}))$

   f. Which gives a sensible operation: the function in part (d) or the function in part (e)?

44. *Function Composition* Think of each of the following processes as a function designated by the indicated letter: *f*, putting on a sock; *g*, taking off a sock. Describe each of the functions in parts (a) to (c).

   a. f(left foot)   **b.** $f(f$(left foot))

   c. $(g \circ f)$(right foot)

45. *Shoe Sizes* A woman's shoe that is size x in Japan is size $s(x)$ in the United States, where $s(x) = x - 17$. A woman's shoe that is size x in the United States is size $p(x)$ in Britain, where $p(x) = x - 1.5$. Find a function that will convert Japanese shoe size to British shoe size.
(Source: Kuru International Exchange Association)

46. *Shoe Sizes* A man's shoe that is size x in Britain is size $d(x)$ in the United States, where $d(x) = x + 0.5$. A man's shoe that is size x in the United States is size $t(x)$ in Continental size, where $t(x) = x + 34.5$. Find a function that will convert British shoe size to Continental shoe size.
(Source: Kuru International Exchange Association)

47. *Exchange Rates* On March 11, 2011, each Japanese yen was worth 0.34954 Russian ruble and each Chilean peso was worth 0.171718 Japanese yen. Find the value of 100 Chilean pesos in Russian rubles on March 11, 2011. Round the answer to two decimal places.
(Source: *x*-rates.com)

48. *Exchange Rates* On March 11, 2011, each euro was worth 1.3773 U.S. dollars and each Mexican peso was worth 0.06047 euro. Find the value of 100 Mexican pesos in U.S. dollars.

49. *Facebook* If $f(x)$ represents the number of Facebook users x years after 2000 and $g(x)$ represents the number of MySpace users x years after 2000, what function represents the percent of Facebook users who are MySpace users x years after 2000?

50. *Home Computers* If $f(x)$ represents the percent of American homes with computers and $g(x)$ represents the number of American homes, with x equal to the number of years after 1990, then what function represents the number of American homes with computers, with x equal to the number of years after 1990?

51. *Education* If the function $f(x)$ gives the number of female PhDs produced by American universities x years after 2005 and the function $g(x)$ gives the number of male PhDs produced by American universities x years after 2005, what function gives the total number of PhDs produced by American universities x years after 2005?

52. *Wind Chill* If the air temperature is 25°F, the wind chill temperature C is given by $C = 59.914 - 2.35s - 20.14\sqrt{s}$, where s is the wind speed in miles per hour.

   a. State two functions whose difference gives this function.

   b. Graph this function for $3 \leq s \leq 12$.

   c. Is the function increasing or decreasing on this domain?

53. *Discount Prices* Half-Price Books has a sale with an additional 20% off the regular (1/2) price of their books. What percent of the retail price is charged during this sale?
(Source: Half-Price Books, Cleveland, Ohio)

54. *Average Cost* The monthly average cost of producing 42-inch plasma TVs is given by $\overline{C}(x) = \dfrac{100{,}000}{x} + 150$ dollars, where x is the number of sets produced.

   a. Graph this function for $x > 0$.

   b. Will the average cost function decrease or increase as the number of sets produced increases?

   c. What transformations of the graph of the reciprocal function give the graph of this function?

SECTION PREVIEW **Loans**

A business property is purchased with a promise to pay off a $60,000 loan plus the $16,500 interest on this loan by making 60 monthly payments of $1275. The amount of money remaining to be paid on the loan plus interest is given by the function

$$f(x) = 76{,}500 - 1275x$$

where x is the number of months for which payments have been made. If we know how much remains to be paid and want to find how many months remain to make payments, we can find the inverse of the above function. (See Example 6.)

In this section, we determine if two functions are inverses of each other and find the inverse of a function if it has one. We also solve applied problems by using inverse functions. ■

Inverse Functions

We have seen that we can convert a temperature of $x°$ Celsius to a Fahrenheit temperature by using the formula

$$F(x) = \frac{9}{5}x + 32$$

The function that can be used to convert a temperature of $x°$ Fahrenheit back to a Celsius temperature is

$$C(x) = \frac{5x - 160}{9}$$

To see that one of these functions "undoes" what the other one does, see Table 4.8, which gives selected inputs and the resulting outputs for both functions. Observe that if we input the number 50 into F, the output after it is operated on by F is 122, and if 122 is then operated on by C, the output is 50, which means that $C(F(50)) = 50$. In addition, if 122 is operated on by C, the output is 50, and if 50 is operated on by F, the output is 122, so $F(C(122)) = 122$.

Table 4.8

x° C	−20	0	50	100
F(x)	−4	32	122	212

x° F	−4	32	122	212
C(x)	−20	0	50	100

In fact, $C(F(x)) = x$ and $F(C(x)) = x$ for any input x, as we will see in Example 1. This means that both of these composite functions act as identity functions because their outputs are the same as their inputs. This happens because the second function performs the inverse operations of the first function. Because of this, we say that $C(x)$ and $F(x)$ are **inverse functions**.

Inverse Functions

Functions f and g for which $f(g(x)) = x$ for all x in the domain of g and $g(f(x)) = x$ for all x in the domain of f are called inverse functions. In this case, we denote g by f^{-1}, read as "f inverse."

EXAMPLE 1 ▶   Temperature Measurement

The function that can be used to convert a temperature of $x°$ Celsius to a Fahrenheit temperature is

$$F(x) = \frac{9}{5}x + 32$$

The function that can be used to convert a temperature of $x°$ Fahrenheit back to a Celsius temperature is

$$C(x) = \frac{5x - 160}{9}$$

To see how the two conversion formulas for temperature are related, find $C(F(x))$ and $F(C(x))$ and determine if the functions are inverse functions.

SOLUTION

We compute $C(F(x))$ and $F(C(x))$. Evaluating C at $F(x)$ gives

$$C(F(x)) = C\left(\frac{9}{5}x + 32\right) = \frac{5\left(\frac{9}{5}x + 32\right) - 160}{9} = \frac{9x + 160 - 160}{9} = x$$

and evaluating F at $C(x)$ gives

$$F(C(x)) = F\left(\frac{5x - 160}{9}\right) = \frac{9}{5} \cdot \left(\frac{5x - 160}{9}\right) + 32 = \frac{45x}{45} - \frac{1440}{45} + 32 = x$$

Because $C(F(x)) = x$ and $F(C(x)) = x$, the two functions F and C are inverses.

Consider a function that doubles each input. Its inverse takes half of each input. For example, the discrete function f with equation $f(x) = 2x$ and domain $\{1, 4, 5, 7\}$ has the range $\{2, 8, 10, 14\}$. The inverse of this function has the equation $f^{-1}(x) = \frac{x}{2}$. The domain of the inverse function is the set of outputs of the original function, $\{2, 8, 10, 14\}$. The outputs of the inverse function form the set $\{1, 4, 5, 7\}$, which is the domain of the original function. Figure 4.32 illustrates the relationship between the domains and ranges of the function f and its inverse f^{-1}. In this case and in every case, the domain of the inverse function is the range of the original function, and the range of the inverse function is the domain of the original function.

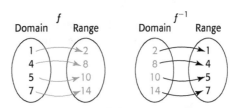

Figure 4.32

It is very important to note that the "$^{-1}$" used in f^{-1} and g^{-1} is *not* an exponent, but rather a symbol used to denote the inverse of the function. The expression f^{-1} *always* refers to the inverse function of f and *never* to the reciprocal $\frac{1}{f}$ of f; that is,

$$f^{-1}(x) \neq \frac{1}{f(x)}.$$

One-to-One Functions

In general, we can show that a function has an inverse if it is a **one-to-one function**. We define a one-to-one function as follows.

One-to-One Function

A function f is a one-to-one function if each output of the function corresponds to exactly one input in the domain of the function. This means there is a one-to-one correspondence between the elements of the domain and the elements of the range.

This statement means that for a one-to-one function f, $f(a) \neq f(b)$ if $a \neq b$.

EXAMPLE 2 ▶ **One-to-One Functions**

Determine if each of the functions is a one-to-one function.

a. $f(x) = 3x^4$ **b.** $f(x) = x^3 - 1$

SOLUTION

a. Clearly $a = -2$ and $b = 2$ are different inputs, but they both give the same output for $f(x) = 3x^4$.

$$f(-2) = 3(-2)^4 = 48 \quad \text{and} \quad f(2) = 3(2)^4 = 48$$

Thus, $y = 3x^4$ is not a one-to-one function.

b. Suppose that $a \neq b$. Then $a^3 \neq b^3$ and $a^3 - 1 \neq b^3 - 1$, so if $f(x) = x^3 - 1$, $f(a) \neq f(b)$. This satisfies the condition $f(a) \neq f(b)$ if $a \neq b$, so the function $f(x) = x^3 - 1$ is one-to-one.

Recall that no vertical line can intersect the graph of a function in more than one point. The definition of a one-to-one function means that if a function is one-to-one, a horizontal line will intersect its graph in at most one point.

Horizontal Line Test

A function is one-to-one if no horizontal line can intersect the graph of the function in more than one point.

EXAMPLE 3 ▶ **Horizontal Line Test**

Determine if each of the functions is one-to-one by using the horizontal line test.

a. $y = 3x^4$ **b.** $y = x^3 - 1$

SOLUTION

a. We can see that $y = 3x^4$ is not a one-to-one function because we can observe that the graph of this function does not pass the horizontal line test (Figure 4.33(a)). Note that when $x = 2$, $y = 48$, and when $x = -2$, $y = 48$.

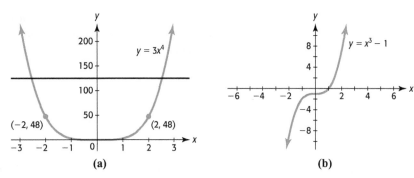

Figure 4.33

b. The graph in Figure 4.33(b) and the horizontal line test indicate that the function $y = x^3 - 1$ is one-to-one. That is, no horizontal line will intersect this graph in more than one point.

We summarize the information about inverse functions as follows.

Inverse Functions

The functions f and g are inverse functions if, whenever the pair (a, b) satisfies $y = f(x)$, the pair (b, a) satisfies $y = g(x)$. Note that when this happens,

$$f(g(x)) = x \quad \text{and} \quad g(f(x)) = x$$

for all x in the domain of g and f, respectively. The domain of the function f is the range of its inverse g, and the domain of g is the range of f.

EXAMPLE 4 ▶ **Inverse Functions**

a. Determine if $f(x) = x^3 - 1$ has an inverse function.

b. Verify that $g(x) = \sqrt[3]{x + 1}$ is the inverse function of $f(x) = x^3 - 1$.

c. Find the domain and range of each function.

SOLUTION

a. Because each output of the function $f(x) = x^3 - 1$ corresponds to exactly one input, the function is one-to-one. Thus, it has an inverse function.

b. To verify that the functions $f(x) = x^3 - 1$ and $g(x) = \sqrt[3]{x + 1}$ are inverse functions, we show that the compositions $f(g(x))$ and $g(f(x))$ are each equal to the identity function, x.

$$f(g(x)) = f(\sqrt[3]{x + 1}) = (\sqrt[3]{x + 1})^3 - 1 = (x + 1) - 1 = x \quad \text{and}$$
$$g(f(x)) = g(x^3 - 1) = \sqrt[3]{(x^3 - 1) + 1} = \sqrt[3]{x^3} = x$$

Thus, we have verified that these functions are inverses.

c. The cube of any real number decreased by 1 is a real number, so the domain of f is the set of all real numbers. The cube root of any real number plus 1 is also a real number, so the domain of g is the set of all real numbers. The ranges of f and g are the set of real numbers.

By using the definition of inverse functions, we can find the equation for the inverse function of f by interchanging x and y in the equation $y = f(x)$ and solving the new equation for y.

Finding the Inverse of a Function

To find the inverse of the function f that is defined by the equation $y = f(x)$:

1. Rewrite the equation replacing $f(x)$ with y.

2. Interchange x and y in the equation defining the function.

3. Solve the new equation for y. If this equation cannot be solved uniquely for y, the original function has no inverse function.

4. Replace y with $f^{-1}(x)$.

EXAMPLE 5 ▶   **Finding an Inverse Function**

a. Find the inverse function of $f(x) = \dfrac{2x - 1}{3}$.

b. Graph $f(x) = \dfrac{2x - 1}{3}$ and its inverse function on the same axes.

SOLUTION

a. Using the steps for finding the inverse of a function, we have

$$y = \frac{2x - 1}{3} \qquad \text{Replace } f(x) \text{ with } y.$$

$$x = \frac{2y - 1}{3} \qquad \text{Interchange } x \text{ and } y.$$

$$3x = 2y - 1 \qquad \text{Solve for } y.$$

$$3x + 1 = 2y$$

$$\frac{3x + 1}{2} = y$$

$$f^{-1}(x) = \frac{3x + 1}{2} \qquad \text{Replace } y \text{ with } f^{-1}(x).$$

b. The graphs of $f(x) = \dfrac{2x - 1}{3}$ and its inverse $f^{-1}(x) = \dfrac{3x + 1}{2}$ are shown in Figure 4.34.

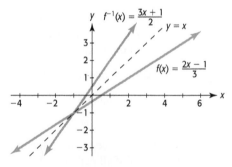

Figure 4.34

Notice that the graphs of $y = f(x)$ and $y = f^{-1}(x)$ in Figure 4.34 appear to be reflections of each other about the line $y = x$. This occurs because, if you were to choose any point (a, b) on the graph of the function f and interchange the x- and y-coordinates, the new point (b, a) would be on the graph of the inverse function f^{-1}. This should make sense because the inverse function is formed by interchanging x and y in the equation defining the original function. In fact, this relationship occurs for every function and its inverse.

Graphs of Inverse Functions

The graphs of a function and its inverse are symmetric with respect to the line $y = x$.

We again illustrate the symmetry of graphs of inverse functions in Figure 4.35, which shows the graphs of the inverse functions $f(x) = x^3 - 2$ and $f^{-1}(x) = \sqrt[3]{x + 2}$. Note that the point $(0, -2)$ is on the graph of $f(x) = x^3 - 2$ and the point $(-2, 0)$ is on the graph of $f^{-1}(x) = \sqrt[3]{x + 2}$.

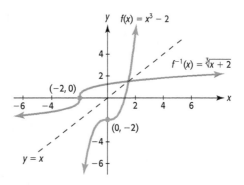

Figure 4.35

EXAMPLE 6 ▶ **Loan Repayment**

A business property is purchased with a promise to pay off a $60,000 loan plus the $16,500 interest on this loan by making 60 monthly payments of $1275. The amount of money remaining to be paid on the loan plus interest is given by the function

$$f(x) = 76{,}500 - 1275x$$

where x is the number of monthly payments remaining.

a. Find the inverse of this function.

b. Use the inverse to determine how many monthly payments remain if $35,700 remains to be paid.

SOLUTION

a. Replacing $f(x)$ with y gives $y = 76{,}500 - 1275x$.

Interchanging x and y gives the equation $x = 76{,}500 - 1275y$.

Solving this new equation for y gives $y = \dfrac{76{,}500 - x}{1275}$.

Replacing y with $f^{-1}(x)$ gives the inverse function $f^{-1}(x) = \dfrac{76{,}500 - x}{1275}$.

b. The inverse function gives the number of months remaining to make payments if x dollars remain to be paid.

$$f^{-1}(35,700) = \frac{76,500 - 35,700}{1275} = 32$$

so 32 monthly payments remain.

Inverse Functions on Limited Domains

As we have stated, a function cannot have an inverse function if it is not a one-to-one function. However, if there is a limited domain over which such a function is a one-to-one function, then it has an inverse function for this domain. Consider the function $f(x) = x^2$. The horizontal line test on the graph of this function (Figure 4.36(a)) indicates that this function is not a one-to-one function and that the function f does not have an inverse function. To see why, consider the attempt to find the inverse function:

$$y = x^2$$
$$x = y^2 \qquad \text{Interchange } x \text{ and } y.$$
$$\pm\sqrt{x} = y \qquad \text{Solve for } y.$$

This equation is not the inverse function because it is not a function (one value of x gives two values for y). Because we cannot solve $x = y^2$ uniquely for y, $f(x) = x^2$ does not have an inverse function.

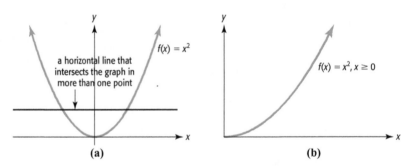

(a)   **(b)**

Figure 4.36

However, if we limit the domain of the function to $x \geq 0$, no horizontal line intersects the graph of $f(x) = x^2$ more than once, so the function is one-to-one on this limited domain (Figure 4.36(b)), and the function has an inverse. If we restrict the domain of the original function by requiring that $x \geq 0$, then the range of the inverse function is restricted to $y \geq 0$, and the equation $y = \sqrt{x}$, or $f^{-1}(x) = \sqrt{x}$, defines the inverse function. Figure 4.37 shows the graph of $f(x) = x^2$ for $x \geq 0$ and its inverse $f^{-1}(x) = \sqrt{x}$. Note that the graphs are symmetric about the line $y = x$.

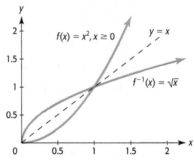

Figure 4.37

EXAMPLE 7 ▶ Velocity of Blood

Because of friction from the walls of an artery, the velocity of blood is greatest at the center of the artery and decreases as the distance x from the center increases. The (nonnegative) velocity in centimeters per second of a blood corpuscle in an artery can be modeled by the function

$$v = k(R^2 - x^2), k > 0, 0 \le x \le R$$

where R is the radius of the artery and k is a constant that is determined by the pressure, the viscosity of the blood, and the length of the artery. In the case where $k = 2$ and $R = 0.1$ centimeter, the velocity can be written as a function of the distance x from the center as

$$v(x) = 2(0.01 - x^2) \text{ centimeter/second}$$

a. Assuming the velocity is nonnegative, does the inverse of this function exist on a domain limited by the context of the application?

b. What is the inverse function?

c. What does the inverse function mean in the context of this application?

SOLUTION

a. Because x represents distance, it is nonnegative, and because $R = 0.1$ and x equals the distance of blood in an artery from the center of the artery, $0 \le x \le 0.1$. Hence, the function is one-to-one for $0 \le x \le 0.1$ and we can find its inverse.

b. We find the inverse function as follows.

$$f(x) = 2(0.01 - x^2)$$
$$y = 2(0.01 - x^2) \qquad \text{Replace } f(x) \text{ with } y.$$
$$x = 2(0.01 - y^2) \qquad \text{Interchange } y \text{ and } x.$$
$$x = 0.02 - 2y^2 \qquad \text{Solve for } y.$$
$$2y^2 = 0.02 - x$$
$$y^2 = \frac{0.02 - x}{2}$$
$$y = \sqrt{\frac{0.02 - x}{2}} \qquad \text{Only nonnegative values of } y \text{ are possible.}$$
$$f^{-1}(x) = \sqrt{\frac{0.02 - x}{2}} \qquad \text{Replace } y \text{ with } f^{-1}(x).$$

c. The inverse function gives the distance from the center of the artery as a function of the velocity of a blood corpuscle.

Skills CHECK 4.3

In Exercises 1 and 2, determine if the function f defined by the arrow diagram has an inverse. If it does, create an arrow diagram that defines the inverse. If not, explain why not.

1.
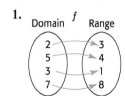

2.
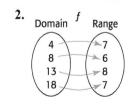

In Exercises 3 and 4, determine if the function f defined by the set of ordered pairs has an inverse. If so, find the inverse.

3. $\{(5, 2), (4, 1), (3, 7), (6, 2)\}$

4. $\{(2, 8), (3, 9), (4, 10), (5, 11)\}$

5. If $f(x) = 3x$ and $g(x) = \dfrac{x}{3}$,

    a. What are $f(g(x))$ and $g(f(x))$?

    b. Are $f(x)$ and $g(x)$ inverse functions?

6. If $f(x) = 4x - 1$ and $g(x) = \dfrac{x + 1}{4}$,

    a. What are $f(g(x))$ and $g(f(x))$?

    b. Are $f(x)$ and $g(x)$ inverse functions?

7. If $f(x) = x^3 + 1$ and $g(x) = \sqrt[3]{x - 1}$, are $f(x)$ and $g(x)$ inverse functions?

8. If $f(x) = (x - 2)^3$ and $g(x) = \sqrt[3]{x} + 2$, are $f(x)$ and $g(x)$ inverse functions?

9. For the function f defined by $f(x) = 3x - 4$, complete the tables below for f and f^{-1}.

x	f(x)
−1	−7
0	
1	
2	
3	

x	f⁻¹(x)
−7	−1
−4	
−1	
2	
5	

10. For the function g defined by $g(x) = 2x^3 - 1$, complete the tables below for g and g^{-1}.

x	g(x)
−2	−17
−1	
0	
1	
2	

x	g⁻¹(x)
−17	−2
−3	
−1	
1	
15	

In Exercises 11–14, determine whether the function is one-to-one.

11. $\{(1, 5), (2, 6), (3, 7), (4, 5)\}$

12. $\{(2, -4), (5, -8), (8, -12), (11, -16)\}$

13. $f(x) = (x - 3)^3$ **14.** $f(x) = \dfrac{1}{x}$

In Exercises 15 and 16, determine whether each graph is the graph of a one-to-one function.

15.

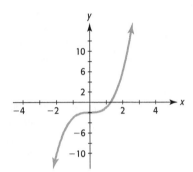

16.

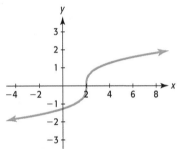

In Exercises 17 and 18, determine whether the function is one-to-one.

17. $y = -2x^4$ **18.** $y = \sqrt{x + 3}$

19. a. Write the inverse of $f(x) = 3x - 4$.

    b. Do the values for f^{-1} in the table of Exercise 9 fit the equation for f^{-1}?

20. a. Write the inverse of $g(x) = 2x^3 - 1$.

    b. Do the values for g^{-1} in the table of Exercise 10 fit the equation for g^{-1}?

21. If function h has an inverse and $h^{-1}(-2) = 3$, find $h(3)$.

22. Find the inverse of $f(x) = \dfrac{1}{x}$.

23. Find the inverse of $g(x) = 4x + 1$.

24. Find the inverse of $f(x) = 4x^2$ for $x \geq 0$.

25. Find the inverse of $g(x) = x^2 - 3$ for $x \geq 0$.

26. Graph $g(x) = \sqrt{x}$ and its inverse $g^{-1}(x)$ for $x \geq 0$ on the same axes.

27. Graph $g(x) = \sqrt[3]{x}$ and its inverse $g^{-1}(x)$ on the same axes.

28. $f(x) = (x - 2)^2$ and $g(x) = \sqrt{x} + 2$ are inverse functions for what values of x?

29. Is the function $f(x) = 2x^3 + 1$ a one-to-one function? Does it have an inverse?

30. Sketch the graph of $y = f^{-1}(x)$ on the axes with the graph of $y = f(x)$, shown below.

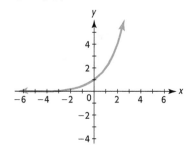

31. Sketch the graph of $y = f^{-1}(x)$ on the axes with the graph of $y = f(x)$, shown below.

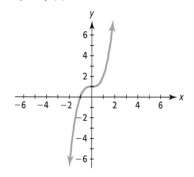

EXERCISES 4.3

32. *Shoe Sizes* A man's shoe that is size x in Britain is size $d(x)$ in the United States, where $d(x) = x + 0.5$.

   a. Find the inverse of the function.

   b. Use the inverse function to find the British size of a shoe if it is U.S. size $8\frac{1}{2}$.
   (Source: Kuru International Exchange Association)

33. *Shoe Sizes* A man's shoe that is size x in the United States is size $t(x)$ in Continental size, where $t(x) = x + 34.5$.

   a. Find a function that will convert Continental shoe size to U.S. shoe size.

   b. Use the inverse function to find the U.S. size if the Continental size of a shoe is 43.
   (Source: Kuru International Exchange Association)

34. *Investments* If x dollars are invested at 10% for 6 years, the future value of the investment is given by $S(x) = x + 0.6x$.

   a. Find the inverse of this function.

   b. What do the outputs of the inverse function represent?

   c. Use this function to find the amount of money that must be invested for 6 years at 10% to have a future value of $24,000.

35. *Cigarettes* For the years 1997–2009, the percent of high school seniors who have tried cigarettes is given by $f(t) = 82.074 - 2.087t$, where t is the number of years after 1990. Find the inverse of this function and use it to find the year in which the percent fell below 41%.

   (Source: Monitoring the Future)

36. *Apparent Temperature* If the outside temperature is 90°F, the apparent temperature is given by $A(x) = 82.35 + 29.3x$, where x is the humidity written as a decimal. Find the inverse of this function and use it to find the percent humidity that will give an apparent temperature of 97° if the temperature is 90°F.
   (Source: "Temperature-Humidity Indices," *The UMAP Journal*, Fall 1989)

37. *Antidepressants* The function that models the percent of children ages 0–19 taking antidepressants from 2004 to 2009 is $f(x) = -0.085x + 2.97$, where x is the number of years after 2000.

   a. Find the inverse of this function. What do the outputs of the inverse function represent?

   b. Use the inverse function to find when the percentage is 2.3%.
   (Source: Medco Health Solutions)

38. *Body-Heat Loss* The model for body-heat loss depends on the coefficient of convection $K = f(x)$, which depends on wind speed x according to the equation $f(x) = 4\sqrt{4x + 1}$.

   a. What are the domain and range of this function without regard to the context of this application?

   b. Find the inverse of this function.

   c. What are the domain and range of the inverse function?

   d. In the context of the application, what are the domain and range of the inverse function?

39. *Algorithmic Relationship* For many species of fish, the weight W is a function of the length x, given by

$W = kx^3$, where k is a constant depending on the species. Suppose $k = 0.002$, W is in pounds, and x is in inches, so the weight is $W(x) = 0.002x^3$.

a. Find the inverse function of this function.

b. What does the inverse function give?

c. Use the inverse function to find the length of a fish that weighs 2 pounds.

d. In the context of the application, what are the domain and range of the inverse function?

40. *Decoding Messages* If we assign numbers to the letters of the alphabet as follows and assign 27 to a blank space, we can convert a message to a numerical sequence. We can "encode" a message by adding 3 to each number that represents a letter in a message.

A	B	C	D	E	F	G	H	I	J	K	L	M
1	2	3	4	5	6	7	8	9	10	11	12	13

N	O	P	Q	R	S	T	U	V	W	X	Y	Z
14	15	16	17	18	19	20	21	22	23	24	25	26

Thus, the message "Go for it" can be encoded by using the numbers to represent the letters and further encoded by using the function $C(x) = x + 3$. The coded message would be 10 18 30 9 18 21 30 12 23. Find the inverse of the function and use it to decode 23 11 8 30 21 8 4 15 30 23 11 12 17 10.

41. *Decoding Messages* Use the numerical representation from Exercise 40 and the inverse of the encoding function $C(x) = 3x + 2$ to decode 41 5 35 17 83 41 77 83 14 5 77.

42. *Social Security Numbers and Income Taxes* Consider the function that assigns each person who pays federal income tax his or her Social Security number. Is this a one-to-one function? Explain.

43. *Checkbook Balance* Consider the function with the check number in your checkbook as input and the dollar amount of the check as output. Is this a one-to-one function? Explain.

44. *Volume of a Cube* The volume of a cube is $f(x) = x^3$ cubic inches, where x is the length of the edge of the cube in inches.

a. Is this function one-to-one?

b. Find the inverse of this function.

c. What are the domain and range of this inverse function in the context of the application?

d. How could this inverse function be used?

45. *Volume of a Sphere* The volume of a sphere is $f(x) = \frac{4}{3}\pi x^3$ cubic inches, where x is the radius of the sphere in inches.

a. Is this function one-to-one?

b. Find the inverse of this function.

c. What are the domain and range of the inverse function for this application?

d. How could this inverse function be used?

e. What is the radius of a sphere if its volume is 65,450 cubic inches?

46. *Surface Area* The surface area of a cube is $f(x) = 6x^2$ cm^2, where x is the length of the edge of the cube in centimeters.

a. For what values of x does this model make sense? Is the model a one-to-one function for these values of x?

b. What is the inverse of this function on this interval?

c. How could the inverse function be used?

47. *Currency Conversion* The function that converts Canadian dollars to U.S. dollars according to the January 19, 2011, values is $f(x) = 1.0136x$, where x is the number of Canadian dollars and $f(x)$ is the number of U.S. dollars.

a. Find the inverse function for f and interpret its meaning.

b. Use f and f^{-1} to determine the money you will have if you take 500 U.S. dollars to Canada, convert them to Canadian dollars, don't spend any, and then convert them back to U.S. dollars. (Assume that there is no fee for conversion and the conversion rate remains the same.)

(Source: Expedia.com)

48. *Supply* The supply function for a product is $p(x) = \frac{1}{4}x^2 + 20$, where x is the number of thousands of units a manufacturer will supply if the price is $p(x)$ dollars.

a. Is this function a one-to-one function?

b. What is the domain of this function in the context of the application?

c. Is the function one-to-one for the domain in part (b)?

d. Find the inverse of this function and use it to find how many units the manufacturer is willing to supply if the price is $101.

49. *Illumination* The intensity of illumination of a light is a function of the distance from the light. For a given light, the intensity is given by $I(x) = \dfrac{300,000}{x^2}$ candle-power, where x is the distance in feet from the light.

a. Is this function a one-to-one function if it is not limited by the context of the application?

b. What is the domain of this function in the context of the application?

c. Is the function one-to-one for the domain in part (b)?

d. Find the inverse of this function on the domain from part (b) and use it to find the distance at which the intensity of the light is 75,000 candle-power.

50. *First-Class Postage* The postage charged for first-class mail is a function of its weight. The U.S. Postal Service uses the following table to describe the rates for 2011.

Weight Increment, x	Postal Rate
First ounce or fraction of an ounce	$0.44
Each additional ounce or fraction	$0.20

(Source: pe.usps.gov/text)

a. Convert this table to a piecewise-defined function $P(x)$ that represents postage for letters weighing more than 0 and no more than 3 ounces, using x as the weight in ounces and $P(x)$ as the postage in cents.

b. Does P have an inverse function? Why or why not?

51. Suppose the function that converts United Kingdom (U.K.) pounds to U.S. dollars is $f(x) = 1.6249x$, where x is the number of pounds and $f(x)$ is the number of U.S. dollars.

a. Find the inverse function for f and interpret its meaning.

b. Use f and f^{-1} to determine the money you will have if you take 1000 U.S. dollars to the United Kingdom, convert them to pounds, don't spend any, and then convert them back to U.S. dollars. (Assume that there is no fee for conversion and the conversion rate remains the same.)

(Source: International Monetary Fund)

52. *Path of a Ball* If a ball is thrown into the air at a velocity of 96 feet per second from a building that is 256 feet high, the height of the ball after x seconds is $f(x) = 256 + 96x - 16x^2$ feet.

a. For how many seconds will the ball be in the air?

b. Is this function a one-to-one function over this time interval?

c. Give an interval over which the function is one-to-one.

d. Find the inverse of the function over the interval $0 \le x \le 3$. What does it give?

4.4 Additional Equations and Inequalities

KEY OBJECTIVES

- Solve radical equations
- Solve equations with rational powers
- Solve quadratic inequalities
- Solve power inequalities
- Solve inequalities involving absolute value

SECTION PREVIEW **Profit**

The daily profit from the production and sale of x units of a product is given by

$$P(x) = -0.01x^2 + 20.25x - 500$$

Because a profit occurs when $P(x)$ is positive, there is a profit for those values of x that make $P(x) > 0$. Thus, the values of x that give a profit are the solutions to

$$-0.01x^2 + 20.25x - 500 > 0$$

In this section, we solve quadratic inequalities by using both analytical and graphical methods. We also solve equations and inequalities involving radicals, rational powers, and absolute values. ∎

Radical Equations; Equations Involving Rational Powers

An equation containing a radical can frequently be converted to an equation that does not contain a radical by raising both sides of the equation to an appropriate power. For example, an equation containing a square root radical can usually be converted by squaring both sides of the equation. It is possible, however, that raising both sides of an equation to a power may produce an *extraneous* solution (a value that does not satisfy the original equation). For this reason, we must check solutions when using this technique. To solve an equation containing a radical, we use the following steps.

Solving Radical Equations

1. Isolate a single radical on one side of the equation.

2. Square both sides of the equation. (Or raise both sides to a power that is equal to the index of the radical.)

3. If a radical remains, repeat steps 1 and 2.

4. Solve the resulting equation.

5. All solutions must be checked in the original equation, and only those that satisfy the original equation are actual solutions.

EXAMPLE 1 ▶ Radical Equation

Solve $\sqrt{x + 5} + 1 = x$.

SOLUTION

To eliminate the radical, we isolate it on one side and square both sides:

$$\sqrt{x + 5} = x - 1$$
$$x + 5 = x^2 - 2x + 1$$
$$0 = x^2 - 3x - 4$$
$$0 = (x - 4)(x + 1)$$
$$x = 4 \quad \text{or} \quad x = -1$$

Checking these values shows that 4 is a solution and -1 is not.

$x = 4$	$x = -1$
$\sqrt{4 + 5} + 1 \stackrel{?}{=} 4$	$\sqrt{-1 + 5} + 1 \stackrel{?}{=} -1$
$\sqrt{9} + 1 \stackrel{?}{=} 4$	$\sqrt{4} + 1 \stackrel{?}{=} -1$
$3 + 1 \stackrel{?}{=} 4$	$2 + 1 \stackrel{?}{=} -1$
$4 = 4$	$3 \neq -1$

We can also check graphically by the intersection method. Graphing $y = \sqrt{x + 5} + 1$ and $y = x$, we find the point of intersection to be $(4, 4)$ (Figure 4.38). Note that there is *not* a point of intersection at $x = -1$.

Thus, the solution to the equation $\sqrt{x + 5} + 1 = x$ is $x = 4$.

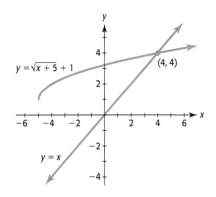

Figure 4.38

EXAMPLE 2 ▶ **Radical Equation**

Solve $\sqrt{4x - 8} - 1 = \sqrt{2x - 5}$.

SOLUTION

We begin by confirming that one of the radicals is isolated. Then we square both sides of the equation:

$$(\sqrt{4x - 8} - 1)^2 = (\sqrt{2x - 5})^2$$
$$4x - 8 - 2\sqrt{4x - 8} + 1 = 2x - 5$$
$$4x - 7 - 2\sqrt{4x - 8} = 2x - 5$$

Now we must isolate the remaining radical and square both sides of the equation again:

$$2x - 2 = 2\sqrt{4x - 8}$$
$$(2x - 2)^2 = (2\sqrt{4x - 8})^2$$
$$4x^2 - 8x + 4 = 4(4x - 8)$$
$$4x^2 - 8x + 4 = 16x - 32$$
$$4x^2 - 24x + 36 = 0$$
$$4(x^2 - 6x + 9) = 0$$
$$4(x - 3)^2 = 0$$
$$x = 3$$

We can check by substituting or graphing. Graphing $y = \sqrt{4x - 8} - 1$ and $y = \sqrt{2x - 5}$, we find the point of intersection at $x = 3$ (Figure 4.39). Thus, $x = 3$ is the solution.

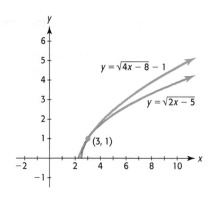

Figure 4.39

Equations Containing Rational Powers

Some equations containing rational powers can be solved by writing the equation as a radical equation.

EXAMPLE 3 ▶ **Equation with Rational Powers**

Solve the equation $(x - 3)^{2/3} - 4 = 0$.

SOLUTION

Rewriting the equation as a radical equation and with the radical isolated gives

$$(x - 3)^{2/3} = 4$$
$$\sqrt[3]{(x - 3)^2} = 4$$

Cubing both sides of the equation and solving gives

$$(x - 3)^2 = 64$$
$$\sqrt{(x - 3)^2} = \pm\sqrt{64}$$
$$x - 3 = \pm 8$$
$$x = 3 \pm 8$$
$$x = 11 \quad \text{or} \quad x = -5$$

Entering each solution in the original equation shows that each checks there.

$x = 11$	$x = -5$
$(11 - 3)^{2/3} - 4 \overset{?}{=} 0$	$(-5 - 3)^{2/3} - 4 \overset{?}{=} 0$
$8^{2/3} - 4 \overset{?}{=} 0$	$(-8)^{2/3} - 4 \overset{?}{=} 0$
$(\sqrt[3]{8})^2 - 4 \overset{?}{=} 0$	$(\sqrt[3]{-8})^2 - 4 \overset{?}{=} 0$
$2^2 - 4 \overset{?}{=} 0$	$(-2)^2 - 4 \overset{?}{=} 0$
$0 = 0$	$0 = 0$

Quadratic Inequalities

A **quadratic inequality** is an inequality that can be written in the form

$$ax^2 + bx + c > 0$$

where a, b, and c are real numbers and $a \neq 0$ (or with $>$ replaced by $<$, $\geq$, or $\leq$).

Algebraic Solution of Quadratic Inequalities

To solve a quadratic inequality $f(x) > 0$ or $f(x) < 0$ algebraically, we first need to find the zeros of $f(x)$. The zeros can be found by factoring or the quadratic formula. If the quadratic function has two real zeros, these two zeros divide the real number line into three intervals. Within each interval, the value of the quadratic function is either always positive or always negative, so we can use one value in each interval to test the function on the interval. We can use the following steps, which summarize these ideas, to solve quadratic inequalities.

Solving a Quadratic Inequality Algebraically

1. Write an equivalent inequality with 0 on one side and with the function $f(x)$ on the other side.

2. Solve $f(x) = 0$.

3. Create a sign diagram that uses the solutions from step 2 to divide the number line into intervals. Pick a test value in each interval and determine whether $f(x)$ is positive or negative in that interval to create a sign diagram.*

4. Identify the intervals that satisfy the inequality in step 1. The values of x that define these intervals are solutions to the original inequality.

EXAMPLE 4 ▶ Solve the inequality $x^2 - 3x > 3 - 5x$.

SOLUTION

Rewriting the inequality with 0 on the right side of the inequality gives $f(x) > 0$ with $f(x) = x^2 + 2x - 3$:

$$x^2 + 2x - 3 > 0$$

Writing the equation $f(x) = 0$ and solving for x gives

$$x^2 + 2x - 3 = 0$$
$$(x + 3)(x - 1) = 0$$
$$x = -3 \quad \text{or} \quad x = 1$$

The two values of x, -3 and 1, divide the number line into three intervals: the numbers less than -3, the numbers between -3 and 1, and the numbers greater than 1. We need only find the signs of $(x + 3)$ and $(x - 1)$ in each interval and then find the sign of their product to find the solution to the original inequality. Testing a value in each interval determines the sign of each factor in each interval. (See the sign diagram in Figure 4.40.)

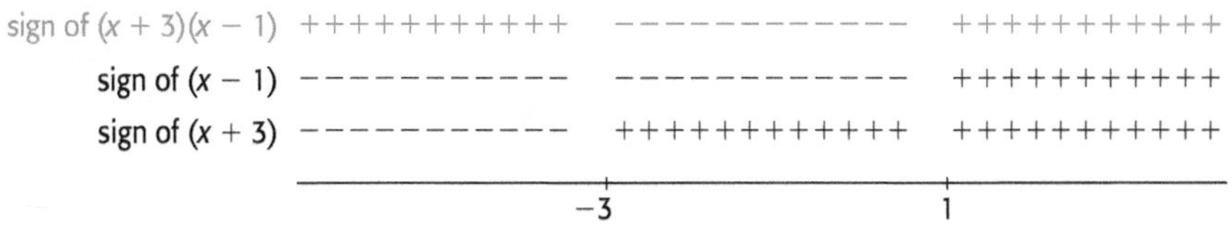

sign of $(x + 3)(x - 1)$ +++++++++++   ------------   +++++++++++
sign of $(x - 1)$ ------------   ------------   +++++++++++
sign of $(x + 3)$ ------------   +++++++++++   +++++++++++

                        −3   1

Figure 4.40

The function $f(x) = (x + 3)(x - 1)$ is positive on the intervals $(-\infty, -3)$ and $(1, \infty)$, so the solution to $x^2 + 2x - 3 > 0$ and thus the original inequality is

$$x < -3 \quad \text{or} \quad x > 1$$

———————

*The numerical feature of your graphing utility can be used to test the x-values.

Graphical Solution of Quadratic Inequalities

Recall that we solved a linear inequality $f(x) > 0$ (or $f(x) < 0$) graphically by graphing the related equation $f(x) = 0$ and observing where the graph is above (or below) the x-axis—that is, where $f(x)$ is positive (or negative).

Similarly, we can solve a quadratic inequality $f(x) > 0$ (or $f(x) < 0$) graphically by graphing the related equation $f(x) = 0$ and observing the x-values of the intervals where $f(x)$ is positive (or negative).

For example, we can graphically solve the inequality

$$x^2 - 3x - 4 \leq 0$$

by graphing

$$y = x^2 - 3x - 4$$

and observing the part of the graph that is below or on the x-axis. Using the x-intercept method, we find that $x = -1$ and $x = 4$ give $y = 0$, and we see that the graph is below or on the x-axis for values of x satisfying $-1 \leq x \leq 4$ (Figure 4.41). Thus, the solution to the inequality is $-1 \leq x \leq 4$.

Table 4.9 shows the possible graphs of quadratic functions and the solutions to the related inequalities. Note that if the graph of $y = f(x)$ lies entirely above the x-axis, the solution to $f(x) > 0$ is the set of all real numbers, and there is no solution to $f(x) < 0$. (The graph of $y = f(x)$ can also touch the x-axis in one point.)

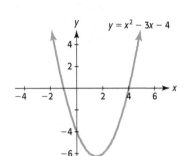

Figure 4.41

Table 4.9 Quadratic Inequalities

Orientation of Graph of $y = f(x)$	Inequality to Be Solved	Part of Graph That Satisfies the Inequality	Solution to Inequality
	$f(x) > 0$	where the graph is above the x-axis	$x < a$ or $x > b$
	$f(x) < 0$	where the graph is below the x-axis	$a < x < b$
	$f(x) > 0$	where the graph is above the x-axis	$a < x < b$
	$f(x) < 0$	where the graph is below the x-axis	$x < a$ or $x > b$
	$f(x) > 0$	the entire graph	all real numbers
	$f(x) < 0$	none of the graph	no solution
	$f(x) > 0$	none of the graph	no solution
	$f(x) < 0$	the entire graph	all real numbers

EXAMPLE 5 ▶ Height of a Model Rocket

A model rocket is projected straight upward from ground level according to the equation

$$h = -16t^2 + 192t, \; t \geq 0$$

where h is the height in feet and t is the time in seconds. During what time interval will the height of the rocket exceed 320 feet?

ALGEBRAIC SOLUTION

To find the time interval when the rocket is higher than 320 feet, we find the values of t for which

$$-16t^2 + 192t > 320$$

Getting 0 on the right side of the inequality, we have

$$-16t^2 + 192t - 320 > 0$$

First we determine when $-16t^2 + 192t - 320$ is equal to 0:

$$-16t^2 + 192t - 320 = 0$$
$$-16(t^2 - 12t + 20) = 0$$
$$-16(t - 2)(t - 10) = 0$$
$$t = 2 \quad \text{or} \quad t = 10$$

The values of $t = 2$ and $t = 10$ divide the number line into three intervals: the numbers less than 2, the numbers between 2 and 10, and the numbers greater than 10 (Figure 4.42). To solve the inequality, we find the sign of the product of -16, $(t - 2)$, and $(t - 10)$ in each interval. We do this by testing a value in each interval.

Figure 4.42

This shows that the product $-16(t - 2)(t - 10)$ is positive only for $2 < t < 10$, so the solution to $-16t^2 + 192t - 320 > 0$ is $2 < t < 10$. This solution is also a solution to the original inequality, so the rocket exceeds 320 feet between 2 and 10 seconds after launch.

Note that if the inequality problem is applied, we must check that the solution makes sense in the context of the problem.

GRAPHICAL SOLUTION

To find the interval when the height exceeds 320 feet, we can also solve $-16t^2 + 192t > 320$ using a graphing utility. To use the x-intercept method, we rewrite the inequality with 0 on the right side:

$$-16t^2 + 192t - 320 > 0$$

Using the variable x in place of the variable t, we enter $y_1 = -16x^2 + 192x - 320$ and graph the function (Figure 4.43). The x-intercepts of the graph are $x = 2$ and $x = 10$.

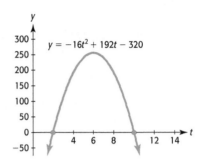

Figure 4.43

The solution to $-16t^2 + 192t - 320 > 0$ is the interval where the graph is above the t-axis, which is

$$2 < t < 10$$

Thus, the height of the rocket exceeds 320 feet between 2 and 10 seconds.

EXAMPLE 6 ▶   **Internet Use**

According to Internet World Stats, the number of worldwide Internet users, in millions, during the years 1995 to 2007 can be modeled by the equation $y = 3.604x^2 + 27.626x - 261.714$, where x is the number of years after 1990. For what years from 1995 through 2015 does this model indicate the number of worldwide Internet users to be at least 1103 million?

SOLUTION

To find the years from 1995 through 2015 when the number of Internet users is at least 1103 million, we solve the inequality

$$3.604x^2 + 27.626x - 261.714 \geq 1103$$

We begin solving this inequality by graphing $y = 3.604x^2 + 27.626x - 261.714$. We set the viewing window from $x = 5$ (1995) to $x = 25$ (2015). The graph of this function is shown in Figure 4.44(a).

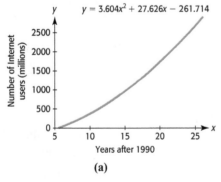 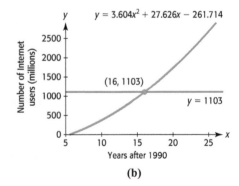

(a)   (b)

Figure 4.44

We seek the point(s) where the graph of $y = 1103$ intersects the graph of $y = 3.604x^2 + 27.626x - 261.714$. Graphing the line $y = 1103$ and intersecting it with the graph of the quadratic function in the given window gives the point of intersection $(16, 1103)$, as shown in Figure 4.44(b). This graph shows that the number of users is at or above 1103 million for $16 \le x \le 25$. Because x represents the number of years after 1990, the number of Internet users is at least 1103 million for the years 2006 through 2015.

Power Inequalities

To solve a power inequality, we will use a combination of analytical and graphical methods.

Power Inequalities

To solve a power inequality, first solve the related equation. Then use graphical methods to find the values of the variable that satisfy the inequality.

EXAMPLE 7 ▶ Investment

The future value of $3000 invested for 3 years at rate r, compounded annually, is given by $S = 3000(1 + r)^3$. What interest rate will give a future value of at least $3630?

SOLUTION

To solve this problem, we solve the inequality

$$3000(1 + r)^3 \ge 3630$$

We begin by solving the related equation using the root method.

$$3000(1 + r)^3 = 3630$$

$$(1 + r)^3 = 1.21 \qquad \text{Divide both sides by 3000.}$$

$$1 + r = \sqrt[3]{1.21} \qquad \text{Take the cube root} \left(\frac{1}{3} \text{power}\right) \text{of both sides.}$$

$$1 + r \approx 1.0656$$

$$r \approx 0.0656$$

This tells us that the investment will have a future value of $3630 in 3 years if the interest rate is approximately 6.56%, and it helps us set the window to solve the inequality graphically. Graphing $y_1 = 3000(1 + r)^3$ and $y_2 = 3630$ on the same axes shows that $3000(1 + r)^3 \ge 3630$ if $r \ge 0.0656$ (Figure 4.45). Thus, the investment will result in at least $3630 if the interest rate is 6.56% or higher. Note that in the context of this problem, only values of r from 0 (0%) to 1 (100%) make sense.

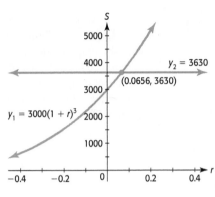

Figure 4.45

Inequalities Involving Absolute Values

Recall that if u is an algebraic expression, then for $a \geq 0$, $|u| = a$ means that $u = a$ or $u = -a$. We can represent inequalities involving absolute values as follows.

For $a \geq 0$:

$|u| < a$ means that $-a < u < a$.

$|u| \leq a$ means that $-a \leq u \leq a$.

$|u| > a$ means that $u < -a$ or $u > a$.

$|u| \geq a$ means that $u \leq -a$ or $u \geq a$.

We can solve inequalities involving absolute values algebraically or graphically.

EXAMPLE 8 ▶   Solve the following inequalities and verify the solutions graphically.

a. $|2x - 3| \leq 5$

b. $|3x + 4| - 5 > 0$

SOLUTION

a. This inequality is equivalent to $-5 \leq 2x - 3 \leq 5$. If we add 3 to all three parts of this inequality and then divide all parts by 2, we get the solution.

$$-5 \leq 2x - 3 \leq 5$$
$$-2 \leq 2x \leq 8$$
$$-1 \leq x \leq 4$$

Figure 4.46(a) shows that the graph of $y = |2x - 3|$ is on or below the graph of $y = 5$ for $-1 \leq x \leq 4$.

b. This inequality is equivalent to $|3x + 4| > 5$, which is equivalent to $3x + 4 < -5$ or $3x + 4 > 5$. We solve each of the inequalities as follows:

$$3x + 4 < -5 \quad \text{or} \quad 3x + 4 > 5$$
$$3x < -9 \qquad\qquad 3x > 1$$
$$x < -3 \qquad\qquad x > \frac{1}{3}$$

Figure 4.46(b) shows that the graph of $y = |3x + 4| - 5$ is above the x-axis for $x < -3$ or $x > \frac{1}{3}$.

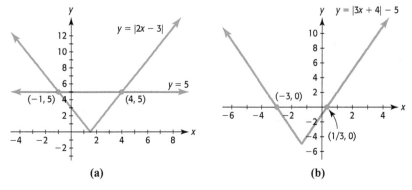

(a)   **(b)**

Figure 4.46

Skills CHECK 4.4

In Exercises 1–12, solve the equations algebraically and check graphically or by substitution.

1. $\sqrt{2x^2 - 1} - x = 0$ **2.** $\sqrt{3x^2 + 4} - 2x = 0$

3. $\sqrt[3]{x - 1} = -2$ **4.** $\sqrt[3]{4 - x} = 3$

5. $\sqrt{3x - 2} + 2 = x$ **6.** $\sqrt{x - 2} + 2 = x$

7. $\sqrt[3]{4x + 5} = \sqrt[3]{x^2 - 7}$

8. $\sqrt{5x - 6} = \sqrt{x^2 - 2x}$

9. $\sqrt{x} - 1 = \sqrt{x - 5}$

10. $\sqrt{x} - 10 = -\sqrt{x - 20}$

11. $(x + 4)^{2/3} = 9$ **12.** $(x - 5)^{3/2} = 64$

In Exercises 13–22, use algebraic methods to solve the inequalities.

13. $x^2 + 4x < 0$ **14.** $x^2 - 25x < 0$

15. $9 - x^2 \geq 0$ **16.** $x > x^2$

17. $-x^2 + 9x - 20 > 0$

18. $2x^2 - 8x < 0$

19. $2x^2 - 8x \geq 24$ **20.** $t^2 + 17t \leq 8t - 14$

21. $x^2 - 6x < 7$ **22.** $4x^2 - 4x + 1 > 0$

In Exercises 23–26, use graphical methods to solve the inequalities.

23. $2x^2 - 7x + 2 \geq 0$ **24.** $w^2 - 5w + 4 > 0$

25. $5x^2 \geq 2x + 6$ **26.** $2x^2 \leq 5x + 6$

In Exercises 27–34, solve the inequalities by using algebraic and graphical methods.

27. $(x + 1)^3 < 4$ **28.** $(x - 2)^3 \geq -2$

29. $(x - 3)^5 < 32$ **30.** $(x + 5)^4 > 16$

31. $|2x - 1| < 3$ **32.** $|3x + 1| \leq 5$

33. $|x - 6| \geq 2$ **34.** $|x + 8| > 7$

In Exercises 35–38, you are given the graphs of several functions of the form $f(x) = ax^2 + bx + c$ for different values of a, b, and c. For each function, (a) solve $f(x) \geq 0$, and (b) solve $f(x) < 0$.

35.

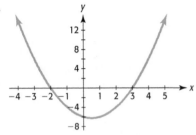

36.

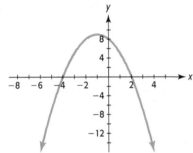

37.

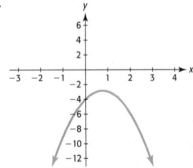

38.

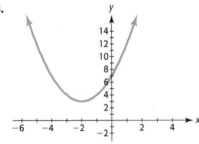

In Exercises 39 and 40, you are given the graphs of two functions f(x) and g(x). Solve $f(x) \leq g(x)$.

39.

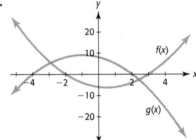

40.

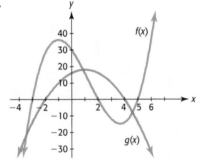

EXERCISES 4.4

Use algebraic and/or graphical methods to solve Exercises 41–50.

41. **Profit** The monthly profit from producing and selling x units of a product is given by

$$P(x) = -0.3x^2 + 1230x - 120,000$$

Producing and selling how many units will result in a profit for this product?

42. **Profit** The monthly profit from producing and selling x units of a product is given by

$$P(x) = -0.01x^2 + 62x - 12,000$$

Producing and selling how many units will result in a profit for this product?

43. **Profit** The revenue from sales of x units of a product is given by $R(x) = 200x - 0.01x^2$, and the cost of producing and selling the product is $C(x) = 38x + 0.01x^2 + 16,000$. Producing and selling how many units will result in a profit?

44. *Profit* The revenue from sales of x units of a product is given by $R(x) = 600x - 0.01x^2$, and the cost of producing and selling the product is $C(x) = 77x + 0.02x^2 + 52,000$. Producing and selling how many units will result in a profit?

45. *Projectiles* Two projectiles are fired into the air over a lake, with the height of the first projectile given by $y = 100 + 130t - 16t^2$ and the height of the second projectile given by $y = -16t^2 + 180t$, where y is in feet and t is in seconds. Over what time interval, before the lower one hits the lake, is the second projectile above the first?

46. *Projectile* A rocket shot into the air has height $s = 128t - 16t^2$ feet, where t is the number of seconds after the rocket is shot. During what time after the rocket is shot is it at least 240 feet high?

47. *Tobacco Sales* Sales of fine-cut cigarettes (in millions) in Canada are given by $y = -13.898x^2 + 255.467x + 5425.618$, where x is the number of years after 1980. During what years does this model indicate that the sales would be at least 6 billion?
(Source: Health Canada)

48. *World Population* The low long-range world population numbers and projections for the years 1995–2150 are given by the equation $y = -0.00036x^2 + 0.0385x + 5.823$, where x is the number of years after 1990 and y is in billions. During what years does this model estimate the population to be above 6 billion?
(Source: U.N. Department of Economic and Social Affairs)

49. *Foreign-Born Population* Suppose the percent of the U.S. population that is foreign born is given by $y = 0.003x^2 - 0.438x + 20.18$, where x is the number of years after 1900. During what years does this model indicate that the percent is at most 12.62?
(Source: U.S. Census Bureau)

50. *Gross Domestic Product* The U.S. gross domestic product (in billions of constant dollars) can be modeled by the equation $y = 3.99x^2 - 432.50x + 12,862.21$, where x is the number of years after 1900. During what years prior to 2010 was the gross domestic product less than $4 trillion?
(Source: Microsoft Encarta 98)

Use graphical and/or numerical methods to solve Exercises 51–56.

51. *Wind Chill* The wind chill temperature when the outside temperature is 20°F is given by $y = 0.0052x^2 - 0.62x + 15.0$, where x is the wind speed in mph. For what wind speeds between 0 mph and 90 mph is the wind chill temperature −3°F or below?

52. *Hotel Room Supply* The percentage change p (from the previous year) in the hotel room supply is given by $p = 0.025x^2 - 0.645x + 4.005$, where x is the number of years after 1998. If this model is accurate, during what years after 1998 will the percentage change be less than 3%?
(Source: Hotel and Motel Management)

53. *Drug Use* The percent of high school students who have ever used marijuana is given by $y = -0.1967x^2 + 4.063x + 27.7455$, where x is the number of years after 1990. Use the model to estimate the years when the percent who ever used marijuana is greater than 43.1%.

54. *Airplane Crashes* The number of all airplane crashes (in thousands) is given by the equation $y = 0.0057x^2 - 0.197x + 3.613$, where x is the number of years after 1980. During what years from 1980 through 2015 does the model indicate the number of crashes as being below 3146?

55. *High School Smokers* The percent of high school students who smoked cigarettes on 1 or more of the 30 days preceding the Youth Risk Behavior Survey is given by $y = -0.061x^2 + 0.275x + 33.698$, where x is the number of years after 1990. During what years from 1990 on is the percent greater than 20%?

56. *Car Design* Sports cars are designed so that the driver's seat is comfortable for people of height 5 feet 8 inches, plus or minus 8 inches.

a. Write an absolute value inequality that gives the height x of a person who will be uncomfortable.

b. Solve this inequality for x to identify the heights of people who are uncomfortable.

57. *Retail Sales* November and December retail sales, excluding autos, for the years 2001–2010 can be modeled by the function $S(x) = -1.751x^2 + 38.167x + 388.997$ billion dollars, where x is the number of years after 2000. During what years does the model estimate retail sales to be above \$500 billion?

58. *Voltage* Required voltage for an electric oven is 220 volts, but it will function normally if the voltage varies from 220 by 10 volts.

   a. Write an absolute value inequality that gives the voltage x for which the oven will work normally.

   b. Solve this inequality for x.

59. *Purchasing Power* Inflation causes a decrease in the value of money used to purchase goods and services. The purchasing power of a 1983 dollar based on consumer prices for 1968–2010 can be modeled by the function $y = 34.394x^{-1.109}$, where x is the number of years after 1960. For what years through 2012 is the purchasing power of a 1983 dollar less than \$1.00, according to the model?

---

chapter 4 ▸ SUMMARY

In this chapter, we presented the building blocks for function construction. We can construct new functions by horizontal or vertical transformation, stretching or compressing, or reflection. We can also construct new functions by combining functions algebraically or with function composition. Inverse functions can be used to "undo" the operations in a function. We finished the chapter by solving additional equations and inequalities.

Key Concepts and Formulas

4.1 Transformations of Graphs and Symmetry

Vertical shifts of graphs	If k is a positive real number, • The graph of $g(x) = f(x) + k$ can be obtained by shifting the graph of $f(x)$ upward k units. • The graph of $g(x) = f(x) - k$ can be obtained by shifting the graph of $f(x)$ downward k units.
Horizontal shifts of graphs	If h is a positive real number, • The graph of $g(x) = f(x - h)$ can be obtained by shifting the graph of $f(x)$ to the right h units. • The graph of $g(x) = f(x + h)$ can be obtained by shifting the graph of $f(x)$ to the left h units.
Stretching and compressing graphs	The graph of $y = af(x)$ is obtained by vertically stretching the graph of $f(x)$ using a factor of $|a|$ if $|a| > 1$ and vertically compressing the graph of $f(x)$ using a factor of $|a|$ if $|a| < 1$.
Reflections of graphs across the coordinate axes	The graph of $y = -f(x)$ can be obtained by reflecting the graph of $y = f(x)$ across the x-axis. The graph of $y = f(-x)$ can be obtained by reflecting the graph of $y = f(x)$ across the y-axis.

Symmetry with respect to the y-axis	The graph of $y = f(x)$ is symmetric with respect to the y-axis if $f(-x) = f(x)$. Such a function is called an even function.
Symmetry with respect to the origin	The graph of $y = f(x)$ is symmetric with respect to the origin if $f(-x) = -f(x)$. Such a function is called an odd function.
Symmetry with respect to the x-axis	The graph of an equation is symmetric with respect to the x-axis if, for every point (x, y) on the graph, the point $(x, -y)$ is also on the graph.

4.2 Combining Functions; Composite Functions

Operations with functions	*Sum*: $(f + g)(x) = f(x) + g(x)$ *Difference*: $(f - g)(x) = f(x) - g(x)$ *Product*: $(f \cdot g)(x) = f(x) \cdot g(x)$ *Quotient*: $\left(\dfrac{f}{g}\right)(x) = \dfrac{f(x)}{g(x)}, g(x) \neq 0$ The domain of the sum, difference, and product of f and g consists of all real numbers of the input variable for which f and g are defined. The domain of the quotient function consists of all real numbers for which f and g are defined and $g \neq 0$.
Average cost	The quotient of the cost function $C(x)$ and the identity function $I(x) = x$ gives the average cost function $\overline{C}(x) = \dfrac{C(x)}{x}$.
Composite functions	The notation for the composite function "f of g" is $(f \circ g)(x) = f(g(x))$. The domain of $f \circ g$ is the subset of the domain of g for which $f \circ g$ is defined.

4.3 One-to-One and Inverse Functions

Inverse functions	If $f(g(x)) = x$ and $g(f(x)) = x$ for all x in the domain of g and f, then f and g are inverse functions. In addition, the functions f and g are inverse functions if, whenever the pair (a, b) satisfies $y = f(x)$, the pair (b, a) satisfies $y = g(x)$. We denote g by f^{-1}, read as "f inverse."
One-to-one functions	For a function f to have an inverse, f must be one-to-one. A function f is one-to-one if each output of f corresponds to exactly one input of f.
Horizontal line test	If no horizontal line can intersect the graph of a function in more than one point, then the function is one-to-one.
Finding the inverse functions	To find the inverse of the function f that is defined by the equation $y = f(x)$: 1. Rewrite the equation with y replacing $f(x)$. 2. Interchange x and y in the equation defining the function. 3. Solve the new equation for y. If this equation cannot be solved uniquely for y, the function has no inverse. 4. Replace y by $f^{-1}(x)$.
Graphs of inverse functions	The graphs of a function and its inverse are symmetric with respect to the line $y = x$.
Inverse functions on limited domains	If a function is one-to-one on a limited domain, then it has an inverse function on that domain.

4.4 Additional Equations and Inequalities

Radical equations	An equation containing radicals can frequently be converted to an equation that does not contain radicals by raising both sides of the equation to a power that is equal to the index of the radical. The solutions must be checked when using this technique.
Equations with rational powers	Some equations with rational powers can be solved by writing the equation as a radical equation.
Quadratic inequality	A quadratic inequality is an inequality that can be written in the form $ax^2 + bx + c > 0$, where a, b, c are real numbers and $a \neq 0$ (or with $>$ replaced by $<$, $\geq$, or $\leq$).
Solving a quadratic inequality algebraically	To solve a quadratic inequality $f(x) < 0$ or $f(x) > 0$ algebraically, solve $f(x) = 0$ and use the solutions to divide the number line into intervals. Pick a test value in each interval to determine whether $f(x)$ is positive or negative in that interval and identify the intervals that satisfy the original inequality.
Solving a quadratic inequality graphically	To solve a quadratic inequality $f(x) < 0$ (or $f(x) > 0$) graphically, graph the related equation $y = f(x)$ and observe the x-values of the intervals where the graph of $f(x)$ is below (or above) the x-axis.
Power inequalities	To solve a power inequality, first solve the related equation. Then use graphical methods to find the values of the variable that satisfy the inequality.
Absolute value inequalities	If $a > 0$, the solution to $|u| < a$ is $-a < u < a$. If $a > 0$, the solution to $|u| > a$ is $u < -a$ or $u > a$.

chapter 4 ▸ SKILLS CHECK

1. How is the graph of $g(x) = (x - 8)^2 + 7$ transformed from the graph of $f(x) = x^2$?

2. How is the graph of $g(x) = -2(x + 1)^3$ transformed from the graph of $f(x) = x^3$?

3. **a.** Graph the functions $f(x) = \sqrt{x}$ and $g(x) = \sqrt{x + 2} - 3$.

   b. How are the graphs related?

4. What is the domain of the function $g(x) = \sqrt{x + 2} - 3$?

5. Suppose the graph of $f(x) = x^{1/3}$ is shifted up 4 units and to the right 6 units. What is the equation that gives the new graph?

6. Suppose the graph of $f(x) = x^{1/3}$ is vertically stretched by a factor of 3 and then shifted down 5 units. What is the equation that gives the new graph?

For Exercises 7–9, match each graph with the correct equation.

   a. $y = |x| + 2$

   b. $y = x^3$

   c. $y = \dfrac{3}{x - 1} + 1$

   d. $y = \dfrac{-3}{x + 1} + 1$

   e. $y = 2x^3$

   f. $y = |x + 2|$

7.

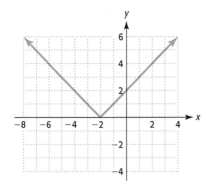

8.

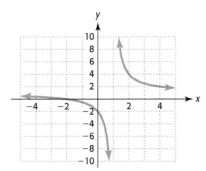

9.

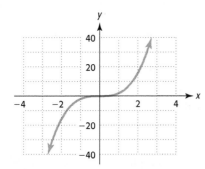

In Exercises 10 and 11, determine algebraically whether the graph of the function is symmetric with respect to the x-axis, y-axis, and/or origin. Confirm graphically.

10. $f(x) = x^3 - 4x$ **11.** $f(x) = -x^2 + 5$

12. Determine whether the function $f(x) = -\dfrac{2}{x}$ is even, odd, or neither.

For Exercises 13–20, use the functions $f(x) = 3x^2 - 5x$, $g(x) = 6x - 4$, $h(x) = 5 - x^3$ to find the following.

13. $(f + g)(x)$ **14.** $(h - g)(x)$

15. $(g \cdot f)(x)$ **16.** $\left(\dfrac{h}{g}\right)(x)$

17. $(f - g)(-2)$ **18.** $(f \circ g)(x)$

19. $(g \circ f)(x)$ **20.** $(g \circ h)(-3)$

21. For the functions $f(x) = 2x - 5$ and $g(x) = \dfrac{x + 5}{2}$,

   a. Find $f(g(x))$ and $g(f(x))$.

   b. What is the relationship between $f(x)$ and $g(x)$?

22. Find the inverse of $f(x) = 3x - 2$.

23. Find the inverse of $g(x) = \sqrt[3]{x - 1}$.

24. Graph $f(x) = (x + 1)^2$ and its inverse $f^{-1}(x)$ on the domain $[-1, 10]$.

25. Is the function $f(x) = \dfrac{x}{x - 1}$ a one-to-one function?

26. Does the function $f(x) = x^2 - 3x$ have an inverse?

27. Solve $\sqrt{4x^2 + 1} = 2x + 2$.

28. Solve $\sqrt{3x^2 - 8} + x = 0$.

29. Solve the inequality $x^2 - 7x \le 18$.

30. Solve the inequality $2x^2 + 5x \ge 3$.

31. Solve $|2x - 4| \le 8$.

32. Solve $|4x - 3| \ge 15$.

33. Solve $(x - 4)^3 < 4096$.

34. Solve $(x + 2)^2 \ge 512$.

35. *Ballistics* Ballistic experts are able to identify the weapon that fired a certain bullet by studying the markings on the bullet. If a test is conducted by firing a bullet into a bale of paper, the distance that the bullet will travel is given by $s = 64 - (4 - 10t)^3$ inches, for $0 \le t \le 0.4$, t in seconds.

   a. Graph this function for $0 \le t \le 0.4$.

   b. How far does the bullet travel during this time period?

36. *Projectile* A toy rocket is fired into the air from the top of a building, with its height given by $S = -16(t - 4)^2 + 380$ feet, with t in seconds.

   a. In how many seconds will the rocket reach its maximum height?

   b. What is the maximum possible height?

   c. Describe the transformations needed to obtain the graph of $S(t)$ from the graph of $y = t^2$.

37. *Airline Traffic* The number of millions of passengers traveling on U.S. airlines can be modeled by $f(x) = -7.232x^2 + 95.117x + 441.138$, where x is the number of years after 2000.

   a. During what year does the model indicate that the number of passengers was at a minimum?

   b. Determine the years during which the number of passengers was above 700 million.

38. *Pet Industry* U.S. pet industry expenditures can be modeled by $y = 0.037x^2 + 1.018x + 12.721$ million dollars, where x equals the number of years after 1990. During what years from 1990 through 2010 did Americans spend more than $40 million on their pets?

39. *Degrees Earned* The number of postsecondary degrees earned, in thousands (including associate's, bachelor's, master's, first professional, and doctoral degrees), can be modeled by $y = 3.980x^2 - 17.597x + 2180.899$, where x is the number of years after 1990. If this model is accurate, during what years from 1990 on would the number of postsecondary degrees earned be less than 3500 thousand? (Source: National Center for Education Statistics, *Digest of Education Statistics*)

40. *Average Cost* Suppose the total cost function for a product is determined to be $C(x) = 30x + 3150$, where x is the number of units produced and sold.

   a. Write the average cost function, which is the quotient of two functions.

   b. Graph this function for $0 < x \le 20$.

41. *Average Cost* The monthly average cost of producing x sofas is $\overline{C}(x) = \dfrac{50{,}000}{x} + 120$ dollars.

   a. Graph this function for $x > 0$.

   b. Will the average cost function decrease or increase as the number of units produced increases?

   c. What transformations of the graph of the reciprocal function give the graph of this function?

42. *Supply and Demand* The price per unit of a product is $\$p$, and the number of units of the product is denoted by q. Suppose the supply function for a product is given by $p = \dfrac{180 + q}{6}$, and the demand for the product is given by $p = \dfrac{300}{q} - 20$.

   a. Which of these functions is a shifted reciprocal function?

   b. Graph the supply and demand functions on the same axes.

43. *Supply and Demand* The supply function for a commodity is given by $p = 58 + \dfrac{q}{2}$, and the demand function for this commodity is given by $p = \dfrac{2555}{q + 5}$.

   a. Which of these functions is a shifted reciprocal function?

   b. Graph the supply and demand functions on the same axes.

44. *Marijuana Use* For the years 1991 through 2006, the percent M of high school seniors who have tried marijuana can be considered as a function of the time x according to the model

$$M(x) = -0.2(x - 10.3)^2 + 48.968$$

where x is the number of years after 1990.

a. Describe the transformations of the graph of $f(x) = x^2$ needed to obtain the graph of M.

b. Find and interpret $M(1)$, $M(2)$, $M(4)$, and $M(6)$.

c. Using the information in parts (a) and (b), sketch a graph of M.

d. The survey on which M is based was taken once during each of the selected years. State the domain of the related discrete function.

45. *Southwest Airlines* The number of employees, in thousands, of Southwest Airlines can be modeled by the function $E(t) = 0.017t^2 + 2.164t + 8.061$, where t is the number of years after 1990. Southwest Airlines' revenue during the same time period is given by the model $R(E) = 0.165E - 0.226$ billion dollars when there are E thousand employees of the company.

a. Find and interpret the meaning of the function $R(E(t))$.

b. Use the result of part (a) to find $R(E(3))$. Interpret this result.

c. How many employees worked for Southwest Airlines in 1997?

d. What was the 1997 revenue for this airline?
(Source: Hoover's Online Capsules)

46. *Prison Sentences* The mean time in prison y for certain crimes can be found as a function of the mean sentence length x, using $f(x) = 0.554x - 2.886$, where x and y are measured in months.

a. Find the inverse of this function.

b. Interpret the inverse function from part (a).
(Source: Index of Leading Cultural Indicators)

47. *Education Spending* Personal expenditures for higher education rose dramatically from 1990 to 2008 and can be modeled by $f(x) = 5.582x + 28.093$ billion dollars, where x is the number of years after 1990.

a. Find a formula for the inverse of function f.

b. Interpret your answer to part (a).

c. If the inverse is f^{-1}, find $f^{-1}(f(10))$.
(Source: Bureau of Economic Analysis, U.S. Department of Commerce)

48. *Purchasing Power* The purchasing power of a 1983 dollar based on consumer prices for 1968–2010 can be modeled by the function $P(x) = 34.394x^{-1.109}$, where x is the number of years after 1960.

a. Another function that might be used to model purchasing power is $f(x) = \dfrac{50}{1.6x} - 0.2$. What is the basic function that can be transformed to obtain f?

b. The purchasing power of a 1983 dollar in 2010 was $0.456. Which model gives the better estimate of the actual value?

49. *Profit* The monthly profit from producing and selling x units of a product is given by

$$P(x) = -0.01x^2 + 62x - 12{,}000$$

Producing and selling how many units will result in profit for this product ($P(x) > 0$)?

50. *Personal Savings Rate* The personal savings rate for Americans for certain years from 1960 to 2006 can be modeled by the equation

$$y = -0.008x^2 + 0.21x + 7.04$$

where x is the number of years after 1960 and y is the rate as a percent. Use graphical or numerical methods to find the years from 1960 on during which the model indicates that the personal savings rate was above 4%.

Group Activities
▶ EXTENDED APPLICATIONS

1. Cost, Revenue, and Profit

The following table gives the weekly revenue and cost, respectively, for selected numbers of units of production and sale of a product by the Quest Manufacturing Company.

Number of Units	Revenue ($)	Number of Units	Cost ($)
100	6800	100	32,900
300	20,400	300	39,300
500	34,000	500	46,500
900	61,200	900	63,300
1400	95,200	1400	88,800
1800	122,400	1800	112,800
2500	170,000	2500	162,500

Provide the information requested and answer the questions.

1. Use technology to determine the functions that model revenue and cost functions for this product, using x as the number of units produced and sold.
2. **a.** Combine the revenue and cost functions with the correct operation to create the profit function for this product.
   b. Use the profit function to complete the following table:

x (units)	Profit, P(x) ($)
0	
100	
600	
1600	
2000	
2500	

3. Find the number of units of this product that must be produced and sold to break even.
4. Find the maximum possible profit and the number of units that gives the maximum profit.
5. **a.** Use operations with functions to create the average cost function for the product.
   b. Complete the following table.

x (units)	Average Cost, C(x) ($/unit)
1	
100	
300	
1400	
2000	
2500	

   c. Graph this function using the viewing window [0, 2500] by [0, 400].
6. Graph the average cost function using the viewing window [0, 4000] by [0, 100]. Determine the number of units that should be produced to minimize the average cost and the minimum average cost.
7. Compare the number of units that produced the minimum average cost with the number of units that produced the maximum profit. Are they the same number of units? Discuss which of these values is more important to the manufacturer and why.

2. Cell Phone Revenue

The table in the right column gives the number of U.S. cellular telephone subscribers and the average monthly bill for the years 1998–2008. To investigate the average monthly revenue for cell phone companies over the time period from 1998 to 2008, answer the following.

1. Find a quadratic function $S = f(t)$ that models the number of subscribers, in millions, as a function of the number of years after 1995. Write the model with three decimal places.
2. Graph the data and the model to visually determine if the model is a good fit for the subscriber data.
3. Find a quadratic function $D = g(t)$ that models the average monthly cell phone bill, in dollars, as a function of the number of years after 1995. Write the model with three decimal places.
4. Graph the data and the model to visually determine if the model is a good fit for the billing data.
5. Using models from steps 3 and 4 with coefficients rounded to one decimal place, take the product of these two functions to find a function that models the average monthly revenue for the cell phone companies.
6. Graph the data points representing the years and average revenue on the same axes as the function found in step 5. Is it a good fit? Explain.
7. Create a new column of revenue data by multiplying the number of subscribers in each year times the average monthly bill for that year.

8. Find a quadratic function that is a good fit for the data representing the years after 1995 and average revenue. Is your model a better fit than the function found in step 5?
9. Find a 4th degree (quartic on a graphing calculator) function that models the data representing the years after 1995 and average revenue. Does this model or the model found in step 5 fit the data better?

Year	Number of Subscribers (millions)	Average Monthly Bill ($)
1998	69.209	39.43
1999	86.047	41.24
2000	109.478	45.27
2001	128.375	47.37
2002	140.766	48.40
2003	158.722	49.91
2004	182.140	50.64
2005	207.896	49.98
2006	233.041	50.56
2007	255.396	49.94
2008	270.334	49.79

(Source: CTIA–The Wireless Association)

5

Exponential and Logarithmic Functions

The intensities of earthquakes like the one in Japan in 2011 are measured with the Richter scale, which uses logarithmic functions. Logarithmic functions are also used in measuring loudness (in decibels) and stellar magnitude and in calculating pH values. Exponential functions are used in many real-world settings, such as the growth of investments, the growth of populations, sales decay, and radioactive decay.

Algebra TOOLBOX

KEY OBJECTIVES

- Use properties of exponents with integers
- Use properties of exponents with real numbers
- Simplify exponential expressions
- Write numbers in scientific notation
- Convert numbers in scientific notation to standard notation

Additional Properties of Exponents

In this Toolbox, we discuss properties of integer and real exponents and exponential expressions. These properties are useful in the discussion of exponential functions and of logarithmic functions, which are related to exponential functions. Integer exponents and rational exponents were discussed in Chapter 3, as well as the Product Property and Quotient Property.

For real numbers a and b and integers m and n,

1. $a^m \cdot a^n = a^{m+n}$ (Product Property)
2. $\dfrac{a^m}{a^n} = a^{m-n}, a \neq 0$ (Quotient Property)

Additional properties of exponents, which can be developed using the properties above, follow:

For real numbers a and b and integers m and n,

3. $(ab)^m = a^m b^m$ (Power of a Product Property)
4. $\left(\dfrac{a}{b}\right)^m = \dfrac{a^m}{b^m}$, for $b \neq 0$ (Power of a Quotient Property)
5. $(a^m)^n = a^{mn}$ (Power of a Power Property)
6. $a^{-m} = \dfrac{1}{a^m}$, for $a \neq 0$

For example, we can prove Property 3 for positive integer m as follows:

$$(ab)^m = \underbrace{(ab)(ab) \cdots (ab)}_{m \text{ times}} = \underbrace{(a \cdot a \cdots a)}_{m \text{ times}} \underbrace{(b \cdot b \cdots b)}_{m \text{ times}} = a^m b^m$$

EXAMPLE 1 ▶ Using the Properties of Exponents

Use properties of exponents to simplify each of the following. Assume that denominators are nonzero.

a. $\dfrac{5^6}{5^4}$ **b.** $\dfrac{y^2}{y^5}$ **c.** $(3xy)^3$ **d.** $\left(\dfrac{y}{z}\right)^4$ **e.** $3^{15-2m} \cdot 3^{2m}$

SOLUTION

a. $\dfrac{5^6}{5^4} = 5^{6-4} = 5^2 = 25$ **b.** $\dfrac{y^2}{y^5} = y^{2-5} = y^{-3} = \dfrac{1}{y^3}$ **c.** $(3xy)^3 = 3^3 x^3 y^3 = 27 x^3 y^3$

d. $\left(\dfrac{y}{z}\right)^4 = \dfrac{y^4}{z^4}$ **e.** $3^{15-2m} \cdot 3^{2m} = 3^{(15-2m)+2m} = 3^{15-2m+2m} = 3^{15}$

We can also simplify and evaluate expressions involving powers of powers.

EXAMPLE 2 ▶ Powers of Powers

a. Simplify $(x^4)^5$.   **b.** Evaluate $(2^3)^5$.   **c.** Simplify $(x^2 y^3)^5$.   **d.** Evaluate 2^{3^2}.

SOLUTION

a. $(x^4)^5 = x^{4 \cdot 5} = x^{20}$

b. We can evaluate $(2^3)^5$ in two ways:

$$(2^3)^5 = 2^{15} = 32{,}768 \quad \text{or} \quad (2^3)^5 = (8)^5 = 32{,}768$$

c. $(x^2y^3)^5 = (x^2)^5(y^3)^5 = x^{10}y^{15}$ d. $2^{3^2} = 2^9 = 512$

EXAMPLE 3 ▶   Applying Exponent Properties

Compute the following products and quotients and write the answer with positive exponents.

a. $(-2x^{-2}y)(5x^{-2}y^{-3})$ b. $\dfrac{8xy^{-2}}{2x^4y^{-6}}$ c. $\dfrac{\dfrac{2x^{-1}y}{3a}}{\dfrac{6xy^{-2}}{5a}}$ d. $(4^0x^3y^{-2})^{-3}$

SOLUTION

a. $(-2x^{-2}y)(5x^{-2}y^{-3}) = -10x^{-2+(-2)}y^{1+(-3)} = -10x^{-4}y^{-2}$

$$= -10 \cdot \frac{1}{x^4} \cdot \frac{1}{y^2} = \frac{-10}{x^4y^2}$$

b. $\dfrac{8xy^{-2}}{2x^4y^{-6}} = 4x^{1-4}y^{-2-(-6)} = 4x^{-3}y^4 = 4 \cdot \dfrac{1}{x^3} \cdot y^4 = \dfrac{4y^4}{x^3}$

c. To perform this division, we invert and multiply.

$$\frac{\dfrac{2x^{-1}y}{3a}}{\dfrac{6xy^{-2}}{5a}} = \frac{2x^{-1}y}{3a} \cdot \frac{5a}{6xy^{-2}} = \frac{10ax^{-1}y}{18axy^{-2}} = \frac{5x^{-1-1}y^{1-(-2)}}{9} = \frac{5x^{-2}y^3}{9} = \frac{5y^3}{9x^2}$$

d. $(4^0x^3y^{-2})^{-3} = 4^0x^{-9}y^6 = 1 \cdot \dfrac{1}{x^9} \cdot y^6 = \dfrac{y^6}{x^9}$

Real Number Exponents

All the above properties also hold for rational exponents so long as no negative numbers in even powered roots result, and they also apply for all real numbers that give expressions that are real numbers.

EXAMPLE 4 ▶   Operations with Real Exponents

Perform the indicated operations.

a. $(x^{1/3})(x^{3/4})$ b. $\dfrac{y^{1/2}}{y^{2/5}}$ c. $(c^{1/3})^{3/4}$

SOLUTION

a. $(x^{1/3})(x^{3/4}) = x^{1/3+3/4} = x^{13/12}$ b. $\dfrac{y^{1/2}}{y^{2/5}} = y^{1/2-2/5} = y^{1/10}$

c. $(c^{1/3})^{3/4} = c^{1/3 \cdot 3/4} = c^{1/4}$

Scientific Notation

Evaluating exponential and logarithmic functions in this chapter may result in outputs that are very large or very close to 0. **Scientific notation** is a convenient way to write very large (positive or negative) numbers or numbers close to 0. Numbers in scientific notation have the form

$$N \times 10^p \text{ (where } 1 \le N < 10 \text{ and } p \text{ is an integer)}$$

For instance, the scientific notation form of 2,654,000 is 2.654×10^6. When evaluating functions or solving equations with a calculator, we may see an expression that looks like 5.122E-8 appear on the screen. This calculator expression represents 5.122×10^{-8}, which is scientific notation for $0.00000005122 \approx 0$. Figure 5.1 shows two numbers expressed in standard notation and in scientific notation on a calculator display.

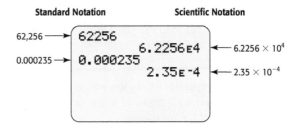

Figure 5.1

A number written in scientific notation can be converted to standard notation by multiplying (when the exponent on 10 is positive) or by dividing (when the exponent on 10 is negative). For example, we convert 7.983×10^5 to standard notation by multiplying 7.983 by $10^5 = 100,000$:

$$7.983 \times 10^5 = 7.983 \cdot 100,000 = 798,300$$

We convert 4.563×10^{-7} to standard notation by dividing 4.563 by $10^7 = 10,000,000$:

$$4.563 \times 10^{-7} = \frac{4.563}{10,000,000} = 0.0000004563$$

Multiplying two numbers in scientific notation involves adding the powers of 10, and dividing them involves subtracting the powers of 10.

EXAMPLE 5 ▶ ## Scientific Notation

Compute the following and write the answers in scientific notation.

a. $(7.983 \times 10^5)(4.563 \times 10^{-7})$ **b.** $(7.983 \times 10^5)/(4.563 \times 10^{-7})$

SOLUTION

a. $(7.983 \times 10^5)(4.563 \times 10^{-7}) = 36.426429 \times 10^{5+(-7)}$

$$= 36.426429 \times 10^{-2} = (3.6426429 \times 10^1) \times 10^{-2} = 3.6426429 \times 10^{-1}$$

The calculation using technology is shown in Figure 5.2(a).

b. $(7.983 \times 10^5)/(4.563 \times 10^{-7}) = 1.749506903 \times 10^{5-(-7)}$

$$= 1.749506903 \times 10^{12}. \text{ Figure 5.2(b) shows the calculation using technology.}$$

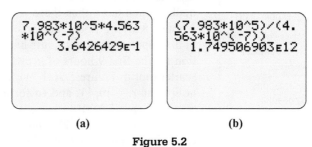

(a)   (b)

Figure 5.2

Toolbox EXERCISES

In Exercises 1 and 2, use the properties of exponents to simplify.

1. a. $x^4 \cdot x^3$ **b.** $\dfrac{x^{12}}{x^7}$ **c.** $(4ay)^4$ **d.** $\left(\dfrac{3}{z}\right)^4$

 e. $2^3 \cdot 2^2$ **f.** $(x^4)^2$

2. a. $y^5 \cdot y$ **b.** $\dfrac{w^{10}}{w^4}$ **c.** $(6bx)^3$ **d.** $\left(\dfrac{5z}{2}\right)^3$

 e. $3^2 \cdot 3^3$ **f.** $(2y^3)^4$

In Exercises 3–18, use the properties of exponents to simplify the expressions and remove all zero and negative exponents. Assume that all variables are nonzero.

3. 10^{5^0} **4.** 4^{2^2}

5. $x^{-4} \cdot x^{-3}$ **6.** $y^{-5} \cdot y^{-3}$

7. $(c^{-6})^3$ **8.** $(x^{-2})^4$

9. $\dfrac{a^{-4}}{a^{-5}}$ **10.** $\dfrac{b^{-6}}{b^{-8}}$

11. $(x^{-1/2})(x^{2/3})$ **12.** $(y^{-1/3})(y^{2/5})$

13. $(3a^{-3}b^2)(2a^2b^{-4})$ **14.** $(4a^{-2}b^3)(-2a^4b^{-5})$

15. $\left(\dfrac{2x^{-3}}{x^2}\right)^{-2}$ **16.** $\left(\dfrac{3y^{-4}}{2y^2}\right)^{-3}$

17. $\dfrac{28a^4b^{-3}}{-4a^6b^{-2}}$ **18.** $\dfrac{36x^5y^{-2}}{-6x^6y^{-4}}$

In Exercises 19–22, write the numbers in scientific notation.

19. 46,000,000   **20.** 862,000,000,000

21. 0.000094   **22.** 0.00000278

In Exercises 23–26, write the numbers in standard form.

23. 4.372×10^5 **24.** 7.91×10^6

25. 5.6294×10^{-4} **26.** 6.3478×10^{-3}

In Exercises 27 and 28, multiply or divide, as indicated, and write the result in scientific notation.

27. $(6.250 \times 10^7)(5.933 \times 10^{-2})$

28. $\dfrac{2.961 \times 10^{-2}}{4.583 \times 10^{-4}}$

In Exercises 29–34, use the properties of exponents to simplify the expressions. Assume that all variables are nonzero.

29. $x^{1/2} \cdot x^{5/6}$ **30.** $y^{2/5} \cdot y^{1/4}$

31. $(c^{2/3})^{5/2}$ **32.** $(x^{3/2})^{3/4}$

33. $\dfrac{x^{3/4}}{x^{1/2}}$ **34.** $\dfrac{y^{3/8}}{y^{1/4}}$

5.1 | Exponential Functions

KEY OBJECTIVES

- Graph and apply exponential functions
- Find horizontal asymptotes
- Graph and apply exponential growth functions
- Graph and apply exponential decay functions
- Compare transformations of graphs of exponential functions

SECTION PREVIEW **Paramecia**

The primitive single-cell animal called a paramecium reproduces by splitting into two pieces (a process called binary fission), so the population doubles each time there is a split. If we assume that the population begins with 1 paramecium and doubles each hour, then there will be 2 paramecia after 1 hour, 4 paramecia after 2 hours, 8 after 3 hours, and so on. If we let y represent the number of paramecia in the population after x hours have passed, the points (x, y) that satisfy this function for the first 9 hours of growth are described by the data in Table 5.1 and the scatter plot in Figure 5.3(a). We can show that each of these points and the data points for $x = 10, 11$, and so forth lie on the graph of the function $y = 2^x$, with $x \geq 0$ (Figure 5.3(b)).

Functions like $y = 2^x$, which have a constant base raised to a variable power, are called exponential functions. We discuss exponential functions and their applications in this section.

Table 5.1

x (hours)	0	1	2	3	4	5	6	7	8	9
y (paramecia)	1	2	4	8	16	32	64	128	256	512

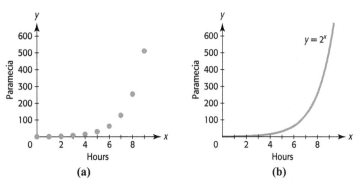

Figure 5.3

Exponential Functions

Recall that a linear function has a constant rate of change; that is, the outputs change by the same amount for each unit increase in the inputs. An exponential function has outputs that are *multiplied* by a fixed number for each unit increase in the inputs.

Note that in Example 1, the number y of paramecia is multiplied by 2 for each hour increase in time x, so the number is an exponential function of time.

EXAMPLE 1 ▶ Paramecia

As discussed in the Section Preview, the primitive single-cell animal called a paramecium reproduces by splitting into two pieces, so the population doubles each time there is a split. The number y of paramecia in the population after x hours have passed fits the graph of $y = 2^x$.

a. In the context of the paramecium application, what inputs can be used?

b. Is the function $y = 2^x$ discrete or continuous?

c. Graph the function $y = 2^x$, without regard to restricting inputs to those that make sense for the paramecium application.

d. What are the domain and range of this function?

SOLUTION

a. Not every point on the graph of $y = 2^x$ in Figure 5.3(b) describes a number of paramecia. For example, the point $(7.1, 2^{7.1})$ is on this graph, but $2^{7.1} \approx 137.187$ does not represent a number of paramecia because it is not a whole number. Since only whole number inputs will give whole number outputs, only whole number inputs can be used for this application.

b. Every point describing a number of paramecia is on the graph of $y = 2^x$, so the function $y = 2^x$ is a continuous model for this discrete paramecium growth function.

c. Some values satisfying the function $y = 2^x$ are shown in Table 5.2, and the graph of the function is shown in Figure 5.4. The function that models the growth of paramecia was restricted because of the physical setting, but if the domain of the function $y = 2^x$ is not restricted, it is the set of real numbers. Note that the range of this function is the set of all positive real numbers because there is no power of 2 that results in a value of 0 or a negative number.

Table 5.2

x	y
−3	$2^{-3} = 0.125$
−1.5	$2^{-1.5} \approx 0.35$
−0.5	$2^{-0.5} \approx 0.71$
0	$2^0 = 1$
0.5	$2^{0.5} \approx 1.41$
1.5	$2^{1.5} \approx 2.83$
3	$2^3 = 8$
5	$2^5 = 32$

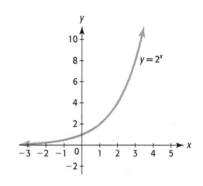

Figure 5.4

The graph in Figure 5.4 approaches but never touches the x-axis as x becomes more negative (approaching $-\infty$), so the x-axis is a **horizontal asymptote** for the graph. As previously mentioned, the function $y = 2^x$ is an example of a special class of functions called **exponential functions**. In general, we define an exponential function as follows.

Exponential Function

If b is a positive real number, $b \neq 1$, then the function $f(x) = b^x$ is an exponential function. The constant b is called the *base* of the function, and the variable x is the *exponent*.

If we graph another exponential function $y = b^x$ that has a base b greater than 1, the graph will have the same basic shape as the graph of $y = 2^x$. (See the graphs of $y = 1.5^x$ and $y = 12^x$ in Figure 5.5.)

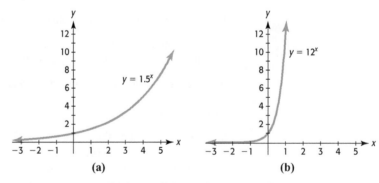

(a)   (b)

Figure 5.5

The exponential function $y = b^x$ with base b satisfying $0 < b < 1$ is a *decreasing* function. See the graph of $y = \left(\dfrac{1}{3}\right)^x$ in Figure 5.6(a). We can also write the function $y = \left(\dfrac{1}{3}\right)^x$ in the form $y = 3^{-x}$ because $3^{-x} = \dfrac{1}{3^x} = \left(\dfrac{1}{3}\right)^x$. (See Figure 5.6(b).) In general, a function of the form $y = b^{-x}$ with $b > 1$ can also be written as $y = c^x$ with $0 < c < 1$ and $c = \dfrac{1}{b}$, and their graphs are identical.

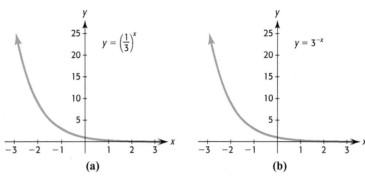

Figure 5.6

Graphs of Exponential Functions

Equation: $y = b^x$

x-intercept: none

y-intercept: (0, 1)

Domain: all real numbers

Range: all real numbers $y > 0$

Horizontal asymptote: *x*-axis

Shape: an increasing function if $b > 1$, $y = b^x$
a decreasing function if $y = b^x$ with $0 < b < 1$
or if $y = b^{-x}$ with $b > 1$

Transformations of Graphs of Exponential Functions

Graphs of exponential functions, like graphs of other types of functions, can be shifted, reflected, or stretched. Transformations, which were introduced in Chapter 4, can be applied to graphs of exponential functions.

EXAMPLE 2 ▶ **Transformations of Graphs of Exponential Functions**

Explain how the graph of each of the following functions compares with the graph of $y = 3^x$, and graph each function on the same axes as $y = 3^x$.

a. $y = 3^{x-4}$ **b.** $y = 2 + 3^{x-4}$ **c.** $y = 5(3^x)$

SOLUTION

a. The graph of $y = 3^{x-4}$ has the same shape as the graph of $y = 3^x$, but it is shifted 4 units to the right. The graph of $y = 3^x$ and the graph of $y = 3^{x-4}$ are shown in Figure 5.7(a).

b. The graph of $y = 2 + 3^{x-4}$ has the same shape as the graph of $y = 3^x$, but it is shifted 4 units to the right and 2 units up. The graph of $y = 2 + 3^{x-4}$ is shown in Figure 5.7(b). The horizontal asymptote is $y = 2$.

c. As we saw in Chapter 4, multiplication of a function by a constant greater than 1 stretches the graph of the function by a factor equal to that constant. Each of the *y*-values of $y = 5(3^x)$ is 5 times the corresponding *y*-value of $y = 3^x$. The graph of $y = 5(3^x)$ is shown in Figure 5.7(c).

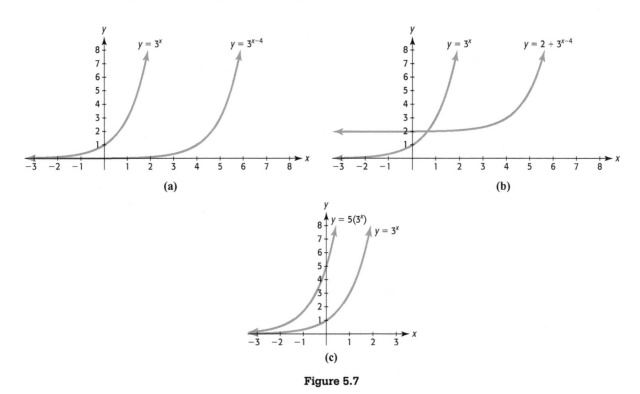

Figure 5.7

Exponential Growth Models

Functions of the form $y = b^x$ with $b > 1$ and, more generally, functions of the form $y = a(b^{kx})$ with $b > 1$, $a > 0$, and $k > 0$ are increasing and can be used to model growth.

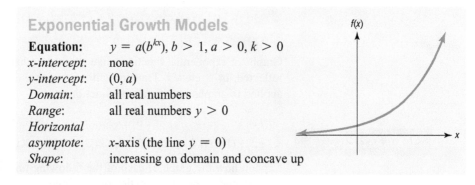

Exponential Growth Models

Equation: $y = a(b^{kx})$, $b > 1$, $a > 0$, $k > 0$
x-intercept: none
y-intercept: $(0, a)$
Domain: all real numbers
Range: all real numbers $y > 0$
Horizontal asymptote: x-axis (the line $y = 0$)
Shape: increasing on domain and concave up

As we will see in Section 5.5, the future value S of an investment of P dollars invested for t years at interest rate r, compounded annually, is given by the exponential growth function

$$S = P(1 + r)^t$$

EXAMPLE 3 ▶   **Inflation**

Suppose that inflation is predicted to average 4% per year for each year from 2012 to 2025. This means that an item that costs $10,000 one year will cost $10,000(1.04) the next year and $10,000(1.04)(1.04) = $10,000(1.04^2)$ the following year.

a. Write the function that gives the cost of a $10,000 item t years after 2012.

b. Graph the growth model found in part (a) for $t = 0$ to $t = 13$.

c. If an item costs $10,000 in 2012, use the model to predict its cost in 2025.

SOLUTION

a. The function that gives the cost of a $10,000 item t years after 2012 is

$$y = 10,000(1.04^t)$$

b. The graph of $y = 10,000(1.04^t)$ is shown in Figure 5.8.

c. The year 2025 is 13 years from 2012, so the predicted cost of this item in 2025 is

$$y = 10,000(1.04^{13}) = 16,650.74 \text{ dollars}$$

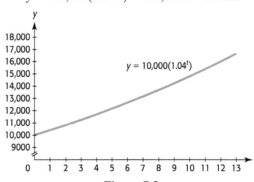

Figure 5.8

EXAMPLE 4 ▶ **Personal Income**

Total personal income in the United States (in billions of dollars) for selected years from 1960 and projected to 2018 can be modeled by

$$y = 492.4(1.07^x)$$

with x equal to the number of years after 1960.

a. What does the model predict the total U.S. personal income to be in 2014?

b. Graphically determine the year during which the model predicts that total U.S. personal income will reach $28.5 trillion.

SOLUTION

a. Evaluating the model at $x = 54$ gives total U.S. personal income in 2014 to be $19,013 billion, or $19.013 trillion.

b. Figure 5.9 shows the graph of the model and the graph of $y = 28,500$. Using Intersect gives $x = 59.98 \approx 60$, so the model predicts that total U.S. personal income will reach $28.5 trillion ($28,500 billion) in 2020.

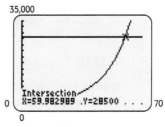

Figure 5.9

Spreadsheet ▶ SOLUTION As with linear and quadratic functions, Excel can be used to graph exponential functions and solve exponential equations. Table 5.3 shows the output of $y = 492.4(1.07^x)$ at several values of x, including $x = 60$, and Figure 5.10 displays the graph of this function.

Table 5.3

	A	B
1	0	492.4
2	10	968.6253
3	20	1905.433
4	30	3748.274
5	40	7373.423
6	50	14504.64
7	59	26666.19
8	60	28532.82

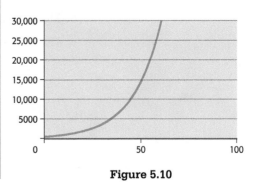

Figure 5.10

Exponential Decay Models

Functions of the form $y = a(b^{kx})$ with $b > 1$, $a > 0$, and $k < 0$ or $y = a(b^{kx})$ with $0 < b < 1$, $a > 0$, and $k > 0$ are decreasing and can be used to model decay. For example, suppose a couple retires with a fixed income of $40,000 per year. If inflation averages 4% per year for each of the next 25 years, the value of what the couple can purchase with the $40,000 will decrease each year, with its *purchasing power* in the xth year after they retire given by the *exponential decay model*

$$P(x) = 40,000(0.96^x)$$

Using this model gives the value of what the $40,000 purchases in the 25th year after they retire to be

$$P(x) = 40,000(0.96^{25}) = 14,416 \text{ dollars}$$

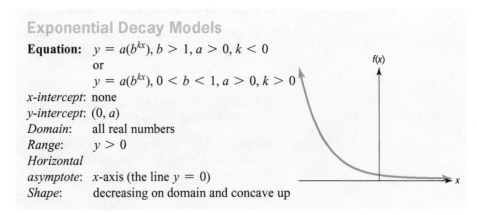

Exponential Decay Models

Equation: $y = a(b^{kx})$, $b > 1$, $a > 0$, $k < 0$
or
$y = a(b^{kx})$, $0 < b < 1$, $a > 0$, $k > 0$

x-intercept: none
y-intercept: $(0, a)$
Domain: all real numbers
Range: $y > 0$
Horizontal asymptote: *x*-axis (the line $y = 0$)
Shape: decreasing on domain and concave up

EXAMPLE 5 ▶   **Sales Decay**

It pays to advertise, and it is frequently true that weekly sales will drop rapidly for many products after an advertising campaign ends. This decline in sales is called *sales decay*. Suppose that the decay in the sales of a product is given by

$$S = 1000(2^{-0.5x}) \text{ dollars}$$

where x is the number of weeks after the end of a sales campaign. Use this function to answer the following.

a. What is the level of sales when the advertising campaign ends?

b. What is the level of sales 1 week after the end of the campaign?

c. Use a graph of the function to estimate the week in which sales equal $500.

d. According to this model, will sales ever fall to zero?

SOLUTION

a. The campaign ends when $x = 0$, so $S = 1000(2^{-0.5(0)}) = 1000(2^0) = 1000(1) = \$1000$.

b. At 1 week after the end of the campaign, $x = 1$. Thus, $S = 1000(2^{-0.5(1)}) = \$707.11$.

c. The graph of this sales decay function is shown in Figure 5.11(a). One way to find the x-value for which $S = 500$ is to graph $y_1 = 1000(2^{-0.5x})$ and $y_2 = 500$ and find the point of intersection of the two graphs. See Figure 5.11(b), which shows that $y = 500$ when $x = 2$. Thus, sales fall to half their original amount after 2 weeks.

d. The graph of this sales decay function approaches the positive x-axis as x gets large, but it never reaches the x-axis. Thus, sales will never reach a value of $0.

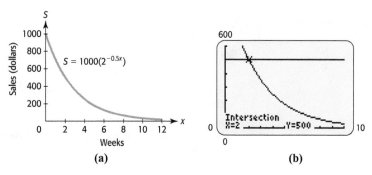

(a)   (b)

Figure 5.11

The Number e

Many real applications involve exponential functions with the base e. The number e is an irrational number with decimal approximation 2.718281828 (to nine decimal places).

$$e \approx 2.718281828$$

The exponential function with base e occurs frequently in biology and in finance. We discuss the derivation of e and how it is used in finance in Section 5.5. Because e is close in value to 3, the graph of the exponential function $y = e^x$ has the same basic shape as the graph of $y = 3^x$. (See Figure 5.12, which compares the graphs of $y = e^x$, $y = 3^x$, and $y = 2^x$.)

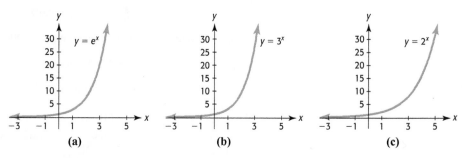

(a)   (b)   (c)

Figure 5.12

The growth in the value of an investment is another important application that can be modeled by an exponential function. As we prove in Section 5.5, the future value S of an investment of P dollars for t years at interest rate r, compounded continuously, is given by

$$S = Pe^{rt} \text{ dollars}$$

EXAMPLE 6 ▶ **Growth of an Investment**

If $10,000 is invested for 15 years at 10%, compounded continuously, what is the future value of the investment?

SOLUTION

The future value of this investment is

$$S = Pe^{rt} = 10{,}000e^{0.10(15)} = 10{,}000e^{1.5} = 44{,}816.89 \text{ dollars}$$

EXAMPLE 7 ▶ **Carbon-14 Dating**

During the life of an organism, the ratio of carbon-14 (C-14) atoms to carbon-12 (C-12) atoms remains about the same. When the organism dies, the number of C-14 atoms within its carcass begins to gradually decrease, but the number of C-12 atoms does not, so they can be measured to determine how many C-14 atoms were present at death. The number of C-14 atoms in an organism when it is discovered can then be compared to the number present at death to find how long ago the organism lived. The ratio of the present amount (y) of C-14 to the original amount (y_0) t years ago is given by

$$\frac{y}{y_0} = e^{-0.00012097t}$$

producing a function

$$y = f(t) = y_0 e^{-0.00012097t}$$

that gives the amount remaining after t years.
(Source: Thomas Jefferson National Accelerator Facility, http://www.jlab.org/)

a. If the original amount of carbon-14 present in an artifact is 100 grams, how much remains after 2000 years?

b. What percent of the original amount of carbon-14 remains after 4500 years?

SOLUTION

a. The initial amount y_0 is 100 grams and the time t is 2000 years, so we substitute these values in $f(t) = y_0 e^{-0.00012097t}$ and calculate:

$$f(2000) = 100e^{-0.00012097(2000)} \approx 78.51$$

Thus, the amount of carbon-14 present after 2000 years is approximately 78.5 grams.

b. If y_0 is the original amount, the amount that will remain after 4500 years is

$$f(4500) = y_0 e^{-0.00012097(4500)} = 0.58y_0$$

Thus, approximately 58% of the original amount will remain after 4500 years. (Note that you do not need to know the original amount of carbon-14 in order to answer this question.)

Skills CHECK 5.1

1. Which of the following functions are exponential functions?

   a. $y = 3x + 3$ **b.** $y = x^3$ **c.** $y = 3^x$

   d. $y = \left(\dfrac{2}{3}\right)^x$ **e.** $y = e^x$ **f.** $y = x^e$

2. Determine if each of the following functions models exponential growth or exponential decay.

   a. $y = 2^{0.1x}$ **b.** $y = 3^{-1.4x}$

   c. $y = 4e^{-5x}$ **d.** $y = 0.8^{3x}$

3. **a.** Graph the function $f(x) = e^x$ on the window $[-5, 5]$ by $[-10, 30]$.

   b. Find $f(1)$, $f(-1)$, and $f(4)$, rounded to three decimal places.

   c. What is the horizontal asymptote of the graph?

   d. What is the y-intercept?

4. **a.** Graph the function $f(x) = 5^x$ on the window $[-5, 5]$ by $[-10, 30]$.

   b. Find $f(1)$, $f(3)$, and $f(-2)$.

   c. What is the horizontal asymptote of the graph?

   d. What is the y-intercept?

5. Graph the function $y = 3^x$ on $[-5, 5]$ by $[-10, 30]$.

6. Graph $y = 5e^{-x}$ on $[-4, 4]$ by $[-1, 20]$.

In Exercises 7–12, graph each function.

7. $y = 5^{-x}$ 8. $y = 3^{-2x}$

9. $y = 3^x + 5$ 10. $y = 2(1.5)^{-x}$

11. $y = 3^{(x-2)} - 4$ 12. $y = 3^{(x-1)} + 2$

In Exercises 13–18, match the graph with its equation.

13. $y = 4^{x+3}$ **A.**

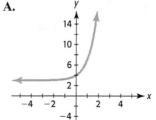

14. $y = 2 \cdot 4^x$ **B.**

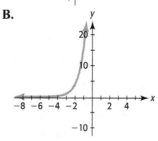

15. $y = 4^x + 3$ **C.**

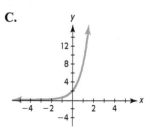

16. $y = 4^{-x}$ **D.**

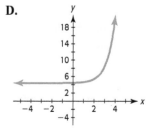

17. $y = -4^x$ **E.**

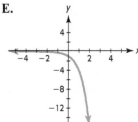

18. $y = 4^{x-2} + 4$ **F.**

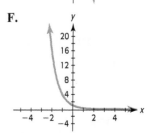

In Exercises 19–24, use your knowledge of transformations to compare the graph of the function with the graph of $f(x) = 4^x$. Then graph each function for $-4 \le x \le 4$.

19. $y = 4^x + 2$ 20. $y = 4^{x-1}$ 21. $y = 4^{-x}$

22. $y = -4^x$ 23. $y = 3(4^x)$ 24. $y = 3 \cdot 4^{(x-2)} - 3$

25. How does the graph of Exercise 24 compare with the graph of Exercise 23?

26. Which of the functions in Exercises 19–24 are increasing functions, and which are decreasing functions?

27. **a.** Graph $f(x) = 12e^{-0.2x}$ for $-5 \le x \le 15$.

   b. Find $f(10)$ and $f(-10)$.

   c. Does this function represent growth or decay?

28. If $y = 200(2^{-0.01x})$,

   a. Find the value of y (to two decimal places) when $x = 20$.

   b. Use graphical or numerical methods to determine the value of x that gives $y = 100$.

EXERCISES 5.1

29. *Sales Decay* At the end of an advertising campaign, weekly sales of smartphones declined according to the equation $y = 12,000(2^{-0.08x})$ dollars, where x is the number of weeks after the end of the campaign.

   a. Determine the sales at the end of the ad campaign.

   b. Determine the sales 6 weeks after the end of the campaign.

   c. Does this model indicate that sales will eventually reach $0?

30. *Sales Decay* At the end of an advertising campaign, weekly sales of camcorders declined according to the equation $y = 10,000(3^{-0.05x})$ dollars, where x is the number of weeks after the campaign ended.

   a. Determine the sales at the end of the ad campaign.

   b. Determine the sales 8 weeks after the end of the campaign.

   c. How do we know, by inspecting the equation, that this function is decreasing?

31. *Investment*
   a. Graph the function $S = 80,000(1.05^t)$, which gives the future value of $80,000 invested at 5%, compounded annually, for t years, $0 \le t \le 12$.

   b. Find the future value of $80,000 invested for 10 years at 5%, compounded annually.

32. *Investment*
   a. Graph the function $S = 56,000(1.09^t)$, which gives the future value of $56,000 invested at 9%, compounded annually, for t years, $0 \le t \le 15$.

   b. Find the future value of $56,000 invested for 13 years at 9%, compounded annually.

33. *Continuous Compounding* If $8000 is invested for t years at 8% interest, compounded continuously, the future value is given by $S = 8000e^{0.08t}$ dollars.

   a. Graph this function for $0 \le t \le 15$.

   b. Use the graph to estimate when the future value will be $20,000.

   c. Complete the following table.

t (year)	S ($)
10	
20	
	46,449.50

34. *Continuous Compounding* If $35,000 is invested for t years at 9% interest, compounded continuously, the future value is given by $S = 35,000e^{0.09t}$ dollars.

   a. Graph this function for $0 \le t \le 8$.

   b. Use the graph to estimate when the future value will be $60,000.

   c. Complete the following table.

t (year)	S ($)
10	
15	
	253,496

35. *Radioactive Decay* The amount of radioactive isotope thorium-234 present at time t is given by $A(t) = 500e^{-0.02828t}$ grams, where t is the time in years that the isotope decays. The initial amount present is 500 grams.

   a. How many grams remain after 10 years?

   b. Graph this function for $0 \le t \le 100$.

   c. If the half-life is the time it takes for half of the initial amount to decay, use graphical methods to estimate the half-life of this isotope.

36. *Radioactive Decay* A breeder reactor converts stable uranium-238 into the isotope plutonium-239. The decay of this isotope is given by $A(t) = 100e^{-0.00002876t}$, where $A(t)$ is the amount of the isotope at time t (in years) and 100 grams is the original amount.

   a. How many grams remain after 100 years?

   b. Graph this function for $0 \le t \le 50,000$.

   c. The half-life is the time it takes for half of the initial amount to decay; use graphical methods to estimate the half-life of this isotope.

37. *Sales Decay* At the end of an advertising campaign, weekly sales at an electronics store declined according to the equation $y = 2000(2^{-0.1x})$ dollars, where x is the number of weeks after the end of the campaign.

   a. Graph this function for $0 \le x \le 60$.

   b. Use the graph to find the weekly sales 10 weeks after the campaign ended.

   c. Comment on "It pays to advertise" for this store.

38. *Sales Decay* At the end of an advertising campaign, weekly retail sales of a product declined according to the equation $y = 40,000(3^{-0.1x})$ dollars, where x is the number of weeks after the campaign ended.

a. Graph this function for $0 \leq x \leq 50$.

b. Find the weekly sales 10 weeks after the campaign ended.

c. Should the retailers consider another advertising campaign even if it costs $5000?

39. *Purchasing Power* The purchasing power (real value of money) decreases if inflation is present in the economy. For example, the purchasing power of R dollars after t years of 5% inflation is given by the model

$$P = R(0.95^t) \text{ dollars}$$

a. What will be the purchasing power of $40,000 after 20 years of 5% inflation?

b. How does this affect people planning to retire at age 50?
(Source: *Viewpoints*, VALIC, 1993)

40. *Purchasing Power* If a retired couple has a fixed income of $60,000 per year, the purchasing power (real value of money) after t years of 5% inflation is given by the equation $P = 60,000(0.95^t)$ dollars.

a. What is the purchasing power of their income after 4 years?

b. Test numerical values in this function to determine the years when their purchasing power will be less than $30,000.

41. *Real Estate Inflation* During a 5-year period of constant inflation, the value of a $100,000 property will increase according to the equation $v = 100,000e^{0.05t}$.

a. What will be the value of this property in 4 years?

b. Use a table or graph to estimate when this property will double in value.

42. *Inflation* An antique table increases in value according to the function $v(x) = 850(1.04^x)$ dollars, where x is the number of years after 1990.

a. How much was the table worth in 1990?

b. If the pattern indicated by the function remains valid, what was the value of the table in 2005?

c. Use a table or graph to estimate the year when this table would reach double its 1990 value.

43. *Population* The population in a certain city was 53,000 in 2000, and its future size is predicted to be $P(t) = 53,000e^{0.015t}$ people, where t is the number of years after 2000.

a. Does this model indicate that the population is increasing or decreasing?

b. Use this function to estimate the population of the city in 2005.

c. Use this function to estimate the population of the city in 2010.

d. What is the average rate of growth between 2000 and 2010?

44. *Population* The population in a certain city was 800,000 in 2003, and its future size is predicted to be $P = 800,000e^{-0.020t}$ people, where t is the number of years after 2003.

a. Does this model indicate that the population is increasing or decreasing?

b. Use this model to estimate the population of the city in 2010.

c. Use this model to predict the population of the city in 2020.

d. What is the average rate of change in population between 2010 and 2020?

45. *Carbon-14 Dating* An exponential decay function can be used to model the number of grams of a radioactive material that remain after a period of time. Carbon-14 decays over time, with the amount remaining after t years given by $y = 100e^{-0.00012097t}$ if 100 grams is the original amount.

a. How much remains after 1000 years?

b. Use graphical methods to estimate the number of years until 10 grams of carbon-14 remain.

46. *Drugs in the Bloodstream* If a drug is injected into the bloodstream, the percent of the maximum dosage that is present at time t is given by

$$y = 100(1 - e^{-0.35(10-t)})$$

where t is in hours, with $0 \leq t \leq 10$.

a. What percent of the drug is present after 2 hours?

b. Graph this function.

c. When is the drug totally gone from the bloodstream?

47. *Normal Curve* The "curve" on which many students like to be graded is the bell-shaped normal curve. The equation $y = \dfrac{1}{\sqrt{2\pi}}e^{-(x-50)^2/2}$ describes the normal curve for a standardized test, where x is the test score before curving.

a. Graph this function for x between 47 and 53 and for y between 0 and 0.5.

b. The average score for the test is the score that gives the largest output y. Use the graph to find the average score.

48. *IQ Measure* The frequency of IQ measures x follows a bell-shaped normal curve with equation $y = \dfrac{1}{\sqrt{2\pi}}e^{-(x-100)^2/20}$. Graph this function for x between 80 and 120 and for y between 0 and 0.4.

SECTION PREVIEW **Diabetes**

Projections from 2010 to 2050 indicate that the fraction of U.S. adults with diabetes (diagnosed and undiagnosed) will eventually reach one-third of the adult population. The percent can be modeled by the logarithmic function

$$p(x) = -12.975 + 11.851 \ln x$$

where x is the number of years after 2000. To find the year when the percent reaches 33%, we can graph the equations $y_1 = -12.975 + 11.851 \ln x$ and $y_2 = 33$ on the same axes and find the point of intersection. (See Example 8.) (Source: Centers for Disease Control and Prevention)

In this section, we introduce logarithmic functions, discuss the relationship between exponential and logarithmic functions, and use the definition of logarithmic functions to rewrite exponential equations in a new form, called the logarithmic form of the equation. We also use logarithmic functions in real applications. ∎

Logarithmic Functions

Recall that we discussed inverse functions in Section 4.3. The inverse of a function is a second function that "undoes" what the original function does. For example, if the original function doubles each input, its inverse takes half of each input. If a function is defined by $y = f(x)$, the inverse of this function can be found by interchanging x and y in $y = f(x)$ and solving the new equation for y. As we saw in Section 4.3, a function has an inverse if it is a one-to-one function (that is, there is a one-to-one correspondence between the inputs and outputs defining the function).

Every exponential function of the form $y = b^x$, with $b > 0$ and $b \neq 1$, is one-to-one, so every exponential function of this form has an inverse function. The inverse function of

$$y = b^x$$

is found by interchanging x and y and solving the new equation for y. Interchanging x and y in $y = b^x$ gives

$$x = b^y$$

In this new function, y is the power to which we raise b to get the number x. To solve an expression like this for the exponent, we need new notation. We define y, the power to which we raise the base b to get the number x, as

$$y = \log_b x$$

This inverse function is called a **logarithmic function** with base b.

Logarithmic Function

For $x > 0, b > 0$, and $b \neq 1$, the logarithmic function with base b is

$$y = \log_b x$$

which is defined by $x = b^y$. That is, $\log_b x$ is the exponent to which we must raise the base b to get the number x.

This function is the inverse function of the exponential function $y = b^x$.

Note that, according to the definition of the logarithmic function, $y = \log_b x$ and $x = b^y$ are two different forms of the same equation. We call $y = \log_b x$ the *logarithmic form* of the equation and $x = b^y$ the *exponential form* of the equation. The number b is called the **base** in both $y = \log_b x$ and $x = b^y$, and y is the **logarithm** in $y = \log_b x$ and the **exponent** in $x = b^y$. Thus, a logarithm is an exponent.

EXAMPLE 1 ▶ **Converting from Exponential to Logarithmic Form**

Write each of the following exponential equations in logarithmic form.

a. $3^2 = 9$

b. $4^{-1} = \dfrac{1}{4}$

c. $5^1 = 5$

d. $x = 3^y$

SOLUTION

To write an exponential equation in its equivalent logarithmic form, remember that the base of the logarithm is the same as the base of the exponent in the exponential form. Also, the exponent in the exponential form is the value of the logarithm in the logarithmic form.

a. To write $3^2 = 9$ in logarithmic form, note that the base of the exponent is 3, so the base of the logarithm will also be 3. The exponent, 2, will be the value of the logarithm. Thus,

$$3^2 = 9 \quad \text{is equivalent to} \quad \log_3 9 = 2$$

We can refer to 9 as the **argument** of the logarithm.

b. To write $4^{-1} = \dfrac{1}{4}$ in logarithmic form, note that the base of the exponent is 4, so the base of the logarithm will also be 4. The exponent, -1, will be the value of the logarithm. Thus,

$$4^{-1} = \frac{1}{4} \quad \text{is equivalent to} \quad \log_4 \frac{1}{4} = -1$$

c. The base of the exponent in $5^1 = 5$ is 5, so the base of the logarithm will be 5, and the value of the logarithm will be 1. Thus,

$$5^1 = 5 \quad \text{is equivalent to} \quad \log_5 5 = 1$$

d. Similarly, the base of the exponent in $x = 3^y$ is 3, so the base of the logarithm will be 3, and the value of the logarithm will be y. Thus,

$$x = 3^y \quad \text{is equivalent to} \quad \log_3 x = y$$

EXAMPLE 2 ▶ **Converting from Logarithmic to Exponential Form**

Write each of the following logarithmic equations in exponential form.

a. $\log_2 16 = 4$

b. $\log_{10} 0.0001 = -4$

c. $\log_6 1 = 0$

SOLUTION

To write a logarithmic equation in its equivalent exponential form, remember that the base of the logarithm will be the base of the exponent in the exponential form. Also, the value of the logarithm will be the exponent in the exponential form. The argument of the logarithm will be the value of the exponential expression.

a. To write $\log_2 16 = 4$ in exponential form, note that the base of the exponent will be 2, the exponent will be 4, and the value of the exponential expression will be 16. Thus,

$$\log_2 16 = 4 \quad \text{is equivalent to} \quad 2^4 = 16$$

b. To write $\log_{10} 0.0001 = -4$ in exponential form, note that the base of the exponent will be 10, the exponent will be -4, and the value of the exponential expression will be 0.0001. Thus,

$$\log_{10} 0.0001 = -4 \quad \text{is equivalent to} \quad 10^{-4} = 0.0001$$

c. The base of the exponent in $\log_6 1 = 0$ will be 6, the exponent will be 0, and the value of the exponential expression will be 1. Thus,

$$\log_6 1 = 0 \quad \text{is equivalent to} \quad 6^0 = 1$$

We can sometimes more easily evaluate logarithmic functions for different inputs by changing the function from logarithmic form to exponential form.

EXAMPLE 3 ▶ **Evaluating Logarithms**

a. Write $y = \log_2 x$ in exponential form.

b. Use the exponential form from part (a) to find y when $x = 8$.

c. Evaluate $\log_2 8$. **d.** Evaluate $\log_4 \dfrac{1}{16}$. **e.** Evaluate $\log_{10} 0.001$.

SOLUTION

a. To rewrite $y = \log_2 x$ in exponential form, note that the base of the logarithm, 2, becomes the base of the exponential expression, and y equals the exponent in the equivalent exponential form. Thus, the exponential form is

$$2^y = x$$

b. To find y when x is 8, we solve $2^y = 8$. We can easily see that $y = 3$ because $2^3 = 8$.

c. To evaluate $\log_2 8$, we write $\log_2 8 = y$ in its equivalent exponential form $2^y = 8$. From part (b), $y = 3$. Thus, $\log_2 8 = 3$.

d. To evaluate $\log_4 \dfrac{1}{16}$, we write $\log_4 \dfrac{1}{16} = y$ in its equivalent exponential form $4^y = \dfrac{1}{16}$. Because $4^{-2} = \dfrac{1}{16}$, $y = -2$. Thus, $\log_4 \dfrac{1}{16} = -2$.

e. To evaluate $\log_{10} 0.001$, we write $\log_{10} 0.001 = y$ in its equivalent exponential form $10^y = 0.001$. Because $10^{-3} = 0.001$, $y = -3$. Thus, $\log_{10} 0.001 = -3$.

The relationship between the logarithmic and exponential forms of an equation is used in the following example.

EXAMPLE 4 ▶ Graphing a Logarithmic Function

Graph $y = \log_2 x$.

SOLUTION

We can graph $y = \log_2 x$ by graphing the equivalent equation $x = 2^y$. Table 5.4 gives the values (found by substituting values of y and calculating x), and the graph is shown in Figure 5.13.

Table 5.4

$x = 2^y$	y
$\frac{1}{8}$	-3
$\frac{1}{4}$	-2
$\frac{1}{2}$	-1
1	0
2	1
4	2
8	3

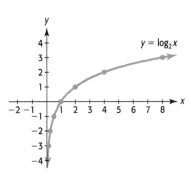

Figure 5.13

Figures 5.14(a) and (b) show graphs of the logarithmic function $y = \log_b x$ for a base $b > 1$ and a base $0 < b < 1$. Note that the graph of the function in Figure 5.14(a) is increasing and the graph of the function in Figure 5.14(b) is decreasing.

Logarithmic Function

Equation:	$y = \log_b x$
x-intercept:	$(1, 0)$
y-intercept:	none
Domain:	$x > 0$
Range:	all real numbers
Vertical asymptote:	y-axis (the line $x = 0$)

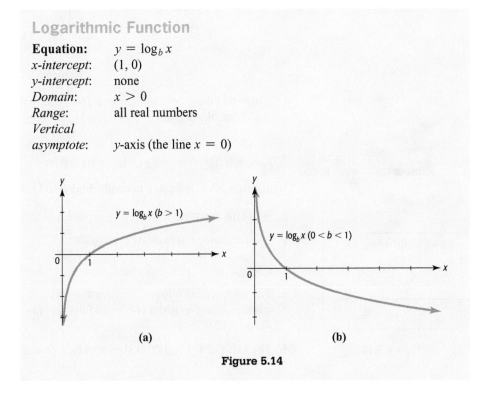

(a)   (b)

Figure 5.14

Common Logarithms

We can use technology to evaluate logarithmic functions if the base of the logarithm is 10 or e. Because our number system uses base 10, logarithms with a base of 10 are called **common logarithms**, and, by convention, $\log_{10} x$ is simply denoted $\log x$. The graph of $y = \log x$ can be drawn by plotting points satisfying $x = 10^y$. Table 5.5 shows some points on the graph of $y = \log x$, using powers of 10 as x-values, and Figure 5.15 shows the graph.

Table 5.5

x	y = log x
1000	3
100	2
10	1
1	0
0.1	−1
0.01	−2
0.001	−3

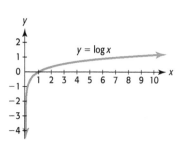

Figure 5.15

Graphing utilities and spreadsheets can be used to evaluate and graph $f(x) = \log x$, using the key or command $\boxed{\log}$. Figure 5.16(a) shows the graph of $y = \log x$ on a graphing calculator. The graph of a logarithmic function with any base can be created with a spreadsheet such as Excel by entering the formula $= \log(x, base)$. The graph in Figure 5.16(b) results from the Excel command "$= \log(x, 10)$."

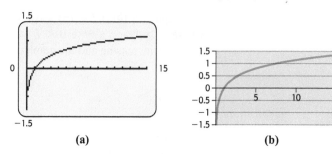

(a)   (b)

Figure 5.16

Note that the graphs in Figures 5.16(a) and (b) appear to touch the y-axis, although we know that the domain of $y = \log x$ is $x > 0$.

EXAMPLE 5 ▶ **Evaluating Base 10 Logarithms**

Using each of the indicated methods, find $f(10,000)$ if $f(x) = \log x$.

a. Write the equation $y = \log x$ in exponential form and find y when $x = 10,000$.

b. Use technology to evaluate $\log 10,000$.

```
log(10000)
              4
```

Figure 5.17

SOLUTION

a. The exponential form of $y = \log x = \log_{10} x$ is $10^y = x$. Substituting 10,000 for x in this equation gives $10^y = 10,000$. Because $10^4 = 10,000$, we see that $y = 4$. Thus, $f(10,000) = 4$.

b. The value of $\log 10,000$ is shown to be 4 on the calculator screen in Figure 5.17, so $f(10,000) = 4$.

EXAMPLE 6 ▶   pH Scale

To simplify measurement of the acidity or basicity of a solution, the pH (hydrogen potential) scale was developed. The pH is given by the formula $pH = -\log[H^+]$, where $\log[H^+]$ is the concentration of hydrogen ions in moles per liter in the solution. The pH scale ranges from 0 to 14. A pH of 7 is neutral; a pH less than 7 is acidic; a pH greater than 7 is basic.

a. Using Table 5.6, find the pH level of bleach.

b. Using Table 5.6, find the pH level of lemon juice.

c. Lower levels of pH indicate a more acidic solution. How much more acidic is lemon juice than bleach?

Table 5.6

Concentration of Hydrogen Ions	Examples
10^{-14}	Liquid drain cleaner
10^{-13}	Bleach, oven cleaner
10^{-12}	Soapy water
10^{-11}	Ammonia
10^{-10}	Milk of magnesia
10^{-9}	Toothpaste
10^{-8}	Baking soda
10^{-7}	"Pure" water
10^{-6}	Milk
10^{-5}	Black coffee
10^{-4}	Tomato juice
10^{-3}	Orange juice
10^{-2}	Lemon juice
10^{-1}	Sulfuric acid
1	Battery acid

SOLUTION

a. The concentration of hydrogen ions for bleach is 10^{-13}. So the pH will be $-\log(10^{-13}) = 13$. See Figure 5.18(a).

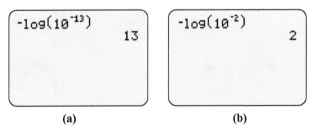

(a)   (b)

Figure 5.18

b. The concentration of hydrogen ions for lemon juice is 10^{-2}. So the pH will be $-\log(10^{-2}) = 2$. See Figure 5.18(b).

c. To find the ratio of the acidity of the two solutions, we divide the concentrations of hydrogen ions. Thus, the ratio is

$$\frac{10^{-2}}{10^{-13}} = \frac{10^{13}}{10^2} = 10^{11} = 100,000,000,000$$

indicating that lemon juice is 100,000,000,000 times as acidic as bleach.

EXAMPLE 7 ▶ ### Transformations of Graphs of Logarithmic Functions

Explain how the graph of each of the following functions compares with the graph of $y = \log x$, find the domain, and graph each function.

a. $y = \log(x + 3)$ **b.** $y = 4 + \log(x - 2)$ **c.** $y = \frac{1}{3}\log x$

SOLUTION

a. The graph of $y = \log(x + 3)$ has the same shape as the graph of $y = \log x$, but it is shifted 3 units to the left. Note that the vertical asymptote of the graph of $y = \log x$, $x = 0$, is also shifted 3 units left, so the vertical asymptote of $y = \log(x + 3)$ is $x = -3$. This means that, as the x-values approach -3, the y-values decrease without bound. See Figure 5.19(a). Because the graph has been shifted left 3 units, the domain is $(-3, \infty)$.

b. The graph of $y = 4 + \log(x - 2)$ has the same shape as the graph of $y = \log x$, but it is shifted 2 units to the right and 4 units up. The vertical asymptote is $x = 2$, and the graph is shown in Figure 5.19(b). The domain is $(2, \infty)$.

c. As we saw in Chapter 4, multiplication of a function by a constant less than 1 compresses the graph by a factor equal to that constant. So each of the y-values of $y = \frac{1}{3}\log x$ is $\frac{1}{3}$ times the corresponding y-value of $y = \log x$, and the graph is compressed. See Figure 5.19(c). The domain is $(0, \infty)$.

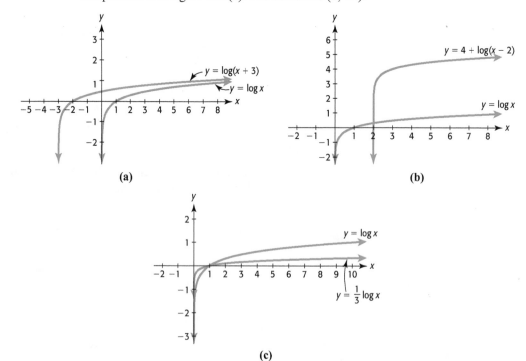

(a)

(b)

(c)

Figure 5.19

Natural Logarithms

As we have seen in the previous section, the number e is important in many business and life science applications. Likewise, logarithms to base e are important in many applications. Because they are used so frequently, especially in science, they are called **natural logarithms**, and we use the special notation $\ln x$ to denote $\log_e x$.

> ### Natural Logarithms
>
> The logarithmic function with base e is $y = \log_e x$, defined by $x = e^y$ for all positive numbers x and denoted by
>
> $$\ln x = \log_e x$$

Most graphing utilities and spreadsheets have the key or command $\boxed{\ln}$ to use in evaluating functions involving natural logarithms. Figure 5.20(a) shows the graph of $y = \ln x$, which is similar to the graph of $y = \log_3 x$ (Figure 5.20(b)) because the value of e, the base of $\ln x$, is close to 3.

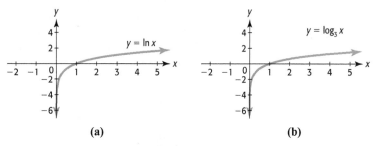

(a)   (b)

Figure 5.20

EXAMPLE 8 ▶   **Diabetes**

Projections from 2010 to 2050 indicate that the percent of U.S. adults with diabetes (diagnosed and undiagnosed) can be modeled by $p(x) = -12.975 + 11.851 \ln x$, where x is the number of years after 2000.

a. Graph this function.

b. Is the function increasing or decreasing? What does this mean in the context of the application?

c. What does this model predict the percent of U.S. adults with diabetes will be in 2022?

d. Use the graph to estimate the year in which this model predicts the percent will reach 33%.
(Source: Centers for Disease Control and Prevention)

SOLUTION

a. Using the window $0 \le x \le 50$ and $-1 \le x \le 50$, the graph of $p(x) = -12.975 + 11.851 \ln x$ is as shown in Figure 5.21(a).

b. The function is increasing, which means the percent of U.S. adults with diabetes is predicted to increase from 2010 to 2050.

c. The year 2022 is 22 years after 2000, so the percent in 2022 is estimated to be

$$p(22) = -12.975 + 11.851 \ln 22 \approx 23.7$$

d. To find the x-value for which $y = p(x) = 33$, we graph $y_1 = -12.975 + 11.851 \ln x$ and $y_2 = 33$ on the same axes and find the point of intersection. This shows that the percent is 33% when $x \approx 48.4$, during 2049. (See Figure 5.21(b).)

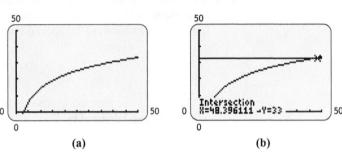

(a)   (b)

Figure 5.21

Logarithmic Properties

We have learned that logarithms are exponents and that a logarithmic function with base a is the inverse of an exponential function with base a. Thus, the properties of logarithms can be derived from the properties of exponents. These properties are useful in solving equations involving exponents and logarithms.

The basic properties of logarithms are easy to derive from the definition of a logarithm.

Basic Properties of Logarithms

For $b > 0, b \neq 1$,

1. $\log_b b = 1$

2. $\log_b 1 = 0$

3. $\log_b b^x = x$

4. $b^{\log_b x} = x$

5. For positive real numbers M and N, if $M = N$, then $\log_b M = \log_b N$.

These properties are easily justified by rewriting the logarithmic form as an exponential form or vice versa. In particular, $\log_b 1 = 0$ because $b^0 = 1$, $\log_b b = 1$ because $b^1 = b$, $\log_b b^x = x$ because $b^x = b^x$, and $b^{\log_b x} = x$ because $\log_b x = \log_b x$.

EXAMPLE 9 ▶ **Logarithmic Properties**

Use the basic properties of logarithms to simplify the following.

a. $\log_5 5^{10}$ **b.** $\log_4 4$ **c.** $\log_4 1$ **d.** $\log 10^7$ **e.** $\ln e^3$ **f.** $\log\left(\dfrac{1}{10^3}\right)$

SOLUTION

a. $\log_5 5^{10} = 10$ by Property 3   **b.** $\log_4 4 = 1$ by Property 1

c. $\log_4 1 = 0$ by Property 2   **d.** $\log 10^7 = \log_{10} 10^7 = 7$ by Property 3

e. $\ln e^3 = \log_e e^3 = 3$ by Property 3   **f.** $\log\left(\dfrac{1}{10^3}\right) = \log 10^{-3} = \log_{10} 10^{-3} = -3$

Other logarithmic properties are helpful in simplifying logarithmic expressions, which in turn help us solve some logarithmic equations. These properties are also directly related to the properties of exponents.

Additional Logarithmic Properties

For $b > 0, b \neq 1, k$ a real number, and M and N positive real numbers,

6. $\log_b(MN) = \log_b M + \log_b N$ (Product Property)

7. $\log_b\left(\dfrac{M}{N}\right) = \log_b M - \log_b N$ (Quotient Property)

8. $\log_b M^k = k \log_b M$ (Power Property)

To prove Property 6, we let $x = \log_b M$ and $y = \log_b N$, so that $M = b^x$ and $N = b^y$. Then

$$\log_b(MN) = \log_b(b^x \cdot b^y) = \log_b(b^{x+y}) = x + y = \log_b M + \log_b N$$

To prove Property 8, we let $x = \log_b M$, so that $M = b^x$. Then

$$\log_b M^k = \log_b(b^x)^k = \log_b b^{kx} = kx = k \log_b M$$

EXAMPLE 10 ▶ **Rewriting Logarithms**

Rewrite each of the following expressions as a sum, difference, or product of logarithms, and simplify if possible.

a. $\log_4 5(x - 7)$ **b.** $\ln[e^2(e + 3)]$ **c.** $\log\left(\dfrac{x - 8}{x}\right)$

d. $\log_4 y^6$ **e.** $\ln\left(\dfrac{1}{x^5}\right)$

SOLUTION

a. By Property 6, $\log_4 5(x - 7) = \log_4 5 + \log_4(x - 7)$.

b. By Property 6, $\ln[e^2(e + 3)] = \ln e^2 + \ln(e + 3)$.
By Property 3, this equals $2 + \ln(e + 3)$.

c. By Property 7, $\log\left(\dfrac{x - 8}{x}\right) = \log(x - 8) - \log x$.

d. By Property 8, $\log_4 y^6 = 6 \log_4 y$.

e. By Property 7, Property 2, and Property 8,

$$\ln\left(\frac{1}{x^5}\right) = \ln 1 - \ln x^5 = 0 - 5 \ln x = -5 \ln x$$

Caution: There is no property of logarithms that allows us to rewrite a logarithm of a sum or a difference. That is why in Example 10(a) the expression $\log_4(x - 7)$ was **not** written as $\log_4 x - \log_4 7$.

EXAMPLE 11 ▶ **Rewriting Logarithms**

Rewrite each of the following expressions as a single logarithm.

a. $\log_3 x + 4 \log_3 y$

b. $\dfrac{1}{2} \log a - 3 \log b$

c. $\ln(5x) - 3 \ln z$

SOLUTION

a. $\log_3 x + 4 \log_3 y = \log_3 x + \log_3 y^4$ Use Logarithmic Property 8.

$\qquad\qquad\qquad\qquad = \log_3 xy^4$ Use Logarithmic Property 6.

b. $\dfrac{1}{2} \log a - 3 \log b = \log a^{1/2} - \log b^3$ Use Logarithmic Property 8.

$\qquad\qquad\qquad\qquad = \log\left(\dfrac{a^{1/2}}{b^3}\right)$ Use Logarithmic Property 7.

c. $\ln(5x) - 3 \ln z = \ln(5x) - \ln z^3$ Use Logarithmic Property 8.

$\qquad\qquad\qquad\qquad = \ln\left(\dfrac{5x}{z^3}\right)$ Use Logarithmic Property 7.

EXAMPLE 12 ▶ **Applying the Properties of Logarithms**

a. Use logarithmic properties to estimate $\ln(5e)$ if $\ln 5 \approx 1.61$.

b. Find $\log_a\left(\dfrac{15}{8}\right)$ if $\log_a 15 = 2.71$ and $\log_a 8 = 2.079$.

c. Find $\log_a\left(\dfrac{15^2}{\sqrt{8}}\right)^3$ if $\log_a 15 = 2.71$ and $\log_a 8 = 2.079$.

SOLUTION

a. $\ln(5e) = \ln 5 + \ln e \approx 1.61 + 1 = 2.61$

b. $\log_a\left(\dfrac{15}{8}\right) = \log_a 15 - \log_a 8 = 2.71 - 2.079 = 0.631$

c. $\log_a\left(\dfrac{15^2}{\sqrt{8}}\right)^3 = 3 \log_a\left(\dfrac{15^2}{\sqrt{8}}\right) = 3[\log_a 15^2 - \log_a 8^{1/2}]$

$\qquad = 3\left[2 \log_a 15 - \dfrac{1}{2} \log_a 8\right] = 3\left[2(2.71) - \dfrac{1}{2}(2.079)\right] = 13.1415$

Richter Scale

Because the intensities of earthquakes are so large, the Richter scale was developed to provide a "scaled down" measuring system that permits earthquakes to be more easily measured and compared. The Richter scale gives the magnitude R of an earthquake using the formula

$$R = \log\left(\dfrac{I}{I_0}\right)$$

where I is the intensity of the earthquake and I_0 is a certain minimum intensity used for comparison.

Table 5.7 describes the typical effects of earthquakes of various magnitudes near the epicenter.

Table 5.7

Description	Richter Magnitudes	Earthquake Effects
Micro	Less than 2.0	Microearthquakes, not felt
Very minor	2.0–2.9	Generally not felt, but recorded
Minor	3.0–3.9	Often felt, but rarely causes damage
Light	4.0–4.9	Noticeable shaking of indoor items, rattling noises; significant damage unlikely
Moderate	5.0–5.9	Can cause damage to poorly constructed buildings over small regions
Strong	6.0–6.9	Can be destructive in areas up to about 100 miles across in populated areas
Major	7.0–7.9	Can cause serious damage over larger areas
Great	8.0–8.9	Can cause serious damage in areas several hundred miles across
Rarely, great	9.0–9.9	Devastating in areas several thousand miles across
Meteoric	10.0+	Never recorded

(Source: U.S. Geological Survey)

EXAMPLE 13 ▶ Richter Scale

a. If an earthquake has an intensity of 10,000 times I_0, what is the magnitude of the earthquake? Describe the effects of this earthquake.

b. Show that if the Richter scale reading of an earthquake is k, the intensity of the earthquake is $I = 10^k I_0$.

c. An earthquake that measured 9.0 on the Richter scale occurred in the Indian Ocean in December 2004, causing a devastating tsunami that killed thousands of people. Express the intensity of this earthquake in terms of I_0.

d. If an earthquake measures 7.0 on the Richter scale, give the intensity of this earthquake in terms of I_0. How much more intense is the earthquake in part (c) than the one with the Richter scale measurement of 7.0?

e. If one earthquake has intensity 320,000 times I_0 and a second has intensity 3,200,000 times I_0, what is the difference in their Richter scale measurements?

SOLUTION

a. If the intensity of the earthquake is $10{,}000 I_0$, we substitute $10{,}000 I_0$ for I in the equation $R = \log\left(\dfrac{I}{I_0}\right)$, getting $R = \log\dfrac{10{,}000 I_0}{I_0} = \log 10{,}000 = 4$. Thus, the magnitude of the earthquake is 4, signifying a light earthquake that would not likely cause significant damage.

b. If $R = k$, $k = \log\left(\dfrac{I}{I_0}\right)$, and the exponential form of this equation is $10^k = \dfrac{I}{I_0}$, so
$I = 10^k I_0$.

c. The Richter scale reading for this earthquake is 9.0, so its intensity is
$$I = 10^{9.0} I_0 = 1{,}000{,}000{,}000 \; I_0$$
Thus, the earthquake had an intensity of 1 billion times I_0.

d. The Richter scale reading for this earthquake is 7.0, so its intensity is $I = 10^{7.0} I_0$.
Thus, the ratio of intensities of the two earthquakes is
$$\frac{10^{9.0} I_0}{10^{7.0} I_0} = \frac{10^{9.0}}{10^{7.0}} = 10^2 = 100$$

Therefore, the Indian Ocean earthquake is 100 times as intense as the other earthquake.

e. Recall that the Richter scale measurement is given by $R = \log\left(\dfrac{I}{I_0}\right)$. The difference of the Richter scale measurements is
$$R_2 - R_1 = \log\frac{3{,}200{,}000 I_0}{I_0} - \log\frac{320{,}000 I_0}{I_0} = \log 3{,}200{,}000 - \log 320{,}000$$

Using Property 6 gives us
$$\log 3{,}200{,}000 - \log 320{,}000 = \log\frac{3{,}200{,}000}{320{,}000} = \log 10 = 1$$

Thus, the difference in the Richter scale measurements is 1.

Note from Example 13(d) that the Richter scale measurement of the Indian Ocean earthquake is 2 larger than the Richter scale measurement of the other earthquake, and that the Indian Ocean earthquake is 10^2 times as intense. In general, we have the following.

Richter Scale

1. If the intensity of an earthquake is I, its Richter scale measurement is
$$R = \log\frac{I}{I_0}$$

2. If the Richter scale reading of an earthquake is k, its intensity is
$$I = 10^k I_0$$

3. If the difference of the Richter scale measurements of two earthquakes is the positive number d, the intensity of the larger earthquake is 10^d times that of the smaller earthquake.

Skills CHECK 5.2

In Exercises 1–4, write the logarithmic equations in exponential form.

1. $y = \log_3 x$ **2.** $2y = \log_5 x$

3. $y = \ln(2x)$ **4.** $y = \log(-x)$

In Exercises 5–8, write the exponential equations in logarithmic form.

5. $x = 4^y$ **6.** $m = 3^p$

7. $32 = 2^5$ **8.** $9^{2x} = y$

In Exercises 9 and 10, evaluate the logarithms, if possible. Round each answer to three decimal places.

9. a. $\log 7$ **10. a.** $\log 456$

     b. $\ln 86$ **b.** $\log(-12)$

     c. $\log 63{,}980$ **c.** $\ln 10$

In Exercises 11–13, find the value of the logarithms without using a calculator.

11. a. $\log_2 32$ **b.** $\log_9 81$

     c. $\log_3 27$ **d.** $\log_4 64$

     e. $\log_5 625$

12. a. $\log_2 64$ **b.** $\log_9 27$

     c. $\log_4 2$ **d.** $\ln (e^3)$

     e. $\log 100$

13. a. $\log_3 \dfrac{1}{27}$ **b.** $\ln 1$

     c. $\ln e$ **d.** $\log 0.0001$

14. Graph the functions by changing to exponential form.

     a. $y = \log_3 x$ **b.** $y = \log_5 x$

In Exercises 15–18, graph the functions with technology.

15. $y = 2 \ln x$ **16.** $y = 4 \log x$

17. $y = \log(x + 1) + 2$ **18.** $y = \ln(x - 2) - 3$

19. a. Write the inverse of $y = 4^x$ in logarithmic form.

     b. Graph $y = 4^x$ and its inverse and discuss the symmetry of their graphs.

20. a. Write the inverse of $y = 3^x$ in logarithmic form.

     b. Graph $y = 3^x$ and its inverse and discuss the symmetry of their graphs.

21. Write $\log_a a = x$ in exponential form and find x to evaluate $\log_a a$ for any $a > 0, a \neq 1$.

22. Write $\log_a 1 = x$ in exponential form and find x to evaluate $\log_a 1$ for any $a > 0, a \neq 1$.

In Exercises 23–26, use the properties of logarithms to evaluate the expressions.

23. $\log 10^{14}$ **24.** $\ln e^5$

25. $10^{\log_{10} 12}$ **26.** $6^{\log_6 25}$

In Exercises 27–30, use $\log_a(20) = 1.4406$ and $\log_a(5) = 0.7740$ to evaluate each expression.

27. $\log_a(100)$ **28.** $\log_a(4)$

29. $\log_a 5^3$ **30.** $\log_a \sqrt{20}$

In Exercises 31–34, rewrite each expression as a sum, difference, or product of logarithms, and simplify if possible.

31. $\ln \dfrac{3x - 2}{x + 1}$ **32.** $\log[x^3(3x - 4)^5]$

33. $\log_3 \dfrac{\sqrt[3]{4x + 1}}{4x^2}$ **34.** $\log_3 \dfrac{\sqrt[3]{3x - 1}}{5x^2}$

In Exercises 35–38, rewrite each expression as a single logarithm.

35. $3 \log_2 x + \log_2 y$ **36.** $\log x - \dfrac{1}{3} \log y$

37. $4 \ln(2a) - \ln b$

38. $6 \ln(5y) + 2 \ln x$

EXERCISES 5.2

39. *Life Span* On the basis of data and projections for the years 1910 through 2020 the expected life span of people in the United States can be described by the function $f(x) = 11.027 + 14.304 \ln x$ years, where x is the number of years from 1900 to the person's birth year.

    a. What does this model estimate the life span to be for people born in 1925? In 2007? (Give each answer to the nearest year.)

    b. Explain why these numbers are so different.
    (Source: National Center for Health Statistics)

40. *Japan's Population* The population of Japan for the years 1984–2006 is approximated by the logarithmic function $y = 114.016 + 4.267 \ln x$ million people, with x equal to the number of years after 1980. According to the model, what is the estimated population in 2000? In 2020?
    (Source: www.jinjapan.org/stat/)

41. *Poverty Threshold* A single person is considered as living in poverty if his or her income falls below the federal government's official poverty level, which is adjusted every year for inflation. The function that models the poverty threshold for the years 1987–2008 is $f(x) = -3130.3 + 4056.8 \ln x$ dollars, where x is the number of years after 1980.

    a. What does this model give as the poverty threshold for 2015? For 2020?

    b. Is this function increasing or decreasing?

    c. Graph this function for $x = 5$ to $x = 40$.
    (Source: U.S. Census Bureau)

42. *Loan Repayment* The number of years t that it takes to pay off a $100,000 loan at 10% interest by making annual payments of R dollars is

    $$t = \frac{\ln R - \ln(R - 10,000)}{\ln 1.1}, R > 10,000$$

    If the annual payment is $16,274.54, in how many years will the loan be paid off?

43. *Supply* Suppose that the supply of a product is given by $p = 20 + 6 \ln(2q + 1)$, where p is the price per unit and q is the number of units supplied. What price will give a supply of 5200 units?

44. *Demand* Suppose that the demand function for a product is $p = \dfrac{500}{\ln(q + 1)}$, where p is the price per unit and q is the number of units demanded. What price will give a demand for 6400 units?

45. *Female Workers* For the years 1970 to 2040, the percent of females over 15 years old in the workforce is given or projected by $y = 27.4 + 5.02 \ln x$, where x is the number of years from 1960.

    a. What does the model predict the percent to be in 2011? In 2015?

    b. Is the percent of female workers increasing or decreasing?

46. *Births to Unmarried Women* The percent of live births to unwed mothers for the years 1970–2007 can be modeled by the function $y = -37.016 + 19.278 \ln x$, where x is the number of years from 1960.

    a. What does the model predict the percent to be in 2015? In 2022?

    b. Is the percent increasing or decreasing?

47. *Doubling Time* If $4000 is invested in an account earning 10% annual interest, compounded continuously, then the number of years that it takes for the amount to grow to $8000 is $n = \dfrac{\ln 2}{0.10}$. Find the number of years.

48. *Doubling Time* If $5400 is invested in an account earning 7% annual interest, compounded continuously, then the number of years that it takes for the amount to grow to $10,800 is $n = \dfrac{\ln 2}{0.07}$. Find the number of years.

49. *Doubling Time* The number of quarters (a quarter equals 3 months) needed to double an investment when a lump sum is invested at 8%, compounded quarterly, is given by $n = \dfrac{\log 2}{0.0086}$. In how many years will the investment double?

50. *Doubling Time* The number of periods needed to double an investment when a lump sum is invested at 12%, compounded semiannually, is $n = \dfrac{\log 2}{0.0253}$. How many years pass before the investment doubles in value?

The number of years t it takes for an investment to double if it earns r percent (as a decimal), compounded annually, is $t = \dfrac{\ln 2}{\ln(1 + r)}$. Use this formula in Exercises 51 and 52.

51. *Investing* In how many years will an investment double if it is invested at 8%, compounded annually?

52. *Investing* In how many years will an investment double if it is invested at 12.3%, compounded annually?

53. *Earthquakes* If an earthquake has an intensity of 25,000 times I_0, what is the magnitude of the earthquake?

54. *Earthquakes*
  a. If an earthquake has an intensity of 250,000 times I_0, what is the magnitude of the earthquake?

  b. Compare the magnitudes of two earthquakes when the intensity of one is 10 times the intensity of the other. (See Exercise 53.)

55. *Earthquakes* The Richter scale measurement for the southern Sumatra earthquake of 2007 was 6.4. Find the intensity of this earthquake as a multiple of I_0.

56. *Earthquakes* The Richter scale measurement for the San Francisco earthquake that occurred in 1906 was 8.25. Find the intensity of this earthquake as a multiple of I_0.

57. *Earthquakes* The Richter scale reading for the San Francisco earthquake of 1989 was 7.1. Find the intensity I of this earthquake as a multiple of I_0.

58. *Earthquakes* On January 9, 2008, an earthquake in northern Algeria measuring 4.81 on the Richter scale killed 1 person, and on May 12, 2008, an earthquake in China measuring 7.9 on the Richter scale killed 67,180 people. How many times more intense was the earthquake in China than the one in northern Algeria?

59. *Earthquakes* The largest earthquake ever to strike San Francisco (in 1906) measured 8.25 on the Richter scale, and the second largest (in 1989) measured 7.1. How many times more intense was the 1906 earthquake than the 1989 earthquake?

60. *Earthquakes* The largest earthquake ever recorded in Japan occurred in 2011 and measured 9.0 on the Richter scale. An earthquake measuring 8.25 occurred in Japan in 1983. Calculate the ratio of the intensities of these earthquakes.

61. *Japan Tsunami* In May of 2008, an earthquake measuring 6.8 on the Richter scale struck near the east coast of Honshu, Japan. In March of 2011, an earthquake measuring 9.0 struck that same region, causing a devastating tsunami that killed thousands of people. The 2011 earthquake was how many times as intense as the one in 2008?

Use the following information to answer the questions in Exercises 62–67. To quantify the intensity of sound, the decibel scale (named for Alexander Graham Bell) was developed. The formula for loudness L on the decibel scale is $L = 10 \log \dfrac{I}{I_0}$, where I_0 is the intensity of sound just below the threshold of hearing, which is approximately 10^{-16} watt per square centimeter.

62. *Decibel Scale* Find the decibel reading for a sound with intensity 20,000 times I_0.

63. *Decibel Scale* Compare the decibel readings of two sounds if the intensity of the first sound is 100 times the intensity of the second sound.

64. *Decibel Scale* Use the exponential form of the function $L = 10 \log \dfrac{I}{I_0}$ to find the intensity of a sound if its decibel reading is 40.

65. *Decibel Scale* Use the exponential form of $L = 10 \log \dfrac{I}{I_0}$ to find the intensity of a sound if its decibel reading is 140. (Sound at this level causes pain.)

66. *Decibel Scale* The intensity of a whisper is $115I_0$, and the intensity of a busy street is $9,500,000I_0$. How do their decibel levels compare?

67. *Decibel Scale* If the decibel reading of a sound that is painful is 140 and the decibel reading of rock music is 120, how much more intense is the painful sound than the rock music?

To answer the questions in Exercises 68–72, use the fact that pH is given by the formula $\text{pH} = -\log[\text{H}^+]$.

68. *pH Levels* Find the pH value of beer for which $[\text{H}^+] = 0.0000631$.

69. *pH Levels* The pH value of eggs is 7.79. Find the hydrogen-ion concentration for eggs.

70. *pH Levels* The most common solutions have a pH range between 1 and 14. Use an exponential form of the pH formula to find the values of $[\text{H}^+]$ associated with these pH levels.

71. *pH Levels* Lower levels of pH indicate a more acidic solution. If the pH of ketchup is 3.9 and the pH of peanut butter is 6.3, how much more acidic is ketchup than peanut butter?

72. *pH Levels* Pure sea water in the middle of the ocean has a pH of 8.3. The pH of sea water in an aquarium must vary between 8 and 8.5; beyond these values, animals will experience certain physiological problems. If the pH of water in an aquarium is 8, how much more acidic is this water than pure sea water?

5.3   Exponential and Logarithmic Equations

KEY OBJECTIVES

- Solve an exponential equation by writing it in logarithmic form
- Convert logarithms using the change of base formula
- Solve an exponential equation by using properties of logarithms
- Solve logarithmic equations
- Solve exponential and logarithmic inequalities

SECTION PREVIEW   Carbon-14 Dating

Carbon-14 dating is a process by which scientists can tell the age of many fossils. To find the age of a fossil that originally contained 1000 grams of carbon-14 and now contains 1 gram, we solve the equation

$$1 = 1000e^{-0.00012097t}$$

An approximate solution of this equation can be found graphically, but sometimes a window that shows the solution is hard to find. If this is the case, it may be easier to solve an exponential equation like this one algebraically.

In this section, we consider two algebraic methods of solving exponential equations. The first method involves converting the equation to logarithmic form and then solving the logarithmic equation for the variable. The second method involves taking the logarithm of both sides of the equation and using the properties of logarithms to solve for the variable. ■

Solving Exponential Equations Using Logarithmic Forms

When we wish to solve an equation for a variable that is contained in an exponent, we can remove the variable from the exponent by converting the equation to its logarithmic form. The steps in this solution method follow.

Solving Exponential Equations Using Logarithmic Forms

To solve an exponential equation using logarithmic form:

1. Rewrite the equation with the term containing the exponent by itself on one side.
2. Divide both sides by the coefficient of the term containing the exponent.
3. Change the new equation to logarithmic form.
4. Solve for the variable.

This method is illustrated in the following example.

EXAMPLE 1 ▶ Solving a Base 10 Exponential Equation

a. Solve the equation $3000 = 150(10^{4t})$ for t by converting it to logarithmic form.

b. Solve the equation graphically to confirm the solution.

SOLUTION

a. 1. The term containing the exponent is by itself on one side of the equation.

   2. We divide both sides of the equation by 150:

$$3000 = 150(10^{4t})$$
$$20 = 10^{4t}$$

   3. We rewrite this equation in logarithmic form:

$$4t = \log_{10} 20, \quad \text{or} \quad 4t = \log 20$$

   4. Solving for t gives the solution:

$$t = \frac{\log 20}{4} \approx 0.32526$$

b. We solve the equation graphically by using the intersection method. Entering $y_1 = 3000$ and $y_2 = 150(10^{4t})$, we graph using the window [0, 0.4] by [0, 3500] and find the point of intersection to be about (0.32526, 3000) (Figure 5.22). Thus, the solution to the equation $3000 = 150(10^{4t})$ is $t \approx 0.32526$, as we found in part (a).

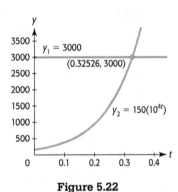

Figure 5.22

EXAMPLE 2 ▶ Doubling Time

a. Prove that the time it takes for an investment to double its value is $t = \dfrac{\ln 2}{r}$ if the interest rate is r, compounded continuously.

b. Suppose $2500 is invested in an account earning 6% annual interest, compounded continuously. How long will it take for the amount to grow to $5000?

SOLUTION

a. If P dollars are invested for t years at an annual interest rate r, compounded continuously, then the investment will grow to a future value S according to the equation

$$S = Pe^{rt}$$

and the investment will be doubled when $S = 2P$, giving

$$2P = Pe^{rt}, \quad \text{or} \quad 2 = e^{rt}$$

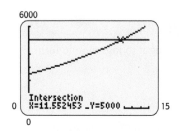

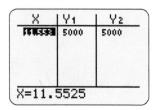

Figure 5.23

Converting this exponential equation to its equivalent logarithmic form gives

$$\log_e 2 = rt, \quad \text{or} \quad \ln 2 = rt$$

and solving this equation for t gives the doubling time,

$$t = \frac{\ln 2}{r}$$

b. We seek the time it takes to double the investment, and the doubling time is given by

$$t = \frac{\ln 2}{0.06} \approx 11.5525$$

Thus, it will take approximately 11.6 years for an investment of $2500 to grow to $5000 if it is invested at an annual rate of 6%, compounded continuously. We can confirm this solution graphically or numerically (Figure 5.23). Note that this is the same time it would take any investment at 6%, compounded continuously, to double.

EXAMPLE 3 ▶ Carbon-14 Dating

Radioactive carbon-14 decays according to the equation

$$y = y_0 e^{-0.00012097t}$$

where y_0 is the original amount and y is the amount of carbon-14 at time t years. To find the age of a fossil if the original amount of carbon-14 was 1000 grams and the present amount is 1 gram, we solve the equation

$$1 = 1000 e^{-0.00012097t}$$

a. Find the age of the fossil by converting the equation to logarithmic form.

b. Find the age of the fossil using graphical methods.

SOLUTION

a. To write the equation so that the coefficient of the exponential term is 1, we divide both sides by 1000, obtaining

$$0.001 = e^{-0.00012097t}$$

Writing this equation in logarithmic form gives the equation

$$-0.00012097t = \log_e 0.001, \quad \text{or} \quad -0.00012097t = \ln 0.001$$

Solving the equation for t gives

$$t = \frac{\ln 0.001}{-0.00012097} \approx 57{,}103$$

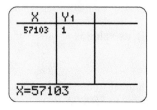

Figure 5.24

Thus, the age of the fossil is calculated to be about 57,103 years.

We can confirm that this is the solution to $1 = 1000 e^{-0.00012097t}$ by using a table of values with $y_1 = 1000 e^{-0.00012097t}$. The result of entering 57,103 for x is shown in Figure 5.24.

b. We can solve the equation graphically by using the intersection method. Entering

$$y_1 = 1000 e^{-0.00012097t} \quad \text{and} \quad y_2 = 1$$

and investigating some y-values shows that we need very large x-values so that y is near 1 (see Figure 5.25 (a)). Graphing using the window [0, 80000] by [−3, 3], we find the x-coordinate of the intersection point to be 57,103.044 (see Figure 5.25(b).) Thus, as we found in part (a), the age of the fossil is about 57,103 years.

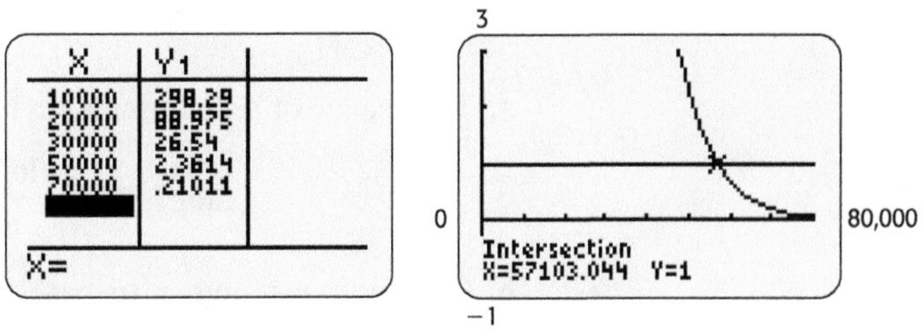

Figure 5.25

While finding this solution from a graph is easy, actually finding a window that contains the intersection point is not. It may be easier to solve the equation algebraically, as we did in part (a), and then use graphical or numerical methods to confirm the solution.

Change of Base

Most calculators and other graphing utilities can be used to evaluate logarithms and graph logarithmic functions if the base is 10 or e. Graphing logarithmic functions with other bases usually requires converting to exponential form to determine outputs and then graphing the functions by plotting points by hand. However, we can use a special formula called the **change of base formula** to rewrite logarithms so that the base is 10 or e. The formula for changing from base a to base b is developed below.

Suppose $y = \log_a x$. Then

$$a^y = x$$

$\log_b a^y = \log_b x$ Take logarithm, base b, of both sides (Property 5 of Logarithms).

$y \log_b a = \log_b x$ Use the Power Property of Logarithms.

$$y = \frac{\log_b x}{\log_b a}$$ Solve for y.

$$\log_a x = \frac{\log_b x}{\log_b a}$$ Substitute $\log_a x$ for y.

The general change of base formula is summarized below.

> **Change of Base Formula**
>
> If $b > 0, b \neq 1, a > 0, a \neq 1$, and $x > 0$, then
>
> $$\log_a x = \frac{\log_b x}{\log_b a}$$
>
> In particular, for base 10 and base e,
>
> $$\log_a x = \frac{\log x}{\log a} \quad \text{and} \quad \log_a x = \frac{\ln x}{\ln a}$$

EXAMPLE 4 ▶   **Applying the Change of Base Formula**

a. Evaluate $\log_8 124$.

Graph the functions in parts (b) and (c) by changing each logarithm to a common logarithm and then by changing the logarithm to a natural logarithm.

b. $y = \log_3 x$ **c.** $y = \log_2(-3x)$

SOLUTION

a. The change of base formula (to base 10) can be used to evaluate this logarithm:

$$\log_8 124 = \frac{\log 124}{\log 8} = 2.318 \text{ approximately}$$

b. Changing $y = \log_3 x$ to base 10 gives $\log_3 x = \dfrac{\log x}{\log 3}$, so we graph $y = \dfrac{\log x}{\log 3}$ (Figure 5.26(a)). Changing to base e gives $\log_3 x = \dfrac{\ln x}{\ln 3}$, so we graph $y = \dfrac{\ln x}{\ln 3}$ (Figure 5.26(b)). Note that the graphs appear to be identical (as they should be because they both represent $y = \log_3 x$).

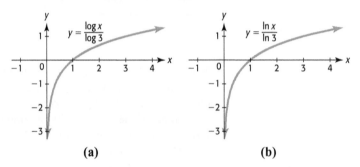

(a)   (b)

Figure 5.26

c. Changing $y = \log_2(-3x)$ to base 10 gives $\log_2(-3x) = \dfrac{\log(-3x)}{\log 2}$, so we graph $y = \dfrac{\log(-3x)}{\log 2}$ (Figure 5.27(a)). Changing to base e gives $\log_2(-3x) = \dfrac{\ln(-3x)}{\ln 2}$, so we graph $y = \dfrac{\ln(-3x)}{\ln 2}$ (Figure 5.27(b)). Note that the graphs are identical. Also notice that because logarithms are defined only for positive inputs, x must be negative so that $-3x$ is positive. Thus, the domain of $y = \log_2(-3x)$ is the interval $(-\infty, 0)$.

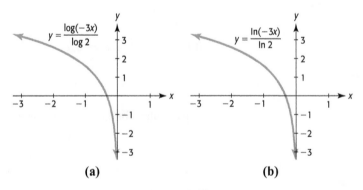

(a)   (b)

Figure 5.27

The change of base formula is also useful in solving exponential equations whose base is neither 10 nor e.

EXAMPLE 5 ▶ Investment

If $10,000 is invested for t years at 10%, compounded annually, the future value is given by

$$S = 10,000(1.10^t)$$

In how many years will the investment grow to $45,950?

SOLUTION

We seek to solve the equation

$$45,950 = 10,000(1.10^t)$$

Dividing both sides by 10,000 gives

$$4.5950 = 1.10^t$$

Writing the equation in logarithmic form solves the equation for t:

$$t = \log_{1.10} 4.5950$$

Using the change of base formula gives the number of years:

$$t = \log_{1.10} 4.5950 = \frac{\log 4.5950}{\log 1.10} = 16$$

Thus, $10,000 will grow to $45,950 in 16 years.

Solving Exponential Equations Using Logarithmic Properties

With the properties of logarithms at our disposal, we consider a second method for solving exponential equations, which may be easier to use. This method of solving an exponential equation involves taking the logarithm of both sides of the equation (Property 5 of Logarithms) and then using properties of logarithms to write the equation in a form that we can solve.

Solving Exponential Equations Using Logarithmic Properties

To solve an exponential equation using logarithmic properties:

1. Rewrite the equation with a base raised to a power on one side.
2. Take the logarithm, base e or 10, of both sides of the equation.
3. Use a logarithmic property to remove the variable from the exponent.
4. Solve for the variable.

This method is illustrated in the following example.

EXAMPLE 6 ▶ Solution of Exponential Equations

Solve the following exponential equations.

a. $4096 = 8^{2x}$

b. $6(4^{3x-2}) = 120$

SOLUTION

a. **1.** This equation has the base 8 raised to a variable power on one side.

2. Taking the logarithm, base 10, of both sides of the equation $4096 = 8^{2x}$ gives

$$\log 4096 = \log 8^{2x}$$

3. Using the Power Property of Logarithms removes the variable x from the exponent:

$$\log 4096 = 2x \log 8$$

4. Solving for x gives the solution:

$$\frac{\log 4096}{2 \log 8} = x$$

$$x = 2$$

b. We first isolate the exponential expression on one side of the equation by dividing both sides by 6:

$$\frac{6(4^{3x-2})}{6} = \frac{120}{6}$$

$$4^{3x-2} = 20$$

Taking the natural logarithm of both sides leads to the solution:

$$\ln 4^{3x-2} = \ln 20$$

$$(3x - 2) \ln 4 = \ln 20$$

$$3x - 2 = \frac{\ln 20}{\ln 4}$$

$$x = \frac{1}{3}\left(\frac{\ln 20}{\ln 4} + 2\right) \approx 1.387$$

An alternative method of solving the equation is to write $4^{3x-2} = 20$ in logarithmic form:

$$\log_4 20 = 3x - 2$$

$$x = \frac{\log_4 20 + 2}{3}$$

The change of base formula can be used to compute this value and verify that this solution is the same as above:

$$x = \frac{\dfrac{\ln 20}{\ln 4} + 2}{3} \approx 1.387$$

Solution of Logarithmic Equations

Some logarithmic equations can be solved by converting to exponential form.

EXAMPLE 7 ▶ **Solving a Logarithmic Equation**

Solve $4 \log_3 x = -8$ by converting to exponential form and verify the solution graphically.

SOLUTION

We first isolate the logarithm by dividing both sides of the equation by 4:

$$\log_3 x = -2$$

Writing $\log_3 x = -2$ in exponential form gives the solution

$$3^{-2} = x$$

$$x = \frac{1}{9}$$

Graphing $y_1 = 4 \log_3 x$ and $y_2 = -8$ with a window that contains $x = \frac{1}{9}$ and $y = -8$, we find the point of intersection to be $\left(\frac{1}{9}, -8 \right)$. See Figures 5.28(a) and (b).

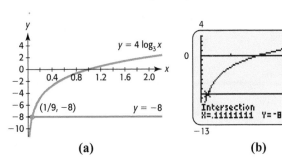

(a)   **(b)**

Figure 5.28

| EXAMPLE 8 ▶ | Solving a Logarithmic Equation |

a. Solve $6 + 3 \ln x = 12$ by writing the equation in exponential form.

b. Solve the equation graphically.

SOLUTION

a. We first solve the equation for $\ln x$:

$$6 + 3 \ln x = 12$$

$$3 \ln x = 6$$

$$\ln x = 2$$

Writing $\log_e x = 2$ in exponential form gives

$$x = e^2$$

b. We solve $6 + 3 \ln x = 12$ graphically by the intersection method. Entering $y_1 = 6 + 3 \ln x$ and $y_2 = 12$, we graph using the window $[-2, 10]$ by $[-10, 15]$ and find the point of intersection to be about $(7.38906, 12)$ (Figure 5.29). Thus, the solution to the equation $6 + 3 \ln x = 12$ is $x \approx 7.38906$. Because $e^2 \approx 7.38906$, the solutions agree.

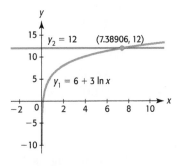

Figure 5.29

The properties of logarithms are frequently useful in solving logarithmic equations. Consider the following example.

EXAMPLE 9 ▶   Solving a Logarithmic Equation

Solve $\ln x + 3 = \ln(x + 4)$ by converting the equation to exponential form and then using algebraic methods.

SOLUTION

We first write the equation with the logarithmic expressions on one side:

$$\ln x + 3 = \ln(x + 4)$$
$$3 = \ln(x + 4) - \ln x$$

Using the Quotient Property of Logarithms gives

$$3 = \ln \frac{x + 4}{x}$$

Writing the equation in exponential form gives

$$e^3 = \frac{x + 4}{x}$$

We can now solve for x:

$$e^3 x = x + 4$$
$$e^3 x - x = 4$$
$$x(e^3 - 1) = 4$$
$$x = \frac{4}{e^3 - 1} \approx 0.21$$

EXAMPLE 10 ▶   Global Warming

In an effort to reduce global warming, it has been proposed that a tax be levied based on the emissions of carbon dioxide into the atmosphere. The cost–benefit equation $\ln(1 - P) = -0.0034 - 0.0053t$ estimates the relationship between the percent reduction of emissions of carbon dioxide P (as a decimal) and the tax t in dollars per ton of carbon.
(Source: W. Clime, *The Economics of Global Warming*)

a. Solve the equation for P, the estimated percent reduction in emissions.

b. Determine the estimated percent reduction in emissions if a tax of \$100 per ton is levied.

SOLUTION

a. We solve $\ln(1 - P) = -0.0034 - 0.0053t$ for P by writing the equation in exponential form.

$$(1 - P) = e^{-0.0034 - 0.0053t}$$
$$P = 1 - e^{-0.0034 - 0.0053t}$$

b. Substituting 100 for t gives $P = 0.4134$, so a \$100 tax per ton of carbon is estimated to reduce carbon dioxide emissions by 41.3%.

Some exponential and logarithmic equations are difficult or impossible to solve algebraically, and finding approximate solutions to real data problems is frequently easier with graphical methods.

Exponential and Logarithmic Inequalities

Inequalities involving exponential and logarithmic functions can be solved by solving the related equation algebraically and then investigating the inequality graphically. Consider the following example.

EXAMPLE 11 ▶   Sales Decay

After the end of an advertising campaign, the daily sales of Genapet fell rapidly, with daily sales given by $S = 3200e^{-0.08x}$ dollars, where x is the number of days from the end of the campaign. For how many days after the campaign ended were sales at least $1980?

SOLUTION

To solve this problem, we find the solution to the inequality $3200e^{-0.08x} \geq 1980$. We begin our solution by solving the related equation:

$$3200e^{-0.08x} = 1980$$

$$e^{-0.08x} = 0.61875 \qquad \text{Divide both sides by 3200.}$$

$$\ln e^{-0.08x} = \ln 0.61875 \qquad \text{Take the logarithm, base } e, \text{ of both sides.}$$

$$-0.08x \approx -0.4801 \qquad \text{Use the Power Property of Logarithms.}$$

$$x \approx 6$$

We now investigate the inequality graphically.

To solve this inequality graphically, we graph $y_1 = 3200e^{-0.08x}$ and $y_2 = 1980$ on the same axes, with nonnegative x-values since x represents the number of days. The graph shows that $y_1 = 3200e^{-0.08x}$ is above $y_2 = 1980$ for $x < 6$ (Figure 5.30). Thus, the daily revenue is at least $1980 for each of the first 6 days after the end of the advertising campaign.

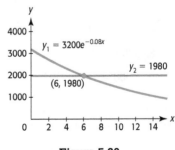

Figure 5.30

EXAMPLE 12 ▶   Cost–Benefit

The cost–benefit equation $\ln(1 - P) = -0.0034 - 0.0053t$ estimates the relationship between the percent reduction of emissions of carbon dioxide P (as a decimal) and the tax t in dollars per ton of carbon. What tax will give a reduction of at least 50%?

SOLUTION

To find the tax, we find the value of t that gives $P = 0.50$:

$$\ln(1 - 0.50) = -0.0034 - 0.0053t$$
$$-0.6931 = -0.0034 - 0.0053t$$
$$0.0053t = 0.6897$$
$$t = 130.14$$

In Example 10, we solved this cost–benefit for P, getting $P = 1 - e^{-0.0034 - 0.0053t}$. The graphs of this equation and $P = 0.50$ on the same axes are shown in Figure 5.31. The point of intersection of the graphs is $(130.14, 0.5)$, and the graph shows that the percent reduction of emissions is more than 50% if the tax is above \$130.14 per ton of carbon.

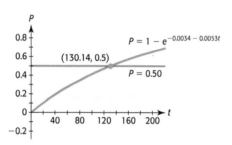

Figure 5.31

Skills CHECK 5.3

In Exercises 1–10, solve the equations algebraically and check graphically. Round to three decimal places.

1. $1600 = 10^x$ **2.** $4600 = 10^x$

3. $2500 = e^x$ **4.** $54.6 = e^x$

5. $8900 = e^{5x}$ **6.** $2400 = 10^{8x}$

7. $4000 = 200e^{8x}$ **8.** $5200 = 13e^{12x}$

9. $8000 = 500(10^x)$ **10.** $9000 = 400(10^x)$

In Exercises 11–14, use a change of base formula to evaluate each logarithm. Give your answers rounded to four decimal places.

11. $\log_6 18$ **12.** $\log_7 215$

13. $\log_8 \sqrt{2}$ **14.** $\log_4 \sqrt[3]{10}$

In Exercises 15–22, solve the equations.

15. $8^x = 1024$ **16.** $9^x = 2187$

17. $2(5^{3x}) = 31{,}250$ **18.** $2(6^{2x}) = 2592$

19. $5^{x-2} = 11.18$ **20.** $3^{x-4} = 140.3$

21. $18{,}000 = 30(2^{12x})$ **22.** $5880 = 21(2^{3x})$

In Exercises 23–36, solve the logarithmic equations.

23. $\log_2 x = 3$ **24.** $\log_4 x = -2$

25. $5 + 2\ln x = 8$ **26.** $4 + 3\log x = 10$

27. $5 + \ln(8x) = 23 - 2\ln x$

28. $3\ln x + 8 = \ln(3x) + 12.18$

29. $2\log x - 2 = \log(x - 25)$

30. $\ln(x - 6) + 4 = \ln x + 3$

31. $\log_3 x + \log_3 9 = 1$

32. $\log_2 x + \log_2(x - 6) = 4$

33. $\log_2 x = \log_2 5 + 3$

34. $\log_2 x = 3 - \log_2 2x$

35. $\log 3x + \log 2x = \log 150$

36. $\ln(x + 2) + \ln x = \ln(x + 12)$

In Exercises 37–40, solve the inequalities.

37. $3^x < 243$ **38.** $7^x \geq 2401$

39. $5(2^x) \geq 2560$ **40.** $15(4^x) \leq 15,360$

EXERCISES 5.3

41. Supply The supply function for a certain size boat is given by $p = 340(2^q)$ boats, where p dollars is the price per boat and q is the quantity of boats supplied at that price. What quantity will be supplied if the price is $10,880 per boat?

42. Demand The demand function for a dining room table is given by $p = 4000(3^{-q})$ dollars per table, where p is the price and q is the quantity, in thousands of tables, demanded at that price. What quantity will be demanded if the price per table is $256.60?

43. Sales Decay After a television advertising campaign ended, the weekly sales of Korbel champagne fell rapidly. Weekly sales in a city were given by $S = 25,000e^{-0.072x}$ dollars, where x is the number of weeks after the campaign ended.

   a. Write the logarithmic form of this function.

   b. Use the logarithmic form of this function to find the number of weeks after the end of the campaign before weekly sales fell to $16,230.

44. Sales Decay After a television advertising campaign ended, sales of Genapet fell rapidly, with daily sales given by $S = 3200e^{-0.08x}$ dollars, where x is the number of days after the campaign ended.

   a. Write the logarithmic form of this function.

   b. Use the logarithmic form of this function to find the number of days after the end of the campaign before daily sales fell to $2145.

45. Sales Decay After the end of an advertising campaign, the daily sales of Genapet fell rapidly, with daily sales given by $S = 3200e^{-0.08x}$ dollars, where x is the number of days from the end of the campaign.

   a. What were daily sales when the campaign ended?

   b. How many days passed after the campaign ended before daily sales were below half of what they were at the end of the campaign?

46. Sales Decay After the end of a television advertising campaign, the weekly sales of Korbel champagne fell rapidly, with weekly sales given by $S = 25,000e^{-0.072x}$ dollars, where x is the number of weeks from the end of the campaign.

   a. What were weekly sales when the campaign ended?

   b. How many weeks passed after the campaign ended before weekly sales were below half of what they were at the end of the campaign?

47. Super Bowl Ads A minute ad during Super Bowl VII in 1973 cost $200,000. The price tag for a 30-second ad slot during the 2011 Super Bowl was $3 million. The cost of a 30-second slot of advertising for the Super Bowl can be modeled by $y = 0.0000966(1.101^x)$, where x is the number of years after 1900 and y is the cost in millions.

   a. According to the model, what was the cost of a 30-second Super Bowl ad in 2000?

   b. If the model remained accurate, when would the cost for a 30-second ad be $5,000,000?

48. Population Growth The population in a certain city was 53,000 in 2000, and its future size was predicted to be $P = 53,000e^{0.015t}$ people, where t is the number of years after 2000. Determine algebraically when the population was predicted to reach 60,000, and verify your solution graphically.

49. Purchasing Power The purchasing power (real value of money) decreases if inflation is present in the economy. For example, the purchasing power of $40,000 after t years of 5% inflation is given by the model

$$P = 40,000e^{-0.05t} \text{ dollars}$$

How long will it take for the value of a $40,000 pension to have a purchasing power of $20,000 under 5% inflation?
(Source: *Viewpoints*, VALIC)

50. *Purchasing Power* If a retired couple has a fixed income of $60,000 per year, the purchasing power of their income (adjusted value of the money) after t years of 5% inflation is given by the equation $P = 60,000e^{-0.05t}$. In how many years will the purchasing power of their income be half of their current income?

51. *Real Estate Inflation* During a 5-year period of constant inflation, the value of a $100,000 property increases according to the equation $v = 100,000e^{0.03t}$ dollars. In how many years will the value of this building be double its current value?

52. *Real Estate Inflation* During a 10-year period of constant inflation, the value of a $200,000 property is given by the equation $v = 200,000e^{0.05t}$ dollars. In how many years will the value of this building be $254,250?

53. *Radioactive Decay* The amount of radioactive isotope thorium-234 present in a certain sample at time t is given by $A(t) = 500e^{-0.02828t}$ grams, where t years is the time since the initial amount was measured.

   a. Find the initial amount of the isotope present in the sample.

   b. Find the half-life of this isotope. That is, find the number of years until half of the original amount of the isotope remains.

54. *Radioactive Decay* The amount of radioactive isotope thorium-234 present in a certain sample at time t is given by $A(t) = 500e^{-0.02828t}$ grams, where t years is the time since the initial amount was measured. How long will it take for the amount of isotope to equal 318 grams?

55. *Drugs in the Bloodstream* The concentration of a drug in the bloodstream from the time the drug is injected until 8 hours later is given by

   $$y = 100(1 - e^{-0.312(8-t)}) \text{ percent, } t \text{ in hours,}$$

   where the drug is administered at time $t = 0$. In how many hours will the drug concentration be 79% of the initial dose?

56. *Drugs in the Bloodstream* If a drug is injected into the bloodstream, the percent of the maximum dosage that is present at time t is given by $y = 100(1 - e^{-0.35(10-t)})$, where t is in hours, with $0 \le t \le 10$. In how many hours will the percent reach 65%?

57. *Gold Prices* The price of an ounce of gold in U.S. dollars for the years 2002–2011 can be modeled by the function $y = 117.911(1.247^x)$, where x is the number of years after 2000.

   a. Graph the function for the given time interval.

   b. According to the model, when during this time period was the price of an ounce of gold $1000?

   c. According to the model, will the price of gold ever be lower than $1500 an ounce? Why or why not? Do you think this is reasonable?

58. *Radioactive Decay* A breeder reactor converts stable uranium-238 into the isotope plutonium-239. The decay of this isotope is given by $A(t) = A_0e^{-0.00002876t}$, where $A(t)$ is the amount of the isotope at time t (in years) and A_0 is the original amount. If the original amount is 100 pounds, find the half-life of this isotope.

59. *Cost* Suppose the weekly cost for the production of x units of a product is given by $C(x) = 3452 + 50 \ln(x + 1)$ dollars. Use graphical methods to estimate the number of units produced if the total cost is $3556.

60. *Supply* Suppose the daily supply function for a product is $p = 31 + \ln(x + 2)$, where p is in dollars and x is the number of units supplied. Use graphical methods to estimate the number of units that will be supplied if the price is $35.70.

61. *Doubling Time* If P dollars are invested at an annual interest rate r, compounded annually for t years, the future value of the investment is given by $S = P(1.07)^t$. Find a formula for the number of years it will take to double the initial investment.

62. *Doubling Time* The future value of a lump sum P that is invested for n years at 10%, compounded annually, is $S = P(1.10)^n$. Show that the number of years it would take for this investment to double is $n = \log_{1.10} 2$.

63. *Investment* At the end of t years, the future value of an investment of $20,000 at 7%, compounded annually, is given by $S = 20,000(1 + 0.07)^t$. In how many years will the investment grow to $48,196.90?

64. *Investment* At the end of t years, the future value of an investment of $30,000 at 9%, compounded annually, is given by $S = 30,000(1 + 0.09)^t$. In how many years will the investment grow to $129,829?

65. *Investing* Find in how many years $40,000 invested at 10%, compounded annually, will grow to $64,420.40.

66. *Investing* In how many years will $40,000 invested at 8%, compounded annually, grow to $86,357?

67. *Life Span* Based on data from 1920 and projected to 2020, the expected life span of people in the United States can be described by the function $f(x) = 11.027 + 14.304 \ln x$, where x is the number of years from 1900 to the person's birth year.

   a. Estimate the birth year for which the expected life span is 78 years.

   b. Use graphical methods to determine the birth year for which the expected life span is 78 years. Does this agree with the solution in part (a)?
   (Source: National Center for Health Statistics)

68. *Supply* Suppose that the supply of a product is given by $p = 20 + 6 \ln(2q + 1)$, where p is the price per unit and q is the number of units supplied. How many units will be supplied if the price per unit is $68.04?

69. *Groupon Valuation* In 2007, the daily deals website Groupon was formed. In 2011, the company was expected to launch an initial public offering that could value the company at $25 billion. The value of Groupon, in millions of dollars, can be modeled by the equation $G(x) = 174.075(1.378^x)$, where x is the number of months after December 2009.

   a. What does the model give as the value of Groupon in December 2009?

   b. What does the model give as the value of Groupon in December 2010?

   c. What is the estimated percent of increase from December 2009 to December 2010?

70. *Demand* Suppose that the demand function for a product is $p = \dfrac{500}{\ln(q + 1)}$, where p is the price per unit and q is the number of units demanded. How many units will be demanded if the price is $61.71 per unit?

71. *Deforestation* The number of square miles per year of rain forest destroyed in Brazil is given by $y = 4899.7601(1.0468^x)$, where x is the number of years from 1990. In what year did 6447 square miles get destroyed, according to the model?
   (Source: National Institute of Space Research [INPE] data)

72. *Global Warming* In an effort to reduce global warming, it has been proposed that a tax be levied based on the emissions of carbon dioxide into the atmosphere. The cost–benefit equation $\ln(1 - P) = -0.0034 - 0.0053t$ estimates the relationship between the percent reduction of emissions of carbon dioxide P (as a decimal) and the tax t in dollars per ton of carbon dioxide.

   a. Solve the equation for t, giving t as a function of P. Graph the function.

   b. Use the equation in part (a) to find what tax will give a 30% reduction in emissions.
   (Source: W. Clime, *The Economics of Global Warming*)

73. *Rule of 72* The "Rule of 72" is a simplified way to determine how long an investment will take to double, given a fixed annual rate of interest. By dividing 72 by the annual interest rate, investors can get a rough estimate of how many years it will take for the initial investment to double. Algebraically we know that the time it takes an investment to double is $\dfrac{\ln 2}{r}$, when the interest is compounded continuously and r is written as a decimal.

   a. Complete the table to compare the exact time it takes for an investment to double to the "Rule of 72" time. (Round to two decimal places.)

Annual Interest Rate	Rule of 72 Years	Exact Years
2%		
3%		
4%		
5%		
6%		
7%		
8%		
9%		
10%		
11%		

   b. Compute the differences between the two sets of outputs. What conclusion can you reach about using the Rule of 72 estimate?

74. *Doubling Time* The number of quarters needed to double an investment when a lump sum is invested at 8%, compounded quarterly, is given by $n = \log_{1.02} 2$.

   a. Use the change of base formula to find n.

   b. In how many years will the investment double?

75. *Doubling Time* The number of periods needed to double an investment when a lump sum is invested at 12%, compounded semiannually, is given by $n = \log_{1.06} 2$.

   a. Use the change of base formula to find n.

   b. How many years pass before the investment doubles in value?

76. *Annuities* If $2000 is invested at the end of each year in an annuity that pays 5%, compounded annually, the number of years it takes for the future value to amount to $40,000 is given by $t = \log_{1.05} 2$. Use the change of base formula to find the number of years until the future value is $40,000.

77. *Annuities* If $1000 is invested at the end of each year in an annuity that pays 8%, compounded annually, the number of years it takes for the future value to amount to $30,000 is given by $t = \log_{1.08} 3.4$. Use the change of base formula to find the number of years until the future value is $30,000.

78. *Deforestation* One of the major causes of rain forest deforestation is agricultural and residential development. The number of hectares (2.47 acres) destroyed in a particular year t can be modeled by $y = -3.91435 + 2.62196 \ln t$, where $t = 0$ in 1950. When will more than 7 hectares be destroyed per year?

79. *Market Share* Suppose that after a company introduces a new product, the number of months before its market share is x percent is given by

$$m = 20 \ln \frac{50}{50 - x}, x < 50$$

After how many months is the market share more than 45%, according to this model?

80. *Drugs in the Bloodstream* The concentration of a drug in the bloodstream from the time the drug is administered until 8 hours later is given by $y = 100(1 - e^{-0.312(8-t)})$ percent, where the drug is administered at time $t = 0$. For what time period is the amount of drug present more than 60%?

81. *Carbon-14 Dating* An exponential decay function can be used to model the number of grams of a radioactive material that remain after a period of time. Carbon-14 decays over time, with the amount remaining after t years given by $y = y_0 e^{-0.00012097t}$, where y_0 is the original amount. If the original amount of carbon-14 is 200 grams, find the number of years until 155.6 grams of carbon-14 remain.

82. *Sales Decay* After a television advertising campaign ended, the weekly sales of Korbel champagne fell rapidly. Weekly sales in a city were given by $S = 25,000e^{-0.072x}$ dollars, where x is the number of weeks after the campaign ended.

a. Use the logarithmic form of this function to find the number of weeks after the end of the campaign that passed before weekly sales fell below $16,230.

b. Check your solution by graphical or numerical methods.

83. *Sales Decay* After a television advertising campaign ended, the weekly sales of Turtledove bars fell rapidly. Weekly sales are given by $S = 600e^{-0.05x}$ thousand dollars, where x is the number of days after the campaign ended.

a. Use the logarithmic form of this function to find the number of weeks after the end of the campaign before weekly sales fell below $269.60.

b. Check your solution by graphical or numerical methods.

5.4 Exponential and Logarithmic Models

KEY OBJECTIVES

- Model data with exponential functions
- Use constant percent change to determine if data fit an exponential model
- Compare quadratic and exponential models of data
- Model data with logarithmic functions

SECTION PREVIEW **Diabetes**

In Example 8 of Section 5.2, we solved problems about diabetes by using the fact that the percent of U.S. adults with diabetes (diagnosed and undiagnosed) can be modeled by the logarithmic function

$$p(x) = -12.975 + 11.851 \ln x$$

where x is the number of years after 2000. In this section, we will use projections from the Centers for Disease Control and Prevention to create this logarithmic model using technology. (See Example 5.)

Many sources provide data that can be modeled by exponential growth and decay functions, and technology can be used to find exponential functions that model data. In this section, we model real data with exponential functions and determine when this type of model is appropriate. We also create exponential functions that model phenomena characterized by constant percent change, and we model real data with logarithmic functions when appropriate. ■

Modeling with Exponential Functions

Exponential functions can be used to model real data if the data exhibit rapid growth or decay. When the scatter plot of data shows a very rapid increase or decrease, it is possible that an exponential function can be used to model the data. Consider the following examples.

EXAMPLE 1 ▶ **Insurance Premiums**

The monthly premiums for $250,000 in term-life insurance over a 10-year term period increase with the age of the men purchasing the insurance. The monthly premiums for nonsmoking males are shown in Table 5.8

Table 5.8

Age (years)	Monthly Premium for 10-Year Term Insurance ($)	Age (years)	Monthly Premium for 10-Year Term Insurance ($)
35	123	60	783
40	148	65	1330
45	225	70	2448
50	338	75	4400
55	500		

(Source: Quotesmith.com)

a. Graph the data in the table with x as age and y in dollars.

b. Create an exponential function that models these premiums as a function of age.

c. Graph the data and the exponential function that models the data on the same axes.

SOLUTION

a. The scatter plot of the data is shown in Figure 5.32(a). The plot shows that the outputs rise rapidly as the inputs increase.

b. Using technology gives the exponential model for the monthly premium, rounded to four decimal places, as

$$y = 4.0389(1.0946^x) \text{ dollars}$$

where x is the age in years.

c. Figure 5.32(b) shows a scatter plot of the data and a graph of the (unrounded) exponential equation used to model it. The model appears to be a good, but not perfect, fit for the data.

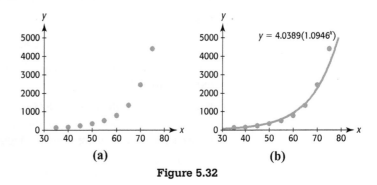

(a)   (b)

Figure 5.32

EXAMPLE 2 ▶ **E-Commerce Sales**

Table 5.9

Year	Sales (millions of $)
1998	4988
1999	15,000
2000	28,885
2001	34,353
2002	45,117
2003	57,861
2004	73,558
2005	92,475
2006	114,445
2007	137,344
2008	141,890
2009	144,100
2010	165,400

(Source: U.S. Census Bureau)

E-commerce is taking a bigger slice of the overall retail sales pie and is growing far faster than retail sales. E-commerce sales in the United States totaled $165.4 billion in 2010, up 14.8% from 2009. Table 5.9 gives retail trade sales by e-commerce in millions of dollars for the years 1998–2010.

a. Graph the data, with x equal to the number of years after 1990.

b. Find an exponential function that models the data, using as input the number of years after 1990.

c. Graph the data and the exponential model on the same axes.

d. Use the reported model from part (b) to estimate e-commerce retail sales for 2015.

e. The recession in 2008 caused retail spending to slow down. Remove the last three data points and model the remaining data with an exponential function.

f. Use the model in part (e) to estimate e-commerce retail sales for 2015, and compare this result to the result from part (d).

SOLUTION

a. The graph of the data is shown in Figure 5.33(a).

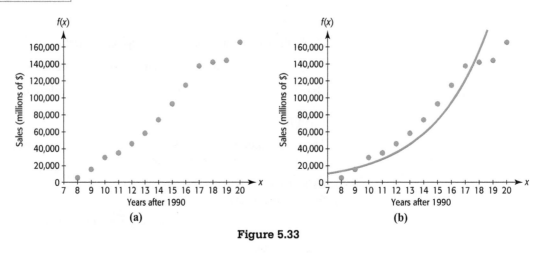

Figure 5.33

b. The exponential equation that models the data, rounded to three decimal places, is $f(x) = 1770.821(1.282^x)$, where x is the number of years after 1990.

c. The graphs of the data and the exponential model are shown in Figure 5.33(b).

d. Evaluating the reported model at $x = 25$ for the year 2015 gives

$$f(25) = 1770.821(1.282^{25}) = 881,811$$

Thus, e-commerce retail sales are expected to be $881,811,000,000 in 2015, according to the model.

e. Removing the last three data points and using exponential regression gives the model $g(x) = 807.519(1.374^x)$, where x is the number of years after 1990. The graphs of the data and the model are shown in Figure 5.34.

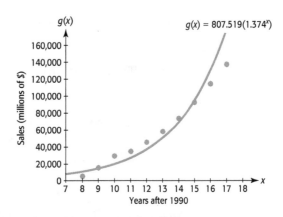

Figure 5.34

f. Evaluating the reported model in part (e) at $x = 25$ gives

$$g(25) = 807.519(1.374^{25}) = 2,274,160.268$$

Thus, e-commerce retail sales are expected to be $2,274,160,268,000 in 2015, according to the model.

The model using all data through 2010 predicts sales of $881,811,000,000 in 2015, whereas the model using data through 2007 predicts sales of $2,274,160,268,000 in 2015. There is a large difference between these amounts!

Spreadsheet ▸ SOLUTION We can use software programs and spreadsheets to find the exponential function that is the best fit for the data. Table 5.10 shows the Excel spreadsheet for the aligned data of Example 2. Selecting the cells containing the data, getting the scatter plot of the data, and selecting the Exponential Trendline gives the equation of the exponential function that is the best fit for the data, along with the scatter plot and the graph of the best-fitting function. Checking Display equation gives the equation on the graph (see Figure 5.35).

Table 5.10

	1	2
A	Years after 1990	Sales (millions)
B	8	4988
C	9	15000
D	10	28885
E	11	34353
F	12	45117
G	13	57861
H	14	73558
I	15	92475
J	16	114445
K	17	137344
L	18	141890
M	19	144100
N	20	165400

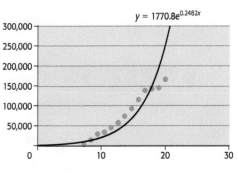

Figure 5.35

The function created by Excel,

$$y = 1770.8e^{0.2482x}$$

is in a different form than the model found in Example 2(b), which is

$$f(x) = 1770.821(1.282^x)$$

However, they are equivalent because $e^{0.2482} \approx 1.282$, which means that

$$y = 1770.8e^{0.2482x} = 1770.8(1.282^x)$$

Constant Percent Change in Exponential Models

How can we determine if a set of data can be modeled by an exponential function? To investigate this, we look at Table 5.11, which gives the growth of paramecia for each hour up to 9 hours.

Table 5.11

x (hours)	0	1	2	3	4	5	6	7	8	9
y (paramecia)	1	2	4	8	16	32	64	128	256	512

If we look at the first differences in the output, we see that they are not constant (Table 5.12). But if we calculate the percent change of the outputs for equally spaced inputs, we see that they are constant. For example, from hour 2 to hour 3, the population grew by 4 units; this represents a 100% increase from the hour-2 population. From hour 3 to hour 4, the population grew by 8, which is a 100% increase over the hour-3 population. In fact, this population increases by 100% each hour. This means that the population 1 hour from now will be the present population plus 100% of the present population.

Table 5.12

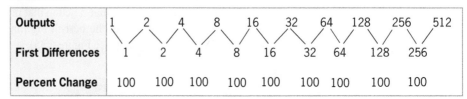

Because the percent change in the outputs is constant for equally spaced inputs in this example, an exponential model ($y = 2^x$) fits the data perfectly.

Constant Percent Changes

If the percent change of the outputs of a set of data is constant for equally spaced inputs, an exponential function will be a perfect fit for the data.

If the percent change of the outputs is approximately constant for equally spaced inputs, an exponential function will be an approximate fit for the data.

EXAMPLE 3 ▶   **Sales Decay**

Suppose a company develops a product that is released with great expectations and extensive advertising, but sales suffer because of bad word of mouth from dissatisfied customers.

a. Use the monthly sales data shown in Table 5.13 to determine the percent change for each of the months given.

Table 5.13

Month	1	2	3	4	5	6	7	8
Sales (thousands of $)	780	608	475	370	289	225	176	137

b. Find the exponential function that models the data.

c. Graph the data and the model on the same axes.

SOLUTION

a. The inputs are equally spaced; the differences of outputs and the percent changes are shown in Table 5.14.

Table 5.14

Outputs	780	608	475	370	289	225	176	137
First Differences	−172	−133	−105	−81	−64	−49	−39	
Percent of Change	−22.1	−21.9	−22.1	−21.9	−22.1	−21.8	−22.2	

The percent change of the outputs is approximately −22%. This means that the sales 1 month from now will be approximately 22% less than the sales now.

b. Because the percent change is nearly constant, an exponential function should fit these data well. Technology gives the model

$$y = 999.781(0.780^x)$$

where x is the month and y is the sales in thousands of dollars.

c. The graphs of the data and the (unrounded) model are shown in Figure 5.36.

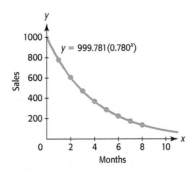

Figure 5.36

Exponential Models

Most graphing utilities give the best-fitting exponential model in the form

$$y = a \cdot b^x$$

It can be shown that the constant percent change of this function is

$$(b - 1) \cdot 100\%$$

Thus, if $b > 1$, the function is increasing (growing), and if $0 < b < 1$, the function is decreasing (decaying).

For the function $y = 2^x$, $b = 2$, so the constant percent change is $r = (2 - 1) \cdot 100\% = 100\%$ (which we found in Table 5.11). For the function $y = 1000(0.78^x)$, the base is $b = 0.78$, and the percent change is $r = (0.78 - 1) \cdot 100\% = -22\%$.

Because $a \cdot b^0 = a \cdot 1 = a$, the value of $y = a \cdot b^x$ is a when $x = 0$, so a is the y-intercept of the graph of the function. Because $y = a$ when $x = 0$, we say that the **initial value** of the function is a. If the constant percent change (as a decimal) is $r = b - 1$, then $b = 1 + r$, and we can write $y = a \cdot b^x$ as

$$y = a(1 + r)^x$$

Exponential Model

If a set of data has initial value a and a constant percent change r (written as a decimal) for equally spaced inputs x, the data can be modeled by the exponential function

$$y = a(1 + r)^x$$

for exponential growth and by the exponential function

$$y = a(1 - r)^x$$

for exponential decay.

To illustrate this, suppose we know that the present population of a city is 100,000 and that it will grow at a constant percent of 10% per year. Then we know that the future population can be modeled by an exponential function with initial value $a = 100,000$ and constant percent change $r = 10\% = 0.10$ per year. That is, the population can be modeled by

$$y = 100,000(1 + 0.10)^x = 100,000(1.10)^x$$

where x is the number of years from the present.

EXAMPLE 4 ▶ Inflation

Suppose inflation averages 4% per year for each year from 2000 to 2010. This means that an item that costs \$1 one year will cost \$1.04 one year later. In the second year, the \$1.04 cost will increase by a factor of 1.04, to $(1.04)(1.04) = 1.04^2$.

a. Write an expression that gives the cost t years after 2000 of an item costing \$1 in 2000.

b. Write an exponential function that models the cost of an item t years from 2000 if its cost was \$100 in 2000.

c. Use the model to find the cost of the item from part (b) in 2010.

SOLUTION

a. The cost of an item is \$1 in 2000, and the cost increases at a constant percent of $4\% = 0.04$ per year, so the cost after t years will be

$$1(1 + 0.04)^t = 1.04^t \text{ dollars}$$

b. The inflation rate is 4% = 0.04. Thus, if an item costs $100 in 2000, the function that gives the cost after t years is

$$f(t) = 100(1 + 0.04)^t = 100(1.04^t) \text{ dollars}$$

c. The year 2010 is 10 years after 2000, so the cost of an item costing $100 in 2000 is

$$f(10) = 100(1.04^{10}) = 148.02 \text{ dollars}$$

Comparison of Models

Sometimes it is hard to determine what type of model is the best fit for a set of data. When a scatter plot exhibits a single curvature without a visible high or low point, it is sometimes difficult to determine whether a power, quadratic, or exponential function should be used to model the data. Suppose after graphing the data points we are unsure which of these models is most appropriate. We can find the three models, graph them on the same axes as the data points, and inspect the graphs to see which is the best visual fit.*

Consider the exponential model for insurance premiums as a function of age that we found in Example 1. The curvature indicated by the scatter plot suggests that either a power, quadratic, or exponential model may fit the data. To compare the goodness of fit of the models, we find each model and compare its graph with the scatter plot (Figure 5.37).

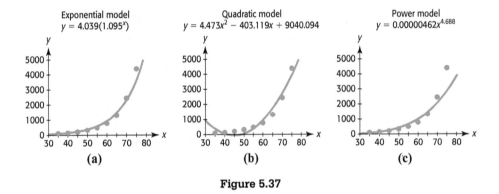

Figure 5.37

The exponential function appears to provide a better fit than the quadratic function because the data points seem to be approaching the x-axis as the input becomes closer to zero from the right. (Recall that the x-axis is a horizontal asymptote for the basic exponential function.) The exponential function also appears to be a better fit than the power function, especially for larger values of x.

When fitting exponential functions to data using technology, it is important to know that problems arise if the input values are large. For instance, when the input values are years (like 2000 and 2010), a technology-determined model may appear to be of the form $y = 0 \cdot b^x$, a function that certainly does not make sense as a model for exponential data. At other times when large inputs are used, your calculator or computer software may give an error message and not even produce an equation. To avoid these problems, it is helpful to align the inputs by converting years to the number of years after a specified year (for example, inputting years from 1990 reduces the size of the inputs).

---

* Statistical measures of goodness of fit exist, but caution must be used in applying these measures. For example, the correlation coefficient r found for linear fits to data is different from the coefficient of determination R^2 found for quadratic fits to data.

> **Technology Note**
>
> When using technology to fit an exponential model to data, you should align the inputs to reasonably small values.

Logarithmic Models

As with other functions we have studied, we can create models involving logarithms. Data that exhibit an initial rapid increase and then have a slow rate of growth can often be described by the function

$$f(x) = a + b \ln x \qquad (\text{for } b > 0)$$

Note that the parameter a is the vertical shift of the graph of $y = \ln x$ and the parameter b affects how much the graph of $y = \ln x$ is stretched. If $b < 0$, the graph of $f(x) = a + b \ln x$ decreases rather than increases. Most graphing utilities will create logarithmic models in the form $y = a + b \ln x$.

EXAMPLE 5 ▶ **Diabetes**

As Table 5.15 shows, projections indicate that the percent of U.S. adults with diabetes could dramatically increase.

a. Find a logarithmic model that fits the data in Table 5.15, with $x = 0$ in 2000.

b. Use the reported model to predict the percent of U.S. adults with diabetes in 2027.

c. In what year does this model predict the percent to be 26.9%?

Table 5.15

Year	Percent	Year	Percent	Year	Percent
2010	15.7	2025	24.2	2040	31.4
2015	18.9	2030	27.2	2045	32.1
2020	21.1	2035	29.0	2050	34.3

(Source: Centers for Disease Control and Prevention)

SOLUTION

a. Entering the aligned input data (number of years after 2000) as the x-values and the percents as the y-values, we use logarithmic regression on a graphing utility to find the function that models the data. This function, rounded to three decimal places, is

$$y = -12.975 + 11.851 \ln x$$

where x is the number of years after 2000. The graphs of the aligned data and the function are shown in Figure 5.38.

b. Evaluating the reported model at $x = 27$ gives $y = -12.975 + 11.851 \ln 27 \approx 26.1$ percent in 2027.

c. Setting $y = 26.9$ gives

$$26.9 = -12.975 + 11.851 \ln x$$
$$39.875 = 11.851 \ln x$$
$$3.3647 = \ln x$$
$$x = e^{3.3647} = 28.9$$

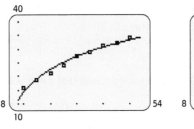

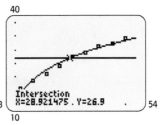

Figure 5.38   **Figure 5.39**

We could also solve graphically by intersecting the graphs of $y_1 = -12.975 + 11.851 \ln x$ and $y_2 = 26.9$, as shown in Figure 5.39. Thus, the percent reaches 27% in 2029.

Note that the input data in Example 5 were not aligned as the number of years after 2010 because the first aligned input value would be 0 and the logarithm of 0 does not exist.

Technology Note

When using technology to fit a logarithmic model to data, you must align the data so that all input values are positive.

EXAMPLE 6 ▶   Women in the Labor Force

The percents of women in the labor force for selected years from 1940 to 2008 are shown in Table 5.16.

Table 5.16   Percent of Women in the Labor Force

Year	Percent of Labor Force Population Aged 16 and Over	Year	Percent of Labor Force Population Aged 16 and Over
1940	24.3	1998	46.3
1950	29.6	1999	46.5
1960	33.4	2000	46.6
1970	38.1	2001	46.5
1980	42.5	2002	46.5
1990	45.2	2003	47.0
1993	45.5	2004	46.0
1994	46.0	2005	46.4
1995	46.1	2006	46.0
1996	46.2	2007	46.4
1997	46.2	2008	44.0

(Source: U.S. Department of Labor, Women's Bureau)

a. Find a logarithmic function that models these data. Align the input to be the number of years after 1900.

b. Graph the equation and the aligned data points. Comment on how the model fits the data.

c. Assuming that the model is valid in 2015, use it to estimate the percent of women in the labor force in 2015.

SOLUTION

a. A logarithmic function that models these data is

$$f(x) = -60.235 + 23.095 \ln x$$

where x is the number of years after 1900.

b. The scatter plot of the data and the graph of the logarithmic function that models the data are shown in Figure 5.40. The model appears to be a good fit for the data.

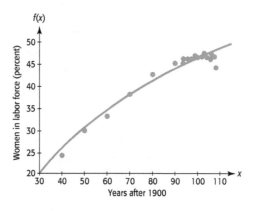

Figure 5.40

c. The percent of women in the labor force in 2015 can be estimated by evaluating $f(115) = 49.3$.

Table 5.17

	A	B
	Years after 1900	Percent of Women
1	40	24.3
2	50	29.6
3	60	33.4
4	70	38.1
5	80	42.5
6	90	45.2
7	93	45.5
8	94	46
9	95	46.1
10	96	46.2
11	97	46.2
12	98	46.3
13	99	46.5
14	100	46.6
15	101	46.5
16	102	46.5
17	103	47
18	104	46
19	105	46.4
20	106	46
21	107	46.4
22	108	44

Spreadsheet ▸ SOLUTION We can use software programs and spreadsheets to find the logarithmic function that is the best fit for data. Table 5.17 shows the Excel spreadsheet for the data of Example 6. Selecting the cells containing the data, getting the scatter plot of the data, and selecting Logarithmic Trendline gives the equation of the logarithmic function that is the best fit for the data, along with the scatter plot and the graph of the best-fitting function. Checking Display equation gives the equation on the graph (see Figure 5.41).

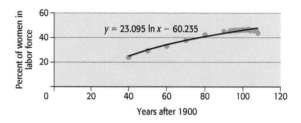

Figure 5.41

Exponents, Logarithms, and Linear Regression

Now that we have knowledge of the properties of logarithms and how exponential functions are related to logarithmic functions, we can show that linear regression can be used to create exponential models that fit data. (The development of linear regression is discussed in Section 2.2.)

Logarithmic Property 3 states that $\log_b b^x = x$ and, in particular, that $\ln e^x = x$, so if we have data that can be approximated by an exponential function, we can convert the data to a linear form by taking the logarithm, base e, of the outputs. We can use linear regression to find the linear function that is the best fit for the converted data and then use the linear function as the exponent of e, which gives the exponential function that is the best fit for the original data. Consider Table 5.18, which gives the number y of paramecia in a population after x hours. We showed in Section 5.1 that the population can be modeled by $y = 2^x$.

Table 5.18

x (hours)	0	1	2	3	4	5	6	7	8	9
y (paramecia)	1	2	4	8	16	32	64	128	256	512
ln y	0	.6931	1.3863	2.0794	2.7726	3.4657	4.1589	4.8520	5.5452	6.2383

The third row of Table 5.18 has the logarithms, base e, of the numbers (y-values) in the second row rounded to four decimal places, and the relationship between x and $\ln y$ is linear. The first differences of the $\ln y$ values are constant, with the difference approximately equal to 0.6931. Using linear regression on the x and $\ln y$ values (to ten decimal places) gives the equation of the linear model:*

$$\ln y = 0.6931471806x$$

Because we seek the equation solved for y, we can write this equation in its exponential form, getting

$$y = e^{0.6931471806x}$$

To show that this model for the data is equal to $y = 2^x$, which we found in Section 5.1, we use properties of exponents and note that $e^{0.6931471806} \approx 2$:

$$y = e^{0.6931471806x} = (e^{0.6931471806})^x \approx 2^x$$

Fortunately, most graphing utilities have combined these steps to give the exponential model directly from the data.

Skills CHECK 5.4

1. Find the exponential function that models the data in the table below.

x	−2	−1	0	1	2	3	4	5
y	2/9	2/3	2	6	18	54	162	486

2. The following table has input x and output $f(x)$. Test the percent change of the outputs to determine if the function is exactly exponential, approximately exponential, or not exponential.

x	1	2	3	4	5	6
f(x)	4	16	64	256	1024	4096

---

* This equation could be found directly, in the form $\ln y = ax + b$, where $b = 0$ and a equals the constant difference. This is because the exponential function is a perfect fit for the data in this application.

3. The following table has input x and output $g(x)$. Test the percent change of the outputs to determine if the function is exactly exponential, approximately exponential, or not exponential.

x	1	2	3	4	5	6
g(x)	2.5	6	8.5	10	8	6

4. The following table has input x and output $h(x)$. Test the percent change of the outputs to determine if the function is exactly exponential, approximately exponential, or not exponential.

x	1	2	3	4	5	6
h(x)	1.5	2.25	3.8	5	11	17

5. Find the exponential function that is the best fit for $f(x)$ defined by the table in Exercise 2.

6. Find the exponential function that is the best fit for $h(x)$ defined by the table in Exercise 4.

7. **a.** Make a scatter plot of the data in the table below.

   b. Does it appear that a linear model or an exponential model is the better fit for the data?

x	1	2	3	4	5	6
y	2	3.1	4.3	5.4	6.5	7.6

8. Find the exponential function that models the data in the table below. Round the model with three-decimal place accuracy.

x	−2	−1	0	1	2	3
y	5	30	150	1000	4000	20,000

9. Compare the first differences and the percent change of the outputs to determine if the data in the table below should be modeled by a linear or an exponential function.

x	1	2	3	4	5
y	2	6	14	34	81

10. Use a scatter plot to determine if a linear or exponential function is the better fit for the data in Exercise 9.

11. Find the linear *or* exponential function that is the better fit for the data in Exercise 9. Round the model to three-decimal-place accuracy.

12. Find the logarithmic function that models the data in the table below. Round the model to two-decimal-place accuracy.

x	1	2	3	4	5	6	7
y	2	4.08	5.3	6.16	6.83	7.38	7.84

13. **a.** Make a scatter plot of the data in the table below.

   b. Does it appear that a linear model or a logarithmic model is the better fit for the data?

x	1	3	5	7	9
y	−2	1	3	4	5

14. **a.** Find a logarithmic function that models the data in the table in Exercise 13. Round the model to three-decimal-place accuracy.

   b. Find a linear function that models the data.

   c. Visually determine which model is the better fit for the data.

15. **a.** Make a scatter plot of the data in the table below.

   b. Find a power function that models the data. Round to three decimal places.

   c. Find a quadratic function that models the data. Round to three decimal places.

   d. Find a logarithmic function that models the data. Round to three decimal places.

x	1	2	3	4	5	6
y	3.5	5.5	6.8	7.2	8	9

16. Let $y = f(x)$ represent the power model found in Exercise 15(b), $y = g(x)$ represent the quadratic model found in Exercise 15(c), and $y = h(x)$ represent the logarithmic model found in Exercise 15(d). Graph each function on the same axes as the scatter plot, using the window [0, 12] by [0, 15]. Which model appears to be the best fit?

EXERCISES 5.4

Report models accurate to three decimal places unless otherwise specified. Use the unrounded function to calculate and to graph the function.

Use the exponential form $y = a(1 + r)^x$ to model the information in Exercises 17–20.

17. *Inflation* Suppose that the retail price of an automobile is $30,000 in 2000 and that it increases at 4% per year.

   a. Write the equation of the exponential function that models the retail price of the automobile t years after 2000.

   b. Use the model to predict the retail price of the automobile in 2015.

18. *Population* Suppose that the population of a city is 190,000 in 2000 and that it grows at 3% per year.

   a. Write the equation of the exponential function that models the annual growth.

   b. Use the model to find the population of this city in 2010.

19. *Sales Decay* At the end of an advertising campaign, weekly sales amounted to $20,000. They then decreased by 2% each week after the end of the campaign.

   a. Write the equation of the exponential function that models the weekly sales.

   b. Find the sales 5 weeks after the end of the advertising campaign.

20. *Inflation* The average price of a house in a certain city was $220,000 in 2008, and it increases at 3% per year.

   a. Write the equation of the exponential function that models the average price of a house t years after 2008.

   b. Use the model to predict the average price of a house in 2013.

21. *Personal Income* Total personal income in the United States (in billions of dollars) for selected years from 1960 and projected to 2018 is given in the following table.

   a. The data can be modeled by an exponential function. Write the equation of this function, with x as the number of years after 1960.

Year	Personal Income (billions of $)
1960	411.5
1970	838.8
1980	2307.9
1990	4878.6
2000	8429.7
2008	12,100.7
2018	19,129.6

(Source: Bureau of Economic Analysis, U.S. Department of Commerce)

   b. If this model is accurate, what will be the total U.S. personal income in 2015?

   c. In what year does the model predict the total personal income will reach $19 trillion?

22. *Cohabiting Households* The table below gives the number of cohabiting without marriage households (in thousands) for selected years from 1960 to 2008.

Year (x)	Cohabiting Households (thousands)	Year (x)	Cohabiting Households (thousands)
1960	439	1994	3661
1970	523	1995	3668
1980	1589	1996	3958
1985	1983	1997	4130
1990	2856	1998	4236
1991	3039	2000	5457
1992	3308	2004	5841
1993	3510	2008	6214

   a. Find the exponential function $y = f(x)$ that models the data, with $x = 0$ in 1960.

   b. Estimate the number of cohabiting households in 2014 using the model.

23. *National Debt* The table that follows gives the U.S. national debt for selected years from 1900 to 2010.

Year	U.S. Debt (billions of $)	Year	U.S. Debt (billions of $)	Year	U.S. Debt (billions of $)
1900	1.2	1985	1823.1	2002	6228.2
1910	1.1	1990	3233.3	2003	6783.2
1920	24.2	1992	4064.6	2004	7379.1
1930	16.1	1994	4692.8	2005	7932.7
1940	43.0	1996	5224.8	2006	8680.2
1945	258.7	1998	5526.2	2007	9229.2
1955	272.8	1999	5656.3	2008	10,699.8
1965	313.8	2000	5674.2	2009	12,311.3
1975	533.2	2001	5807.5	2010	14,025.2

(Source: Bureau of Public Debt, U.S. Treasury)

a. Using a function of the form $y = a(b^x)$, with $x = 0$ in 1900 and y equal to the national debt in billions, model the data.

b. Use the model to predict the debt in 2013.

c. Predict when the debt will be $25 trillion ($25,000 billion).

d. Look at a graph of both the data and the model. What events may affect the accuracy of this model as a predictor of future public debt? Explain.

24. Insurance Premiums The table below gives the annual premiums required for a $250,000 20-year term-life insurance policy on female nonsmokers of different ages.

a. Find an exponential function that models the monthly premium as a function of the age of the female nonsmoking policyholder.

b. Find the quadratic function that is the best fit for the data.

c. Graph each function on the same axes with the data points to determine visually which model is the better fit for the data. Use the window [30, 80] by [−10, 6800].

Age	Monthly Premium for a 20-Year Policy ($)	Age	Monthly Premium for a 20-Year Policy ($)
35	145	60	845
40	185	65	1593
45	253	70	2970
50	363	75	5820
55	550		

(Source: Quotesmith.com)

25. Consumer Price Index The consumer price index (CPI) is calculated by averaging the prices of various items after assigning a weight to each item. The following table gives the consumer price indices for selected years from 1940 through 2010, reflecting buying patterns of all urban consumers, with x representing years past 1900.

a. Find an equation that models these data.

b. Use the model to predict the consumer price index in 2013.

c. According to the model, during what year will the consumer price index pass 300?

Year	Consumer Price Index	Year	Consumer Price Index
1940	14	2004	188.9
1950	24.1	2005	195.3
1960	29.6	2006	201.6
1970	38.8	2007	207.3
1980	82.4	2008	215.3
1990	130.7	2009	214.5
2000	172.2	2010	218.1
2002	179.9		

(Source: U.S. Census Bureau)

26. Facebook The numbers of millions of users of Facebook for the years 2004–2010 are shown in the figure below.

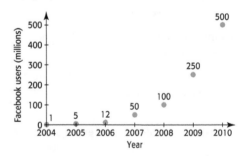

(Source: www.facebook.com)

a. Find an exponential function that models the number of millions of Facebook users. Use the number of years after 2000 as the input.

b. Find the power function that is the best fit for these data.

c. Graph each function on the same axes with the data points and discuss which model is the better fit for the data.

27. *Life Span* The table below gives the life expectancy for people in the United States for the birth years from 1920 and projected to 2020.

a. Find the logarithmic function that models the data, with x equal to 0 in 1900.

b. Find the quadratic function that is the best fit for the data, with $x = 0$ in 1900. Round the quadratic coefficient to four decimal places.

c. Graph each of these functions on the same axes with the data points to determine visually which function is the better model for the data for the years 1920–2020.

d. Evaluate both models for the birth year 2016.

Year	Life Span (years)	Year	Life Span (years)	Year	Life Span (years)
1920	54.1	1988	74.9	2001	77.2
1930	59.7	1989	75.2	2002	77.3
1940	62.9	1990	75.4	2003	77.5
1950	68.2	1992	75.8	2004	77.8
1960	69.7	1994	75.7	2005	77.9
1970	70.8	1996	76.1	2010	78.1
1975	72.6	1998	76.7	2015	78.9
1980	73.7	1999	76.7	2020	79.5
1987	75.0	2000	77.0		

(Source: National Center for Health Statistics)

28. *Poverty Threshold* The following table gives the average poverty thresholds for one person for selected years from 1990 to 2005.

a. Use a logarithmic equation to model the data, with x as the number of years after 1980.

b. Find an exponential model for the data, with $x = 0$ in 1980.

c. Which model is the better fit for the data?

Year	Income ($)	Year	Income ($)	Year	Income ($)
1990	6652	2000	8794	2003	9573
1995	7763	2001	9039	2004	9827
1998	8316	2002	9182	2005	10,160
1999	8501				

(Source: U.S. Census Bureau)

29. *Female Workers* The percents of females in the workforce for selected years from 1970 to 2008 are shown in the table below.

a. Find a logarithmic function that models the data, with y the percent and x the number of years from 1960.

b. When will the percent reach 48%, according to the model?

Year	Percent	Year	Percent
1970	38.1	2000	46.6
1980	42.5	2005	46.4
1990	45.2	2008	44.0
1995	46.1		

30. *Sales Abroad* Because of the weakening of the U.S. dollar, U.S.-based corporations are generating a growing share of their sales overseas. The figure shows the percent of sales made abroad.

a. Find a logarithmic function that is the best fit for the data, with x equal to the number of years after 2000 and y equal to the percent.

b. If the model found in part (a) is accurate for 2007, what would you expect the percent of sales made abroad to be in 2007?

Percent of Sales Made Abroad

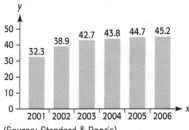

(Source: Standard & Poor's)

31. *College Tuition* Average tuition and required fees at degree-granting institutions, for graduate and first-professional fields of study, for the years 1988–89 through 2009–10 are given in the table.

a. Find an exponential function to model the data, with x equal to the number of years from 1980 to the end of the academic year (1988–89 would be $x = 9$).

b. Use the model to estimate the tuition for 2015–16. Is this interpolation or extrapolation?

Year	Current Dollars
1988–89	3728
1989–90	4135
1990–91	4488
1991–92	5116
1992–93	5475
1993–94	5973
1994–95	6247
1995–96	6741
1996–97	7111
1997–98	7246
1998–99	7685
1999–2000	8071
2000–01	8429
2001–02	8857
2002–03	9226
2003–04	10,312
2004–05	11,004
2005–06	11,621
2006–07	12,312
2007–08	12,962
2008–09	13,634
2009–10	14,537

(Source: *Digest of Educational Statistics*)

32. *Health Services Expenditures* The following table gives total U.S. expenditures (in billions of dollars) for health services and supplies for selected years from 2000 and projected to 2018.

Year	$ (billions)	Year	$ (billions)	Year	$ (billions)
2000	1264	2008	2227	2014	3107
2002	1498	2010	2458	2016	3556
2004	1733	2012	2746	2018	4086
2006	1976				

(Source: U.S. Centers for Medicare and Medicaid Services)

a. Find an exponential function to model these data, with *x* equal to the number of years after 2000.

b. Use the model to estimate the U.S. expenditures for health services and supplies in 2020.

33. *Sexually Active Girls* The percents of girls age *x* or younger who have been sexually active are given in the table below.

a. Create a logarithmic function that models the data, using an input equal to the age of the girls.

b. Use the model to estimate the percent of girls age 17 or younger who have been sexually active.

c. Find the quadratic function that is the best fit for the data.

d. Graph each of these functions on the same axes with the data points to determine which function is the better model for the data.

Age	Cumulative Percent Sexually Active Girls	Cumulative Percent Sexually Active Boys
15	5.4	16.6
16	12.6	28.7
17	27.1	47.9
18	44.0	64.0
19	62.9	77.6
20	73.6	83.0

(Source: "National Longitudinal Survey of Youth," *Risking the Future*, Washington D.C.: National Academy Press)

34. *Sexually Active Boys* The percents of boys age *x* or younger who have been sexually active are given in the table in Exercise 33.

a. Create a logarithmic function that models the data, using an input equal to the age of the boys.

b. Use the model to estimate the percent of boys age 17 or younger who have been sexually active.

c. Compare the percents that are sexually active for the two genders (see Exercise 33). What do you conclude?

35. *Flex Fuel Vehicles* Flexible fuel vehicles (FFVs) are designed to run on gasoline or a blend of up to 85% ethanol (E85). The numbers of millions of FFVs in use in the United States from 1998 to 2009 are given in the figure.

a. Find an exponential function to model the data, with *x* equal to the number of years after 1990.

b. Use the model to estimate the number of FFVs in use in 2015.

E85 FFVs in Use in the United States

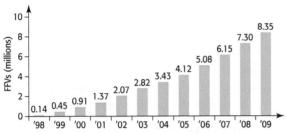

(Source: U.S. Department of Energy, Energy Efficiency, and Renewable Energy)

36. *Fuel Economy* The lifetime gasoline use of light-duty vehicles is a function of the fuel economy, as shown in the figure below.

a. Should the data be modeled by an exponential growth or an exponential decay function?

b. Find an exponential function to model the data.

c. Find a power function to model the data.

d. Which function is the better fit to the data?

e. If a vehicle had a fuel economy of 100 mpg, what would its lifetime gasoline use be, according to the power model?

Light-Duty Vehicle Fuel Consumption

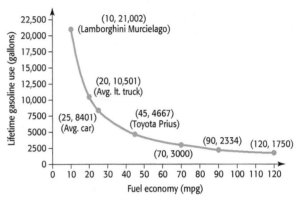

(Source: U.S. Department of Energy, Energy Efficiency, and Renewable Energy)

5.5 Exponential Functions and Investing

KEY OBJECTIVES

- Find future value of investments when interest is compounded k times per year
- Find future value of investments when interest is compounded continuously
- Find the present value of an investment
- Use graphing utilities to model investment data

SECTION PREVIEW Investments

If $1000 is invested in an account that earns 6% interest, with the interest added to the account at the end of each year, the interest is said to be **compounded annually**. The amount to which an investment grows over a period of time is called its **future value**. In this section, we use general formulas to find the future value of money invested when the interest is compounded at regular intervals of time and when it is compounded continuously. We also find the lump sum that must be invested to have it grow to a specified amount in the future. This is called the **present value** of the investment. ∎

Compound Interest

In Section 5.4, we found that if a set of data has initial value a and a constant percent change r for equally spaced inputs x, the data can be modeled by the exponential function

$$y = a(1 + r)^x$$

Thus, if an investment of $1000 earns 6% interest, compounded annually, the future value at the end of t years will be

$$S = \$1000(1.06)^t \text{ dollars}$$

EXAMPLE 1 ▶ Future Value of an Account

To numerically and graphically view how $1000 grows at 6% interest, compounded annually, do the following:

a. Construct a table that gives the future values $S = 1000(1 + 0.06)^t$ for $t = 0, 1, 2, 3, 4,$ and 5 years after the $1000 is invested.

b. Graph $S = 1000(1 + 0.06)^t$ for the values of t given in part (a).

Table 5.19

t (years)	Future Value, S ($)
0	1000.00
1	1060.00
2	1123.60
3	1191.02
4	1262.48
5	1338.23

c. Graph the function $S = 1000(1 + 0.06)^t$ as a continuous function of t for $0 \le t \le 5$.

d. Is there another graph that more accurately represents the amount that would be in the account at any time during the first 5 years that the money is invested?

SOLUTION

a. Substitute the values of t into the equation $S = 1000(1 + 0.06)^t$. Because the output units for S are dollars, we round to the nearest cent to obtain the values in Table 5.19.

b. Figure 5.42 shows the scatter plot of the data in Table 5.19.

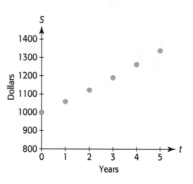

Figure 5.42

c. Figure 5.43(a) shows a continuous graph of the function $S = 1000(1 + 0.06)^t$ for $0 \le t \le 5$.

d. The scatter plot in Figure 5.42 shows the amount in the account at the end of each year, but not at any other time during the 5-year period. The continuous graph in Figure 5.43(a) shows the amount continually increasing, but interest is paid only at the end of each year and the amount in the account is constant at the previous level until more interest is added. So the graph in Figure 5.43(b) is a more accurate graph of the amount that would be in this account because it shows the amount remaining constant until the end of each year, when interest is added.

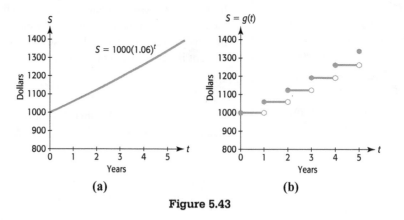

(a)   (b)

Figure 5.43

The function g graphed in Figure 5.43(b) is the step function defined by

$$g(t) = \begin{cases} 1000.00 & \text{if } 0 \le t < 1 \\ 1060.00 & \text{if } 1 \le t < 2 \\ 1123.60 & \text{if } 2 \le t < 3 \\ 1191.02 & \text{if } 3 \le t < 4 \\ 1262.48 & \text{if } 4 \le t < 5 \\ 1338.23 & \text{if } t = 5 \end{cases} \quad \text{dollars in } t \text{ years}$$

This step function is rather cumbersome to work with, so we will find the future value by using the graph in Figure 5.43(a) and the function $S = 1000(1 + 0.06)^t$ *interpreted discretely*. That is, we draw the graph of this function as it appears in Figure 5.43(a) and evaluate the formula for S with the understanding that the only inputs that make sense in this investment example are nonnegative integers. In general, we have the following.

Future Value of an Investment with Annual Compounding

If P dollars are invested at an interest rate r per year, compounded annually, the future value S at the end of t years is

$$S = P(1 + r)^t$$

The **annual interest rate** r is also called the **nominal interest rate**, or simply the **rate**. Remember that the interest rate is usually stated as a percent and that it is converted to a decimal when computing future values of investments.

If the interest is compounded more than once per year, then the additional compounding will result in a larger future value. For example, if the investment is compounded twice per year (semiannually), there will be twice as many compounding periods, with the interest rate in each period equal to

$$r\left(\frac{1}{2}\right) = \frac{r}{2}$$

Thus, the future value is found by doubling the number of periods and halving the interest rate if the compounding is done twice per year. In general, we use the following model, interpreted discretely, to find the amount of money that results when P dollars are invested and earn compound interest.

Future Value of an Investment with Periodic Compounding

If P dollars are invested for t years at the annual interest rate r, where the interest is compounded k times per year, then the interest rate per period is $\frac{r}{k}$, the number of compounding periods is kt, and the future value that results is given by

$$S = P\left(1 + \frac{r}{k}\right)^{kt} \text{ dollars}$$

For example, if $1000 is placed in an account that earns 6% interest per year, with the interest added to the account at the end of every month (compounded monthly), the future value of this investment in 5 years is given by

$$S = 1000\left(1 + \frac{0.06}{12}\right)^{12(5)} = \$1348.85$$

Recall from Table 5.19 that compounding annually gives $1338.23, so compounding monthly rather than yearly results in an additional $1348.85 - \$1338.23 = \$10.62$ from the 5-year investment.

EXAMPLE 2 ▶ Daily versus Annual Compounding of Interest

a. Write the equation that gives the future value of $1000 invested for t years at 8%, compounded annually.

b. Write the equation that gives the future value of $1000 invested for t years at 8%, compounded daily.

c. Graph the equations from parts (a) and (b) on the same axes, with t between 0 and 30.

d. What is the additional amount of interest earned in 30 years from compounding daily rather than annually?

SOLUTION

a. Substituting $P = 1000$ and $r = 0.08$ in $S = P(1 + r)^t$ gives

$$S = 1000(1 + 0.08)^t, \quad \text{or} \quad S = 1000(1.08)^t$$

b. Substituting $P = 1000$, $r = 0.08$, and $k = 365$ in $S = P\left(1 + \dfrac{r}{k}\right)^{kt}$ gives

$$S = 1000\left(1 + \frac{0.08}{365}\right)^{365t}$$

c. The graphs of both functions are shown in Figure 5.44.

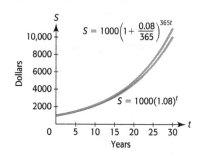

Figure 5.44

d. Evaluating the two functions in parts (a) and (b) at $t = 30$, we calculate the future value of $1000, compounded annually at 8%, to be $10,062.66 and the future value of $1000 at 8%, compounded daily, to be $11,020.28. The additional amount earned in 30 years from compounding daily rather than annually is the difference between these two amounts:

$$\$11,020.28 - \$10,062.66 = \$957.62.$$

Continuous Compounding and the Number e

In Section 5.1, we stated that if P dollars are invested for t years at interest rate r, compounded continuously, the future value of the investment is

$$S = Pe^{rt}$$

To see how the number e becomes part of this formula, we consider the future value of $1 invested at an annual rate of 100%, compounded for 1 year with different compounding periods. If we denote the number of periods per year by k, the model that gives the future value is

$$S = \left(1 + \frac{1}{k}\right)^k \text{ dollars}$$

Table 5.20 shows the future value of this investment for different compounding periods.

Table 5.20

Type of Compounding	Number of Compounding Periods per Year	Future Value ($)
Annually	1	$\left(1 + \dfrac{1}{1}\right)^1 = 2$
Quarterly	4	$\left(1 + \dfrac{1}{4}\right)^4 = 2.44140625$
Monthly	12	$\left(1 + \dfrac{1}{12}\right)^{12} \approx 2.61303529022$
Daily	365	$\left(1 + \dfrac{1}{365}\right)^{365} \approx 2.71456748202$
Hourly	8760	$\left(1 + \dfrac{1}{8760}\right)^{8760} \approx 2.71812669063$
Each minute	525,600	$\left(1 + \dfrac{1}{525,600}\right)^{525,600} \approx 2.7182792154$
x times per year	x	$\left(1 + \dfrac{1}{x}\right)^x$

As Table 5.20 indicates, the future value increases (but not very rapidly) as the number of compounding periods during the year increases. As x gets very large, the future value approaches the number e, which is 2.718281828 to nine decimal places. In general, the outputs that result from larger and larger inputs in the function

$$f(x) = \left(1 + \frac{1}{x}\right)^x$$

approach the number e. (In calculus, e is defined as the *limit* of $\left(1 + \dfrac{1}{x}\right)^x$ as x approaches ∞.)

This definition of e permits us to create a new model for the future value of an investment when interest is compounded continuously. We can let $m = \dfrac{k}{r}$ to rewrite the formula

$$S = P\left(1 + \frac{r}{k}\right)^{kt} \quad \text{as} \quad S = P\left(1 + \frac{1}{m}\right)^{mrt}, \quad \text{or} \quad S = P\left[\left(1 + \frac{1}{m}\right)^m\right]^{rt}$$

Now as the compounding periods k increase without bound, $m = \dfrac{k}{r}$ increases without bound and

$$\left(1 + \frac{1}{m}\right)^m$$

approaches e, so $S = P\left[\left(1 + \dfrac{1}{m}\right)^m\right]^{rt}$ approaches Pe^{rt}.

Future Value of an Investment with Continuous Compounding

If P dollars are invested for t years at an annual interest rate r, compounded continuously, then the future value S is given by

$$S = Pe^{rt} \text{ dollars}$$

For a given principal and interest rate, the function $S = Pe^{rt}$ is a continuous function whose domain consists of real numbers greater than or equal to 0. Unlike the other compound interest functions, this one is not discretely interpreted because of the continuous compounding.

EXAMPLE 3 ▶ Future Value and Continuous Compounding

a. What is the future value of $2650 invested for 8 years at 12%, compounded continuously?

b. How much interest will be earned on this investment?

SOLUTION

a. The future value of this investment is $S = 2650e^{0.12(8)} = 6921.00$.

b. The interest earned on this investment is the future value minus the original investment:

$$\$6921 - \$2650 = \$4271$$

EXAMPLE 4 ▶ Continuous versus Annual Compounding of Interest

a. For each of 9 years, compare the future value of an investment of $1000 at 8%, compounded annually, and of $1000 at 8%, compounded continuously.

b. Graph the functions for annual compounding and for continuous compounding for $t = 30$ years on the same axes.

c. What conclusion can be made regarding compounding annually and compounding continuously?

SOLUTION

a. The future value of $1000, compounded annually, is given by the function $S = 1000(1 + 0.08)^t$, and the future value of $1000, compounded continuously, is given by $S = 1000e^{0.08t}$. By entering these formulas in an Excel spreadsheet and evaluating them for each of 9 years (Table 5.21), we can compare the future value for each of these 9 years. At the end of the 9 years, we see that compounding continuously results in $2054.43 - \$1999.00 = \$55.43$ more than compounding annually yields.

b. The graphs are shown in Figure 5.45.

c. The value of the investment increases more under continuous compounding than under annual compounding.

Table 5.21

	A	B	C
1	YEAR	$S = 1000*(1.08)^t$	$S = 1000*e^{(.08t)}$
2	1	1080	1083.287068
3	2	1166.4	1173.510871
4	3	1259.712	1271.249150
5	4	1360.48896	1377.127764
6	5	1469.328077	1491.824698
7	6	1586.874323	1616.07440
8	7	1713.824269	1750.672500
9	8	1850.93021	1896.480879
10	9	1999.004627	2054.433211

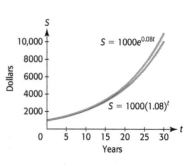

Figure 5.45

Present Value of an Investment

We can write the formula for the future value S when a lump sum P is invested for $n = kt$ periods at interest rate $i = \dfrac{r}{k}$ per compounding period as

$$S = P(1 + i)^n$$

We sometimes need to know the lump sum investment P that will give an amount S in the future. This is called the **present value** P, and it can be found by solving $S = P(1 + i)^n$ for P. Solving for P gives the present value as follows.

Present Value

The lump sum that will give future value S in n compounding periods at rate i per period is the present value

$$P = \frac{S}{(1 + i)^n} \quad \text{or, equivalently,} \quad P = S(1 + i)^{-n}$$

EXAMPLE 5 ▶ **Present Value**

What lump sum must be invested at 10%, compounded semiannually, for the investment to grow to $15,000 in 7 years?

SOLUTION

We have the future value $S = 15{,}000$, $i = \dfrac{0.10}{2} = 0.05$, and $n = 2 \cdot 7 = 14$, so the present value of this investment is

$$P = \frac{15{,}000}{(1 + 0.05)^{14}} = 7576.02 \text{ dollars}$$

Investment Models

We have developed and used investment formulas. We can also use graphing utilities to model actual investment data. Consider the following example.

EXAMPLE 6 ▶ **Mutual Fund Growth**

Table 5.22

Time (years)	Average Annual Total Return ($)
1	1.23
3	2.24
5	3.11
10	6.82

(Source: FundStation)

The data in Table 5.22 give the annual return on an investment of $1.00 made on March 31, 1990, in the AIM Value Fund, Class A Shares. The values reflect reinvestment of all distributions and changes in net asset value but exclude sales charges.

a. Use an exponential function to model these data.

b. Use the model to find what the fund would amount to on March 31, 2001 if $10,000 was invested March 31, 1990 and the fund continued to follow this model after 2000.

c. Is it likely that this fund continued to grow at this rate in 2002?

SOLUTION

a. Using technology to create an exponential model for the data gives

$$A(x) = 1.1609(1.2004)^x$$

Thus, $1.00 invested in this fund grows to $A(x) = 1.1609(1.2004)^x$ dollars x years after March 31, 1990.

   b. March 31, 2001 is 11 years after March 31, 1990, so using the equation from part (a) with $x = 11$ gives $A(11) = 1.1609(1.2004)^{11}$ as the future value of an investment of $1.00. Thus, the future value of an investment of $10,000 in 11 years is

$$\$10{,}000[1.1609(1.2004)^{11}] = \$86{,}572.64$$

   c. No. Most stocks and mutual funds decreased in value in 2002 because of terrorist attacks in 2001.

Skills CHECK 5.5

Evaluate the expressions in Exercises 1–10. Write approximate answers rounded to two decimal places.

1. $15{,}000e^{0.06(20)}$ **2.** $8000e^{0.05(10)}$

3. $3000(1.06^x)$ for $x = 30$

4. $20{,}000(1.07^x)$ for $x = 20$

5. $12{,}000\left(1 + \dfrac{0.10}{k}\right)^{kn}$ for $k = 4$ and $n = 8$

6. $23{,}000\left(1 + \dfrac{0.08}{k}\right)^{kn}$ for $k = 12$ and $n = 20$

7. $P\left(1 + \dfrac{r}{k}\right)^{kn}$ for $P = 3000, r = 8\%, k = 2, n = 18$

8. $P\left(1 + \dfrac{r}{k}\right)^{kn}$ for $P = 8000, r = 12\%, k = 12,$ $n = 8$

9. $300\left(\dfrac{1.02^n - 1}{0.02}\right)$ for $n = 240$

10. $2000\left(\dfrac{1.10^n - 1}{0.10}\right)$ for $n = 12$

11. Find $g(2.5)$, $g(3)$, and $g(3.5)$ if

$$g(x) = \begin{cases} 1000.00 & \text{if } 0 \le x < 1 \\ 1060.00 & \text{if } 1 \le x < 2 \\ 1123.60 & \text{if } 2 \le x < 3 \\ 1191.00 & \text{if } 3 \le x < 4 \\ 1262.50 & \text{if } 4 \le x < 5 \\ 1338.20 & \text{if } x = 5 \end{cases}$$

12. Find $f(2)$, $f(1.99)$, and $f(2.1)$ if

$$f(x) = \begin{cases} 100 & \text{if } 0 \le x \le 1 \\ 200 & \text{if } 1 < x < 2 \\ 300 & \text{if } 2 \le x \le 3 \end{cases}$$

13. Solve $S = P\left(1 + \dfrac{r}{k}\right)^{kn}$ for P.

14. Solve $S = P(1 + i)^n$ for P.

EXERCISES 5.5

15. *Future Value* If $8800 is invested for x years at 8% interest, compounded annually, find the future value that results in

   a. 8 years.   **b.** 30 years.

16. *Investments* Suppose $6400 is invested for x years at 7% interest, compounded annually. Find the future value of this investment at the end of

   a. 10 years.   **b.** 30 years.

17. *Future Value* If $3300 is invested for x years at 10% interest, compounded annually, the future value that results is $S = 3300(1.10)^x$ dollars.

   a. Graph the function for $x = 0$ to $x = 8$.

   b. Use the graph to estimate when the money in the account will double.

18. *Investments* If $5500 is invested for x years at 12% interest, compounded quarterly, the future value that results is $S = 5500(1.03)^{4x}$ dollars.

a. Graph this function for $0 \le x \le 8$.

b. Use the graph to estimate when the money in the account will double.

19. *Future Value* If $10,000 is invested at 12% interest, compounded quarterly, find the future value in 10 years.

20. *Future Value* If $8800 is invested at 6% interest, compounded semiannually, find the future value in 10 years.

21. *Future Value* An amount of $10,000 is invested at 12% interest, compounded daily.

    a. Find the future value in 10 years.

    b. How does this future value compare with the future value in Exercise 19? Why are they different?

22. *Future Value* A total of $8800 is invested at 6% interest, compounded daily.

    a. Find the future value in 10 years.

    b. How does this future value compare with the future value in Exercise 20? Why are they different?

23. *Compound Interest* If $10,000 is invested at 12% interest, compounded monthly, find the interest earned in 15 years.

24. *Compound Interest* If $20,000 is invested at 8% interest, compounded quarterly, find the interest earned in 25 years.

25. *Continuous Compounding* Suppose $10,000 is invested for t years at 6% interest, compounded continuously. Give the future value at the end of

    a. 12 years.

    b. 18 years.

26. *Continuous Compounding* If $42,000 is invested for t years at 7% interest, compounded continuously, find the future value in

    a. 10 years.

    b. 20 years.

27. *Continuous versus Annual Compounding*
    a. If $10,000 is invested at 6%, compounded annually, find the future value in 18 years.

    b. How does this compare with the result from Exercise 25(b)?

28. *Continuous versus Annual Compounding*
    a. If $42,000 is invested at 7%, compounded annually, find the future value in 20 years.

    b. How does this compare with the result from Exercise 26(b)?

29. *Doubling Time* Use a spreadsheet, a table, or a graph to estimate how long it takes for an amount to double if it is invested at 10% interest

    a. Compounded annually.

    b. Compounded continuously.

30. *Doubling Time* Use a spreadsheet, a table, or a graph to estimate how long it takes for an amount to double if it is invested at 6% interest

    a. Compounded annually.

    b. Compounded continuously.

31. *Future Value* Suppose $2000 is invested in an account paying 5% interest, compounded annually. What is the future value of this investment

    a. After 8 years?

    b. After 18 years?

32. *Future Value* If $12,000 is invested in an account that pays 8% interest, compounded quarterly, find the future value of this investment

    a. After 2 quarters.

    b. After 10 years.

33. *Investments* Suppose $3000 is invested in an account that pays 6% interest, compounded monthly. What is the future value of this investment after 12 years?

34. *Investments* If $9000 is invested in an account that pays 8% interest, compounded quarterly, find the future value of this investment

    a. After 0.5 year.

    b. After 15 years.

35. *Doubling Time* If money is invested at 10% interest, compounded quarterly, the future value of the investment doubles approximately every 7 years.

    a. Use this information to complete the table below for an investment of $1000 at 10% interest, compounded quarterly.

Years	0	7	14	21	28
Future Value ($)	1000				

    b. Create an exponential function, rounded to three decimal places, that models the discrete function defined by the table.

    c. Because the interest is compounded quarterly, this model must be interpreted discretely. Use the rounded function to find the value of the investment in 5 years and in $10\frac{1}{2}$ years.

36. *Doubling Time* If money is invested at 11.6% interest, compounded monthly, the future value of the investment doubles approximately every 6 years.

a. Use this information to complete the table below for an investment of $1000 at 11.6% interest, compounded monthly.

Years	0	6	12	18	24
Future Value ($)	1000				

b. Create an exponential function, rounded to three decimal places, that models the discrete function defined by the table.

c. Because the interest is compounded monthly, this model must be interpreted discretely. Use the rounded model to find the value of the investment in 2 months, in 4 years, and in $12\frac{1}{2}$ years.

37. *Present Value* What lump sum must be invested at 10%, compounded monthly, for the investment to grow to $65,000 in 8 years?

38. *Present Value* What lump sum must be invested at 8%, compounded quarterly, for the investment to grow to $30,000 in 12 years?

39. *Present Value* What lump sum investment will grow to $10,000 in 10 years if it is invested at 6%, compounded annually?

40. *Present Value* What lump sum investment will grow to $30,000 if it is invested for 15 years at 7%, compounded annually?

41. *College Tuition* New parents want to put a lump sum into a money market fund to provide $30,000 in 18 years to help pay for college tuition for their child. If the fund averages 10% per year, compounded monthly, how much should they invest?

42. *Trust Fund* Grandparents decide to put a lump sum of money into a trust fund on their granddaughter's 10th birthday so that she will have $1,000,000 on her 60th birthday. If the fund pays 11%, compounded monthly, how much money must they put in the account?

43. *College Tuition* The Toshes need to have $80,000 in 12 years for their son's college tuition. What amount must they invest to meet this goal if they can invest money at 10%, compounded monthly?

44. *Retirement* Hennie and Bob inherit $100,000 and plan to invest part of it for 25 years at 10%, compounded monthly. If they want it to grow to $1 million for their retirement, how much should they invest?

45. *Investment* At the end of t years, the future value of an investment of $10,000 in an account that pays 8%, compounded monthly, is

$$S = 10,000\left(1 + \frac{0.08}{12}\right)^{12t} \text{ dollars}$$

Assuming no withdrawals or additional deposits, how long will it take for the investment to amount to $40,000?

46. *Investment* At the end of t years, the future value of an investment of $25,000 in an account that pays 12%, compounded quarterly, is

$$S = 25,000\left(1 + \frac{0.12}{4}\right)^{4t} \text{ dollars}$$

In how many years will the investment amount to $60,000?

47. *Investment* At the end of t years, the future value of an investment of $38,500 in an account that pays 8%, compounded monthly, is

$$S = 38,500\left(1 + \frac{0.08}{12}\right)^{12t} \text{ dollars}$$

For what time period (in years) will the future value of the investment be more than $100,230?

48. *Investment* At the end of t years, the future value of an investment of $12,000 in an account that pays 8%, compounded monthly, is

$$S = 12,000\left(1 + \frac{0.08}{12}\right)^{12t} \text{ dollars}$$

Assuming no withdrawals or additional deposits, for what time period (in years) will the future value of the investment be less than $48,000?

49. *Investing* If an investment of P dollars earns r percent (as a decimal), compounded m times per year, its future value in t years is $A = P\left(1 + \frac{r}{m}\right)^{mt}$. Prove that the number of years it takes for this investment to double is $t = \dfrac{\ln 2}{m \ln(1 + r/m)}$.

KEY OBJECTIVES

- Find the future value of an ordinary annuity
- Find the present value of an ordinary annuity
- Find the payments needed to amortize a loan

SECTION PREVIEW **Annuities**

To prepare for future retirement, people frequently invest a fixed amount of money at the end of each month into an account that pays interest that is compounded monthly. Such an investment plan or any other characterized by regular payments is called an annuity. In this section, we develop a model to find future values of annuities.

A person planning retirement may also want to know how much money is needed to provide regular payments to him or her during retirement, and a couple may want to know what lump sum to put in an investment to pay future college expenses for their child. The amount of money they need to invest to receive a series of payments in the future is the present value of an annuity.

We can also use the present value concept to determine the payments that are necessary to repay a loan with equal payments made on a regular schedule, which is called amortization. The formulas used to determine the future value and the present value of annuities and the payments necessary to repay loans are applications of exponential functions. ■

Future Value of an Annuity

An **annuity** is a financial plan characterized by regular payments. We can view an annuity as a savings plan where regular payments are made to an account, and we can use an exponential function to determine what the future value of the account will be. One type of annuity, called an **ordinary annuity**, is a financial plan where equal payments are contributed at the end of each period to an account that pays a fixed rate of interest compounded at the same time as payments are made. For example, suppose $1000 is invested at the end of each year for 5 years in an account that pays interest at 10%, compounded annually. To find the future value of this annuity, we can think of it as the sum of 5 investments of $1000, one that draws interest for 4 years (from the end of the first year to the end of the fifth year), one that draws interest for 3 years, and so on (Table 5.23).

Table 5.23

Investment	Future Value
Invested at end of 1st year	$1000(1.10)^4$
Invested at end of 2nd year	$1000(1.10)^3$
Invested at end of 3rd year	$1000(1.10)^2$
Invested at end of 4th year	$1000(1.10)^1$
Invested at end of 5th year	1000

Note that the last investment is at the end of the fifth year, so it earned no interest. The future value of the annuity is the sum of the separate future values. It is

$$S = 1000 + 1000(1.10)^1 + 1000(1.10)^2 + 1000(1.10)^3 + 1000(1.10)^4$$

$$= 6105.10 \text{ dollars}$$

We can find the value of an ordinary annuity like the one above with the following formula.

Future Value of an Ordinary Annuity

If R dollars are contributed at the end of each period for n periods into an annuity that pays interest at rate i at the end of the period, the future value of the annuity is

$$S = R\left(\frac{(1 + i)^n - 1}{i}\right) \text{ dollars}$$

Note that the interest rate used in the above formula is the rate per period, not the rate for a year. We can use this formula to find the future value of an annuity in which $1000 is invested at the end of each year for 5 years into an account that pays interest at 10%, compounded annually. This future value, which we found earlier without the formula, is

$$S = R\left(\frac{(1 + i)^n - 1}{i}\right) = 1000\left(\frac{(1 + 0.10)^5 - 1}{0.10}\right) = 6105.10 \text{ dollars}$$

EXAMPLE 1 ▶ Future Value of an Ordinary Annuity

Find the 5-year future value of an ordinary annuity with a contribution of $500 per quarter into an account that pays 8% per year, compounded quarterly.

SOLUTION

The payments and interest compounding occur quarterly, so the interest rate per period is $\frac{0.08}{4} = 0.02$ and the number of compounding periods is $4(5) = 20$. Substituting the information into the formula for the future value of an annuity gives

$$S = 500\left(\frac{(1 + 0.02)^{20} - 1}{0.02}\right) = 12{,}148.68$$

Thus, the future value of this investment is $12,148.68.

EXAMPLE 2 ▶ Future Value of an Annuity

Harry deposits $200 at the end of each month into an account that pays 12% interest per year, compounded monthly. Find the future value for every 4-month period, for up to 36 months.

SOLUTION

To find the future value of this annuity, we note that the interest rate per period (month) is $\frac{12\%}{12} = 0.01$, so the future value at the end of n months is given by the function

$$S = 200\left(\frac{1.01^n - 1}{0.01}\right) \text{ dollars}$$

This model for the future value of an ordinary annuity is a continuous function with discrete interpretation because the future value changes only at the end of each period. A graph of the function and the table of future values for every 4 months up to 36 months are shown in Figure 5.46 and Table 5.24, respectively. Keep in mind that the graph should be interpreted discretely; that is, only positive integer inputs and corresponding outputs make sense.

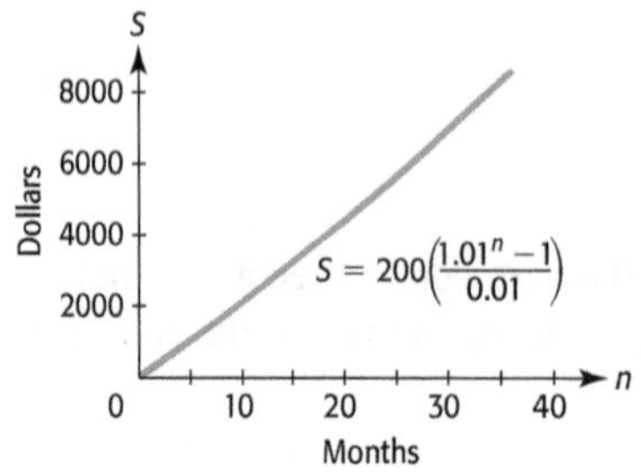

Table 5.24

Number of Months	Value of Annuity ($)
4	812.08
8	1657.13
12	2536.50
16	3451.57
20	4403.80
24	5394.69
28	6425.82
32	7498.81
36	8615.38

$$S = 200\left(\frac{1.01^n - 1}{0.01}\right)$$

Figure 5.46

Present Value of an Annuity

Many retirees purchase annuities that give them regular payments through a period of years. Suppose a lump sum of money is invested to return payments of R dollars at the end of each of n periods, with the investment earning interest at rate i per period, after which $0 will remain in the account. This lump sum is the **present value** of this ordinary annuity, and it is equal to the sum of the present values of each of the future payments. Table 5.25 shows the present value of each payment of this annuity.

Table 5.25

Present	Present Value of This Payment ($)
Paid at end of 1st period	$R(1 + i)^{-1}$
Paid at end of 2nd period	$R(1 + i)^{-2}$
Paid at end of 3rd period	$R(1 + i)^{-3}$
$\vdots$	$\vdots$
Paid at end of $(n - 1)$st period	$R(1 + i)^{-(n-1)}$
Paid at end of nth period	$R(1 + i)^{-n}$

We must have enough money in the account to provide the present value of each of these payments, so the present value of the annuity is

$$A = R(1 + i)^{-1} + R(1 + i)^{-2} + R(1 + i)^{-3} + \cdots + R(1 + i)^{-(n-1)} + R(1 + i)^{-n}$$

This sum can be written in the form

$$A = R\left(\frac{1 - (1 + i)^{-n}}{i}\right)$$

Present Value of an Ordinary Annuity

If a payment of R dollars is to be made at the end of each period for n periods from an account that earns interest at a rate of i per period, then the account is an **ordinary annuity**, and the **present value** is

$$A = R\left(\frac{1 - (1 + i)^{-n}}{i}\right)$$

EXAMPLE 3 ▶ **Present Value of an Annuity**

Suppose a retiring couple wants to establish an annuity that will provide $2000 at the end of each month for 20 years. If the annuity earns 6%, compounded monthly, how much must the couple put in the account to establish the annuity?

SOLUTION

We seek the present value of an annuity that pays $2000 at the end of each month for $12(20) = 240$ months, with interest at $\frac{6\%}{12} = \frac{0.06}{12} = 0.005$ per month. The present value is

$$A = 2000\left(\frac{1 - (1 + 0.005)^{-240}}{0.005}\right) = 279{,}161.54 \text{ dollars}$$

Thus, the couple can receive a payment of $2000 at the end of each month for 20 years if they put a lump sum of $279,161.54 in an annuity. (Note that the sum of the money they will receive over the 20 years is $2000 \cdot 20 \cdot 12 = \$480{,}000$.)

EXAMPLE 4 ▶   Home Mortgage

A couple who wants to purchase a home has $30,000 for a down payment and wants to make monthly payments of $2200. If the interest rate for a 25-year mortgage is 6% per year on the unpaid balance, what is the price of a house they can buy?

SOLUTION

The amount of money that they can pay for the house is the sum of the down payment and the present value of the payments that they can afford. The present value of the $12(25) = 300$ monthly payments of $2200 with interest at $\dfrac{6\%}{12} = \dfrac{0.06}{12} = 0.005$ per month is

$$A = 2200\left(\frac{1 - (1 + 0.005)^{-300}}{0.005}\right) = 341{,}455.10$$

They can buy a house costing $341,455 + 30,000 = \$371,455.10$.

Loan Repayment

When money is borrowed, the borrower must repay the total amount that was borrowed (the debt) plus interest on that debt. Most loans (including those for houses, for cars, and for other consumer goods) require regular payments on the debt plus payment of interest on the unpaid balance of the loan. That is, loans are paid off by a series of partial payments with interest charged on the unpaid balance at the end of each period. The stated interest rate (the **nominal rate**) is the *annual rate*. The rate per period is the nominal rate divided by the number of payment periods per year.

There are two popular repayment plans for these loans. One plan applies an equal amount to the debt each payment period plus the interest for the period, which is the interest on the unpaid balance. For example, a loan of $240,000 for 10 years at 12% could be repaid with 120 monthly payments of $2000 plus 1% $\left(\dfrac{1}{12} \text{ of } 12\%\right)$ of the unpaid balance each month. When this payment method is used, the payments will decrease as the unpaid balance decreases. A few sample payments are shown in Table 5.26.

Table 5.26

Payment Number (month)	Unpaid Balance ($)	Interest on the Unpaid Balance ($)	Balance Reduction ($)	Monthly Payment ($)	New Balance ($)
1	240,000	2400	2000	4400	238,000
2	238,000	2380	2000	4380	236,000
⋮					
51	140,000	1400	2000	3400	138,000
⋮					
101	40,000	400	2000	2400	38,000
⋮					
118	6000	60	2000	2060	4000

A loan of this type can also be repaid by making all payments (including the payment on the balance and the interest) of equal size. The process of repaying the loan in this way is called **amortization**. If a bank makes a loan of this type, it uses an amortization formula that gives the size of the equal payments. This formula is developed by considering the loan as an annuity purchased from the borrower by the bank, with the annuity paying a fixed return to the bank each payment period. The lump sum that the bank gives to the borrower (the principal of the loan) is the present value of this annuity, and each payment that the bank receives from the borrower is a payment from this annuity. To find the size of these equal payments, we solve the formula for the present value of

an annuity, $A = R\left(\dfrac{1 - (1 + i)^{-n}}{i}\right)$, for R. This gives the following formula.

Amortization Formula

If a debt of A dollars, with interest at a rate of i per period, is amortized by n equal periodic payments made at the end of each period, then the size of each payment is

$$R = A\left(\frac{i}{1 - (1 + i)^{-n}}\right)$$

We can use this formula to find the *equal* monthly payments that will amortize the loan of $240,000 for 10 years at 12%. The interest rate per month is 0.01, and there are 120 periods.

$$R = A\left(\frac{i}{1 - (1 + i)^{-n}}\right) = 240,000\left(\frac{0.01}{1 - (1 + 0.01)^{-120}}\right) \approx 3443.303$$

Thus, repaying this loan requires 120 equal payments of $3443.31. If we compare this payment with the selected payments in Table 5.26, we see that the first payment in Table 5.26 is nearly $1000 more than the equal amortization payment and that every payment after the 101st is $1000 less than the amortization payment. The advantage of the amortization method of repaying the loan is that the borrower can budget better by knowing what the payment will be each month. Of course, the balance of the loan will not be reduced as fast with this payment method, because most of the monthly payment will be needed to pay interest in the first few months. The partial spreadsheet in Table 5.27 shows how the outstanding balance is reduced.

Table 5.27

	A	B	C	D	E	F
1	Payment	Unpaid		Monthly	Balance	New
2	Number	Balance	Interest	Payment	Reduction	Balance
3	1	240,000	2400	3443.31	1043.31	238,956.69
4	2	238,956.69	2389.57	3443.31	1053.74	237,902.95
	⋮					
53	51	172,745.38	1727.45	3443.31	1715.86	171,029.52
	⋮					
103	101	63,136.30	631.36	3443.31	2811.95	60,324.35

EXAMPLE 5 ▶ Home Mortgage

A couple that wants to purchase a home with a price of $230,000 has $50,000 for a down payment. If they can get a 25-year mortgage at 9% per year on the unpaid balance,

a. What will be their equal monthly payments?

b. What is the total amount they will pay before they own the house outright?

c. How much interest will they pay?

SOLUTION

a. The amount of money that they must borrow is $230,000 - $50,000 = $180,000. The number of monthly payments is $12(25) = 300$, and the interest rate is $\dfrac{9\%}{12} = \dfrac{0.09}{12} = 0.0075$ per month. The monthly payment is

$$R = 180,000 \left(\frac{0.0075}{1 - (1.0075)^{-300}} \right) \approx 1510.553$$

so the required payment would be $1510.56.

b. The amount that they must pay before owning the house is the down payment plus the total of the 300 payments:

$$\$50,000 + 300(\$1510.56) = \$503,168$$

c. The total interest is the total amount paid minus the price paid for the house. The interest is

$$\$503,168 - \$230,000 = \$273,168$$

When the interest paid is based on the periodic interest rate on the unpaid balance of the loan, the nominal rate is also called the **annual percentage rate (APR)**. You will occasionally see an advertisement for a loan with a stated interest rate and a different, larger APR. This can happen if a lending institution charges fees (sometimes called points) in addition to the stated interest rate. Federal law states that all these charges must be included in computing the APR, which is the true interest rate that is being charged on the loan.

Skills CHECK 5.6

Give answers to two decimal places.

1. Solve $S = P(1 + i)^n$ for P with positive exponent n.

2. Solve $S = P(1 + i)^n$ for P, with no denominator.

3. Solve $Ai = R[1 - (1 + i)^{-n}]$ for A.

4. Evaluate $2000 \left(\dfrac{1 - (1 + 0.01)^{-240}}{0.01} \right)$.

5. Solve $A = R \left(\dfrac{1 - (1 + i)^{-n}}{i} \right)$ for R.

6. Evaluate $240,000 \left(\dfrac{0.01}{1 - (1 + 0.01)^{-120}} \right)$.

EXERCISES 5.6

7. *IRA* Anne Wright decides to invest $4000 in an IRA CD at the end of each year for 10 years. If she makes these payments and the certificates all pay 6%, compounded annually, how much will she have at the end of the 10 years?

8. *Future Value* Find the future value of an annuity of $5000 paid at the end of each year for 20 years, if interest is earned at 9%, compounded annually.

9. *Down Payment* To start a new business, Beth deposits $1000 at the end of each 6-month period in an account that pays 8%, compounded semiannually. How much will she have at the end of 8 years?

10. *Future Value* Find the future value of an annuity of $2600 paid at the end of each 3-month period for 5 years, if interest is earned at 6%, compounded quarterly.

11. *Retirement* Mr. Lawrence invests $600 at the end of each month in an account that pays 7%, compounded monthly. How much will be in the account in 25 years?

12. *Down Payment* To accumulate money for the down payment on a house, the Kings deposit $800 at the end of each month into an account paying 7%, compounded monthly. How much will they have at the end of 5 years?

13. *Dean's List* Parents agree to invest $1000 (at 10%, compounded semiannually) for their daughter on the December 31 or June 30 following each semester that she makes the Dean's List during her 4 years in college. If she makes the Dean's List in each of the 8 semesters, how much will the parents have to give her when she graduates?

14. *College Tuition* To help pay for tuition, parents deposited $1000 into an account on each of their child's birthdays, starting with the first birthday. If the account pays 6%, compounded annually, how much will they have for tuition on her 19th birthday?

15. *Annuities* Find the present value of an annuity that will pay $1000 at the end of each year for 10 years if the interest rate is 7%, compounded annually.

16. *Annuities* Find the present value of an annuity that will pay $500 at the end of each year for 20 years if the interest rate is 9%, compounded annually.

17. *Lottery Winnings* The winner of a "million dollar" lottery is to receive $50,000 plus $50,000 at the end of each year for 19 years or the present value of this annuity in cash. How much cash would she receive if money is worth 8%, compounded annually?

18. *College Tuition* A couple wants to establish a fund that will provide $3000 for tuition at the end of each 6-month period for 4 years. If a lump sum can be placed in an account that pays 8%, compounded semiannually, what lump sum is required?

19. *Insurance Payment* A man is disabled in an accident and wants to receive an insurance payment that will provide him with $3000 at the end of each month for 30 years. If the payment can be placed in an account that pays 9%, compounded monthly, what size payment should he seek?

20. *Auto Leasing* A woman wants to lease rather than buy a car but does not want to make monthly payments. A dealer has the car she wants, which leases for $400 at the end of each month for 48 months. If money is worth 8%, compounded monthly, what lump sum should she offer the dealer to keep the car for 48 months?

21. *Business Sale* A man can sell his Thrifty Electronics business for $800,000 cash or for $100,000 plus $122,000 at the end of each year for 9 years.

   a. Find the present value of the annuity that is offered if money is worth 10%, compounded annually.

   b. If he takes the $800,000, spends $100,000 of it, and invests the rest in a 9-year annuity at 10%, compounded annually, what size annuity payment will he receive at the end of each year?

   c. Which is better, taking the $100,000 and the annuity or taking the cash settlement? Discuss the advantages of your choice.

22. *Sale of a Practice* A physician can sell her practice for $1,200,000 cash or for $200,000 plus $250,000 at the end of each year for 5 years.

   a. Find the present value of the annuity that is offered if money is worth 7%, compounded annually.

   b. If she takes the $1,200,000, spends $200,000 of it, and invests the rest in a 5-year annuity at 7%, compounded annually, what size annuity payment will she receive at the end of each year?

   c. Which is better, taking the $200,000 and the annuity or taking the cash settlement? Discuss the advantages of your choice.

23. *Home Mortgage* A couple wants to buy a house and can afford to pay $1600 per month.

   a. If they can get a loan for 30 years with interest at 9% per year on the unpaid balance and make monthly payments, how much can they pay for a house?

   b. What is the total amount paid over the life of the loan?

   c. What is the total interest paid on the loan?

24. *Auto Loan* A man wants to buy a car and can afford to pay $400 per month.

   a. If he can get a loan for 48 months with interest at 12% per year on the unpaid balance and make monthly payments, how much can he pay for a car?

   b. What is the total amount paid over the life of the loan?

   c. What is the total interest paid on the loan?

25. *Loan Repayment* A loan of $10,000 is to be amortized with quarterly payments over 4 years. If the interest on the loan is 8% per year, paid on the unpaid balance,

   a. What is the interest rate charged each quarter on the unpaid balance?

b. How many payments are made to repay the loan?

c. What payment is required quarterly to amortize the loan?

26. *Loan Repayment* A loan of $36,000 is to be amortized with monthly payments over 6 years. If the interest on the loan is 6% per year, paid on the unpaid balance,

   a. What is the interest rate charged each month on the unpaid balance?

   b. How many payments are made to repay the loan?

   c. What payment is required each month to amortize the loan?

27. *Home Mortgage* A couple who wants to purchase a home with a price of $350,000 has $100,000 for a down payment. If they can get a 30-year mortgage at 6% per year on the unpaid balance,

   a. What will be their monthly payments?

   b. What is the total amount they will pay before they own the house outright?

   c. How much interest will they pay over the life of the loan?

28. *Business Loan* Business partners want to purchase a restaurant that costs $750,000. They have $300,000 for a down payment, and they can get a 25-year business loan for the remaining funds at 8% per year on the unpaid balance, with quarterly payments.

   a. What will be the payments?

   b. What is the total amount that they will pay over the 25-year period?

   c. How much interest will they pay over the life of the loan?

5.7   Logistic and Gompertz Functions

KEY OBJECTIVES

- Graph and apply logistic functions
- Find limiting values of logistic functions
- Model with logistic functions
- Graph and apply Gompertz functions

SECTION PREVIEW   iPods

Apple launched the iPod on October 23, 2001, and the iPod line has dominated digital music player sales in the United States ever since. The high rate of sales caused the iPod's U.S. market share to climb from 31% in 2004 to over 90% by 2008. However, with the release of new technology such as the iPad in April 2010, sales of iPods began to slow down. The annual sales of iPods, in thousands, from 2002 to 2010 are shown in Figure 5.47. (See Example 1.)

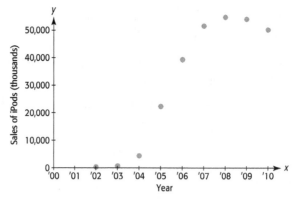

Figure 5.47

    Notice that the graph is an S-shaped curve with a relatively slow start, then a steep climb followed by a leveling off. It is reasonable that sales would eventually level off, as new technology (such as the iPad) is introduced. Very seldom will exponential growth continue indefinitely, so functions of this type better represent many types of sales and organizational growth. Two functions that are characterized by rapid growth followed by leveling off are logistic functions and Gompertz functions. We consider applications of these functions in this section. ∎

Logistic Functions

When growth begins slowly, then increases at a rapid rate, and finally slows over time to a rate that is almost zero, the amount (or number) present at any given time frequently fits on an S-shaped curve. Growth of business organizations and the spread of a virus or disease sometimes occur according to this pattern. For example, the total number y of people on a college campus infected by a virus can be modeled by

$$y = \frac{10{,}000}{1 + 9999e^{-0.99t}}$$

where t is the number of days after an infected student arrives on campus. Table 5.28 shows the number infected on selected days. Note that the number infected grows rapidly and then levels off near 10,000. The graph of this function is shown in Figure 5.48.

Table 5.28

Days	1	3	5	7	9	11	13	15	17
Number Infected	3	19	139	928	4255	8429	9749	9965	9995

A function that can be used to model growth of this type is called a **logistic function**; its graph is similar to the one shown in Figure 5.48.

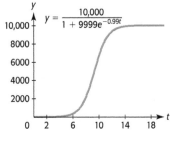

Figure 5.48

Logistic Function

For real numbers a, b, and c, the function

$$f(x) = \frac{c}{1 + ae^{-bx}}$$

is a logistic function.

If $a > 0$, a logistic function increases when $b > 0$ and decreases when $b < 0$.

The parameter c is often called the **limiting value**,* or **upper limit**, because the line $y = c$ is a horizontal asymptote for the logistic function. As seen in Figure 5.49, the line $y = 0$ is also a horizontal asymptote. Logistic growth begins at a rate that is near zero, rapidly increases in an exponential pattern, and then slows to a rate that is again near zero.

Logistic Growth Function

Equation: $y = \dfrac{c}{1 + ae^{-bx}}$ $(a > 0, b > 0)$

x-intercept: none

y-intercept: $\left(0, \dfrac{c}{1 + a}\right)$

Horizontal asymptotes: $y = 0$, $y = c$

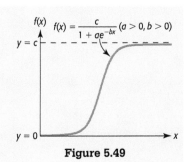

Figure 5.49

Even though the graph of an increasing logistic function has the elongated S-shape indicated in Figure 5.49, it may be the case that we have data showing only a portion of the S.

---

* The logistic function fit by most technologies is a best-fit (least-squares) logistic equation. There may therefore be data values that are larger than the value of c. To avoid confusion, we talk about the limiting value of the logistic function or the upper limit in the problem context.

EXAMPLE 1 ▶ **Sales of iPods**

Table 5.29

Year	iPods Sold
2002	376,000
2003	937,000
2004	4,416,000
2005	22,497,000
2006	39,409,000
2007	51,630,000
2008	54,828,000
2009	54,132,000
2010	50,315,000

(Source: Apple, Inc.)

Apple's iPod line has dominated digital music player sales in the United States ever since 2001. However, with the release of new technology such as the iPad in April 2010, sales of iPods began to slow down. The numbers of iPods sold from 2002 to 2010 are given in Table 5.29.

a. Create a scatter plot of the data, with x equal to the number of years after 2000 and y equal to the number of millions of iPods sold.

b. Find a logistic function to model the data and graph the function with a scatter plot.

c. Use the model to estimate the number of iPods sold in 2015.

d. What is the upper limit for the number of iPods sold, according to the model? Does this seem reasonable?

SOLUTION

a. A scatter plot of the data, with x equal to the number of years after 2000 and y equal to the number of millions of iPods sold, is shown in Figure 5.50(a).

b. A logistic function that models the data is $f(x) = \dfrac{53.35}{1 + 6365e^{-1.664}}$, rounded to two decimal places. (See Figure 5.50(b).) The graphs of the data and the logistic model are shown in Figure 5.50(c).

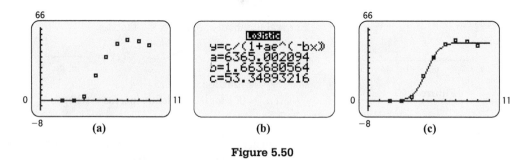

(a)   **(b)**   **(c)**

Figure 5.50

c. According to the model, $f(15) = 53.348927$, so the number of iPods sold in 2015 will be 53,348,927. See Figure 5.51.

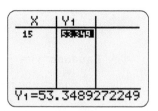

Figure 5.51

d. The upper limit of the function is the value of c in the equation, or 53.348932, which means that the number of iPods sold will never be more than 53,348,932. This sounds reasonable, since new technologies such as the iPad and iPhone have been introduced in the market and have been very successful.

A logistic function can decrease; this behavior occurs when $b < 0$. Demand functions for products often follow such a pattern. Decreasing logistic functions can also be used to represent decay over time. Figure 5.52 shows a decreasing logistic function.

Logistic Decay Function

Equation: $y = \dfrac{c}{1 + ae^{-bx}} (a > 0, b < 0)$

x-intercept: none

y-intercept: $\left(0, \dfrac{c}{1 + a}\right)$

Horizontal asymptotes: $y = 0, y = c$

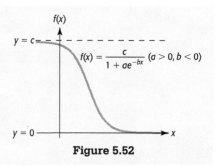

Figure 5.52

The properties of logarithms can also be used to solve for a variable in an exponent in a logistic function.

EXAMPLE 2 ▶ Expected Life Span

The expected life span at birth of people born in the United States can be modeled by the equation

$$y = \frac{81.837}{1 + 0.8213e^{-0.0255x}}$$

where x is the number of years from 1900.

Use this model to estimate the birth year after 1900 that gives an expected life span of 65 years with

a. Graphical methods.

b. Algebraic methods.

SOLUTION

a. To solve the equation

$$65 = \frac{81.837}{1 + 0.8213e^{-0.0255x}}$$

with the *x*-intercept method, we solve

$$0 = \frac{81.837}{1 + 0.8213e^{-0.0255x}} - 65$$

Graphing the function

$$y = \frac{81.837}{1 + 0.8213e^{-0.0255x}} - 65$$

on the window [30, 55] by [−2, 3] shows a point where the graph crosses the *x*-axis (Figure 5.53(a)). Finding the *x*-intercept gives the approximate solution

$$x \approx 45.25$$

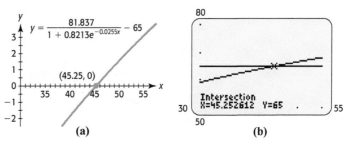

(a)   **(b)**

Figure 5.53

Since $x = 45.25$ represents the number of years from 1900, we conclude that people born in 1946 have an expected life span of 65 years.

We can also use the intersection method to solve the equation graphically. To do this, we graph the equations $y_1 = 65$ and $y_2 = \dfrac{81.837}{1 + 0.8213e^{-0.0255x}}$ and find the point of intersection of the graphs (Figure 5.53(b)).

b. We can also solve the equation

$$65 = \frac{81.837}{1 + 0.8213e^{-0.0255x}}$$

algebraically. To do this, we first multiply both sides by the denominator and then solve for $e^{-0.0255x}$:

$$65(1 + 0.8213e^{-0.0255x}) = 81.837$$
$$65 + 53.3845e^{-0.0255x} = 81.837$$
$$53.3845e^{-0.0255x} = 16.837$$
$$e^{-0.0255x} = \frac{16.837}{53.3845}$$

Taking the natural logarithm of both sides of the equation and using a logarithmic property are the next steps in the solution:

$$\ln e^{-0.0255x} = \ln \frac{16.837}{53.3845}$$

$$-0.0255x = \ln \frac{16.837}{53.3845}$$

$$x = \frac{\ln \dfrac{16.837}{53.3845}}{-0.0255} \approx 45.252612$$

This solution is the same solution that was found (more easily) by using the graphical method in Figure 5.53(a); 1946 is the birth year for people with an expected life span of 65 years.

EXAMPLE 3 ▶   **Miles per Gallon**

The average miles per gallon of gasoline for passenger cars for selected years from 1975 through 2006 are shown in Table 5.30. Find a logistic function that models these data, with x equal to the number of years after 1970. Would this model apply today? Why or why not?

Table 5.30

Year	Miles per Gallon	Year	Miles per Gallon	Year	Miles per Gallon
1975	13.5	1997	24.3	2002	24.5
1980	20.0	1998	24.4	2003	24.7
1985	23.0	1999	24.1	2004	24.7
1990	23.7	2000	24.1	2005	25.0
1995	24.2	2001	24.3	2006	24.6
1996	24.2				

(Source: Environmental Protection Agency)

SOLUTION

The scatter plot of the data is shown in Figure 5.54(a), with y representing miles per gallon and x equal to the number of years after 1970. Technology gives the equation of the best-fitting logistic function as

$$y = \frac{24.458}{1 + 2.805e^{-0.250x}}$$

Figure 5.54(b) shows the graph of this model and the scatter plot on the same axes. The model would not apply today, as sales of unconventional vehicles with better gas mileage, such as hybrids, have increased.

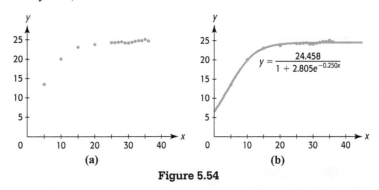

(a)　　　　　　　　(b)

Figure 5.54

Gompertz Functions

Another type of function that models rapid growth that eventually levels off is called a **Gompertz function**. This type of function can also be used to describe human growth and development and the growth of organizations in a limited environment. These functions have equations of the form

$$N = Ca^{R^t}$$

where t represents the time, R $(0 < R < 1)$ is the expected rate of growth of the population, C is the maximum possible number of individuals, and a represents the proportion of C present when $t = 0$.

For example, the equation

$$N = 3000(0.2)^{0.6^t}$$

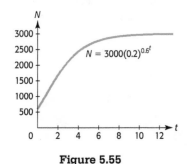

Figure 5.55

could be used to predict the number of employees t years after the opening of a new facility. Here the maximum number of employees C would be 3000, the proportion of 3000 present when $t = 0$ is $a = 0.2$, and R is 0.6. The graph of this function is shown in Figure 5.55. Observe that the initial number of employees, at $t = 0$, is

$$N = 3000(0.2)^{0.6^0} = 3000(0.2)^1 = 3000(0.2) = 600$$

To verify that the maximum possible number of employees is 3000, observe the following. Because $0.6 < 1$, higher powers of t make 0.6^t smaller, with 0.6^t approaching 0 as t approaches ∞. Thus, $N \to 3000(0.2)^0 = 3000(1) = 3000$ as $t \to \infty$.

EXAMPLE 4 ▶ Deer Population

The Gompertz equation

$$N = 1000(0.06)^{0.2^t}$$

predicts the size of a deer herd on a small island t decades from now.

a. What is the number of deer on the island now ($t = 0$)?

b. How many deer are predicted to be on the island 1 decade from now ($t = 1$)?

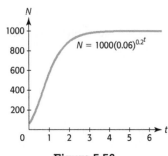

Figure 5.56

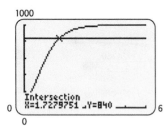

Figure 5.57

c. Graph the function.

d. What is the maximum number of deer predicted by this model?

e. In how many years will there be 840 deer on the island?

SOLUTION

a. If $t = 0$, $N = 1000(0.06)^{0.2^0} = 1000(0.06)^1 = 60$.

b. If $t = 1$, $N = 1000(0.06)^{0.2^1} = 1000(0.06)^{0.2} = 570$ (approximately).

c. The graph is shown in Figure 5.56.

d. According to the model, the maximum number of deer is predicted to be 1000. Evaluating N as t gets large, we see the values of N approach, but never reach, 1000. We say that $N = 1000$ is an asymptote for this graph.

e. To find the number of years from now when there will be 840 deer on the island, we graph $y = 840$ with $y = 1000(0.06)^{0.2^x}$ and find the point of intersection to be (1.728, 840). Thus, after about 17 years there will be 840 deer on the island. (See Figure 5.57.)

EXAMPLE 5 ▶ **Company Growth**

A new dotcom company starts with 3 owners and 5 employees but tells investors that it will grow rapidly, with the total number of people in the company given by the model

$$N = 2000(0.004)^{0.5^t}$$

where t is the number of years from the present. Use graphical methods to determine the year in which the number of employees is predicted to be 1000.

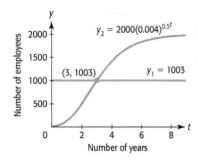

Figure 5.58

SOLUTION

If the company has 1000 employees plus the 3 owners, the total number in the company will be 1003; so we seek to solve the equation

$$1003 = 2000(0.004)^{0.5^t}$$

Using the equations $y_1 = 1003$ and $y_2 = 2000(0.004)^{0.5^t}$ and the intersection method gives $t = 3$ (Figure 5.58). Because t represents the number of years, the owners predict that they will have 1000 employees in 3 years.

Skills CHECK 5.7

Give approximate answers to two decimal places.

1. Evaluate $\dfrac{79.514}{1 + 0.835e^{-0.0298(80)}}$.

2. If $y = \dfrac{79.514}{1 + 0.835e^{-0.0298x}}$ find

   a. y when $x = 10$.

   b. y when $x = 50$.

3. Evaluate $1000(0.06)^{0.2^t}$ for $t = 4$ and for $t = 6$.

4. Evaluate $2000(0.004)^{0.5^t}$ for $t = 5$ and for $t = 10$.

5. a. Graph the function $f(x) = \dfrac{100}{1 + 3e^{-x}}$ for $x = 0$ to $x = 15$.

   b. Find $f(0)$ and $f(10)$.

   c. Is this function increasing or decreasing?

   d. What is the limiting value of this function?

6. a. Graph $f(t) = \dfrac{1000}{1 + 9e^{-0.9t}}$ for $t = 0$ to $t = 15$.

b. Find $f(2)$ and $f(5)$.

c. What is the limiting value of this function?

7. a. Graph $y = 100(0.05)^{0.3^x}$ for $x = 0$ to $x = 10$.

    b. What is the initial value of this function (the y-value when $x = 0$)?

    c. What is the limiting value of this function?

8. a. Graph $N = 2000(0.004)^{0.5^t}$ for $t = 0$ to $t = 10$.

    b. What is the initial value of this function (the y-value when $x = 0$)?

    c. What is the limiting value of this function?

EXERCISES 5.7

9. *Spread of a Disease* The spread of a highly contagious virus in a high school can be described by the logistic function

$$y = \frac{5000}{1 + 999e^{-0.8x}}$$

where x is the number of days after the virus is identified in the school and y is the total number of people who are infected by the virus.

    a. Graph the function for $0 \le x \le 15$.

    b. How many students had the virus when it was first discovered?

    c. What is the upper limit of the number infected by the virus during this period?

10. *Population Growth* Suppose that the size of the population of an island is given by

$$p(t) = \frac{98}{1 + 4e^{-0.1t}} \text{ thousand people}$$

where t is the number of years after 1988.

    a. Graph this function for $0 \le x \le 30$.

    b. Find and interpret $p(10)$.

    c. Find and interpret $p(100)$.

    d. What appears to be an upper limit for the size of this population?

11. *Sexually Active Boys* The percent of boys between ages 15 and 20 who have been sexually active at some time (the cumulative percent) can be modeled by the logistic function

$$y = \frac{89.786}{1 + 4.6531e^{-0.8256x}}$$

where x is the number of years after age 15.

    a. Graph the function for $0 \le x \le 5$.

    b. What does the model estimate the cumulative percent to be for boys whose age is 16?

    c. What cumulative percent does the model estimate for boys of age 21, if it is valid after age 20?

    d. What is the limiting value implied by this model? (Source: "National Longitudinal Survey of Youth," riskingthefuture.com)

12. *Sexually Active Girls* The percent of girls between ages 15 and 20 who have been sexually active at some time (the cumulative percent) can be modeled by the logistic function

$$y = \frac{83.84}{1 + 13.9233e^{-0.9248x}}$$

where x is the number of years after age 15.

    a. Graph the function for $0 \le x \le 5$.

    b. What does the model estimate the cumulative percent to be for girls of age 16?

    c. What cumulative percent does the model estimate for girls of age 20?

    d. What is the upper limit implied by the given logistic model? (Source: "National Longitudinal Survey of Youth," riskingthefuture.com)

13. *Spread of a Rumor* The number of people in a small town who are reached by a rumor about the mayor and an intern is given by

$$N = \frac{10,000}{1 + 100e^{-0.8t}}$$

where t is the number of days after the rumor begins.

    a. How many people will have heard the rumor by the end of the first day?

    b. How many will have heard the rumor by the end of the fourth day?

    c. Use graphical or numerical methods to find the day on which 7300 people in town have heard the rumor.

14. *Sexually Active Girls* The cumulative percent of sexually active girls ages 15 to 20 is given in the table below.

Age (years)	Cumulative Percent Sexually Active
15	5.4
16	12.6
17	27.1
18	44.0
19	62.9
20	73.6

(Source: "National Longitudinal Survey of Youth," riskingthefuture.com)

a. A logarithmic function was used to model this data in Exercises 5.4, problem 33. Find the logistic function that models the data, with x equal to the number of years after age 15.

b. Does this model agree with the model used in Exercise 12?

c. Find the linear function that is the best fit for the data.

d. Graph each function on the same axes with the data points to determine which model appears to be the better fit for the data.

15. *Sexually Active Boys* The cumulative percent of sexually active boys ages 15 to 20 is given in the table.

a. A logarithmic function was used to model this data in Exercises 5.4, problem 34. Create the logistic function that models the data, with x equal to the number of years after age 15.

b. Does this agree with the model used in Exercise 11?

c. Find the linear function that is the best fit for the data.

d. Determine which model appears to be the better fit for the data and graph this function on the same axes with the data points.

Age (years)	Cumulative Percent Sexually Active
15	16.6
16	28.7
17	47.9
18	64.0
19	77.6
20	83.0

(Source: "National Longitudinal Survey of Youth," riskingthefuture.com)

16. *World Internet Usage* The number of worldwide users (in millions) of the Internet for 1995–2007 is given in the following table.

a. Find a logistic function that models the data, with y equal to the number of users in millions and x equal to the number of years after 1990.

b. Use the model to predict the number of users in 2015.

c. What will be the maximum number of worldwide users, according to the model?

d. Graph the aligned data and the model on the same axes. Is the model a good fit?

Year	Users (millions)	Year	Users (millions)
1995	16	2002	598
1996	36	2003	719
1997	70	2004	817
1998	147	2005	1018
1999	248	2006	1093
2000	361	2007	1215
2001	536		

(Source: www.internetworldstats.com)

17. *Births to Unmarried Mothers* The percent of live births to unmarried mothers for selected years from 1970 to 2007 is given in the table below.

Year	Percent	Year	Percent
1970	10.7	2000	33.2
1975	14.3	2003	34.6
1980	18.4	2004	35.8
1985	22.0	2005	36.9
1990	28.0	2006	38.5
1995	32.2	2007	39.7

(Source: *World Almanac and Book of Facts*)

a. Find a logistic function that models the data, with y the percent and x the number of years from 1960.

b. Graph the model and the aligned data on the same axes.

c. What percent does this model predict for 2015?

18. *Internet Usage* The percent of the U.S. population that used the Internet during selected years from 1997 to 2007 is given in the table below.

a. Find the logistic function that models the percent as a function of the years, with x equal to the number of years after 1995.

b. Visually determine if the model is a good fit for the data.

c. What did the model predict would be the percent of Internet users in the United States in 2010?

Year	Percent	Year	Percent
1997	22.2	2003	59.2
2000	44.1	2004	68.8
2001	50.0	2005	68.1
2002	58.0	2007	70.2

(Source: www.internetworldstats.com)

19. *Life Span* The expected life span of people in the United States for certain birth years is given in the table.

Birth Year	Life Span (years)	Year	Life Span (years)
1920	54.1	1994	75.7
1930	59.7	1996	76.1
1940	62.9	1998	76.7
1950	68.2	1999	76.7
1960	69.7	2000	77.0
1970	70.8	2001	77.2
1975	72.6	2002	77.3
1980	73.7	2003	77.5
1987	75.0	2004	77.8
1988	74.9	2005	77.9
1989	75.2	2010	78.1
1990	75.4	2015	78.9
1992	75.8	2020	79.5

(Source: National Center for Health Statistics)

a. Find a logistic function to model the data, with x equal to how many years after 1900 the birth year is.

b. Estimate the expected life span of a person born in the United States in 1955 and a person born in 2006.

c. Find an upper limit for a person's expected life span in the United States, according to this model.

20. *SAT Scores* The average composite SAT scores for entering freshmen at Georgia Southern University are shown in the table below.

Year (x)	1996	1997	1998	1999	2000	2001	2002
SAT Score (y)	967	973	983	987	1008	1028	1052

Year (x)	2003	2004	2005	2006	2007	2008	2009
SAT Score (y)	1056	1080	1098	1104	1108	1111	1106

(Source: Student Information Report System [SIRS])

a. Find a logistic function that models the data, with x equal to the number of years from 1990.

b. Use the model to estimate the composite SAT score for this school in 2015.

21. *Organizational Growth* A new community college predicts that its student body will grow rapidly at first and then begin to level off according to the Gompertz curve with equation $N = 10{,}000(0.4)^{0.2^t}$ students, where t is the number of years after the college opens.

a. What does this model predict the number of students to be when the college opens?

b. How many students are predicted to attend the college after 4 years?

c. Graph the equation for $0 \le t \le 10$ and estimate an upper limit on the number of students at the college.

22. *Organizational Growth* A new technology company started with 6 employees and predicted that its number of employees would grow rapidly at first and then begin to level off according to the Gompertz function

$$N = 150(0.04)^{0.5^t}$$

where t is the number of years after the company started.

a. What does this model predict the number of employees to be in 8 years?

b. Graph the function for $0 \le t \le 10$ and estimate the maximum predicted number of employees.

23. *Sales Growth* The president of a company predicts that sales will increase rapidly after a new product is brought to the market and that the number of units sold monthly can be modeled by $N = 40,000(0.2)^{0.4^t}$, where t represents the number of months after the product is introduced.

  a. How many units will be sold by the end of the first month?

  b. Graph the function for $0 \le t \le 10$.

  c. What is the predicted upper limit on sales?

24. *Company Growth* Because of a new research grant, the number of employees in a firm is expected to grow, with the number of employees modeled by $N = 1600(0.6)^{0.2^t}$, where t is the number of years after the grant was received.

  a. How many employees did the company have when the grant was received?

  b. How many employees did the company have at the end of 3 years after the grant was received?

  c. What is the expected upper limit on the number of employees?

  d. Graph the function.

25. *Company Growth* Suppose that the number of employees in a new company is expected to grow, with the number of employees modeled by $N = 1000(0.01)^{0.5^t}$, where t is the number of years after the company was formed.

  a. How many employees did the company have when it started?

  b. How many employees did the company have at the end of 1 year?

  c. What is the expected upper limit on the number of employees?

  d. Use graphical or numerical methods to find the year in which 930 people are employed.

26. *Sales Growth* The president of a company predicts that sales will increase rapidly after a new advertising campaign, with the number of units sold weekly modeled by $N = 8000(0.1)^{0.3^t}$, where t represents the number of weeks after the advertising campaign begins.

  a. How many units per week will be sold at the beginning of the campaign?

  b. How many units will be sold at the end of the first week?

  c. What is the expected upper limit on the number of units sold per week?

  d. Use graphical or numerical methods to find the first week in which 6500 units will be sold.

27. *Spread of Disease* An employee brings a contagious disease to an office with 150 employees. The number of employees infected by the disease t days after the employees are first exposed to it is given by

$$N = \frac{100}{1 + 79e^{-0.9t}}$$

Use graphical or numerical methods to find the number of days until 99 employees have been infected.

28. *Advertisement* The number of people in a community of 15,360 who are reached by a particular advertisement t weeks after it begins is given by

$$N(t) = \frac{14,000}{1 + 100e^{-0.6t}}$$

Use graphical or numerical methods to find the number of weeks until at least half of the community is reached by this advertisement.

29. *Population Growth* A pair of deer are introduced on a small island, and the population grows until the food supply and natural enemies of the deer on the island limit the population. If the number of deer is

$$N = \frac{180}{1 + 89e^{-0.5554t}}$$

where t is the number of years after the deer are introduced, how long does it take for the deer population to reach 150?

30. *Spread of Disease* A student brings a contagious disease to an elementary school of 1200 students. If the number of students infected by the disease t days after the students are first exposed to it is given by

$$N = \frac{800}{1 + 799e^{-0.9t}}$$

use numerical or graphical methods to find in how many days at least 500 students will be infected.

chapter 5 ▶ SUMMARY

In this chapter, we discussed exponential and logarithmic functions and their applications. We explored the fact that logarithmic and exponential functions are inverses of each other. Many applications in today's world involve exponential and logarithmic functions, including the pH of substances, stellar magnitude, intensity of earthquakes, the growth of bacteria, the decay of radioactive isotopes, and compound interest. Logarithmic and exponential equations are solved by using both algebraic and graphical solution methods.

Key Concepts and Formulas

5.1 Exponential Functions

Exponential function	If a and b are real numbers, $b \neq 1$, then the function $f(x) = a(b^x)$ is an exponential function.
Exponential growth	Whenever the base b of $f(x) = a(b^x)$ is greater than 1 and $a > 0$, the exponential function is increasing and can be used to model exponential growth. The graph of an exponential growth model resembles the graph at the right. The x-axis is a horizontal asymptote, with the graph approaching $-\infty$ on the left.
Exponential decay	Whenever the base b of $f(x) = a(b^{kx})$ is greater than 1 and $a > 0$ and $k < 0$, the exponential function is decreasing and can be used to model exponential decay. The graph of an exponential decay model resembles the graph at the right. The x-axis is a horizontal asymptote, with the graph approaching $+\infty$ on the right. The function $f(x) = a(c^{kx}), a > 0, 0 < c < 1, k > 0$, is also an exponential decay function.
Number e	Many real applications involve exponential functions with the base e. The number e is an irrational number with decimal approximation 2.718281828 (to nine decimal places).

5.2 Logarithmic Functions; Properties of Logarithms

Logarithmic function	For $x > 0, a > 0$, and $a \neq 1$, the logarithmic function with base a is $y = \log_a x$, which is defined by $x = a^y$.

The graph of $y = \log_a x$ for the base $a > 0$ is shown at the right. The y-axis is a vertical asymptote for the graph. |
| **Logarithmic and exponential forms** | These forms are equivalent:

$$y = \log_a x \quad \text{and} \quad x = a^y$$ |

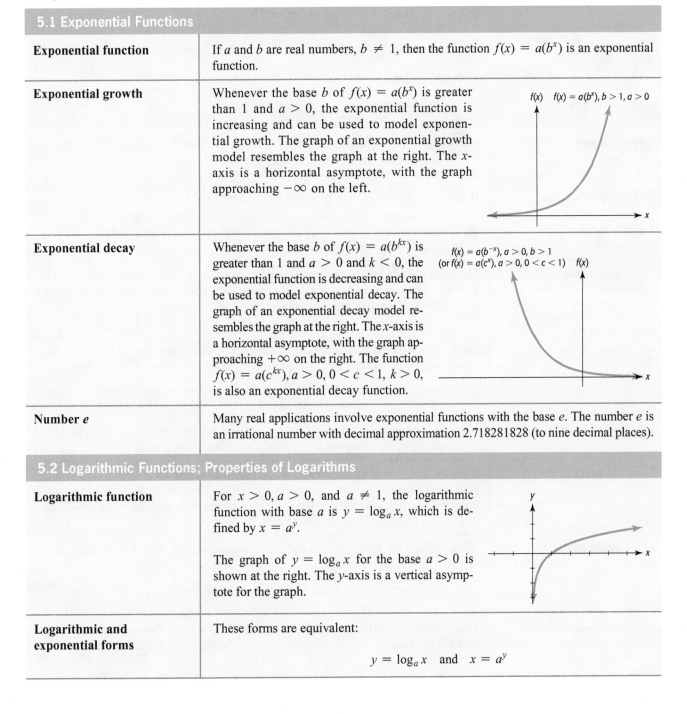

Common logarithms	Logarithms with a base of 10; $\log_{10} x$ is denoted $\log x$.
Natural logarithms	Logarithms with a base of e; $\log_e x$ is denoted $\ln x$.
Properties of logarithms	For $b > 0$, $b \neq 1$, k a real number, and M and N positive real numbers,
	Property 1. $\log_b b = 1$
	Property 2. $\log_b 1 = 0$
	Property 3. $\log_b b^x = x$
	Property 4. $b^{\log_b x} = x$
	Property 5. If $M = N$, then $\log_b M = \log_b N$
	Property 6. $\log_b(MN) = \log_b M + \log_b N$
	Property 7. $\log_b\left(\dfrac{M}{N}\right) = \log_b M - \log_b N$
	Property 8. $\log_b M^k = k \log_b M$

5.3 Exponential and Logarithmic Equations

Solving exponential equations	Many exponential equations can be solved by converting to logarithmic form; that is, $x = a^y$ is equivalent to $y = \log_a x$.
Change of base	If $a > 0$, $a \neq 1$, $b > 0$, $b \neq 1$, and $x > 0$, then $\log_b x = \dfrac{\log_a x}{\log_a b}$.
Solving exponential equations	Some exponential equations can be solved by taking the logarithm of both sides and using logarithmic properties.
Solving logarithmic equations	Many logarithmic equations can be solved by converting to exponential form; that is, $y = \log_a x$ is equivalent to $x = a^y$. Sometimes properties of logarithms can be used to rewrite the equation in a form that can be more easily solved.
Exponential and logarithmic inequalities	Solving inequalities involving exponential and logarithmic functions involves rewriting and solving the related equation. Then graphical methods are used to find the values of the variable that satisfy the inequality.

5.4 Exponential and Logarithmic Models

Exponential models	Many sources provide data that can be modeled by exponential growth and decay functions according to the model $y = ab^x$ for real numbers a and b, $b \neq 1$.
Constant percent change	If the percent change of the outputs of a set of data is constant for equally spaced inputs, an exponential function will be a perfect fit for the data. If the percent change of the outputs is approximately constant for equally spaced inputs, an exponential function will be an approximate fit for the data.
Exponential model	If a set of data has initial value a and has a constant percent change r, the data can be modeled by the exponential function $$y = a(1 + r)^x$$
Comparison of models	To determine which type of model is the best fit for a set of data, we can look at the first differences and second differences of the outputs and the percent change of the outputs.

Logarithmic models	Models can be created involving $\ln x$. Data that exhibit an initial rapid increase and then have a slow rate of growth can often be described by the function $f(x) = a + b \ln x$ for $b > 0$.

5.5 Exponential Functions and Investing

Future value with annual compounding	If P dollars are invested for n years at an interest rate r, compounded annually, then the future value of P is given by $S = P(1 + r)^n$ dollars.
Future value of an investment	If P dollars are invested for t years at the annual interest rate r, where the interest is compounded k times per year, then the interest rate per period is $\dfrac{r}{k}$, the number of compounding periods is kt, and the future value that results is given by $S = P\left(1 + \dfrac{r}{k}\right)^{kt}$ dollars.
Future value with continuous compounding	If P dollars are invested for t years at an annual interest rate r, compounded continuously, then the future value S is given by $S = Pe^{rt}$ dollars.
Present value of an investment	The lump sum P that will give future value S if invested for n periods at interest rate i per compounding period is $$P = \frac{S}{(1 + i)^n}, \quad \text{or} \quad P = S(1 + i)^{-n}$$

5.6 Annuities; Loan Repayment

Future value of an ordinary annuity	If R dollars are contributed at the end of each period for n periods into an annuity that pays interest at rate i at the end of the period, the future value of the annuity is $$S = R\left(\frac{(1 + i)^n - 1}{i}\right)$$
Present value of an ordinary annuity	If a payment of R dollars is to be made at the end of each period for n periods from an account that earns interest at a rate of i per period, then the account is an ordinary annuity, and the present value is $$A = R\left(\frac{1 - (1 + i)^{-n}}{i}\right)$$
Amortization formula	If a debt of A dollars, with interest at a rate of i per period, is amortized by n equal periodic payments made at the end of each period, then the size of each payment is $$R = A\left(\frac{i}{1 - (1 + i)^{-n}}\right)$$

5.7 Logistic and Gompertz Functions

Logistic function	For real numbers a, b, and c, the function $$f(x) = \frac{c}{1 + ae^{-bx}}$$ is a logistic function. A logistic function increases when $a > 0$ and $b > 0$. The graph of a logistic function resembles the graph at right.

Logistic modeling	If data points indicate that growth begins slowly, then increases very rapidly, and finally slows over time to a rate that is almost zero, the data may fit the logistic model $$f(x) = \frac{c}{1 + ae^{-bx}}$$
Gompertz function	For real numbers C, a, and R $(0 < R < 1)$, the function $N = Ca^{R^t}$ is a Gompertz function. A Gompertz function increases rapidly, then levels off as it approaches a limiting value. The graph of a Gompertz function resembles the graph at right.

chapter 5 ▸ SKILLS CHECK

1. a. Graph the function $f(x) = 4e^{-0.3x}$.

    b. Find $f(-10)$ and $f(10)$.

2. Is $f(x) = 4e^{-0.3x}$ an increasing or a decreasing function?

3. Graph $f(x) = 3^x$ on $[-10, 10]$ by $[-10, 20]$.

4. Graph $y = 3^{(x-1)} + 4$ on $[-10, 10]$ by $[0, 15]$.

5. How does the graph of Exercise 4 compare with the graph of Exercise 3?

6. Is $y = 3^{(x-1)} + 4$ an increasing or a decreasing function?

7. a. If $y = 1000(2^{-0.1x})$, find the value of y when $x = 10$.

    b. Use graphical or numerical methods to determine the value of x that gives $y = 250$ if $y = 1000(2^{-0.1x})$.

In Exercises 8 and 9, write the exponential equations in logarithmic form.

8. $x = 6^y$ **9.** $y = 7^{3x}$

In Exercises 10–12, write the logarithmic equations in exponential form.

10. $y = \log_4 x$ **11.** $y = \log(x)$

12. $y = \ln x$

13. Write the inverse of $y = 4^x$ in logarithmic form.

In Exercises 14–16, evaluate the logarithms, if possible. Give approximate solutions to four decimal places.

14. $\log 22$ **15.** $\ln 56$

16. $\log 10$

In Exercises 17–19, find the value of the logarithms without using a calculator.

17. $\log_2 16$ **18.** $\ln(e^4)$

19. $\log 0.001$

In Exercises 20 and 21, use a change of base formula to evaluate each of the logarithms. Give approximate solutions to four decimal places.

20. $\log_3 54$ **21.** $\log_8 56$

In Exercises 22 and 23, graph the functions with technology.

22. $y = \ln(x - 3)$ **23.** $y = \log_3 x$

In Exercises 24–27, solve the exponential equations. Give approximate solutions to four decimal places.

24. $340 = e^x$

25. $1500 = 300e^{8x}$

26. $9200 = 23(2^{3x})$

27. $4(3^x) = 36$

28. Rewrite $\ln\dfrac{(2x - 5)^3}{x - 3}$ as the sum, difference, or product of logarithms, and simplify if possible.

29. Rewrite $6 \log_4 x - 2 \log_4 y$ as a single logarithm.

30. Determine whether a linear or an exponential function is the better fit for the data in the table. Then find the equation of the function with three-decimal-place accuracy.

x	1	2	3	4	5
y	2	5	12	30	75

31. Evaluate $P\left(1 + \dfrac{r}{k}\right)^{kn}$ for $P = 1000$, $r = 8\%$, $k = 12$, $n = 20$, to two decimal places.

32. Evaluate $1000\left(\dfrac{1 - 1.03^{-240+n}}{0.03}\right)$ for $n = 120$, to two decimal places.

33. a. Graph $f(t) = \dfrac{2000}{1 + 8e^{-0.8t}}$ for $t = 0$ to $t = 10$.

    b. Find $f(0)$ and $f(8)$, to two decimal places.

    c. What is the limiting value of this function?

34. a. Graph $y = 500(0.1^{0.2^x})$ for $x = 0$ to $x = 10$.

    b. What is the initial value (the value of y when $x = 0$) of this function?

    c. What is the limiting value of this function?

chapter 5 ▸ REVIEW

35. *Sales of iPads* The iPad has become the most quickly adopted nonphone consumer electronics product in history, topping the DVD player. Apple sold over 300,000 iPads in its first day, and took just 28 days to reach 1 million units sold. Sales of the iPad can be modeled by $y = 0.554(1.455^x)$, where x is the number of months after April 1, 2010 and y is the number of millions of iPads sold. According to this model, how many iPads would be sold by August 2011 ($x = 17$)?

36. *Sales Decay* At the end of an advertising campaign, the daily sales (in dollars) declined, with daily sales given by the equation $y = 2000(2^{-0.1x})$, where x is the number of weeks after the end of the campaign. According to this model, what will sales be 4 weeks after the end of the ad campaign?

37. *Corporate Revenue* Prior to the November 1994 $1.7 billion takeover proposal by Quaker Oats, Snapple Beverage Corporation's annual revenues were given by the function $B(t) = 1.337e^{0.718t}$ million dollars, where t is the number of years after 1985. When does this model indicate that annual revenue was more than $10 million?
(Source: *U.S. News and World Report*, November 14, 1994)

38. *Earthquakes*
    a. If an earthquake has an intensity of 1000 times I_0, what is the magnitude of the earthquake?

    b. An earthquake that measured 6.5 on the Richter scale occurred in Pakistan in February 1991. Express this intensity in terms of I_0.

39. *Earthquakes* On January 23, 2001, an earthquake registering 7.9 hit western India, killing more than 15,000 people. On the same day, an earthquake registering 4.8 hit the United States, killing no one. How much more intense was the Indian earthquake than the American earthquake?

40. *Investments* The number of years needed for an investment of $10,000 to grow to $30,000 when money is invested at 12%, compounded annually, is given by

$$t = \log_{1.12} \frac{30,000}{10,000} = \log_{1.12} 3$$

In how many years will this occur?

41. *Investments* If $1000 is invested at 10%, compounded quarterly, the future value of the investment is given by $S = 1000(2^{x/7})$, where x is the number of years after the investment is made.

    a. Use the logarithmic form of this function to solve the equation for x.

    b. Find when the future value of the investment is $19,504.

42. *Sales Decay* At the end of an advertising campaign, the weekly sales (in dollars) declined, with weekly sales given by the equation $y = 2000(2^{-0.1x})$, where x is the number of weeks after the end of the campaign. In how many weeks will sales be half of the sales at the end of the ad campaign?

43. *Personal Consumption* Using data from the U.S. Bureau of Labor Statistics for selected years from 1998 and projected to 2018, the number of billions of dollars spent for personal consumption in the United States can be modeled by $P = 2969e^{0.051t}$, where t is the number of years after 1985.

a. Is this model one of exponential growth or exponential decay? Explain.

b. Graph this equation with a graphing utility to show the graph from 1985 through the year 2020.

44. *Mobile Home Sales* A company that buys and sells used mobile homes estimates its cost to be given by $C(x) = 2x + 50$ thousand dollars when x mobile homes are purchased. The same company estimates that its revenue from the sale of x mobile homes is given by $R(x) = 10(1.26^x)$ thousand dollars.

a. Combine C and R into a single function that gives the profit for the company when x used mobile homes are bought and sold.

b. Use a graph to find how many mobile homes the company must sell for revenue to be at least $30,000 more than cost.

45. *Sales Decay* At the end of an advertising campaign, the sales (in dollars) declined, with weekly sales given by the equation $y = 40,000(3^{-0.1x})$, where x is the number of weeks after the end of the campaign. In how many weeks will sales be less than half of the sales at the end of the ad campaign?

46. *Carbon-14 Dating* An exponential decay function can be used to model the number of atoms of a radioactive material that remain after a period of time. Carbon-14 decays over time, with the amount remaining after t years given by

$$y = y_0 e^{-0.00012097t}$$

where y_0 is the original amount.

a. If a sample of carbon-14 weighs 100 grams originally, how many grams will be present in 5000 years?

b. If a sample of wood at an archaeological site contains 36% as much carbon-14 as living wood, determine when the wood was cut.

47. *Purchasing Power* If a retired couple has a fixed income of $60,000 per year, the purchasing power of their income (adjusted value of the money) after t years of 5% inflation is given by the equation $P = 60,000e^{-0.05t}$. In how many years will the purchasing power of their income fall below half of their current income?

48. *Investments* If $2000 is invested at 8%, compounded continuously, the future value of the investment after t years is given by $S = 2000e^{0.08t}$ dollars. What is the future value of this investment in 10 years?

49. *Investments* If $3300 is invested for x years at 10%, compounded annually, the future value that will result is $S = 3300(1.10)^x$ dollars. In how many years will the investment result in $13,784.92?

50. *Starbucks* The number of Starbucks stores increased rapidly during 1992–2009, as the following table shows. Find an exponential function that models these data, with x equal to the number of years after 1990 and y equal to the number of U.S. stores.

Year	Number of U.S. Stores	Year	Number of U.S. Stores
1992	113	2001	2925
1993	163	2002	3756
1994	264	2003	4453
1995	430	2004	5452
1996	663	2005	6423
1997	974	2006	7715
1998	1321	2007	9401
1999	1657	2009	11,189
2000	2119		

(Source: starbucks.com)

51. *Students per Computer* The following table gives the average number of students per computer in public schools for the school years that ended in 1985 through 2006.

a. Find an exponential model for the data. Let x be the number of years past 1980.

b. Is this model an exponential growth function or an exponential decay function?

c. How many students per computer in public schools did this model predict for 2010?

Year	Students per Computer	Year	Students per Computer
1985	75	1995	10.5
1986	50	1996	10
1987	37	1997	7.8
1988	32	1998	6.1
1989	25	1999	5.7
1990	22	2000	5.4
1991	20	2001	5.0
1992	18	2002	4.9
1993	16	2004	4.4
1994	14	2006	3.9

(Source: Quality Education Data, Inc., Denver, Colorado)

52. *Cell Phones* The table gives the number of thousands of U.S. cellular telephone subscriberships for 1985–2006.

a. Find the exponential function that is the best fit for the data, with x equal to the number of years after 1980 and y equal to the number of subscriberships in thousands.

b. What does the model estimate for the number of subscriberships in 2007? Is this a good model for the data? Why?

Year	Subscriberships (thousands)	Year	Subscriberships (thousands)
1985	340	1996	44,043
1986	682	1997	55,312
1987	1231	1998	69,209
1988	2069	1999	86,047
1989	3509	2000	109,478
1990	5283	2001	128,375
1991	7557	2002	140,767
1992	11,033	2003	158,722
1993	16,009	2004	182,140
1994	24,134	2005	207,896
1995	33,786	2006	233,000

(Source: Semiannual CTIA Wireless Industry Survey)

53. *World Tourism* The table below shows the receipts (in billions of dollars) for world tourism for 1990–2007. Write an exponential function that models the data, with $x = 0$ in 1980, and predict the receipts in 2015 with the model.

Year	Receipts (billions of $)	Year	Receipts (billions of $)
1990	264	1998	445
1991	278	1999	455
1992	317	2000	473
1993	323	2001	459
1994	356	2002	474
1995	405	2003	524
1996	439	2004	623
1997	443	2007	735

54. *U.S. Internet Users* Historical data on the percent of U.S. residents using the Internet are given in the table below.

a. Find a logarithmic function that models the data, with x equal to the number of years after 1995.

b. Use the model to predict the percent of users in 2015.

c. When will the model no longer be valid? Explain.

Year	Users (percent)	Year	Users (percent)
1997	22.2	2003	59.2
2000	44.1	2004	68.8
2001	50.0	2005	68.1
2002	58.0	2007	70.2

(Source: www.internetworldstats.com)

55. *Corvette Acceleration* The following table shows the times that it takes a 2008 Corvette to reach speeds from 0 mph to 100 mph, in increments of 10 mph after 30 mph. Find a logarithmic function that gives the speed as a function of the time.

Time (sec)	Speed (mph)	Time (sec)	Speed (mph)
1.7	30	5.3	70
2.4	40	6.5	80
3.3	50	7.9	90
4.1	60	9.5	100

56. *Japan's Population* The table gives the population of Japan for selected years from 1984 to 2006.

a. Find the logistic function that models the population N, using an input x equal to the number of years from 1980.

b. Comment on the goodness of fit of the model to the data.

Year	Population (millions)	Year	Population (millions)	Year	Population (millions)
1984	120.235	1992	124.452	1999	126.686
1985	121.049	1993	124.764	2000	126.926
1986	121.672	1994	125.034	2001	127.291
1987	122.264	1995	125.570	2002	127.435
1988	122.783	1996	125.864	2003	127.619
1989	123.255	1997	126.166	2004	127.687
1990	123.611	1998	126.486	2006	127.464
1991	124.043				

(Source: www.jinjapan.org/stat/)

57. *Investment* Suppose $12,500 is invested in an account earning 5% annual interest, compounded continuously. What is the future value in 10 years?

58. *Investments* Find the 7-year future value of an investment of $20,000 placed into an account that pays 6%, compounded annually.

59. *Annuities* At the end of each quarter, $1000 is placed into an account that pays 12%, compounded quarterly. What is the future value of this annuity in 6 years?

60. *Annuities* Find the 10-year future value of an ordinary annuity with a contribution of $1500 at the end of each month, placed into an account that pays 8%, compounded monthly.

61. *Present Value* Find the present value of an annuity that will pay $2000 at the end of each month for 15 years if the interest rate is 8%, compounded monthly.

62. *Present Value* Find the present value of an annuity that will pay $500 at the end of each 6-month period for 12 years if the interest rate is 10%, compounded semiannually.

63. *Loan Amortization* A debt of $2000 with interest at 12%, compounded monthly, is amortized by equal monthly payments for 36 months. What is the size of each payment?

64. *Loan Amortization* A debt of $120,000 with interest at 6%, compounded monthly, is amortized by equal monthly payments for 25 years. What is the size of each monthly payment?

65. *Births to Unmarried Mothers* The percent of live births to unmarried mothers for the years 1970–2007 can be modeled by the logistic function

$$y = \frac{44.742}{1 + 6.870e^{-0.0782x}}$$

where x is the number of years after 1960.

a. Use this model to estimate the percent in 1990 and in 1996.

b. What is the upper limit of the percent of teen mothers who were unmarried, according to this model?

66. *Spread of Disease* A student brings a contagious disease to an elementary school of 2000 students. The number of students infected by the disease is given by

$$n = \frac{1400}{1 + 200e^{-0.5x}}$$

where x is the number of days after the student brings the disease.

a. How many students will be infected in 14 days?

b. How many days will it take for 1312 students to be infected?

67. *Organizational Growth* The president of a new campus of a university predicts that the student body will grow rapidly after the campus is open, with the number of students at the beginning of year t given by

$$N = 4000(0.06^{0.4^{t-1}})$$

a. How many students does this model predict for the beginning of the second year ($t = 2$)?

b. How many students are predicted for the beginning of the tenth year?

c. What is the limit on the number of students that can attend this campus, according to the model?

68. *Sales Growth* The number of units of a new product that were sold each month after the product was introduced is given by

$$N = 18,000(0.03^{0.4^t})$$

where t is the number of months.

a. How many units were sold 10 months after the product was introduced?

b. What is the limit on sales if this model is accurate?

69. *Endangered Species* The following table gives the numbers of species of plants that were endangered in various years from 1980 to 2006.

a. Find the logistic function that models these data. Use x as the number of years past 1975.

b. Use the model to predict the number of species of endangered plants in 2012.

c. When does the model predict that 627 plant species will be endangered?

Year	Endangered Plant Species
1980	50
1985	93
1990	179
1995	432
1998	567
1999	581
2000	593
2001	595
2002	596
2003	599
2006	598

(Source: U.S. Fish and Wildlife Service)

Group Activities
▶ EXTENDED APPLICATIONS

---

1. Chain Letters

Suppose you receive a chain letter asking you to send $1 to the person at the top of the list of six names, then to add your name to the bottom of the list, and finally to send the revised letter to six people. The promise is that when the letters have been sent to each of the people above you on the list, you will be at the top of the list and you will receive a large amount of money. To investigate whether this is worth the dollar that you are asked to spend, create the requested models and answer the following questions.

1. Suppose each of the six people on the original list sent six letters. How much money would the person on the top of the list receive?
2. Consider the person who is second on the original list. How much money would this person receive?
3. Complete the partial table of amounts that will be sent to the person on the top of the list during each "cycle" of the chain letter.

Cycle Number (after original six names)	Money Sent to Person on Top of List ($)
1	6 × 6 = 36
2	6 × 36 = 216
3	
4	
5	

4. If you were going to start such a chain letter (don't, it's illegal), would you put your name at the top of the first 6 names or as number 5?
5. Find the best quadratic, power, and exponential models for the data in the table. Which of these function types gives the best model for the data?
6. Use the exponential function found in part (5) to determine the amount of money the person who was at the bottom of the original list would receive if all people contacted sent the $1 and mailed the letter to 6 additional people.
7. How many people will have to respond to the chain letter for the sixth person on the original list to receive all the money that was promised?
8. If you receive the letter in its tenth cycle, how many other people have been contacted, along with you, assuming that everyone who receives the letter cooperates with its suggestions?
9. Who remains in the United States for you to send your letter to, if no one sends a letter to someone who has already received it?
10. Why do you think the federal government has made it illegal to send chain letters in the U.S. mail?

---

2. Modeling

Your mission is to find real-world data for company sales, a stock price, biological growth, or some sociological trend over a period of years that fit an exponential, logistic, or logarithmic function. To do this, you can look at a statistical graph called a histogram, which describes the situation, or at a graph or table describing it. Once you have found data consisting of numerical values, you are to find the equation that is the best exponential, logistic, or logarithmic fit for the data and then write the model that describes the relationship. Some helpful steps for this process are given in the instructions below.

1. Look in newspapers, in periodicals, on the Internet, or in statistical abstracts for a graph or table of data containing at least six data points. The data must have numerical values for the independent and dependent variables.

2. If you decide to use a relation determined by a graph that you have found, read the graph very carefully to determine the data points or (for full credit) contact the source of the data to obtain the data from which the graph was drawn.
3. If you have a table of values for different years, create a graph of the data to determine the type of function that will be the best fit for the data. To do this, first align the independent variable by letting the input represent the number of years after some convenient year and then enter the data into your graphing calculator and draw a scatter plot. Check to see if the points on the scatter plot lie near some curve that could be described by an exponential, logistic, or logarithmic function. If not, save the data for possible later use and renew your search. You can also test to

see if the percent change of the outputs is nearly constant for equally spaced inputs. If so, the data can be modeled by an exponential function.

4. Use your calculator to create the equation of the function that is the best fit for the data. Graph this equation and the data points on the same axes to see if the equation is reasonable.

5. Write some statements about the data that you have been working with. For example, describe how the quantity is increasing or decreasing over periods of years, why it is changing as it is, or when this model is no longer appropriate and why.

6. Your completed project should include the following:

a. A proper bibliographical reference for the source of your data.

b. An original copy or photocopy of the data being used.

c. A scatter plot of the data and reasons why you chose the model that you did to fit it.

d. The equation that you have created, labeled with appropriate units of measure and variable descriptions.

e. A graph of the scatter plot and the model on the same axes.

f. One or more statements about how you think the model you have created could actually be used. Include any restrictions that should be placed on the model.

Basic Calculator Guide

Operating the TI-83 Plus and TI-84 Plus

Turning the Calculator On and Off

ON		Turns the calculator on
2ND	ON	Turns the calculator off

Adjusting the Display Contrast

2ND	▲	Increases the contrast (darkens the screen)
2ND	▼	Decreases the contrast (lightens the screen)

Note: If the display begins to dim (especially during calculations) and you must adjust the contrast to 8 or 9 in order to see the screen, then batteries are low and you should replace them soon.

The TI-83, TI-84 Plus keyboards are divided into four zones: graphing keys, editing keys, advanced function keys, and scientific calculator keys (Figure 1).

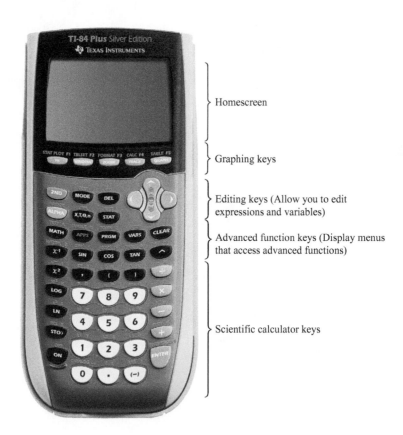

Figure 1

Keystrokes on the TI-83, TI-83 Plus, TI-84 Plus, and TI-Nspire

ENTER	Executes commands or performs a calculation
2ND	Pressing the 2ND key *before* another key accesses the character located above the key and printed in blue (yellow on the TI-83)
ALPHA	Pressing the ALPHA key *before* another key accesses the character located above the key and printed in green
2ND ALPHA	Locks in the ALPHA keyboard
CLEAR	Pressing CLEAR once clears the line
	Pressing CLEAR twice clears the screen
2ND MODE	Returns to the homescreen
DEL	Deletes the character at the cursor
2ND DEL	Inserts characters at the underline cursor
X,T,Θ,n	Enters an X in Function mode, a T in Parametric mode, a θ in Polar mode, or an n in Sequence mode
STO ▶	Stores a value to a variable
^	Raises to an exponent
2ND ^	The number π
(–)	Negative symbol
MATH 1	Converts a rational number to a fraction
MATH ▶ 1: abs(	Computes the absolute value of the number or expression in parentheses
2ND ENTER	Recalls the last entry
ALPHA ·	Used to enter more than one expression on a line
2ND (–)	Recalls the most recent answer to a calculation
x^2	Squares a number or an expression
x^{-1}	Inverse; can be used with a real number or a matrix
2ND x^2	Computes the square root of a number or an expression in parentheses
2ND LN	Returns the constant e raised to a power
ALPHA 0	Space
2ND ◀	Moves the cursor to the beginning of an expression
2ND ▶	Moves the cursor to the end of an expression

Creating Scatter Plots

Clearing Lists

Press STAT and under EDIT press 4:ClrList (Figure 2). Press 2ND 1 (L1) ENTER to clear List 1. Repeat the line by pressing 2ND ENTER and moving the cursor to L1. Press 2ND 2 (L2) and ENTER to clear List 2 (Figure 3).

Another way to clear a list in the STAT menu is, after accessing the list, to press the up arrow until the name of the list is highlighted and then press CLEAR and ENTER. *Caution:* Do not press the DEL key. This will not actually delete the list, but hide it from view. To view the list, press 2ND DEL (INS) and type the list name.

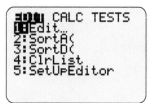

Figure 2

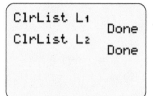

Figure 3

Entering Data into Lists

Press STAT and under EDIT press 1:Edit. This will bring you to the screen where you enter data into lists.

Enter the *x*-values (input) in the column headed L1 and the corresponding *y*-values (output) in the column headed L2 (Figure 4).

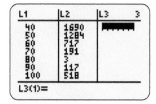

Figure 4

Create a Scatter Plot of Data Points

Go to the Y= menu and turn off or clear any functions entered there. To turn off a function, move the cursor over the = sign and press ENTER.

Press 2ND Y=, 1:Plot 1, highlight On, and then highlight the first graph type (Scatter Plot). Enter Xlist:L1, Ylist:L2, and pick the point plot mark you want (Figure 5).

Choose an appropriate window for the graph and press GRAPH or ZOOM, 9:ZoomStat to plot the data points (Figure 6).

Figure 5

Tracing Along the Plot

Press TRACE and the right arrow to move from point to point. The *x*- and *y*-coordinates will be displayed at the bottom of the screen.

In the upper left-hand corner of the screen, P1:L1,L2 will be displayed. This tells you that you are tracing along the scatter plot (Figure 7).

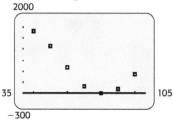

Figure 6

Turning Off Stat Plot

After creating a scatter plot, turn Stat Plot 1 off by pressing Y=, moving up to Plot 1 with the cursor, and pressing ENTER. Stat Plot 1 can be turned on by pressing ENTER with the cursor on Plot 1. It will be highlighted when it is on.

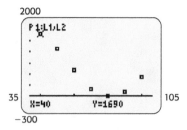

Figure 7

Graphing Equations

Setting Windows

The window defines the highest and lowest values of *x* and *y* on the graph of the function that will be shown on the screen. To set the window manually, press the WINDOW key and enter the values that you want (Figure 8).

The values that define the viewing window can also be set by using ZOOM keys (Figure 9). Frequently the standard window (ZOOM 6) is appropriate. The standard window sets values Xmin=−10, Xmax=10, Ymin=−10, Ymax=10. Often a decimal or integer viewing window (ZOOM 4 or ZOOM 8) gives a better representation of the graph.

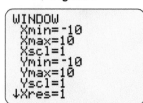

Figure 8

The window should be set so that the important parts of the graph are shown and the unseen parts are suggested. Such a graph is called complete.

Graphing Equations

To graph an equation in the variables x and y, first solve the equation for y in terms of x. If the equation has variables other than x and y, solve for the dependent variable and replace the independent variable with x.

Press the $\boxed{Y=}$ key to access the function entry screen. You can input up to ten functions in the $\boxed{Y=}$ menu (Figure 10). Enter the equation, using the $\boxed{X,T,\Theta,n}$ key to input x. Use the subtraction key $\boxed{-}$ between terms. Use the negation key $\boxed{(-)}$ when the first term is negative. Use parentheses as needed so that what is entered agrees with the order of operations.

The $=$ sign is highlighted to show that Y1 is *on* and ready to be graphed. If the $=$ sign is not highlighted, the equation will remain but its graph is "turned off" and will not appear when $\boxed{GRAPH}$ is pressed. (The graph is "turned on" by repeating the process.)

To erase an equation, press $\boxed{CLEAR}$. To return to the homescreen, press $\boxed{2ND}$ $\boxed{MODE}$ (QUIT).

Determine an appropriate viewing window (Figures 11 and 12).

Pressing $\boxed{GRAPH}$ or a $\boxed{ZOOM}$ key will activate the graph.

Finding Function Values

Using TRACE, VALUE Directly on the Graph

Enter the function to be evaluated in Y1. Choose a window that contains the x-value whose y-value you seek (Figure 13).

Turn all Stat Plots off.

Press $\boxed{TRACE}$ and then enter the selected x-value followed by $\boxed{ENTER}$. The cursor will move to the selected value and give the resulting y-value if the selected x-value is in the window.

If the selected x-value is not in the window, ERR: INVALID will occur.

If the x-value is in the window, the y-value will be given even if it is not in the window (Figures 14 and 15).

Figure 9

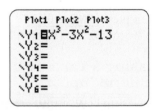

Figure 10 The function $y = x^3 - 3x^2 - 13$ is entered in Y1.

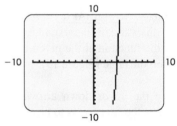

Figure 11 The graph of the function using the standard window.

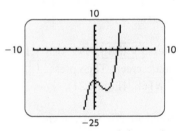

Figure 12 A better graph using the window $[-10, 10]$ by $[-25, 10]$.

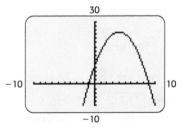

Figure 13 To evaluate $y = -x^2 + 8x + 9$ when $x = 3$ and when $x = -5$, graph the function using the window $[-10, 10]$ by $[-10, 30]$.

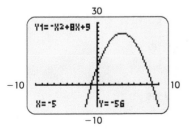

Figure 14

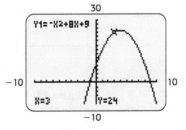

Figure 15

Using the TABLE ASK Feature

Enter the function with the $\boxed{Y=}$ key. (Note: The = sign must be highlighted.) Press $\boxed{2ND}$ $\boxed{WINDOW}$ (TBLSET), move the cursor to Ask opposite Indpnt:, and press $\boxed{ENTER}$ (Figure 16). This allows you to input specific values for x. Pressing $\boxed{DEL}$ will clear entries in the table.

Then press $\boxed{2ND}$ $\boxed{GRAPH}$ (TABLE) and enter the specific values (Figure 17).

Making a Table of Values with Uniform Inputs

Press $\boxed{2ND}$ $\boxed{WINDOW}$ (TBLSET), enter an initial x-value in the table (TblStart), and enter the desired change in the x-value in the table (ΔTbl) (Figure 18).

Enter $\boxed{2ND}$ $\boxed{GRAPH}$ (TABLE) to get the list of x-values and the corresponding y-values. The value of the function at the given value of x can be read from the table (Figure 19).

Use the up or down arrow to find the x-values where the function is to be evaluated (Figure 20).

Using Y-VARS

Use the $\boxed{Y=}$ key to store $Y1 = f(x)$. Press $\boxed{2ND}$ (QUIT).

Press $\boxed{VARS}$, Y-VARS 1,1 to display Y1 on the homescreen. Then press $\boxed{(}$, the x-value, $\boxed{)}$, and $\boxed{ENTER}$ (Figure 21).

Alternatively, you can enter the x-values needed as follows: Y1 ({value 1, value 2, etc.}) $\boxed{ENTER}$ (Figure 22). Values of the function will be displayed.

Finding Intercepts of Graphs

Solve the equation for y. Enter the equation with the $\boxed{Y=}$ key.

Finding the y-Intercept

Press $\boxed{TRACE}$ and enter the value 0. The resulting y-value is the y-intercept of the graph (Figure 23).

Finding the x-Intercept(s)

Set the window so that the intercepts to be located can be seen. The graph of a linear equation will cross the x-axis at most one time; the graph of a quadratic equation will cross the x-axis at most two times; etc.

Figure 16

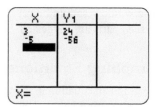

Figure 17

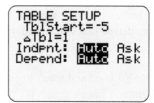

Figure 18

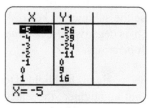

Figure 19

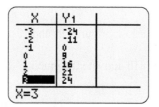

Figure 20

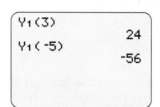

Figure 21

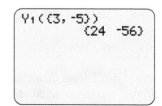

Figure 22

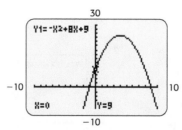

Figure 23 The y-intercept of the graph of the function $y = -x^2 + 8x + 9$ is 9.

To find the point(s) where the graph crosses the *x*-axis, press 2ND TRACE to access the CALC menu. Select 2:zero (Figure 24).

Answer the question "Left Bound?" with ENTER after moving the cursor close to and to the left of an *x*-intercept (Figure 25).

Answer the question "Right Bound?" with ENTER after moving the cursor close to and to the right of this *x*-intercept (Figure 26).

To answer the question "Guess?" press ENTER.

The coordinates of the *x*-intercept will be displayed (Figure 27). Repeat to get all *x*-intercepts (Figure 28). (Some of these values may be approximate.)

Finding Maxima and Minima

To locate a local maximum on the graph of an equation, graph the function using a window that shows all the turning points (Figure 29).

Press 2ND TRACE to access the CALC menu (Figure 30). Select 4:maximum. The calculator will ask the question "Left Bound?" (Figure 31). Move the cursor close to and to the left of the maximum and press ENTER. Next the calculator will ask the question "Right Bound?" (Figure 32). Move the cursor close to and to the right of the maximum and press ENTER. When the calculator asks "Guess?" move the cursor close to the maximum and press ENTER (Figure 33). The coordinates of the maximum will be displayed (Figure 34).

Note: When selecting a left bound or a right bound, you may also type in a value of *x* if you wish.

To locate a local minimum on the graph of an equation, press 2ND TRACE to access the CALC menu. Select 3:minimum, and follow the same steps as above (Figure 35).

Solving Equations Graphically

The *x*-Intercept Method

To solve an equation graphically using the *x*-intercept method, first rewrite the equation with 0 on one side of the equation. Using the Y= key, enter the nonzero side of the equation equal to Y1. Then graph the equation with a window that shows all points where the graph crosses the *x*-axis.

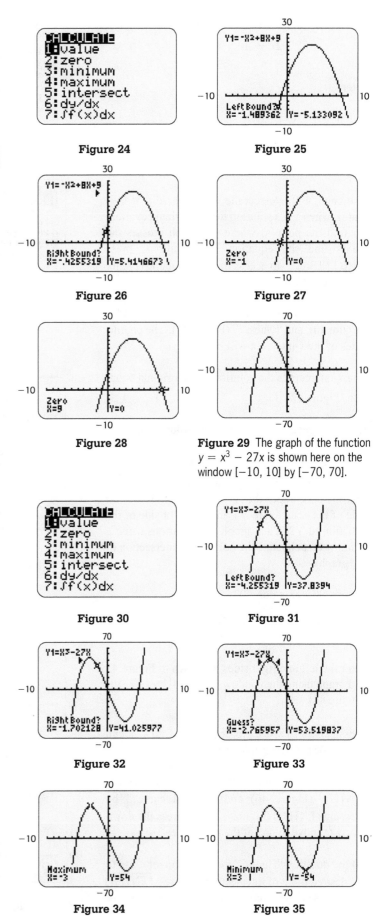

Figure 24   **Figure 25**

Figure 26   **Figure 27**

Figure 28

Figure 29 The graph of the function $y = x^3 - 27x$ is shown here on the window [−10, 10] by [−70, 70].

Figure 30   **Figure 31**

Figure 32   **Figure 33**

Figure 34   **Figure 35**

For example, to solve the equation

$$\frac{2x - 3}{4} = \frac{x}{3} + 1$$

enter

$$Y1 = \frac{2x - 3}{4} - \frac{x}{3} - 1$$

and graph using the window [−10, 15] by [−5, 5] (Figures 36 and 37).

The graph will intersect the x-axis where $y = 0$— that is, when x is a solution to the original equation. To find the point(s) where the graph crosses the x-axis and the equation has solutions, press ⎡2ND⎤ ⎡TRACE⎤ to access the CALC menu (Figure 38). Select 2:zero.

Select "Left Bound?" (Figure 39), "Right Bound?" (Figure 40), and "Guess?" (Figure 41). The coordinates of the x-intercept will be displayed.

The x-intercept is the solution to the original equation (Figure 42).

Repeat to get all x-intercepts (and solutions).

The Intersection Method

To solve an equation graphically using the intersection method, under the Y= menu, assign the left side of the equation to Y1 and the right side of the equation to Y2. Then graph the equations using a window that contains the point(s) of intersection of the graphs.

To solve the equation

$$\frac{2x - 3}{4} = \frac{x}{3} + 1$$

enter the left and right sides as shown in Figure 43 and graph using the window [−10, 20] by [−3, 10] (Figure 44).

Press ⎡2ND⎤ ⎡TRACE⎤ to access the CALC menu (Figure 45). Select 5:intersect.

Answer the question "First curve?" by pressing ⎡ENTER⎤ (Figure 46) and "Second curve?" by pressing ⎡ENTER⎤ (Figure 47). (Or press the down arrow to move to one of the two curves.)

To the question "Guess?" move the cursor close to the desired point of intersection and press ⎡ENTER⎤ (Figure 48). The coordinates of the point of

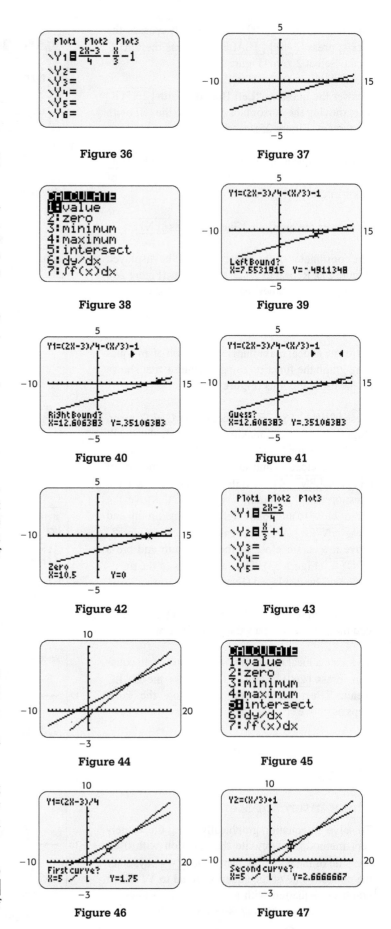

Figure 36

Figure 37

Figure 38

Figure 39

Figure 40

Figure 41

Figure 42

Figure 43

Figure 44

Figure 45

Figure 46

Figure 47

intersection will be displayed. Repeat to get all points of intersection.

The solution(s) to the equation will be the values of x from the points of intersection (Figure 49).

Solving Systems of Equations

To find the solution of a system of equations graphically, first solve each equation for y and use the $\boxed{Y=}$ key with Y1 and Y2 to enter the equations. Graph the equation with an appropriate window.

For example, to solve the system

$$\begin{cases} 4x + 3y = 11 \\ 2x - 5y = -1 \end{cases}$$

graphically, we solve for y:

$$y_1 = \frac{11}{3} - \left(\frac{4}{3}\right)x$$

$$y_2 = \left(\frac{2}{5}\right)x + \frac{1}{5}$$

We graph using ZOOM 4 and then Intersect under the CALC menu to find the point of intersection. If the two lines intersect in one point, the coordinates give the x- and y-values of the solution (Figure 50). The solution of the system above is $x = 2$, $y = 1$. If the two lines are parallel, there is no solution; the system of equations is inconsistent.

For example, to solve the system

$$\begin{cases} 4x + 3y = 4 \\ 8x + 6y = 25 \end{cases}$$

we solve for y (Figure 51) and obtain the graph in Figure 52. This system has no solution. Note that if the lines are parallel, then when we solve for y the equations will show that the lines have the same slope and different y-intercepts.

If the two graphs of the equations give only one line, every point on the line gives a solution to the system and the system is dependent. For example, to solve the system

$$\begin{cases} 2x + 3y = 6 \\ 4x + 6y = 12 \end{cases}$$

we solve for y (Figure 53) and obtain the graph in Figure 54.

This system has many solutions and is dependent. Note that the two graphs will be the same graph if, when we solve for y to use the graphing calculator, the equations are equivalent.

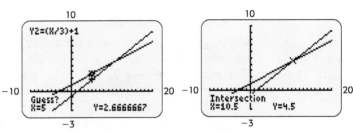

Figure 48

Figure 49 The solution to the equation is $x = 10.5$.

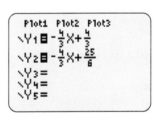

Figure 50

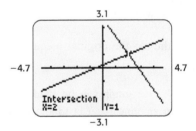

Figure 51

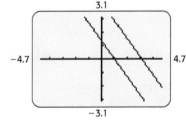

Figure 52

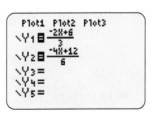

Figure 53

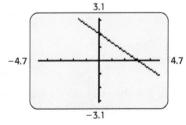

Figure 54

Solving Inequalities Graphically

To solve a linear inequality graphically, first rewrite the inequality with 0 on the right side and simplify. Then, under the Y= menu, assign the left side of the inequality to Y1, so that $Y1 = f(x)$. For example, to solve $3x > 6 + 5x$, rewrite the inequality as $3x - 5x - 6 > 0$ or $-2x - 6 > 0$. Set $Y1 = -2x - 6$.

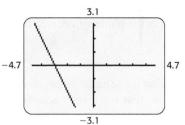

Figure 55

Graph this equation so that the point where the graph crosses the x-axis is visible. Note that the graph will cross the axis in at most one point because the function is of degree 1. (Using ZOOM OUT can help you find this point.)

Using ZOOM 4, the graph of $Y1 = -2x - 6$ is shown in Figure 55.

Use the ZERO command under the CALC menu to find the x-value where the graph crosses the x-axis (the x-intercept). This value can also be found by finding the solution to $0 = f(x)$ algebraically.

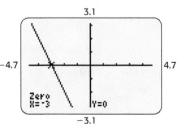

Figure 56

The x-intercept of the graph of $Y1 = -2x - 6$ is $x = -3$ (Figure 56).

Observe the inequality as written with 0 on the right side. If the inequality is $<$, the solution to the original inequality is the interval (bounded by the x-intercept) where the graph is below the x-axis. If the inequality is $>$, the solution to the original inequality is the interval (bounded by the x-intercept) where the graph is above the x-axis.

The solution to $-2x - 6 > 0$, and thus to the original inequality $3x > 6 + 5x$, is $x < -3$.

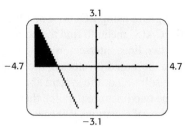

The region above the x-axis and under the graph can be shaded by pressing 2ND PRGM (DRAW), 7:Shade(, and then entering (0, Y1) to shade on the homescreen (Figure 57).

Figure 57

The x-interval where the shading occurs is the solution.

Piecewise-Defined Functions

A piecewise-defined function is defined differently over two or more intervals.

To graph a piecewise-defined function

$$y = \begin{cases} f(x) & \text{if } x \le a \\ g(x) & \text{if } x > a \end{cases}$$

press Y= and enter

$$Y1 = (f(x))/(x \le a)$$

and

$$Y2 = (g(x))/(x > a)$$

Note that the inequality symbols are found by pressing 2ND MATH to access the TEST menu (Figure 58).

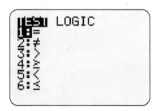

Figure 58

To illustrate, we will enter the function

$$f(x) = \begin{cases} x + 7 & \text{if } x \le -5 \\ -x + 2 & \text{if } x > -5 \end{cases}$$

(Figure 59) and graph using an appropriate window (Figure 60).

Evaluating a piecewise-defined function at a given value of x requires that the correct equation ("piece") be selected. For example, $f(-6)$ and $f(3)$ are shown in Figures 61 and 62, respectively.

Modeling—Regression Equations

Finding an Equation That Models a Set of Data Points

To illustrate, we will find the equation of the line that is the best fit for the average daily numbers of inmates in the Beaufort County, South Carolina, Detention Center for the years 1993–1998, given by the data points (1993, 96), (1994, 109), (1995, 119), (1996, 116), (1997, 137), (1998, 143).

Press $\boxed{\text{STAT}}$ and under EDIT press 1:Edit. Enter the x-values (inputs) in the column headed L1 and the corresponding y-values (outputs) in the column headed L2 (Figure 63).

Press $\boxed{\text{2ND}}$ $\boxed{Y=}$ 1:Plot 1, highlight On, and then highlight the first graph type. Enter Xlist:L1, Ylist:L2, and pick the point plot mark you want (Figure 64).

Press $\boxed{\text{GRAPH}}$ with an appropriate window or $\boxed{\text{ZOOM}}$, 9:ZoomStat to plot the data points (Figure 65).

Observe the point plots to determine what type of function would best model the data.

The graph looks like a line, so use the linear model, with LinReg.

Press $\boxed{\text{STAT}}$, move to CALC, and select the function type to be used to model the data (Figure 66). Enter L1 for Xlist, L2 for Ylist, and for Store RegEQ press the $\boxed{\text{VARS}}$ key, go over to Y-VARS, and select 1:Function and 1:Y1 (Figure 67). Press $\boxed{\text{ENTER}}$. The coefficients of the equation will appear on the screen (Figure 68), and the linear regression equation will appear as Y1 on the Y= screen (see Figure 69).

To see how well the equation models the data, press $\boxed{\text{GRAPH}}$. If the graph does not fit the points well, another type function may be used to model the data.

For some models, r is a diagnostic value that gives the **correlation coefficient**, which determines the strength of the relationship between the independent and dependent variables. To display this value,

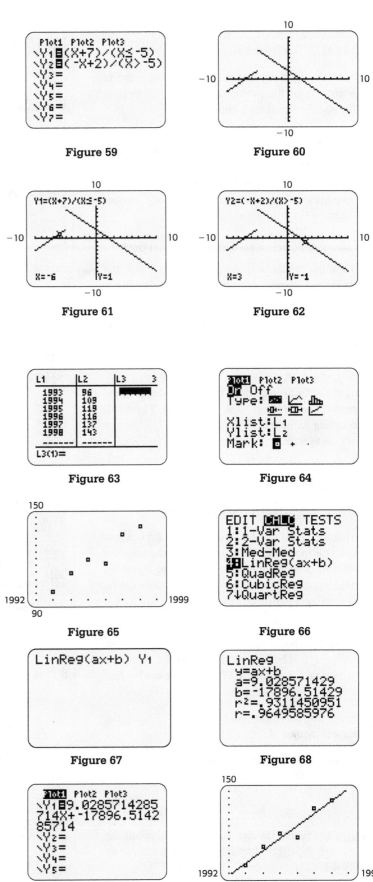

Figure 59

Figure 60

Figure 61

Figure 62

Figure 63

Figure 64

Figure 65

Figure 66

Figure 67

Figure 68

Figure 69

Figure 70

select DiagnosticOn after pressing 2ND 0 (CATALOG) (Figure 68).

Report the equation in a way that makes sense in the context of the problem, with the appropriate units and the variables identified.

The equation that models the average daily number of inmates in the Beaufort County, South Carolina, Detention Center (see page 625) is $y = 9.0286x - 17,896.5143$ (rounded to four decimal places), where x is the year and y is the number of inmates (Figures 69 and 70).

Using a Model to Find an Output

To use the model to find output values inside the data range (*interpolation*) or outside the data range (*extrapolation*), evaluate the function at the desired input value. This may be done using TABLE (Figure 71), Y-VARS (Figure 72), or TRACE (Figure 73). (Note that you may have to change the window to see the x-value to which you are tracing.)

To predict the average daily inmate population in the Beaufort County jail in 2008, we compute Y1 (2008). According to the model, approximately 233 inmates were, on average, in Beaufort County jail in 2008.

When TRACE is used, the window must be changed so that $x = 2008$ is visible (Figure 73).

Using a Model to Find an Input

To use the model to estimate an input for a given output, set the model function equal to Y1 and the desired output equal to Y2 and solve the resulting equations for the input variable. An approximate solution can be found using TABLE or 2ND TRACE (CALC), 5:Intersect.

To use the model to determine in what year the population will be 260, we must solve the equations Y1 = $f(x)$, Y2 = 260. TABLE may be used to find the x-value when y is approximately 260 (Figure 74).

We can also solve the equation Y1 = Y2 using the intersect method (Figures 75 and 76).

According to the model, the population of the jail will reach 260 in the year 2011.

Complex Numbers

Calculations that involve complex numbers of the form $a + bi$ will be displayed as complex numbers only if MODE is set to a + bi (Figure 77).

Complex numbers can be added (Figure 78), subtracted (Figure 79), multiplied (Figure 80), divided

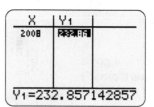

Figure 71 Using TABLE.

Figure 72 Using Y-VARS.

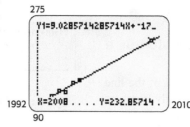

Figure 73 Using TRACE.

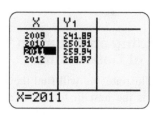

Figure 74

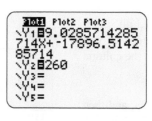

Figure 75

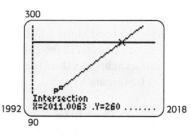

Figure 76

Figure 77

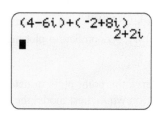

Figure 78

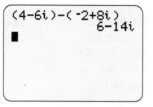

Figure 79

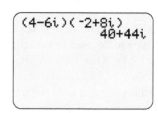

Figure 80

Figure 81

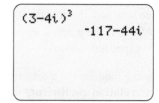

Figure 82

(Figure 81), and raised to powers (Figure 82). The number i is accessed by pressing [2ND] [.].

Care must be taken when writing the quotient of two complex numbers. The answer in Figure 81 appears to have i in the denominator; however, the form of the number is $a + bi$, so the answer is actually $-\dfrac{14}{17} - \dfrac{5}{17}i$.

Operations with Functions

Combinations of Functions

To find the graphs of combinations of two functions $f(x)$ and $g(x)$, enter $f(x)$ as Y1 and $g(x)$ as Y2 under the Y= menu. Consider the example $f(x) = 4x - 8$ and $g(x) = x^2$. To graph $(f + g)(x)$, enter Y1 + Y2 as Y3 under the Y= menu (Figure 83). Place the cursor on the = sign beside Y1 and press [ENTER] to turn off the graph of Y1. Repeat with Y2. Press [GRAPH] with an appropriate window (Figure 84).

To graph $(f - g)(x)$, enter Y1 − Y2 as Y3 under the Y= menu (Figure 85). Place the cursor on the = sign beside Y1 and press [ENTER] to turn off the graph of Y1. Repeat with Y2. Press [GRAPH] (Figure 86).

To graph $(f * g)(x)$, enter Y1*Y2 as Y3 under the Y = menu (Figure 87). Place the cursor on the = sign beside Y1 and press [ENTER] to turn off the graph of Y1. Repeat with Y2. Press [GRAPH] (Figure 88).

To graph $(f/g)(x)$, enter Y1/Y2 as Y3 under the Y = menu (Figure 89). Place the cursor on the = sign beside Y1 and press [ENTER] to turn off the graph of Y1. Repeat with Y2. Press [GRAPH] with an appropriate window (Figure 90).

To evaluate $f + g, f - g, f * g,$ or f/g at a specified value of x, enter Y1, the value of x enclosed in parentheses, the operation to be performed, Y2, and the value of x enclosed in parentheses, and press [ENTER] (Figure 91). Or, if the combination of functions is entered as Y3, enter Y3 and the value of x enclosed in parentheses (Figure 92), or use TABLE.

The value of $(f + g)(3)$ is shown in Figures 91 and 92.

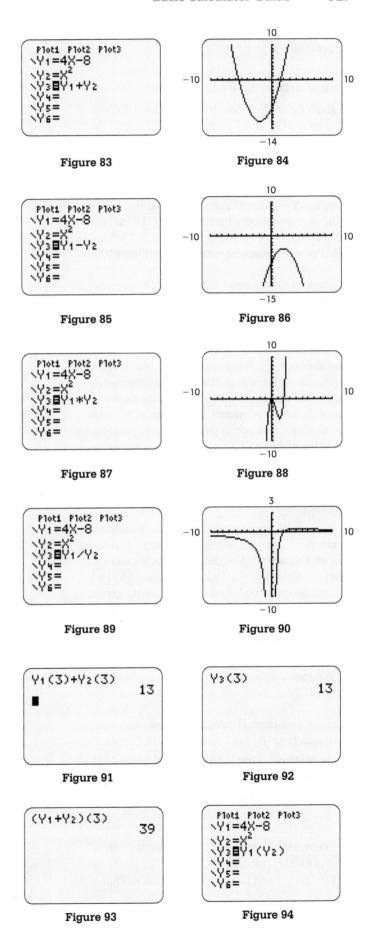

Figure 83

Figure 84

Figure 85

Figure 86

Figure 87

Figure 88

Figure 89

Figure 90

Figure 91

Figure 92

Figure 93

Figure 94

Note: Entering (Y1 + Y2)(3) does not produce the correct result (Figure 93).

Composition of Functions

To graph the composition of two functions $f(x)$ and $g(x)$, enter $f(x)$ as Y1 and $g(x)$ as Y2 under the Y= menu. Consider the example $f(x) = 4x - 8$ and $g(x) = x^2$.

To graph $(f \circ g)(x) = f(g(x))$, enter Y1(Y2) as Y3 under the Y= menu (Figure 94). Place the cursor on the = sign beside Y1 and press $\boxed{\text{ENTER}}$ to turn off the graph of Y1. Repeat with Y2. Press $\boxed{\text{GRAPH}}$ with an appropriate window (Figure 95).

To graph $(g \circ f)(x)$, enter Y2(Y1) as Y4 under the Y= menu (Figure 96). Turn off the graphs of Y1, Y2, and Y3. Press $\boxed{\text{GRAPH}}$ with an appropriate window (Figure 97).

To evaluate $(f \circ g)(x)$ at a specified value of x, enter Y1(Y2(, the value of x, and two closing parentheses (Figure 98), and press $\boxed{\text{ENTER}}$. Or, if the combination of functions is entered as Y3, enter Y3 and the value of x enclosed in parentheses (Figure 99). The value of $(f \circ g)(-5) = f(g(-5))$ is shown in Figures 98 and 99.

Inverse Functions

To show that the graphs of two inverse functions are symmetrical about the line $y = x$, graph the functions on a square window. Enter $f(x)$ as Y1 and the inverse function $g(x)$ as Y2, and press $\boxed{\text{ENTER}}$ with the cursor to the left of Y2 (to make the graph dark). Press $\boxed{\text{GRAPH}}$ with an appropriate window. Enter Y3 = x under the Y= menu and graph using $\boxed{\text{ZOOM}}$, 5:Zsquare. For example, the graphs of $f(x) = x^3 - 3$ and its inverse $g(x) = \sqrt[3]{x + 3}$ are symmetric about the line $y = x$, as shown in Figures 100 and 101.

To graph a function $f(x)$ and its inverse, under the Y= menu, enter $f(x)$ as Y1 (Figure 102). Choose a square window. Press $\boxed{\text{2ND}}$ $\boxed{\text{PRGM}}$ (DRAW) (Figure 103), 8:DrawInv (Figure 104), press $\boxed{\text{VARS}}$, move to Y-VARS, and press $\boxed{\text{ENTER}}$ three times. The graph of $f(x)$ and its inverse will be displayed.

To clear the graph of the inverse, press $\boxed{\text{2ND}}$ $\boxed{\text{PRGM}}$ (DRAW), 1:ClrDraw. The graphs of $f(x) = 2x - 5$ and its inverse are shown in Figure 105.

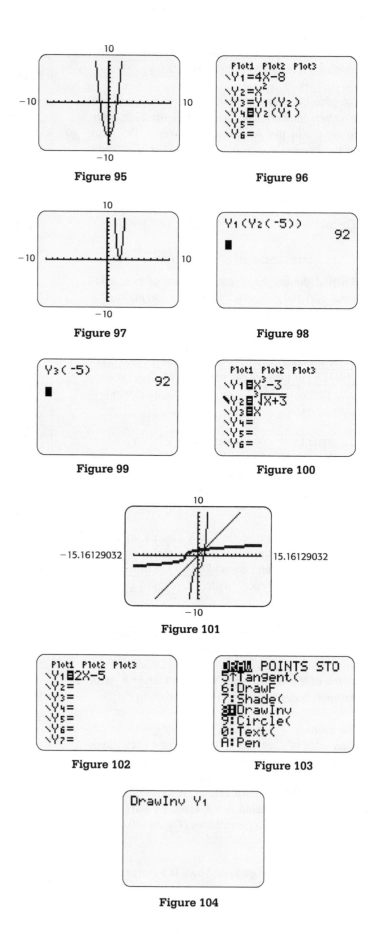

Figure 95   **Figure 96**

Figure 97   **Figure 98**

Figure 99   **Figure 100**

Figure 101

Figure 102   **Figure 103**

Figure 104

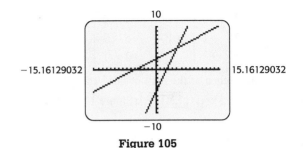

Figure 105

Matrices

Entering Data into Matrices

To enter data into matrices, press 2ND x^{-1} to get the MATRIX menu (Figure 106). Move the cursor to EDIT. Select the number of the matrix into which the data are to be entered. Enter the dimensions of the matrix, and enter the value for each entry of the matrix (Figure 107). Press ENTER after each entry.

For example, to enter as [A] the matrix

$$\begin{bmatrix} 1 & 2 & 3 \\ 2 & -2 & 1 \\ 3 & 1 & -2 \end{bmatrix}$$

enter 3's to set the dimension and enter the elements of the matrix (Figure 108).

To perform operations with the matrix or leave the editor, first press 2ND MODE (QUIT).

To view the matrix, press MATRIX , enter the name of the matrix (Figure 109), and press ENTER (Figure 110).

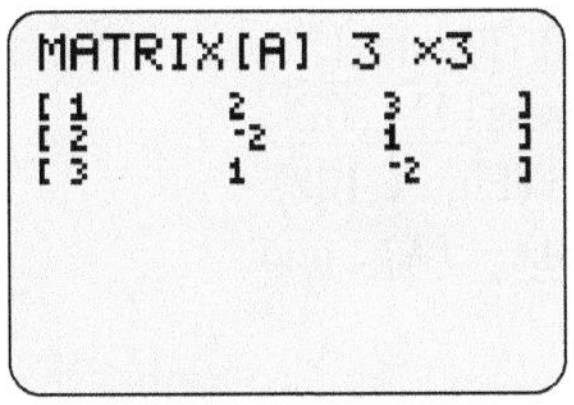

The Identity Matrix

To display an identity matrix of order n (an $n \times n$ matrix consisting of 1's on the main diagonal and 0's elsewhere), press MATRIX , move to MATH, select 5:identity(, and enter the order of the identity matrix desired (Figure 111). The identity matrix of order 2 is shown in Figure 112.

An identity matrix can also be created by entering the numbers directly with MATRIX and EDIT.

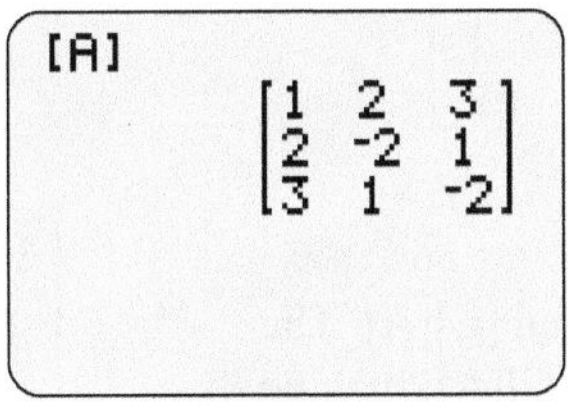

Adding and Subtracting Matrices

To find the sum of two matrices [A] and [D], enter the values of the elements of [A] using MATRIX and EDIT (Figure 113). Press 2ND MODE (QUIT). Enter the values of the elements of [D] using MATRIX and EDIT (Figure 114). Press 2ND MODE (QUIT).

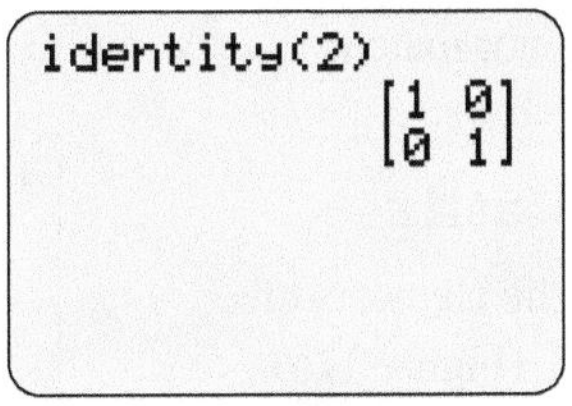

Use MATRIX and NAMES to enter [A] + [D], and press ENTER . If the matrices have the same dimensions, they can be added (or subtracted). If they do not have the same dimensions, an error message will occur.

For example, Figure 115 shows the sum

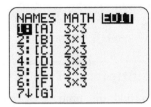

Figure 106

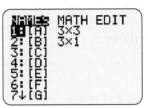

Figure 107

Figure 108

Figure 109

Figure 110

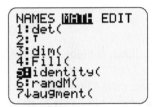

Figure 111

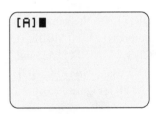

Figure 112

Figure 113

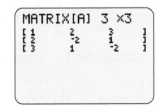

Figure 114

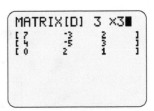

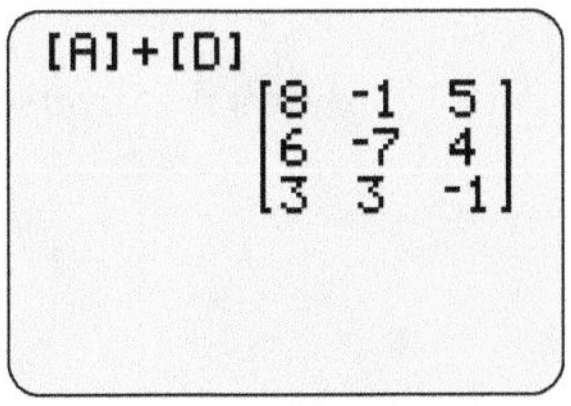

Figure 115

$$\begin{bmatrix} 1 & 2 & 3 \\ 2 & -2 & 1 \\ 3 & 1 & -2 \end{bmatrix} + \begin{bmatrix} 7 & -3 & 2 \\ 4 & -5 & 3 \\ 0 & 2 & 1 \end{bmatrix}$$

To find the difference of the matrices [A] and [D], enter [A] − [D] and press ENTER (Figure 116).

Multiplying a Matrix by a Real Number

We can multiply a matrix [D] by a real number (scalar) k by entering k[D] (or k∗[D]). In Figures 117 and 118, we multiply the matrix [D] by 5.

Multiplying Two Matrices

To find the product of matrices [C][A], press MATRIX, move to EDIT, select 1: [A], enter the dimensions of [A] (Figure 119), and enter the elements of [A] (Figure 120). Press 2ND MODE (QUIT). Do the same for matrix [C] (Figure 121) and press 2ND MODE (QUIT). Press MATRIX [C], ∗, MATRIX [A], and ENTER (Figure 122). Or press MATRIX [C], MATRIX [A], and ENTER (Figure 123).

The product

$$\begin{bmatrix} 1 & 2 & 4 \\ -3 & 2 & -1 \end{bmatrix} \begin{bmatrix} 1 & 2 & 3 \\ 2 & -2 & 1 \\ 3 & 1 & -2 \end{bmatrix}$$

is shown in Figures 122 and 123.

Note that [A][C] does not always equal [C][A]. The product [A][C] may be the same as [C][A], may be different from [C][A], or may not exist.

In Figures 124 and 125, [A][C] cannot be computed because the matrices' dimensions do not match.

Finding the Inverse of a Matrix

To find the inverse of a matrix, enter the elements of the matrix using MATRIX and EDIT (Figure 126). Press 2ND MODE (QUIT). Press MATRIX, enter the name of the matrix, and then press the x^{-1} key and ENTER (Figure 127).

Figure 128 shows the inverse of

$$E = \begin{bmatrix} 2 & 0 & 2 \\ -1 & 0 & 1 \\ 4 & 2 & 0 \end{bmatrix}$$

To see the entries as fractions, press MATH, 1:Frac, and ENTER (Figures 129 and 130).

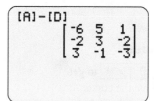

Figure 116

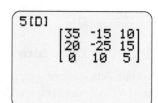

Figure 117

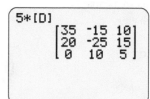

Figure 118

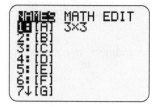

Figure 119

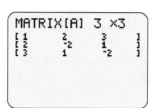

Figure 120

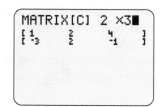

Figure 121

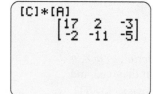

Figure 122

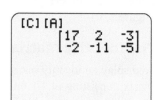

Figure 123

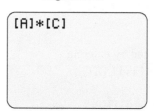

Figure 124

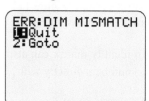

Figure 125

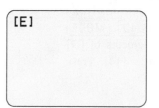

Figure 126

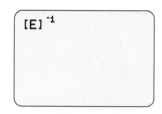

Figure 127

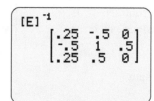

Figure 128

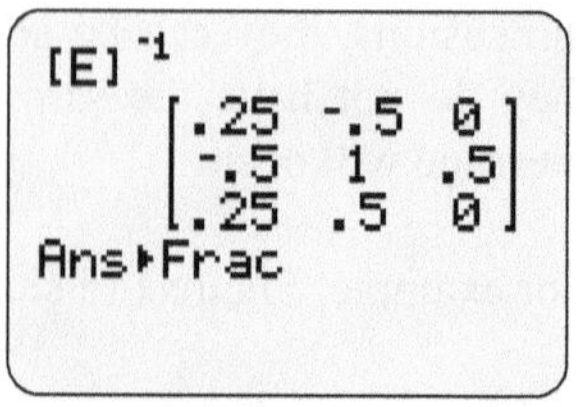

Figure 129

Not all matrices have inverses. Matrices that do not have inverses are called singular matrices (Figures 131–133).

Solving Systems— Reduced Echelon Form

To solve a system of linear equations by using rref under the MATRIX MATH menu, create an augmented matrix [A] with the coefficient matrix augmented by the constants (Figure 134). Use the MATRIX menu to produce a reduced row echelon form of [A], as follows:

1. Press MATRIX and move to the right to MATH (Figure 135).
2. Scroll down to B:rref(and press ENTER, or press ALPHA B (Figure 136). Press MATRIX, 1: [A] to get rref([A]). Press ENTER to get the reduced echelon form.

The system

$$\begin{cases} 2x - y + z = 6 \\ x + 2y - 3z = 9 \\ 3x - 3z = 15 \end{cases}$$

is solved in Figure 137.

If each row in the coefficient matrix (the first three columns) contains a 1 with the other elements 0's, the solution is unique and the number in column 4 of a row is the value of the variable corresponding to a 1 in that row. The solution to the system above is unique: $x = 4$, $y = 1$, and $z = -1$.

If the bottom row contains all zeros, the system has many solutions. The values for the first two variables are found as functions of the third.

If there is a nonzero element in the augment of row 3 and zeros elsewhere in row 3, there is no solution to the system.

Linear Programming

To solve a linear programming problem involving two constraints graphically, write the inequalities as equations, solved for y. Graph the equations. The inequalities $x \geq 0$, $y \geq 0$ limit the graph to Quadrant I, so choose a window with Xmin = 0 and Ymin = 0. Use TRACE or INTERSECT to find the corners of the region, where the borders intersect.

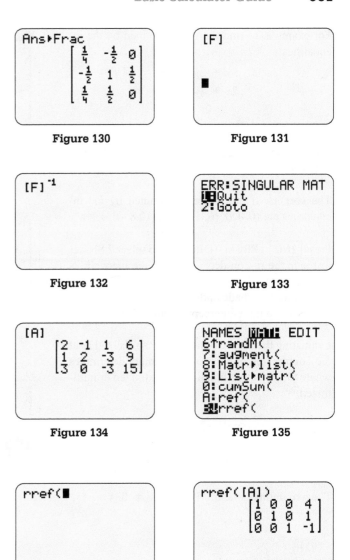

Figure 130

Figure 131

Figure 132

Figure 133

Figure 134

Figure 135

Figure 136

Figure 137

For example, to find the region defined by the inequalities

$$5x + 2y \le 54$$
$$2x + 4y \le 60$$
$$x \ge 0, \; y \ge 0$$

write $y = 27 - 5x/2$ and $y = 15 - x/2$, graph, and find the intersection points as shown in Figures 138 to 140.

The corners of the region determined by the inequalities are $(0, 15)$, $(6, 12)$, and $(10.8, 0)$.

Press 2ND PRGM (DRAW) and select 7:Shade(to shade the region determined by the inequalities (Figure 141). Shade under the border from $x = 0$ to a corner and shade under the second border from the corner to the x-intercept (Figure 142).

Evaluating the objective function $P = 5x + 11y$ at the coordinates of each of the corners determines where the objective function is maximized or minimized.

$$\text{At } (0, 15), f = 165$$
$$\text{At } (6, 12), f = 162$$
$$\text{At } (10.8, 0), f = 54$$

The maximum value of f is 165 at $x = 0$, $y = 15$.

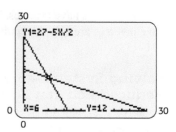

Figure 138

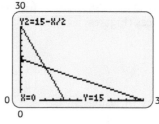

Figure 139

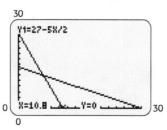

Figure 140

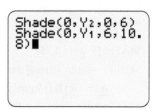

Figure 141

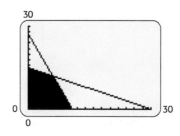

Figure 142

Sequences and Series

Evaluating a Sequence

To evaluate a sequence for different values of n, press MODE and highlight Seq (Figure 143). Press ENTER and 2ND MODE (QUIT). Store the formula for the sequence (in quotes) in u, using STO u. (Press the X,T,Θ,n key to enter n for the formula, and 2ND 7 to get u.) Press ENTER.

For example, to evaluate the sequence with nth term $n^2 + 1$ at $n = 1, 3, 5$, and 9, we store the formula as shown in Figure 144.

Enter u({$a, b, c, \ldots$}) to evaluate the sequence at a, $b, c, \ldots$, and press ENTER (Figure 145).

To generate a sequence after the formula is defined, enter u(nstart, nstop, step) and press ENTER. For the sequence formula above, evaluate every third term beginning with the second term and ending with the eleventh term (Figure 146).

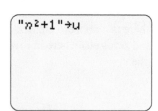

Figure 143

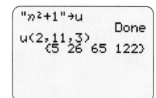

Figure 144

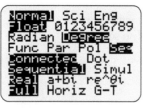

Figure 145

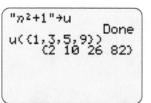

Figure 146

Finding the nth Term of an Arithmetic Sequence

To find the *n*th term of an arithmetic sequence with first term *a* and common difference *d*, press [MODE] and highlight Seq (Figure 147). Press [ENTER] and press [2ND] [MODE] (QUIT).

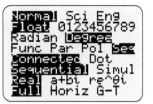

Figure 147

Press [Y=]. At u(*n*) =, enter the formula for the *n*th term of an arithmetic sequence, using the [X,T,Θ,n] key to enter *n*. The formula is $a + (n - 1)d$, where *a* is the first term and *d* is the common difference.

For example, to find the 12th term of the arithmetic sequence with first term 10 and common difference 5, substitute 10 for *a* and 5 for *d*, as shown in Figure 148, to get u(*n*) = $10 + (n - 1)5$.

Figure 148

Press [2ND] [MODE] (QUIT). To find the *n*th term of the sequence, press [2ND] 7, followed by the value of *n* in parentheses, to get u(*n*); then press [ENTER] (Figure 149).

Additional terms can be found in the same manner (Figure 150).

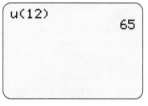

Figure 149 The 12th term.

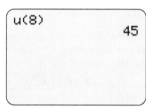

Figure 150 The 8th term.

Finding the Sum of an Arithmetic Sequence

To find the sum of the first *n* terms of an arithmetic sequence, press [MODE] and highlight Seq (Figure 151). Press [ENTER] and press [2ND] [MODE] (QUIT).

Press [Y=]. At v(*n*) =, enter the formula for the sum of the first *n* terms of an arithmetic sequence, using the [X,T,Θ,n] key to enter *n*. The formula is $(n/2)(a + (a + (n - 1)d))$, where *a* is the first term and *d* is the common difference (Figure 152).

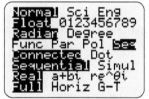

Figure 151

Figure 152

Press [2ND] [MODE] (QUIT). To find the sum of the first *n* terms of the sequence, press [2ND] 8, followed by the value of *n* in parentheses, to get v(*n*); then press [ENTER]. For example, the sum of the first 12 terms of the arithmetic sequence with first term 10 and common difference 5 is shown in Figure 153.

Other sums can be found in the same manner (Figure 154).

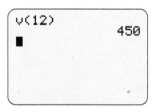

Figure 153 The sum of the first 12 terms.

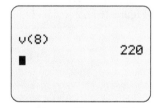

Figure 154 The sum of the first 8 terms.

Finding the nth Term and Sum of a Geometric Sequence

To find the *n*th term of a geometric sequence with first term *a* and common ratio *r*, press [MODE] and highlight Seq (Figure 155). Press [ENTER] and [2ND] [MODE] (QUIT).

Figure 155

Press [Y=]. At u(n) =, enter the formula for the nth term of a geometric sequence, using the [X,T,Θ,n] key to enter n. The formula is ar^{n-1}, where a is the first term and r is the common ratio.

For example, to find the geometric sequence with first term 40 and common ratio 1/2, we substitute 40 for a and (1/2) for r (Figure 156).

Press [2ND] [MODE] (QUIT). To find the nth term of the sequence, press [2ND] 7, followed by the value of n in parentheses, to get u(n); then press [ENTER]. The 8th and 12th terms of the geometric sequence above are shown in Figures 157 and 158, respectively. To get a fractional answer, press [MATH], 1:Frac. Additional terms can be found in the same manner.

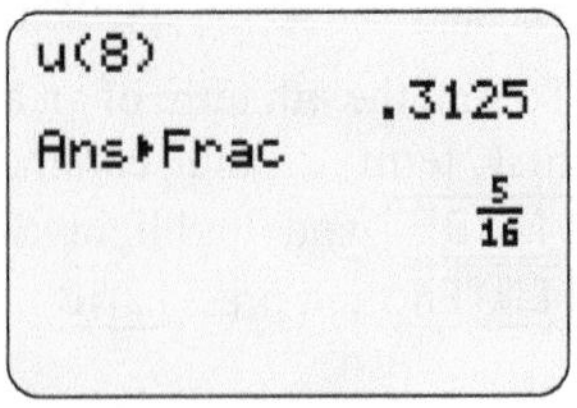

To find the sum of the first n terms of a geometric sequence, press [Y=]. At v(n) =, enter the formula for the sum of the first n terms of a geometric sequence, using the [X,T,Θ,n] key to enter n. The formula is $a(1 - r^n)/(1 - r)$, where a is the first term and r is the common ratio. Press [2ND] [MODE] (QUIT). To find the sum of the first n terms of the sequence, press [2ND] 8, followed by the value of n in parentheses, to get v(n); then press [ENTER].

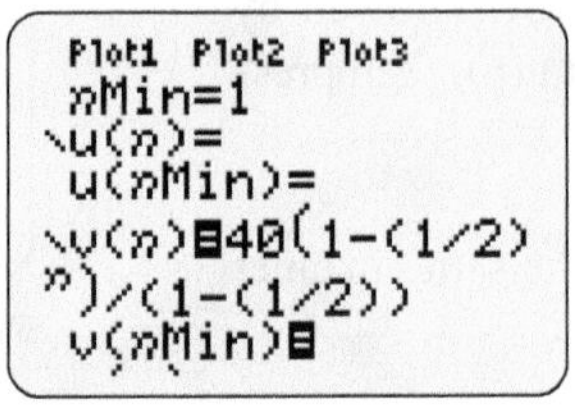

For example, to find the sum of the first 12 terms of the geometric sequence with first term 40 and common ratio 1/2, substitute 40 for a and (1/2) for r (Figure 159).

To get a fractional answer, press [MATH], 1:Frac (Figure 160).

Other sums can be found in the same manner (Figure 161).

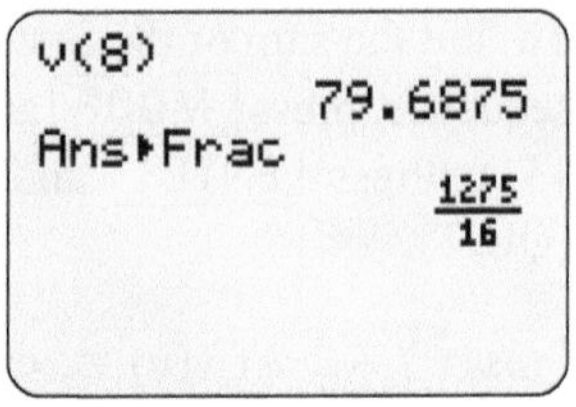

Figure 156

Figure 157

Figure 158

Figure 159

Figure 160 The sum of the first 12 terms.

Figure 161 The sum of the first 8 terms.

Appendix B

Basic Guide to Excel 2003, Excel 2007, and Excel 2010

This appendix describes the basic features and operation of Excel. More details of specific features can be found in the *Graphing Calculator and Excel® Manual*.

Excel 2007 and 2010

The Excel steps shown in this guide are for Excel 2003 and earlier versions, and also for Excel 2007 and 2010. The operations within the spreadsheet are mostly the same for Excel 2003 and Excel 2007 and 2010, but many of the menus and button locations are different. When the Excel 2007 and 2010 steps are significantly different from the Excel 2003 steps, the Excel 2007 and 2010 menus and button locations are printed in color, embedded in or after the Excel 2003 steps in this guide.

To find additional information about how to use Excel 2010 menus and buttons, use 2010 Excel Help as follows:

1. Select File and then Help; then select Microsoft Office Help.

2. Select Getting Started with Excel 2010, and click on WHERE TO GO based on your situation (previous knowledge of Excel).

For Excel 2007, click the ⑦ icon for Microsoft Office Help.

Excel Worksheet

When you start up Excel by using the instructions for your software and computer, the screen in Figure 1 will appear. The components of the **spreadsheet** are shown, and the grid shown is called a **worksheet**. By clicking on the tabs at the bottom, you can move to other worksheets.

Figure 1

Addresses and Operations

Notice the letters at the top of the columns and the numbers identifying the rows. The cell addresses are given by the column and row; for example, the first cell has address A1. You can move from one cell to another with arrow keys, or you can select a cell with a mouse click. After you enter an entry in a cell, press ENTER to accept the entry. To edit the contents of a cell, make the edits in the formula bar at the top. To delete the contents, press the delete key.

File operations such as "open a new file," "saving a file," and "printing a file" are similar to those in Word. For example, <CTRL>S saves a file. You can also format a cell entry by selecting it and using menus similar to those in Word.

Working with Cells

Cell entries, rows containing entries, and columns containing entries can be copied and pasted with the same commands as in Word. For example, a highlighted cell can be copied with <CTRL>C. Sometimes an entry exceeds the width of the cell containing it, especially if it is text. To widen the cells in a column, move the mouse to the right side of the column heading until you see the symbol ↔; then hold down the left button and move the mouse to the right (moving to the left makes the cells narrower). If entering a number results in #####, the number is too long for the cell and the cell should be widened.

You can work with a cell or with a range of cells. To select a range of cells,

1. Click on the beginning of the range of cells, hold down the left mouse button, and drag to the end of the desired range.

2. Release the mouse button. The range of cells will be highlighted. If the range of cells were from A2 through B6, the range would be indicated by A2:B6.

Creating Tables of Numbers in Excel

Entering Independent Values

Type the column heading, such as x, in cell A1, and enter each data value in a cell of the column (Figure 2).

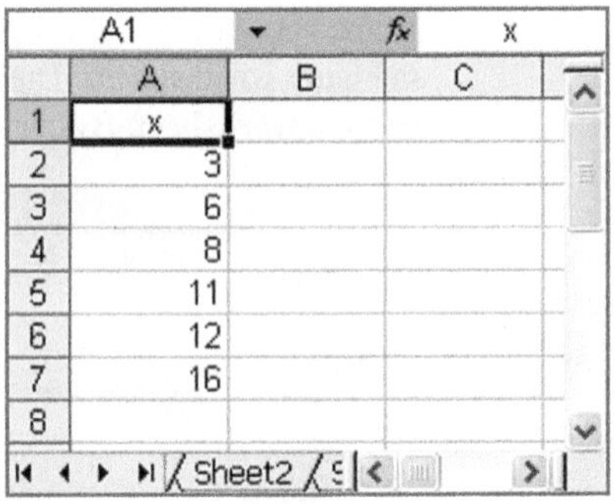

Figure 2

Fill-Down Method for Values That Change by a Constant Increment

1. Type in the first two numbers (in C2 and C3, for example).

2. Select the cells C2 and C3 (C2:C3) (Figure 3).

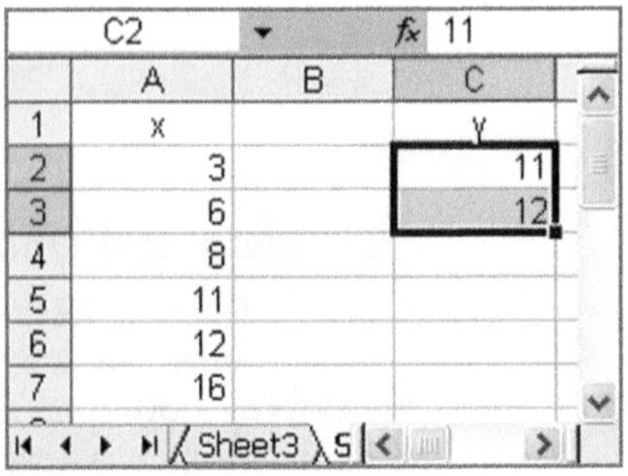

Figure 3

3. Move the mouse to the lower right corner until the mouse becomes a thin + sign.

4. Drag the mouse down to the last cell where data are required (Figure 4). Press the right arrow to remove the highlight.

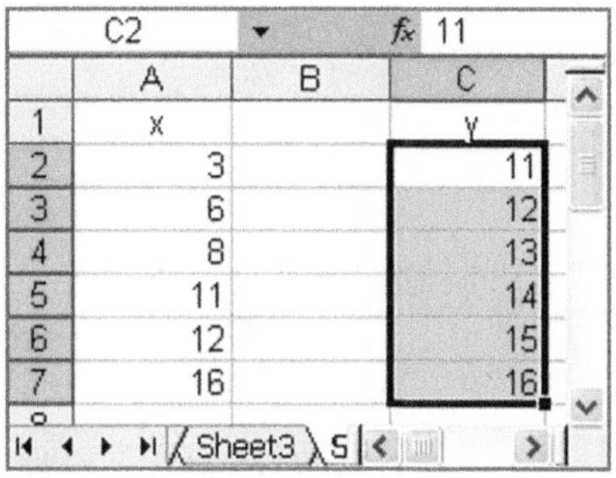

	A	B	C
1	x		y
2	3		11
3	6		12
4	8		13
5	11		14
6	12		15
7	16		16

Figure 4

Evaluating a Function

1. Put headings on the two columns (x and $f(x)$, for example).

2. Begin a formula (in cell B2, for example) by entering =; then enter the function performing operations with data in cells by entering the operation side of the formula with appropriate cell address(es) representing the variable(s) (Figure 5). Use * to indicate multiplication and ^ to indicate power.

3. Enter the value(s) of the variable(s) for which the function is to be evaluated and press ENTER (Figure 6).

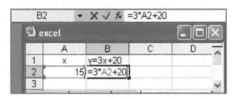

Figure 5

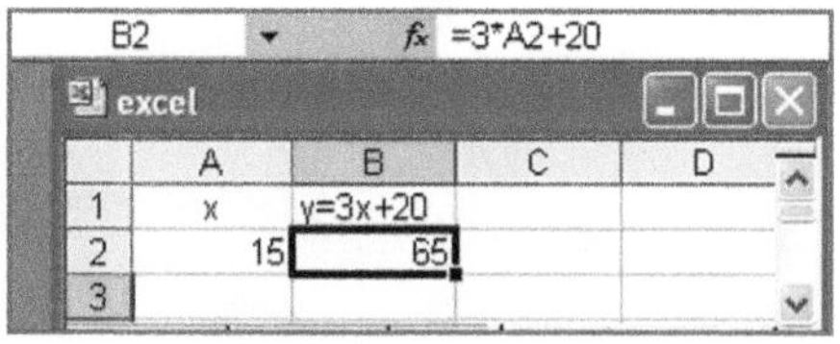

Figure 6

4. Changing the input values will give different function values (Figure 7). Press ENTER to complete the process (Figure 8).

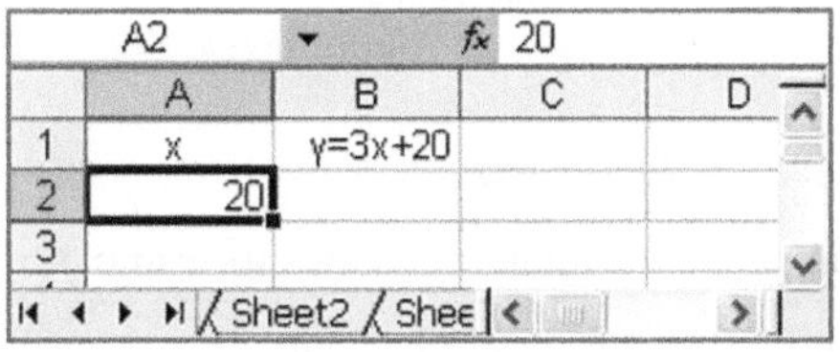

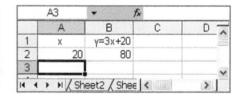

Figure 7

Figure 8

Creating a Table for a Function Using Fill Down

1. Put headings on the two columns (x and $f(x)$, for example).

2. Fill the inputs (x-values) as described on page 636.

3. Enter the formula for the function as described above.

4. Select the cell containing the formula for the function (B2, for example) (Figure 9).

5. Move the mouse to the lower right corner until there is a thin + sign.

6. Drag the mouse down to the last cell where the formula is required, and the values will be displayed (Figure 10). Using an arrow to move to a cell to the right will remove the highlight from the outputs.

B2 =A2-1950

	A	B	C
1	Year	Year-1950	
2	1993	43	
3	1994		
4	1995		
5	1996		
6	1997		
7	1998		
8			

Sheet2

Figure 9

B2 =A2-1950

	A	B	C
1	Year	Year-1950	
2	1993	43	
3	1994	44	
4	1995	45	
5	1996	46	
6	1997	47	
7	1998	48	
8			

Sheet2

Figure 10

The worksheet in Figure 11 has headings typed in, numbers entered with Fill Down in column A, inputs entered in column B, the numbers in column D created with Fill Down using the formula A5 − 1950, and column B copied in Column E.

E14

	A	B	C	D	E	F
1	Cohabiting Households					
2						
3		Households				
4	Year	(thousands)		Year-1950	Thousands	
5	1991	3039		41	3039	
6	1992	3308		42	3308	
7	1993	3510		43	3510	
8	1994	3661		44	3661	
9	1995	3668		45	3668	
10	1996	3958		46	3958	
11	1997	4130		47	4130	
12	1998	4236		48	4236	

Sheet2 / Sheet3 / Sheet4 / She

Figure 11

Graphing a Function of a Single Variable ($y = f(x)$, for example)

(Excel 2007 and 2010 steps are shown below in color.)

1. Create a table containing values for x and $f(x)$ as described on page 637.
2. Highlight the two columns containing the values of x and $f(x)$ (Figure 12).

B2 =3*A2^2-10*A2

	A	B	C
1	x	f(x)=3x^2-10x	
2	-1	13	
3	0	0	
4	1	-7	
5	2	-8	
6	3	-3	
7	4	8	

Sheet1 / She

Figure 12

3. Click the Chart Wizard icon and then select the XY(Scatter) chart type and the chart sub-type with the smooth curve option (Figure 13).

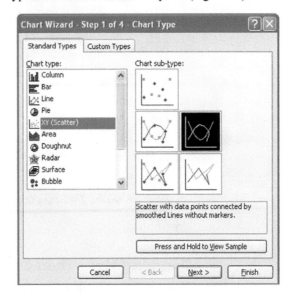

Figure 13

4. Click the Next button to get the Chart Source Data box. Then click Next to get the Chart Options box, and enter your chart title and labels for the *x*- and *y*-axes. (Figure 14).

5. Click Next, select whether the graph should be within the current worksheet or on another, and click Finish.

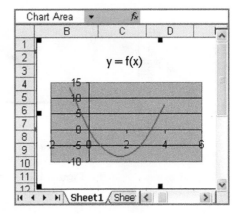

Figure 14

Excel 2007 and 2010

In Excel 2007 and 2010, find the graph of a function as follows:

1. Use the function to create a table containing values for *x* and *f*(*x*).

2. Highlight the two columns containing the values of *x* and *f*(*x*).

3. Select the Insert tab.

4. Select Scatter in the Charts group.

5. Select the smooth curve option under Scatter, and the graph will be as shown in Figure 15 on the next page.

6. To add or change titles or legends, click on the icons in the Layout group under Chart Tools.

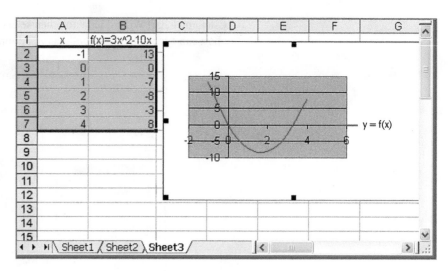

Figure 15

Graphing More Than One Function

1. Create a table with an input column and one column for each function, with multiple function headings.

2. Highlight the columns containing the variable and function values (Figure 16).

3. Proceed with the same steps as are used to graph a single function (Figure 17).

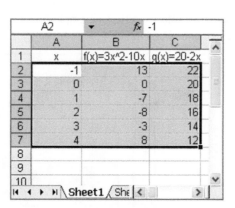

Figure 16

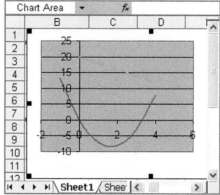

Figure 17

Changing Graphing Windows

1. To change the *x*-scale,

   a. Double click on the *x*-axis, and then click on the Scale tab.

   b. Uncheck the Auto boxes, and change the minimum and maximum values to the desired values.

   c. Click OK.

2. To change the *y*-scale, double click on the *y*-axis and proceed as for the *x*-axis.

3. To delete shading on graphs, double click on the plot area and check "None" for the format of the plot area.

Graphing Discontinuous Functions

An Excel graph will connect all points corresponding to values in the table, so if the function you are graphing is discontinuous for some *x*-value *a*, enter *x*-values near this value and leave (or make) the corresponding $f(a)$ cell blank. Then graph the function using the steps described on page 638 (Figure 18).

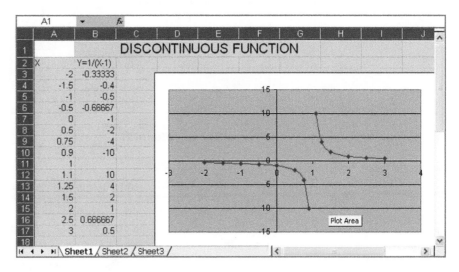

Figure 18

Finding a Minimum or Maximum of a Function with Solver

Finding the Minimum of a Function (if it has a minimum)

(Excel 2007 and 2010 steps are shown below in color.)

1. Enter values for x and for $y = f(x)$, and use them to graph the function over an interval that shows a minimum (Figure 19). (If a minimum exists but does not show, add more points.)

2. Use any value for x in A2 and the same function formula as in Step 1 (Figure 20).

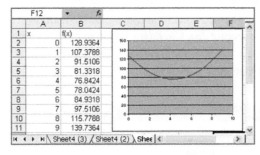

Figure 19   **Figure 20**

3. Invoke Solver by choosing Tools > Solver.

4. Click on Min in the Solver dialog box.

5. Set the Target Cell to B2 by choosing the box and clicking on B2. Then set Changing Cells to A2 with a similar process (Figure 21).

6. Click Solve in the dialog box, and click OK to accept the solution (Figure 22).

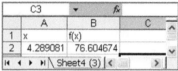

Figure 21   **Figure 22**

Excel 2007 and 2010

In Excel 2007 and 2010, find a minimum of a function (if it exists) as follows:

1. Enter the inputs (*x*-values) in column A and the functional values in column B, and use them to graph the function over an interval that shows a minimum. (If a minimum exists but does not show, add more points.)

2. Use any value for *x* in A2 and the function formula in B2.

3. Select the Data tab, and select ?Solver from the Analysis group.

4. Click on Min in the Solver dialog box.

5. Set the Objective Cell to B2 by choosing the box and clicking on B2. Then set the Changing Cells to A2 with a similar process.

6. Click Solve in the dialog box, and the minimum will show in B2. Click OK to save the solution.

Finding the Maximum of a Function (if it has a maximum)

Use the process above, but in Step 4 click on Max in the Solver dialog box.

Scatter Plots of Data

(Excel 2007 and 2010 steps are shown below in color.)

1. Enter the inputs (*x*-values) in column A and the outputs (*y*-values) in column B.

2. Highlight the two columns, use Chart Wizard to select XY(Scatter), and click Next (Figure 23). Click Next again to move to Step 3.

3. In Step 3 of Chart Wizard, enter the title and the *x*- and *y*-axis labels (Figure 24) and then click Next.

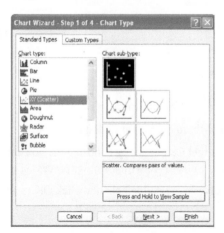

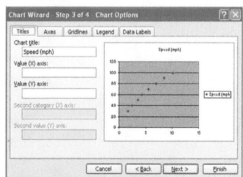

Figure 23   **Figure 24**

4. In Step 4, indicate that the scatter plot should be placed in the current worksheet and then click Finish (Figure 25).

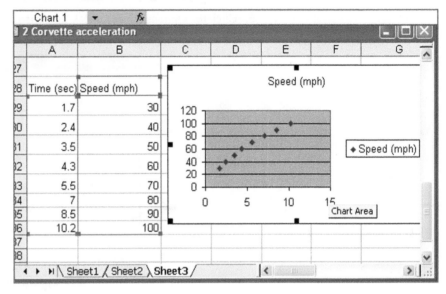

Figure 25

Excel 2007 and 2010

In Excel 2007 and 2010, create a scatter plot of data as follows:

1. Enter the inputs (x-values) in column A and the outputs (y-values) in column B.
2. Highlight the two columns.
3. Select the Insert tab and select Scatter in the Charts group.
4. Select the points option under Scatter. The graph will appear.
5. To add or change titles or legends, click on the icons in the Layout group under Chart Tools.

Finding the Equation of the Line or Curve That Best Fits a Given Set of Data Points

(Excel 2007 and 2010 steps are shown below in color.)

1. Place the scatter plot of the data in the worksheet, as described on page 642.
2. Single click on the scatter plot in the workbook.
3. From the Chart menu choose Add Trendline (Figure 26).

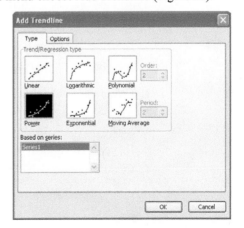

Figure 26

4. Click on the regression type that appears to be the best function fit for the scatter plot. If Polynomial is selected, choose the appropriate Order (degree).

5. Click the Options tab and check the "Display equation on chart" box (Figure 27).

Figure 27

6. Click OK and you will see the graph of the selected function that is the best fit, along with its equation (Figure 28).

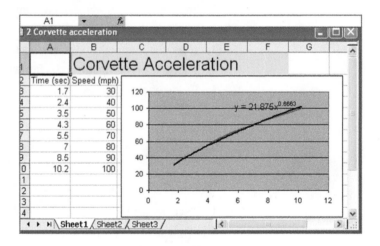

Figure 28

Excel 2007 and 2010

In Excel 2007 and 2010, find the model for a set of data as follows:

1. Create the scatter plot for the data (see above).

2. Right click on one of the data points and select Add Trendline.

3. Select the desired regression type and click Display equation on chart.
   [Note: If Polynomial is selected, choose the appropriate Order (degree).]

4. Close the box and the function and its graph will appear.

Solving a Linear Equation of the Form $f(x) = 0$ Using the Intercept Method

The solution will also be the *x*-intercept of the graph and the zero of the function. (Excel 2007 and 2010 steps are shown below in color.)

1. Enter an *x*-value in A2 and the function formula in B2, as described on page 637 (Figure 29).

2. Highlight B2. From the Tools menu, select Goal Seek.

3. In the dialog box,

    a. Click the Set cell box and click on the B2 cell.

    b. Enter 0 in the To value box.

    c. Click the By changing cell box and click on the A2 cell.

4. Click OK to see the Goal Seek Status box (Figure 30).

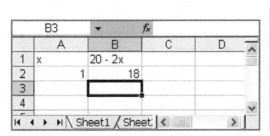
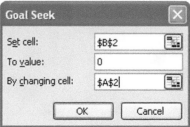

Figure 29   **Figure 30**

5. Click OK again. The solution will be in A2 and 0 will be in B2 (Figure 31).

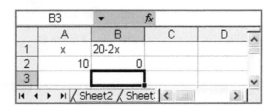

Figure 31

Excel 2007 and 2010

In Excel 2007 and 2010, we solve linear equations of the form $f(x) = 0$ with the intercept method as follows:

1. Enter an *x*-value in A2 and the function formula in B2.

2. Highlight B2.

3. Select the Data tab.

4. Select What-if Analysis in the Data Tools group.

5. Select Goal Seek under What-if Analysis.

6. In the Goal Seek dialog box,

    a. Click the Set cell box and click on the B2 cell.

    b. Enter 0 in the To value box.

    c. Click the By changing box and click on the A2 cell.

7. Click OK to see the Goal Status box.

8. Click OK again. The solution will be in A2 and 0 will be in B2. The solution may be approximate.

Solving a Quadratic Equation of the Form
$f(x) = 0$ Using the Intercept Method

The solutions will also be the x-intercepts of the graph and the zeros of the function. Because the graph of a quadratic function $f(x) = ax^2 + bx + c$ can have two intercepts, it is wise to graph the function using an x-interval with the x-coordinate $\left(h = \dfrac{-b}{2a} \right)$ near the center.

1. Enter x-values centered around h in column A and use the function formula to find the values of $f(x)$ in column B, as described on page 637 (Figure 32).

2. Graph the function as described on page 638, and observe where the function values are at or near 0.

3. Select Goal Seek, enter the address of a cell with a function value in column B at or near 0, enter 0 in To value, and enter the corresponding cell in column A in By changing cell (Figure 33). (In Excel 2007 and 2010, Goal Seek is found under What-if Analysis in the Data Tools group, under the Data tab.) Click OK to find the x-value of the solution in column A. The solution may be approximate.

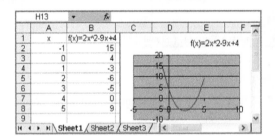   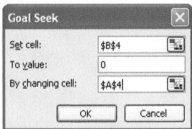

Figure 32   Figure 33

4. One solution is $x = 0.5$ (Figure 34).*

5. After finding the first solution, repeat the process using a second function value at or near 0.

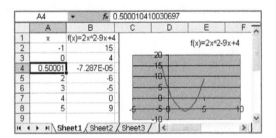

Figure 34

---

*Notes: (1) The solution $x = 0.50001$, giving $f(x) = -7.287\text{E}{-}05$, is an approximation of the exact solution $x = 0.5$, which gives $f(x) = 0$. (2) Solutions of other equations (logarithmic, etc.) can be found with similar solution methods.

Solving an Equation of the Form $f(x) = g(x)$ Using the Intersection Method

If One Solution Is Sought

1. Enter a value for the input variable (x, for example) in cell A2 and the formula for each of the two sides of the equation in cells B2 and C2, respectively.

2. Enter "=B2−C2" in cell D2 (Figure 35).

3. Select Goal Seek to find x when B2−C2=0. (In Excel 2007 and 2010, Goal Seek is found under What-if Analysis in the Data Tools group, under the Data tab.)

4. In the dialog box,

    a. Click the Set cell box and click on the D2 cell.

    b. Enter 0 in the To value box.

    c. Click the By changing cell box and click on the A2 cell (Figure 36).

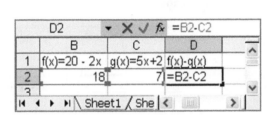
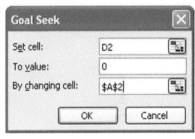

<table>
<tr><td align="center">**Figure 35**</td><td align="center">**Figure 36**</td></tr>
</table>

5. Click OK to find the x-value of the solution in cell A2 (Figure 37). The solution may be approximate.

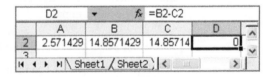

Figure 37

If Multiple Solutions Are Sought (as in the case of a quadratic equation)

1. Proceed as in Steps 1 and 2 above.

2. Enter values for $f(x)$ and for $g(x)$, and use them to graph the function over an interval that shows points of intersection. Input values near those of these points will give column D values that are at or near 0. Entering x-values with $h = \dfrac{-b}{2a}$ near the center is useful when solving quadratic equations.

3. Use Goal Seek to enter in D2 the address of a cell with a function value at or near 0, and complete the process as above. (In Excel 2007 and 2010, Goal Seek is found under What-if Analysis in the Data Tools group, under the Data tab.)

4. After finding the first solution, repeat the process using a second function value at or near 0.

Solving a System of Two Linear Equations in Two Variables

1. Write the two equations as linear functions in the form $y = mx + b$.

2. Enter a value for the input variable (x) in cell A2 and the formula for each of the two equations in cells B2 and C2, respectively.

3. Enter "=B2−C2" in cell D2 (Figure 38).

4. Select Goal Seek to find x when B2−C2=0. (In Excel 2007 and 2010, Goal Seek is found under What-if Analysis in the Data Tools group, under the Data tab.)

5. In the dialog box,
    a. Click the Set cell box and click on the D2 cell.
    b. Enter 0 in the To value box.
    c. Click the By changing cell box and click on the A2 cell.

6. Click OK in the Goal Seek dialog box, getting the solution.

7. The x-value of the solution will be in cell A2, and the y-value will be in both B2 and C2 (Figure 39).

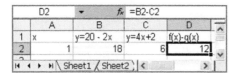
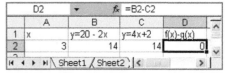

Figure 38   **Figure 39**

Matrices

Excel is useful to add, subtract, multiply, and find inverses of matrices. To perform operations with a matrix, enter each of its elements in a cell of a worksheet. Addition and subtraction of matrices are intuitive with Excel, but finding products and inverses requires special commands.

Adding and Subtracting Matrices (steps for two 3 × 3 matrices)

1. Type a name A in A1 to identify the first matrix.

2. Enter the elements of matrix A in the cells B1:D3.

3. Type a name B in A5 to identify the second matrix.

4. Enter the elements of matrix B in the cells B5:D7.

5. Type a name A+B in A9 to indicate the matrix sum.

6. Type the formula "=B1+B5" in B9 and press ENTER (Figure 40).

Figure 40

7. Use Fill Across to copy this formula across the row to C9 and D9.

8. Use Fill Down to copy the row B9:D9 to B11:D11, which gives the sum (Figure 41).

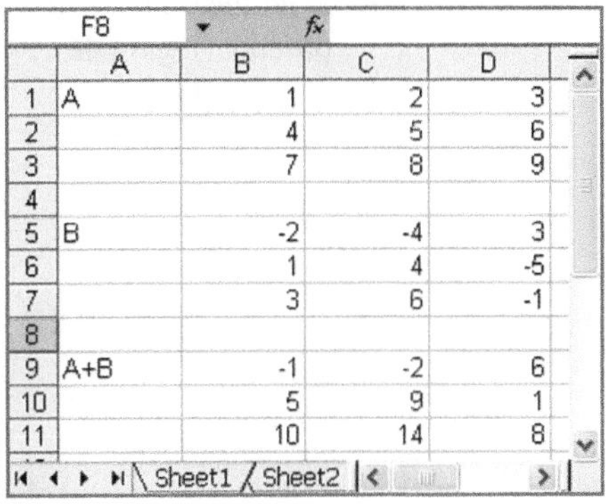

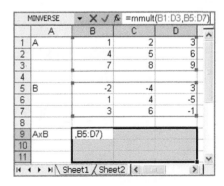

Figure 41

9. To subtract the matrices, change the formula in B9 to "=B1−B5" (Figure 42) and proceed as with addition (Figure 43).

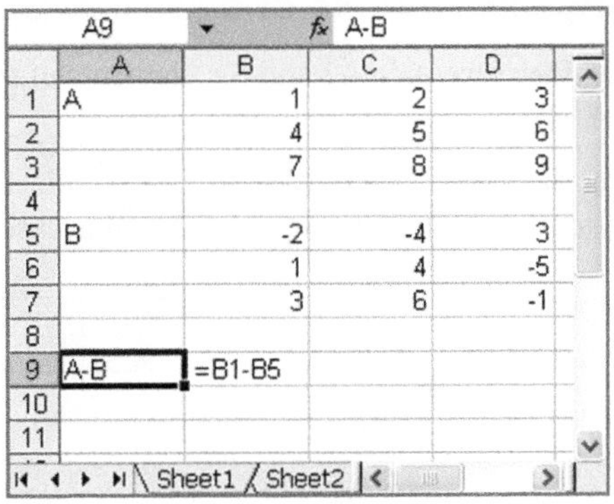

Figure 42 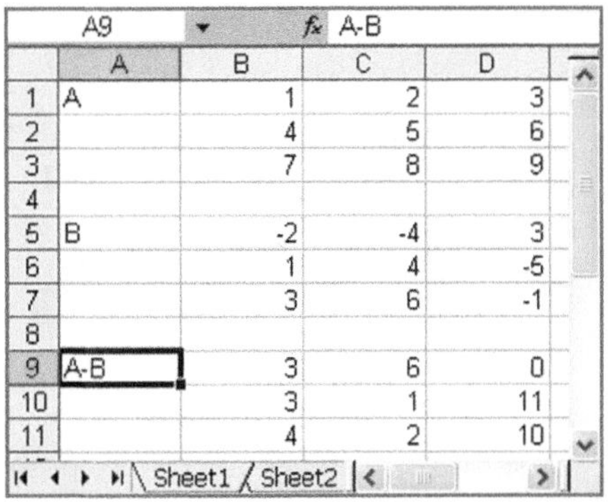   **Figure 43**

Finding the Product of Two Matrices (steps for two 3 × 3 matrices)

1. Enter the names and elements of the matrices as described on page 648.

2. Enter the name A×B in A9 to indicate the matrix product (Figure 44).

3. Select a range of cells that is the correct size to contain the product (B9:D11 in this case).

4. Type "=mmult(" in the formula bar, and then select the cells containing the elements of matrix A (B1:D3 in this case) (Figure 45).

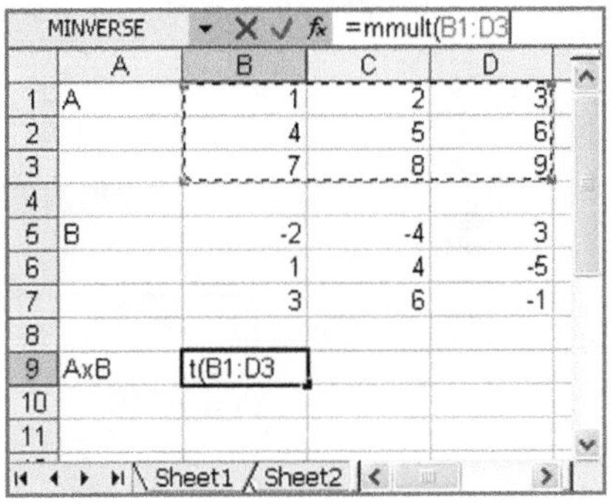

Figure 44   **Figure 45**

5. Staying in the formula bar, type a comma, select the matrix B elements (B5:D7 in this case), and close the parentheses.

6. Hold the CTRL and SHIFT keys down and press ENTER, which will give the product (Figure 46).

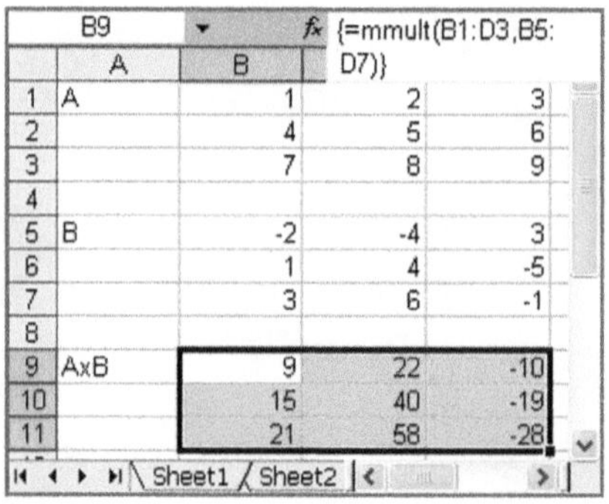

Figure 46

Finding the Inverse of a Matrix (steps for a 3 × 3 matrix)

1. Enter the name A in A1 and the elements of the matrix in B1:D3, as described on page 648.

2. Enter the name "Inverse(A)" in A5 and select a range of cells that is the correct size to contain the inverse (B5:D7 in this case) (Figure 47).

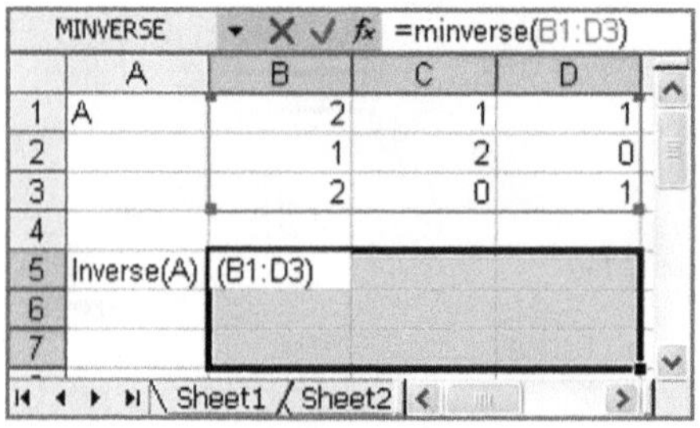

Figure 47

3. In the formula bar, enter "=minverse(", select matrix A (B1:D3), and close the parentheses.

4. Hold the CTRL and SHIFT keys down and press ENTER, which will give the inverse (Figure 48).

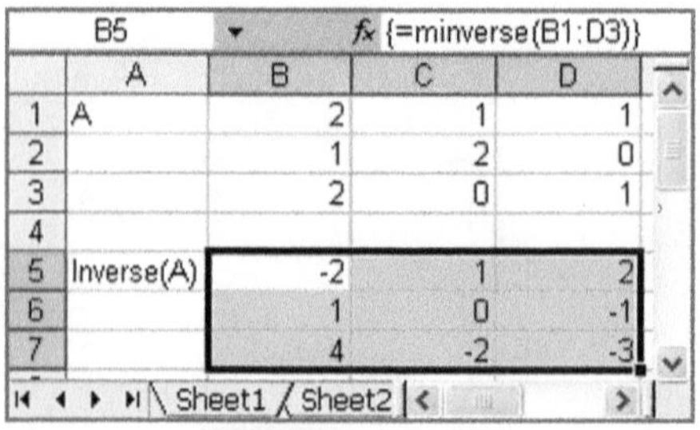

Figure 48

Solving Systems of Linear Equations with Matrix Inverses

A system of linear equations can be solved by multiplying the matrix containing the augment by the inverse of the coefficient matrix. Following are the steps used to solve

$$\begin{cases} 2x + y + z = 8 \\ x + 2y \quad\;\; = 6 \\ 2x \quad\;\;\; + z = 5 \end{cases}$$

1. Enter the coefficient matrix A in B1:D3.

2. Compute the inverse of A in B5:D7, as described on page 650 (Figure 49).

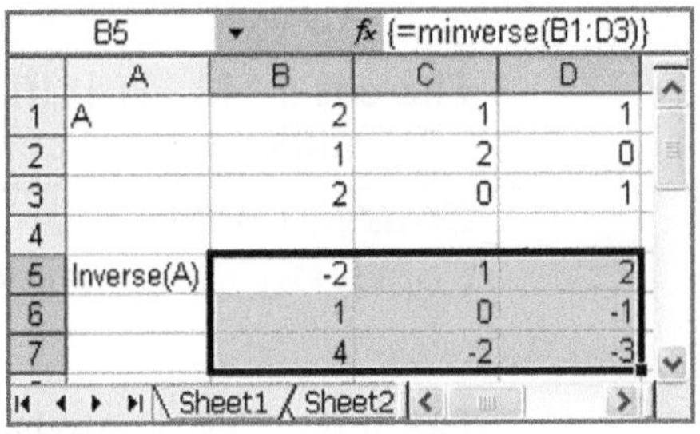

Figure 49

3. Enter B in cell A9 and enter the matrix containing the augment in B9:B11.

4. Enter X in A13 and select the cells B13:B15.

5. In the formula bar, type "=mmult(", select matrix inverse(A) in B5:D7, type a comma, select matrix B in B9:B11, and close the parentheses (Figure 50).

6. HOLD the CTRL and SHIFT keys down and press ENTER, which will give the solution.

7. Matrix X gives the solution $x = 0$, $y = 3$, $z = 5$ (Figure 51).

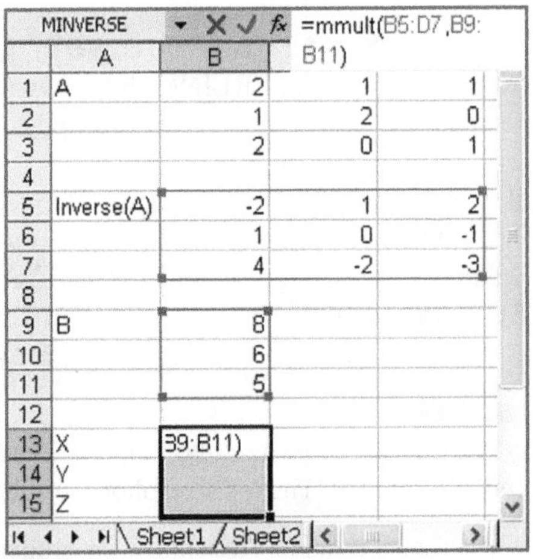

Figure 50

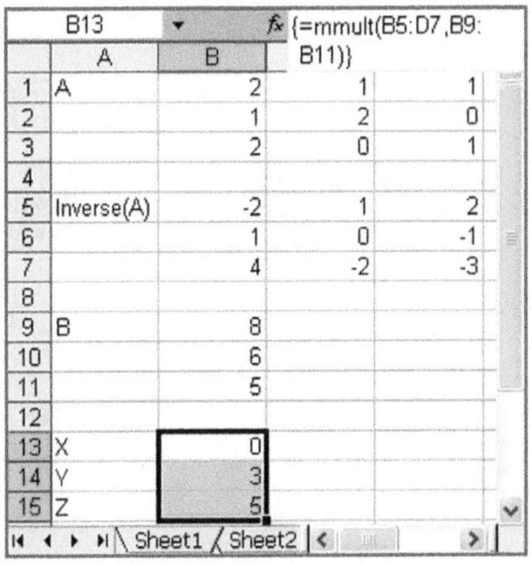

Figure 51

MATH 1101 Introduction to Mathematical Modeling

Instructor: _____

Office: _____COE (College Of Education)

Office phone: 404-413-_____

E-mail: _____@gsu.edu

Office hours: MWF or by appointment

There is a MATH 1101 Academic Assistance (for questions of math 1101 students) held at MW pm-pm in 120 Kell Hall

Course Location & Time: _____ GCB am pm

MyMathLab COURSE ID: _____

Textbook: College Algebra in Context, Georgia State University Edition, by Harshbarger and Yocco, Pearson, and

MyMathLab access code (all homework and quizzes will be delivered online using this software maintained by Pearson)

To register for MyMathLab, you will need:

1. MyMathLab access code. At the GSU bookstore, all new books come bundled with an access code.
2. Course ID which is _____

How do I register for my course?

Please go to the website: HYPERLINK "http://www.coursecompass.com/html/student_getting_started.html"http://www.coursecompass.com/html/student_getting_started.html

It will walk you through the process step by step.

***NOTE** When you register for the course on MyMathLab,

1. You are **expected** to use your **GSU EMAIL.**
2. GSU's zip code is 30303

Important Notes:

1. MyMathLab includes live tutor support available from 5pm to midnight, Sunday through Thursday. The toll free phone number is 888-777-0463.

2. MyMathLab Technical Assistance for Students: 1-800-677-6337

3. Since Georgia State **does not** support MyMathLab (MML) or Course Compass (Pearson Education- the publisher of the textbook supports this software), **it is the responsibility of the student** to use the resources above to resolve all technical issues independently of the University. GSU and its

faculty are not responsible for outcomes due to individual technical issues, nor scheduled MML and Course Compass downtime. It is expected that the student will be responsible for completing their work in a timely fashion as to alleviate any pressures these scheduled downtimes cause. All students will be notified of these downtimes through the announcements page of the course.

Note: This course syllabus only provides a general plan for the course; deviations may be necessary.

1. THE COURSE

a. **Course Description:** We will explore mathematical modeling using graphical, numerical, symbolic, and verbal techniques to describe and explore real-world data and phenomena. Emphasis is on the use of elementary functions to investigate and analyze applied problems and questions, on the use of appropriate supporting technology, and on the effective communication of quantitative concepts and results.

b. **THIS COURSE IS NOT AN APPROPRIATE PREREQUISITE FOR COLLEGE ALGEBRA, PRECALCULUS OR CALCULUS.** You are responsible for understanding the implications of taking MATH 1101. Please discuss your situation with your instructor if necessary.

c. **Prerequitites:** Knowledge of high school algebra II, or equivalent. This includes algebraic expressions, first degree equations and inequalities, exponents, radicals, solving and graphing linear equations, factoring quadratic expressions, and other topics.

d. **Course Coverage:** We will cover the following sections from the text:

    Chapter 1   Functions, Graphs, and Models; Linear Functions (1.1 - 1.4)
    Chapter 2   Linear Models, Equations and Inequalities (2.1 - 2.4)
    Chapter 3   Quadratic and Other Nonlinear Functions (3.1 - 3.4)
    Chapter 4   Additional Topics with Functions (4.1 - 4.3)
    Chapter 5   Exponential and Logarithmic Functions (5.1 - 5.6)

Additional materials needed for the course will be distributed electronically throughu Learn at http://ulearn.gsu.edu, or by email, or as a handout during class.

2. **METHOD OF EVALUATION.** Your course grade will be determined as follows:

a. **Tests (50%).** Four closed book/closed notes tests (12.5% each test) are scheduled during the regular semester (see attached schedule). Tests may not be made up (unless it is a documented University excused absence…see the Student handbook for details).

b. **Homeworks (using MML online) (10%).** There will be mandatory homework offered in an online format for you to work and receive instant feedback!! Of course, as it states, this is a mandatory part of your grade and will be graded by the software. Please note the due dates as there will NOT be any extensions given. There will be an Amnesty day on Friday, in which all online assignments (excluding exams/projects) will be reopened for you to work.

c. **Quizzes (using MML online) (5%)** – There will be mandatory quizzes offered in an online format. Each quiz will be a subset of the homework completed for the specific textbook section. You will be given 60 minutes to complete the quiz and given two attempts. The higher of the two grades will stand. Please note the due dates as there will NOT be any extensions given. There will be an Amnesty day on Friday, December 3 in which all online assignments (excluding exams/projects) will be reopened for you to work.

d. **Semester Project (10%).** Two Excel projects will be assigned during the semester. The Excel software is available on the computers in the GSU computer labs. Basic instructions for using Excel will be given in class. The projects will be collected and graded. The average score of the two projects will be the project grade.

e. **Final Exam (25%).** The two and a half hour final exam will be closed book, closed notes format and will be comprehensive. It will be held in our regular classroom. A missed final exam will result in a score of zero. You may use this final exam grade to replace a missed exam grade. ***Your final exam score may be used to replace the lowest of the four regular semester exams if you take all four exams.***

Example of Course Grade Computation:

Test Grades: T1 = 88, T2 = 72, T3 = 68, T4 = 87

MML homework avg = 95

MML quiz average = 85

Excel Project average = 90

Final Exam=76

Course Grade: 0.125*(88+72+76+87) +0.10*95+ 0.05*85 + 0.10*90 + 0.25*76

3. GRADING SCALE. We will use the following grading scale:

PRIVATE Numeric Ave.	Letter Grade
97 – 100	A+
93 – 96	A
90 – 92	A–
87 – 89	B+
83 – 86	B
80 – 82	B–
77 – 79	C+
70 – 76	C
60 – 69	D
0 – 59	F

4. Makeup Policy: Your final exam grade will replace your missed exam grade. No make-up exams will be given unless in some extreme situations, like university-approved excuses which must be verified in writing. If feasible, written notification in advance is required. Otherwise, it allows two working days for notification. Excuses must have some form of written verification, such as a doctor's note. Absence from the final exam will result in a grade of F for the course unless arrangements are made PRIOR (at least one week before the final exam) to its administration.

5. CALCULATOR Policy. You will be free to use any STAND ALONE calculator (i.e. NOT a part of your cell phone/ipod/pager, etc) or any graphing calculator, but don't forget that you will be asked to provide full working for many questions in your tests and the final. You are not allowed to share calculator with any other party in your class during any in class quiz or exam unless permitted by your instructor.

6. **Academic assistance at GSU:**

   a. Attend academic assistance for MATH 1101 students only, MW pm, 120 Kell Hall.
   b. Visit the Math Assistance Complex (MAC), 122 Kell Hall (phone: 404-413-6462).
   c. Visit the Counseling Center for Learning assistance, Test anxiety classes, and Student support services (phone: 404-413-1641)
   d. African American Student Services (phone: 404-413-1530)
   e. A private tutor list is available at Math Assistance Complex and Math Department

7. **ACADEMIC HONESTY:** Cheating/plagiarism will not be tolerated on any work. A first occurrence will result in a grade of 0 on the assignment for all concerned parties as well as an Academic Dishonesty form being filed with the Dean of Students. A second occurrence will result in a grade of F for the course for the concerned parties and a second Academic Dishonesty form being filed. During in-class quizzes, tests, and the final exam you will be instructed to do your own work, no talking, and not share calculators. Violations of these instructions constitute dishonesty and will be handled in accordance with University policy. The instructor has the option of withholding or denying credit for answers not adequately supported by you.

8. **Inclement Weather Policy:** If the University is closed due to inclement weather, any exam that may have been scheduled for that date will be administered on the next available class date. If an assignment is due that day, it will be due the next class. Please check the website http://www.gsu.edu. This website will communicate whether campus is closed or delayed start.

9. **Attendance Policy:** Attending class is of utmost importance and is your responsibility and yours alone. During class I will clarify important or complex points for you, observe you working problems, and answer your questions.

10. **Conduct Policy:** Please turn off all cell phones, pagers and all other electronic communication devices and keep them off the desk. Text messaging, instant messaging, email, etc. during class is strictly prohibited and is grounds for dismissal. If you are using your cell phone, using your computer for tasks that are not math related, talking, or otherwise disrupting students, you will be asked to leave. After the third incident, you will be administratively removed from the class (as per the Student handbook). Please see the University's Policy on Disruptive Behavior in the *General Catalog*, (www.gsu.edu/images/Downloadables/UG_05_06.pdf) or *On Campus*, the official student handbook (www2.gsu.edu/~wwwdos/codeofconduct_adminpol_a.html).

11. **Withdrawal Policy:** You cannot withdraw from the course simply by ceasing to attend class; you must formally withdraw. If you intend to withdraw, do so before midpoint to avoid a grade of "F" or "WF." See http://www.gsu.edu/es/withdrawals.html

12. COURSE OBJECTIVES.

Algebra. Students will demonstrate the ability to:

a. Graph points.
b. Graph linear, piecewise linear, exponential, logarithmic and quadratic functions, and identify horizontal asymptotes.
c. Determine the equation of a line given two points or one point and the slope.
d. Determine the absolute value of a quantity.
e. Solve and estimate solutions to linear, quadratic, exponential, and logarithmic equations, including use of the properties of exponents and common and natural logarithms.

f. Solve linear systems of two equations by substitution and elimination, including systems that have a unique solution, no solution, or many solutions.

g. Simplify expressions using the laws of exponents and logarithms.

h. Calculate average rate of change of any function.

i. Perform arithmetic calculations to answer questions regarding two-variable data presented in tabular, graphical, or equation form.

j. Express and compare very large and very small numbers using scientific notation and orders of magnitude.

k. Factor quadratic expressions.

l. Complete the square of quadratic expressions.

m. Express the square root of negative numbers in terms of the imaginary unit, i.

n. Given conversion factors, convert units of measure.

o. Use the quadratic formula to solve quadratic equations

Functions. Students will demonstrate:

a. The understanding of the definitions of function, domain, range, independent and dependent variables, and input and output.

b. The ability to determine if tables, graphs, and equations represent functions.

c. The ability to determine the domain and range of functions as mathematical abstractions or in a physical context.

d. The ability to determine from the graph of a function the values of the independent variable for which the function increases, decreases, or remains constant.

Linear and piecewise linear functions. Students will demonstrate the ability to:

a. Determine when two real-world variables are related by a linear or piecewise linear function.

b. Calculate, and interpret average rate of change as slope.

c. Model the behavior of two real-world variables that are directly proportional or are related by a linear or piecewise linear function using tables, graphs, equations.

d. Evaluate linear and piecewise linear functions.

e. Use a linear function to approximate the value of a non-linear function.

f. Interpret the intersection of the graphs of linear functions as equilibrium points.

Quadratic Functions. Students will demonstrate the ability to:

a. Estimate horizontal intercepts of quadratic functions from their graphs.

b. Determine the horizontal intercepts of quadratic functions in factored form.

c. Determine the vertex, axis of symmetry, and horizontal and vertical intercepts of quadratic functions in either the a-b-c or a-h-k forms.

d. Convert quadratic functions from the a-b-c form to the a-h-k form and vice versa.

e. Determine when two real-world variables are related by a quadratic function by calculating the average rate of change of the average rates of change.

f. Model the behavior of two real-world variables that are related by a quadratic function using tables, graphs, equations, or combinations thereof including such applications as maximum area for fixed perimeter, minimum perimeter for fixed area, free fall, maximum profit, and break-even analysis.

Exponential Functions. Students will demonstrate the ability to:

a. Determine when two real-world variables are related by an exponential function.

b. Model the behavior of two real-world variables that are related by an exponential function using tables, graphs, equations, or combinations thereof including such applications as population growth and decay, radioactive decay, simple and compound interest.

c. Change the base of an exponential function to determine rate of growth/decay, growth/decay factor, and effective and nominal interest rate.

d. Express continuous growth/decay in terms of the number e.
e. Evaluate exponential functions.
f. Determine the exponential equation model from the table or graphical model.
g. Compare linear to exponential growth.

Logarithmic Functions. Students will demonstrate:

a. The ability to determine when two real-world variables are related by a logarithmic function.
b. The ability to model the behavior of two real-world variables that are related by a logarithmic function using tables, graphs, equations, or combinations thereof including such applications as Richter scale, pH levels and the decibel system.
c. The understanding of the common and natural logarithms.
d. The ability to graph logarithmic functions.
e. The ability to solve some exponential and logarithmic equations.

Photo Credits

Answers to Selected Exercises

Chapter 1 Functions, Graphs, and Models; Linear Functions

Toolbox Exercises

1. $\{1, 2, 3, 4, 5, 6, 7, 8\}$; $\{x \mid x \in N, x < 9\}$ **2.** Yes
3. No **4.** No **5.** Yes **6.** Yes **7.** Integers, rational
8. Rational **9.** Irrational **10.** $x > -3$ **11.** $-3 \le x \le 3$
12. $x \le 3$ **13.** $(-\infty, 7]$ **14.** $(3, 7]$ **15.** $(-\infty, 4)$
16.
17.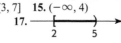
18.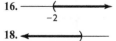
19.

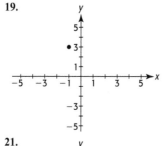

20.

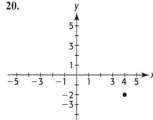

21.

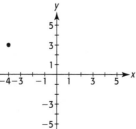

22.

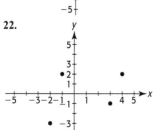

23.

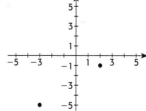

24. Yes; 4 **25.** No **26.** No **27.** Yes; 6
28. $-3x^2, -3; -4x, -4$; constant 8
29. $5x^4, 5; 7x^3, 7$; constant -3
30. $3z^4 + 4z^3 - 27z^2 + 20z - 11$
31. $7y^4 - 2x^3y^4 - 3y^2 - 2x - 9$
32. $4p + 4d$ **33.** $-6x + 14y$ **34.** $-ab - 8ac$
35. $x - 6y$ **36.** $6x + 9xy - 5y$ **37.** $3xyz - 5x$

1.1 Skills Check

1. a. Input **b.** Output **c.** $D = \{-9, -5, -7, 6, 12, 17, 20\}$;
$R = \{5, 6, 7, 4, 9, 10\}$ **d.** Each input gives exactly one output.
3. 5; 9 **5.** No; input 9 for y gives two outputs for x.
7. a. $f(2) = -1$ **b.** $f(2) = -2$ **c.** $f(2) = -3$
9.

x	y
0	2
-2	-4

11. a. -7 **b.** 3 **c.** 18

13. Yes; $D = \{-1, 0, 1, 2, 3\}$; $R = \{5, 7, 2, -1, -8\}$
15. No **17.** Yes **19.** No; one value of x, 3, gives two values of y.
21. b **23.** b **25.** $D = \{-3, -2, -1, 1, 3, 4\}$; $R = \{-8, -4, 2, 4, 6\}$
27. $D = [-10, 8]$; $R = [-12, 2]$ **29.** $x \ge 2$ **31.** All real numbers
except -4 **33.** No **35.** $C = 2\pi r$

1.1 Exercises

37. a. No. **b.** Yes; x, p; there is one closing price p on any given day x.
39. a. Yes; a, p; there is one premium p for each age a. **b.** No
41. Yes; m, r; there is one unemployment rate r for each month m.
43. a. Yes; there is one price for each barcode. **b.** No; numerous
items with the same price have different barcodes. **45.** There is one
output for each input. **47. a.** Yes **b.** The days 1–14 of May
c. $\{171, 172, 173, 174, 175, 176, 177, 178\}$ **d.** May 1, May 3
e. May 14 **f.** 3 days **49. a.** 1096.78; if the car is financed over 3
years, the payment is $1096.78 **b.** $42,580.80; $C(5) = 42,580.80$
c. 4 **d.** $3096.72 **51. a.** 22 million **b.** 11; approximately 11 million
women were in the workforce in 1930. **c.** $\{1930, 1940, 1950, 1960,$
$1970, 1980, 1990, 2000, 2005, 2010, 2015\}$ **d.** For each new indicated
year, the number of women increased. **53. a.** 26.1, 27.1 **b.** 21.5, 25.1
c. 1980; the median age at first marriage was 24.7 in 1980. **d.** Increase
55. a. 41.5 million **b.** 78.0; 78 million U.S. homes used the Internet
in 2008. **c.** 1998 **d.** Increasing; it has increased rapidly.
57. a. 56.0; the birth rate in 1995 for U.S. girls ages 15–19 was 56.0
per 1000. **b.** 2005 **c.** 1990 **d.** The birth rate increased until 2007,
then decreased. **59. a.** 6400; the revenue from the sale of 200 hats is
$6400. **b.** $80,000; $R(2500) = 80,000$ **61. a.** 876.35; the charge
for 1000 kWh is $876.35. **b.** $1304.85 **63. a.** 1200; the daily profit
from the sale of 100 bicycles is $1200. **b.** $1560 **65. a.** All real
numbers except $-\dfrac{3}{7}$ **b.** Positive values of n **67. a.** $0 \le p < 100$
b. 355,500; 2,133,000 **69. a.** 8640; 11,664 **b.** $0 < x < 27$, so volume is positive and box exists **c.** Testing values in the table shows a
maximum volume of 11,664 cubic inches when $x = 18$ inches.

x	y
10	6,800
15	10,800
20	11,200
21	10,584
19	11,552
18	11,664
17	11,560

The dimensions of the box are 18 in. by
18 in. by 36 in.

1.2 Skills Check

1. a.

x	-3	-2	-1	0	1	2	3
$y = x^3$	-27	-8	-1	0	1	8	27

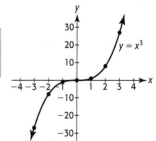

b.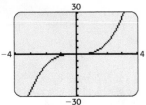

c. Graphs are the same.

13. 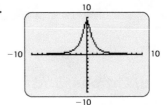 ; yes

3. Answers vary.

x	$f(x)$
-2	-7
-1	-4
0	-1
1	2
2	5

$f(x) = 3x - 1$

15. 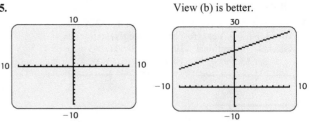   View (b) is better.

5. Answers vary.

x	$f(x)$
-4	8
-2	2
0	0
2	2
4	8

$f(x) = \frac{1}{2}x^2$

17. 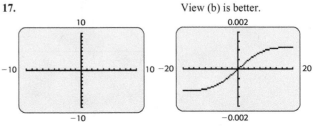   View (b) is better.

19. Letting y vary from -5 to 100 gives one view.

Turning point coordinates: $(0, 50)$

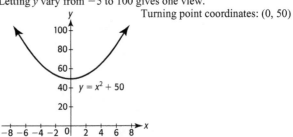

$y = x^2 + 50$

7. Answers vary.

x	$f(x)$
-2	$-1/4$
-1	$-1/3$
0	$-1/2$
1	-1
2	undefined
3	1
4	$1/2$

$f(x) = \dfrac{1}{x - 2}$

21. Setting y from -200 to 300 gives one view.

Turning point coordinates: $(-5, 175), (3, -81)$

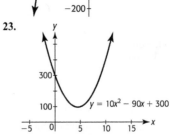

$y = x^3 + 3x^2 - 45x$

9. 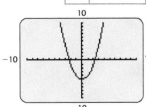 ; yes

23.

$y = 10x^2 - 90x + 300$

11. 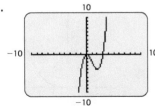 ; yes

25.

t	S
12	51.9
16	72.7
28	135.1
43	213.1

27.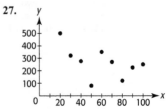

29. a.

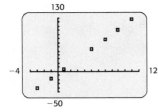

b.

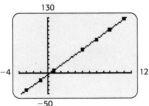

c. Yes; yes

31. a. 300   **b.** 2020

1.2 Exercises

33. a. 16.431; in 1944, there were 16,431,000 women in the workforce.
b. 76,227,000   **35. a.** 1; 19   **b.** 2005; 65.13   **c.** 0; 20

37. a.

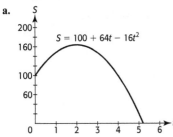

b. 148 ft and 148 ft; ball rising and falling
c. 164 ft, in 2 seconds

39. a.

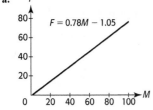

b. $48,090

41. a.

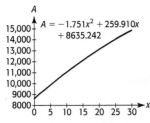

b. $11,931 million, $12,746 million
c. Yes

43. a.

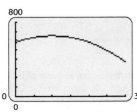

b. 1980 through 2010
c. Decrease
d. 641.24 in 1994 and 396.3 in 2009; decreasing
e. Yes

45.

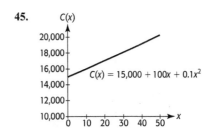

47.

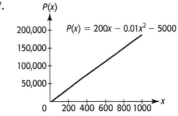

49. a. 0 through 19   **b.** 31,148.869, 54,269.57
c.

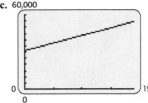

51. a. 299.9 million, or 299,900,000
b.

Years after 2000	Population (millions)
0	275.3
10	299.9
20	324.9
30	351.1
40	377.4
50	403.7
60	432.0

c.

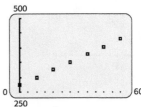

53.

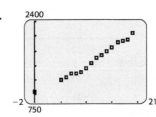

55. a. 6.0%   **b.**

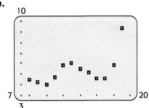

c.

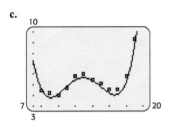

c.

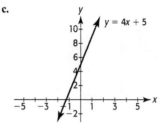

21. a. Slope -100; y-intercept 50,000   **b.** Falling

c.

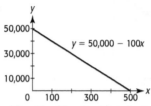

23. 4   **25.** -15 **27.** -2 **29. a.** ii   **b.** i   **31. a.** 0   **b.** 0

1.3 Skills Check

1. b   **3.** $-\dfrac{1}{2}$ **5.** 2   **7. a.** x: 3; y: -5

b.

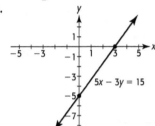

9. a. x: $\dfrac{3}{2}$; y: 3   **b.**

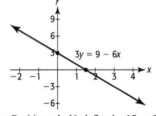

11. 0; undefined   **13. a.** Positive   **b.** Undefined   **15. a.** Slope 4; y-intercept 8   **b.**

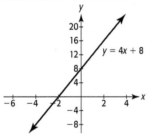

17. a. Slope 0; y-intercept $\dfrac{2}{5}$ **b.**

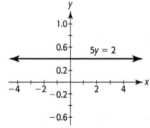

19. a. Slope 4; y-intercept 5   **b.** Rising

1.3 Exercises

33. Yes; it is written in the form $y = ax + b$; x, the number of years after 1990   **35. a.** It is written in the form $y = ax + b$.   **b.** -0.146; it has decreased at a rate of 0.146 per thousand per year since 1980.

37. a. $x = \dfrac{405}{7}$ **b.** $p = 48.6$; in 2000, 48.6% of high school seniors had used marijuana.   **c.** Integer values of $x \geq 0$ on the graph represent years 2000 and after.

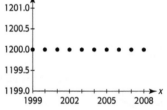

39. a.

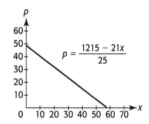

b. Constant function   **c.** $y = 1200$

d.

41. a. Negative   **b.** The percent is decreasing at a rate of 0.504 percentage point per year.   **43. a.** 16,665   **b.** From 2007 to 2010, the number of tweets increased by 16,665 thousand per year.   **c.** No; the model gives an increase of 16,665, which is an increase of 50%.
45. a. Yes   **b.** 0.959   **c.** Minority median annual salaries will increase by $0.959 for each $1 increase in median annual salaries for whites.

47. a. 0.057 **b.** From 1990 to 2050, the percent of the U.S. population that is black increased by 0.057 percentage point per year.
49. x: 50; R: 3500

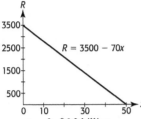

51. a. \$0.975 billion per year **b.** \$6.3 billion
c. No; the amount is negative. **53. a.** $-61,000$ **b.** $-\$61,000$ per year **55.** \$58 per unit **57. a.** 0.56 **b.** \$0.56 per ball **c.** The cost will increase by \$0.56 for each additional ball produced in a month.
59. a. 1.60 **b.** \$1.60 per ball **c.** The revenue will increase by \$1.60 for each additional ball sold in a month. **61.** \$19 per unit

1.4 Skills Check

1. $y = 4x + \dfrac{1}{2}$ **3.** $y = \dfrac{1}{3}x + 3$ **5.** $y = \dfrac{-3}{4}x - 3$ **7.** $x = 9$
9. $y = x + 3$ **11.** $y = 2$ **13.** $y = \dfrac{4}{5}x + 4$ **15.** $y = -3x + 6$
17. $y = \dfrac{3}{2}x + \dfrac{23}{2}$ **19.** $y = 3x + 1$ **21.** $y = -15x + 12$ **23.** 1
25. -3 **27.** -15 **29.** $4x + 2h$ **31. a.** Yes; a scatter plot shows that data fit along a line. **b.** $y = 3x + 555$

1.4 Exercises

33. $y = 12.00 + 0.1034x$ dollars **35.** $y = -3600t + 36,000$
37. a. $y = 2.25x + 22.7$ billion dollars **b.** \$58.7 billion **39. a.** \$25,000
b. \$5000 **c.** $s = 26,000 - 5000t$ **41.** $y = x$; $x = $ deputies;
$y = $ cars **43.** $P(x) = 58x - 12,750$; $x = $ number of units
45. $V = -61,000x + 1,920,000$ **47.** $p = 43.3 - 0.504t$
49. a. 0.045 percentage point per drink **b.** $y = 0.045x$
51. a. $y = 0.504x - 934.46$ **b.** Yes **c.** They are the same.
53. a. 5.572 **b.** \$5.572 billion per year **c.** No **55. a.** -1.1
b. -1.1 births per 1000 girls **c.** The birth rate has decreased over this period. **d.** $y = -1.1x + 104$ **57. a.** 28,664 per year
b. 28,664 **c.** $y = 28,664x - 55,968,047$ **d.** No **e.** Yes
59. a. No; the points in the scatter plot do not lie approximately on a line.
b. 0.2962 million women per year **c.** 0.984 million women per year
d. Yes; since the graph curves, the average rate of change is not constant.
61. a. Increasing at a rate of approximately 2623 thousand people, or approximately 2.623 million people, per year **b.** $y = 2623x + 152,271$ **c.** 217,846 thousand, or 217,846,000; it is close. **d.** The line does not model the data exactly.

Chapter 1 Skills Check

1. Each value of x is assigned exactly one value of y.
2. $D = \{-3, -1, 1, 3, 5, 7, 9, 11, 13\}$;
$R = \{9, 6, 3, 0, -3, -6, -9, -12, -15\}$
3. $f(3) = 0$ **4.** Yes; $y = -\dfrac{3}{2}x + \dfrac{9}{2}$ **5. a.** -2 **b.** 8
c. 14 **6. a.** 1 **b.** -10
7.

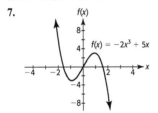

8.

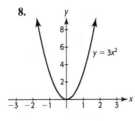

9. Standard window: [0, 40] by [0, 5000] gives a better graph:

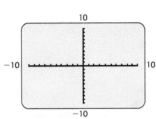 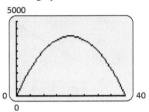

10.

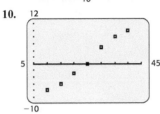

11. a. $x \geq 4$ **b.** All real numbers except 6 **12.** Perpendicular
13. $-\dfrac{7}{6}, \dfrac{6}{7}$ **14.** $-\dfrac{11}{6}$ **15. a.** $(0, -4)$ and $(6, 0)$
b.

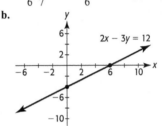

16. $\dfrac{2}{3}$ **17.** Slope -6, y-intercept 3 **18.** -6
19. $y = \dfrac{1}{3}x + 3$ **20.** $y = -\dfrac{3}{4}x - 3$ **21.** $y = x + 4$ **22.** 3
23. a. $5 - 4x - 4h$ **b.** $-4h$ **c.** -4
24. a. $10x + 10h - 50$ **b.** $10h$ **c.** 10

Chapter 1 Review

25. a. Yes **b.** $f(1992) = 82$; in 1992, 82% of black voters supported the Democratic candidate for president. **c.** 1964; in 1964, 94% of black voters supported the Democratic candidate for president.
26. a. $\{1960, 1964, 1968, 1972, 1976, 1980, 1984, 1992, 1996, 2000, 2004, 2008\}$ **b.** No; 1982 was not a presidential election year.
27.

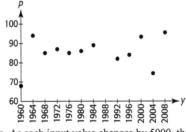

28. a. 0.25
b. 0.25 percentage point per year
c. No

29. a. As each input value changes by 5000, the output value changes by 89.62. **b.** 448.11; the monthly payment to borrow \$25,000 is \$448.11. **c.** $A = 20,000$ **30. a.** $D = \{10,000, 15,000, 20,000, 25,000, 30,000\}$; $R = \{179.25, 268.87, 358.49, 448.11, 537.73\}$
b. No **31. a.** $f(28,000) = 501.882$; if \$28,000 is borrowed, the payment is \$501.88. **b.** Yes **32. a.** $f(1960) = 15.9$; a 65-year-old woman in 1960 was expected to live 15.9 more years, or 80.9 years.
b. $65 + 19.4 = 84.4$ years **c.** 1990 **33. a.** 16.9; a 65-year-old man in 2020 is expected to live 16.9 more years, or 81.9 years. **b.** 77.8 years
c. $g(1990) = 15$ **34. a.** \$42,724.37; $f(10) = 42,724.37$
b. $f(15) = 47,634.67$; the average salary was \$47,634.67 in 2005.
c. Increasing **35. a.** 1.424 **b.** 1,424,000 per year

36. a. 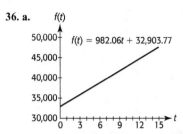 **b.** Years from 1990 to 2005

37. $f(x) = 4500$ **38. a.** $f(x) = 33.8$ **b.** It is a constant function.
39. a. $67,680 **b.** $47,680 **c.** MR: 564; MC: 64 **d.** $m = 64$
e.

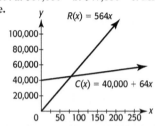

40. a. $P(x) = 500x - 40,000$ **b.** $20,000 **c.** 80 **d.** 500
e. Marginal revenue minus marginal cost **41. a.** $y = 300,000$; the
initial value of the property is $300,000. **b.** $x = 100$; the value of
the property after 100 years is zero dollars. **42. a.** $4.40 per unit
b. Slope 4.4 **c.** $P(x) = 4.4x - 205$ **d.** $4.40 per unit
e. Approximately 47

Chapter 2 Linear Models, Equations, and Inequalities

Toolbox Exercises

1. Division Property; $x = 2$ **2.** Addition Property; $x = 18$
3. Subtraction Property; $x = 5$ **4.** Addition Property; $x = 3$
5. Multiplication Property; $x = 18$ **6.** Division Property; $x = -2$
7. Subtraction Property and Division Property; $x = -10$
8. Addition Property and Multiplication Property; $x = 32$ **9.** $x = 3$
10. $x = \dfrac{3}{5}$ **11.** $x = 16$ **12.** $x = -4$ **13.** $x = 2$ **14.** $x = 1$
15. $x = \dfrac{1}{2}$ **16.** $x = 1$ **17.** $x = 4$ **18.** $x = 5$ **19.** $x = 3$
20. $x = 5$ **21.** Conditional **22.** Contradiction **23.** Identity
24. Conditional **25.** $x > -\dfrac{6}{5}$ **26.** $x \le -2$ **27.** $x > -12$
28. $x > -12$ **29.** $x < \dfrac{-8}{19}$ **30.** $x < \dfrac{6}{11}$ **31.** $x > \dfrac{19}{3}$
32. $x > -16$

2.1 Skills Check

1. $x = -\dfrac{37}{2}$ **3.** $x = 10$ **5.** $x = -\dfrac{13}{24}$ **7.** $x = -63$ **9.** $t = -1$
11. $x = \dfrac{25}{36}$ **13. a.** -20 **b.** -20 **c.** -20 **15. a.** 4 **b.** 4 **c.** 4
17. a. 2 **b.** -34 **c.** 2 **19. a.** 40 **b.** 40 **21. a.** 25 **b.** 25 **c.** 25
23. a. -8.25 **b.** -8.25 **c.** -8.25 **25.** $x = 3$ **27.** $s = -5$
29. $t = -4$ **31.** $t = \dfrac{17}{4}$ **33.** $r = \dfrac{A - P}{Pt}$ **35.** $F = \dfrac{9}{5}C + 32$
37. $n = \dfrac{5m}{2} - \dfrac{P}{4} - \dfrac{A}{2}$

39. $y = \dfrac{5x - 5}{3}$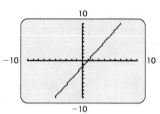

41. $y = \dfrac{6 - x^2}{2}$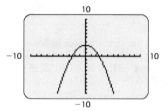

2.1 Exercises

43. 60 months, or 5 years **45.** $6000 **47.** $54,620 **49.** 2020
51. 2010 **53.** 2020 **55.** $x = 55$, so 2015 **57.** 2015 **59.** 1990
61. a. $y = 2.68x + 7.11$ billion dollars **b.** 2020 **63.** 97
65. $78.723 billion **67.** 25% **69.** $1698 **71.** $t = \dfrac{A - P}{Pr}$ **73.** 10%
75. $613.33 **77.** Yes; 2π **79. a.** 4/9 **b.** 72 kg

2.2 Skills Check

1. No; data points do not lie close to a line. **3.** Approximately; all the
data points do not lie on the line, but are close to the line.
5.

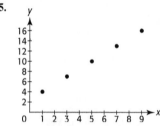

7. Exactly; first differences of the outputs are equal. **9.** $y = 1.5x + 2.5$
11.   **13.** $y = 2.419x - 5.571$

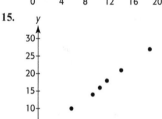

15.   **17.** $y = 1.577x + 1.892$
19. $y = -1.5x + 8$
21. a. Exactly
b. Nonlinear
c. Approximately

2.2 Exercises

23. a. Discrete **b.** Continuous **c.** No; a line will not fit the points well.
25. No; increases from one bar top to the next are not constant.

27. a. Yes; the first differences are constant for uniform inputs.
b. $T = 0.15x - 835$ **c.** They agree.   **d.** $3683.75; interpolation
e. No, just tax due for income in the 15% tax bracket
29. a. $y = 0.465x + 12.049$ **b.** 20.4%   **c.** $x = 28$, so 2028
31. a. $y = 25.568x + 1328.171$ **b.** 1583.9 million (1.5839 billion)
c. 2016   **33. a.** $y = 0.323x + 25.759$ **b.** Not close to an exact fit
35. a.

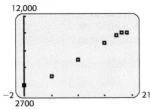

b. Yes

c. $y = 352.933x + 3533.995$ **d.** $11,298.512 billion
37. a.

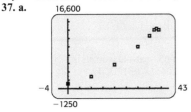

b. $y = 365.910x - 213.561$ **c.** The line is a good fit.

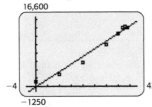

39. a. $S = 122.914x - 477.438$ **b.** $122.914 million   **c.** 16.92%

d. $119.6 million per year   **41. a.** $D = \dfrac{5}{11}W$ **b.** Fits exactly

c. 68.18 mg   **d.** Discretely for integer weights
43. a.

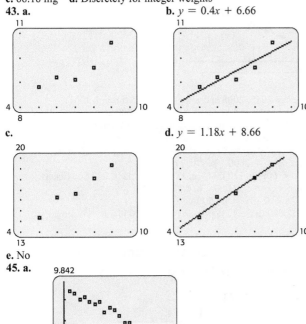

b. $y = 0.4x + 6.66$

c.

d. $y = 1.18x + 8.66$

e. No
45. a.

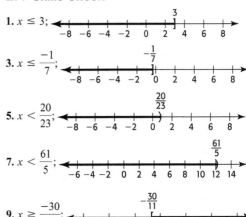

b. $y = -0.139x + 9.633$

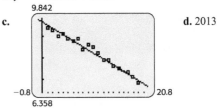

c.

d. 2013

2.3 Skills Check

1. a. No   **b.** Yes   **3.** $(1, 1)$ **5.** $x = -14, y = -54$
7. $x = \dfrac{-4}{7}, y = \dfrac{4}{7}$ **9.** Many solutions; the two equations have the
same graph.   **11.** $x = 2, y = -2$ **13.** $x = 1, y = -1$
15. $x = 2, y = 1$ **17.** $x = \dfrac{4}{7}, y = \dfrac{12}{7}$ **19.** $x = 4, y = 3$
21. Dependent; many solutions   **23.** No solution
25. $x = 2, y = 4$ **27.** $x = \dfrac{32}{11}, y = \dfrac{-14}{11}$ **29.** $x = 3, y = -2$
31. $x = 2, y = 1$ **33.** No solution

2.3 Exercises

35. 60 units   **37.** 682 units
39. a. $260   **b.** 30
41. a. $y = 0.69x + 2.4$ **b.** $y = -0.073x + 7.7$ **c.** 7
43. a. 1987   **b.** Approximately 521,787
45. $x = 19.8$, so 2000
47. $669.8 million in 2008, $805.3 million in 2011
49. a. $x + y = 2400$ **b.** $30x$ **c.** $45y$ **d.** $30x + 45y = 84{,}000$
e. 1600 $30 tickets, 800 $45 tickets
51. a. $75,000 at 8%, $25,000 at 12%   **b.** 12% account is probably
more risky; investor might lose money.   **53.** $200,000 at 6.6%,
$50,000 at 8.6%   **55.** 60 cc of 10% solution, 40 cc of 5% solution
57. 3 glasses of milk, 2 servings of meat   **59.** 2031; 11%
61. a. $p = -\dfrac{1}{2}q + 155$ **b.** $p = \dfrac{1}{4}q + 50$ **c.** $85

63. a. $x = -0.39$ b. No; Peru's is greater after 1990.
65. a. $300x + 200y = 100{,}000$ **b.** 250 in the first group, 125 in
the second group   **67.** 700 units at $30; when the price is $30, the
amount demanded equals the amount supplied equals 700.

2.4 Skills Check

1. $x \le 3$;

3. $x \le \dfrac{-1}{7}$;

5. $x < \dfrac{20}{23}$;

7. $x < \dfrac{61}{5}$;

9. $x \ge \dfrac{-30}{11}$;

11. $x \ge 2.8\overline{6}$;

13. $(-\infty, -2)$ **15.** $[-8, \infty)$ **17. a.** $x = -1$ **b.** $(-\infty, -1)$
19. $\frac{22}{3} \leq x < 12$ **21.** $\frac{5}{2} \leq x \leq 10$ **23.** No solution
25. $x \geq \frac{32}{11}$ or $x \leq \frac{3}{4}$ **27.** $72 \leq x \leq 146$

2.4 Exercises

29. a. $V = 12,000 - 2000t$ **b.** $12,000 - 2000t < 8000$
c. $12,000 - 2000t \geq 6000$ **31.** $C \leq 0$ **33.** More than $22,000
35. 80 through 100 **37. a.** $x = \frac{82.1 - y}{2.1}$ **b.** $42.2 \leq y \leq 61.1$
39. a. 2007 **b.**

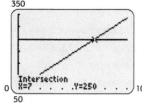

c. Before 2007 **41.** $1350 \leq x \leq 1650$ **43.** Before 1994; after 2001
45. a. $y = 20,000x + 190,000$ **b.** $11 \leq x \leq 14$ **c.** No
47. More than 2000 units **49.** At least 1500 feet
51. $364x - 345,000 > 0$ **53.** 2015 and after **55. a.** About 44.5%
b. $x \leq 12$ **c.** Until 2002

Chapter 2 Skills Check

1. $x = \frac{34}{5}$ **2.** $x = -14.5$ **3.** $x = -138$ **4.** $x = \frac{5}{28}$ **5.** $x = \frac{52}{51}$
6. $x = 5$ **7. a.** 15 **b.** 15 **c.** 15 **8.** $y = \frac{3 + m - 3Pa}{-3P}$
9. $y = \frac{4x - 6}{3}$

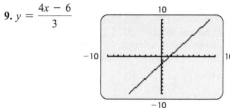

10.

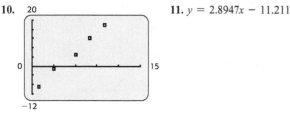

11. $y = 2.8947x - 11.211$

12.

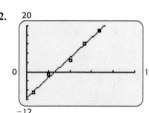

13. No **14.** $x = 2, y = -3$ **15.** $x = \frac{-3}{13}, y = \frac{-15}{13}$ **16.** Many
solutions **17.** No solution **18.** $x = -3, y = 5$
19. $x = \frac{1}{2}, y = -4$ **20.** $x < -\frac{4}{5}$ **21.** $x \leq \frac{25}{28}$ **22.** $6 \leq x < 18$

Chapter 2 Review

23. a. Yes **b.** Yes **c.** $P = f(A) = 0.018A + 0.010$
24. a. 504.01; the predicted monthly payment on a car loan of $28,000
is $504.01. **b.** Yes **c.** $27,895 **25.** 1998 **26.** $f(x) = 4500$
27. a. $f(x) = 200$ **b.** A constant function **28. a.** $22,000 **b.** More
than $22,000 per month **29.** $25,440 each **30.** $300,000 in the safe
account, $120,000 in the risky account **31.** 1994 **32.** More than
120 units **33. a.** $P(x) = 500x - 40,000$ **b.** $x > 80$ **c.** More than
80 units **34. a.** $x > 10$ **b.** 10 years after its purchase
35. a. $P(x) = 4.4x - 205$
b. At least 47 units **36. a.** $y = 0.0638x + 15.702$
b.

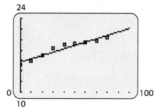

c. 22; the projected life span of a 65-year-old woman is 87 in 2049.
d. 2002 and after **37. a.** $y = 0.0655x + 12.324$
b.

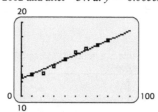

c. 20.8; the projected life span of a 65-year-old man is 86 in 2080.
d. 2144 **e.** Before 2007 **38. a.** A linear equation is reasonable.
b. $y = 271.595x + 16,283.726$ **c.** 21,716,000 **39. a.** A linear
equation is reasonable. **b.** $y = 5.582x + 28.093$ **c.** $150.9 billion
40. a. A linear equation is reasonable. **b.** $y = 0.209x - 0.195$
c. The line seems to fit the data points very well.

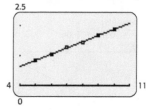

41. Before 2015; 2008 and after **42.** 1995 through 2001
43. Between 17 and 31 months **44.** $140,000 at 8%, $100,000 at
12% **45.** 25 units **46.** Each dosage of medication A contains 2.8
mg, and each dosage of medication B contains 4.2 mg.
47. $p = 100$; $q = 80$ pairs **48.** $p = 79$; $q = 710$ units
49. a. $x + y = 2600$ **b.** $40x$ **c.** $60y$
d. $40x + 60y = 120,000$ **e.** 1800 $40 tickets, 800 $60 tickets
50. a. $x + y = 500,000$ **b.** $0.12x$ **c.** $0.15y$
d. $0.12x + 0.15y = 64,500$
e. $350,000 in the 12% property, $150,000 in the 15% property

Chapter 3 Quadratic, Piecewise-Defined, and Power Functions

Toolbox Exercises

1. $\frac{9}{4}$ **2.** $\frac{8}{27}$ **3.** $\frac{1}{100}$ **4.** $\frac{1}{64}$ **5.** $\frac{1}{8}$ **6.** $\frac{1}{256}$
7. 6 **8.** 4 **9. a.** $x^{3/2}$ **b.** $x^{3/4}$ **c.** $x^{3/5}$

d. $3^{1/2}y^{3/2}$ **e.** $27y^{3/2}$ **10. a.** $\sqrt[4]{a^3}$
b. $-15\sqrt[8]{x^5}$ **c.** $\sqrt[8]{(-15x)^5}$ **11.** $-12a^2x^5y^3$
12. $4x^3y^4 + 8x^2y^3z - 6xy^3z^2$ **13.** $2x^2 - 11x - 21$
14. $k^2 - 6k + 9$ **15.** $16x^2 - 49y^2$ **16.** $3x(x - 4)$
17. $12x^3(x^2 - 2)$ **18.** $(3x - 5m)(3x + 5m)$
19. $(x - 3)(x - 5)$ **20.** $(x - 7)(x + 5)$
21. $(x - 2)(3x + 1)$ **22.** $(2x - 5)(4x - 1)$
23. $3(2n + 1)(n + 6)$ **24.** $3(y - 2)(y + 2)(y^2 + 3)$
25. $(3p + 2)(6p - 1)$ **26.** $(5x - 3)(x - 2y)$
27. a. Imaginary   **b.** Pure imaginary   **c.** Real   **d.** Real
28. a. Imaginary   **b.** Real   **c.** Pure imaginary   **d.** Imaginary
29. $a = 4, b = 0$ **30.** $a = 15, b = -3$ **31.** $a = 2, b = 4$

3.1 Skills Check

1. a. Quadratic   **b.** Up   **c.** Minimum
3. Not quadratic   **5. a.** Quadratic   **b.** Down   **c.** Maximum
7. a.   **b.** Yes

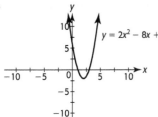

9. a.   **b.** Yes

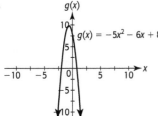

11. a.   **b.** Yes

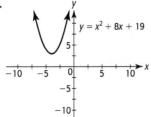

13. a.   **b.** No; the complete graph
                             will be a parabola.

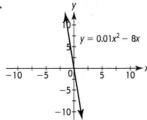

15. $y = (x - 2)^2 - 4$ **17.** y_1
19. $y = -5x^2 + 10x + 8$
21. a. $(1, 3)$ **b.**

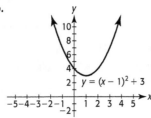

23. a. $(-8, 8)$ **b.**

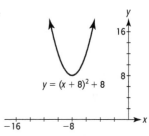

25. a. $(4, -6)$ **b.**

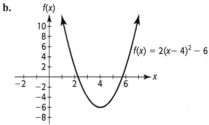

27. a. $(2, 12)$ **b.**

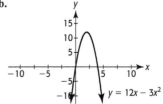

29. a. $(-3, -30)$ **b.**

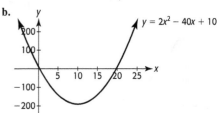

31. a. $x = 10$ **b.**

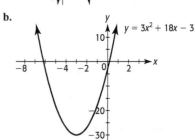

c. $(10, -190)$

33. a. $x = -80$ **b.**

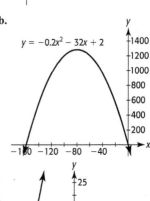

c. $(-80, 1282)$

35.

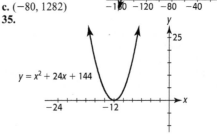

37.

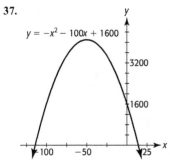

$y = -x^2 - 100x + 1600$

39.

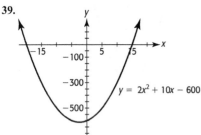

$y = 2x^2 + 10x - 600$

41. $x = 1, x = 3$ **43.** $x = -10, x = 11$
45. $x = -2, x = 0.8$

3.1 Exercises

47. a.

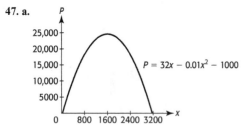

$P = 32x - 0.01x^2 - 1000$

b. It increases.   **c.** Profit decreases.

49. a.   **b.** 577,511

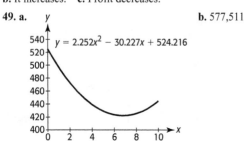

$y = 2.252x^2 - 30.227x + 524.216$

51. a.   **b.** $796.87 billion
                                        c. Extrapolation

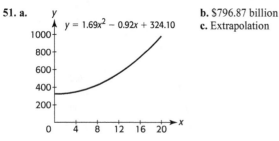

$y = 1.69x^2 - 0.92x + 324.10$

53. a. $t = 2, S = 69.2$ **b.** Two seconds after the ball is thrown, it reaches its maximum height of 69.2 meters.   **c.** The function is increasing until $t = 2$ seconds. The ball rises for two seconds, at which time it reaches its maximum height. After two seconds, the ball falls toward the ground.

55. a.

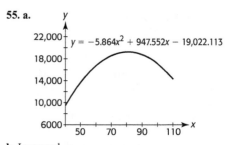

$y = -5.864x^2 + 947.552x - 19,022.113$

b. Increased
c. There are fewer union members when only those employed are included in the data. The graph begins to slope downward.
57. a. 2000   **b.** $37,000   **59. a.** 37,500   **b.** $28,125,000
61. a. Yes   **b.** $A = 100x - x^2$; maximum area of 2500 sq ft when x is 50 ft   **63. a.** (9.38, 29.51)   **b.** 2000   **c.** 29.5%
65. a. Minimum   **b.** (10.184, 33.621); in 2002, the number of people in the United States who lived below the poverty level was at a minimum, 33,621,000.

c.

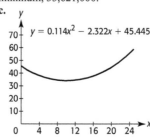

$y = 0.114x^2 - 2.322x + 45.445$

67. a.   **b.** Decreasing   **c.** 25
                                        d. When wind speed
                                        is 0 mph, amount of
                                        pollution is 25 oz per
                                        cubic yard.

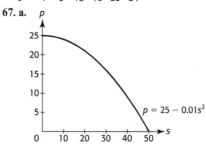

$p = 25 - 0.01s^2$

69. The t-intercepts are 3.5 and -3.75. This means that the ball will strike the pool in 3.5 seconds; -3.75 is meaningless.
71. a. $y = -16t^2 + 32t + 3$ **b.** 19 feet

73. a.

Rent ($)	Number of Apartments	Revenue ($)
1200	100	120,000
1240	98	121,520
1280	96	122,880
1320	94	124,080

b. Yes
c. $1600

75. a. $x = 53.5, y = 6853.3$ **b.** World population will be maximized at 6,853,300,000 in 2044.   **c.** Until 2044

3.2 Skills Check

1. $x = 5, x = -2$ **3.** $x = 8, x = 3$ **5.** $x = -3, x = 2$
7. $t = \dfrac{3}{2}, t = 4$ **9.** $x = -2, x = \dfrac{1}{3}$ **11.** $x = -2, x = 5$
13. $x = \dfrac{2}{3}, x = 2$ **15.** $x = \dfrac{1}{2}, x = -4$ **17.** $w = -\dfrac{1}{2}, w = 3$
19. $x = 8, x = 32$ **21.** $s = 50, s = -15$ **23.** $x = \pm\dfrac{3}{2}$
25. $x = \pm 4\sqrt{2}$ **27.** $x = 2 \pm \sqrt{13}$ **29.** $x = 1, x = 2$
31. $x = \dfrac{5 \pm \sqrt{17}}{2}$ **33.** $x = 1, x = -\dfrac{8}{3}$ **35.** $x = -3, x = 2$

37. $x = \dfrac{2}{3} \approx 0.667, x = -\dfrac{3}{2} = -1.5$

39. $x = \dfrac{-1}{2} = -0.5, x = \dfrac{2}{3} \approx 0.667$

41. $x = \pm 5i$ **43.** $x = 1 \pm 2i$ **45.** $x = -2 \pm 2i$
47. a. Positive **b.** 2 **c.** $x = 3, x = -2$

3.2 Exercises

49. $t = 2$ sec and 4 sec **51.** 20 or 90
53. a. $P(x) = 520x - 10{,}000 - x^2$ **b.** $-\$964$ (loss of \$964)

c. \$5616 **d.** 20 or 500 **55. a.** $s = 50, s = -50$ **b.** There is no
particulate pollution. **c.** $s = 50$; speed is positive. **57. a.** 0 mL or
100 mL **b.** When x is zero, there is no amount of drug in a person's
system, and therefore no sensitivity to the drug. When x is 100 mL,
the amount of drug in a person's system is so high that the person
may be overdosed on the drug and therefore have no sensitivity to the
drug. **59.** \$100 is the price that gives demand = supply = 97 trees

61. a. 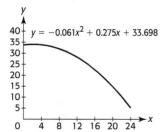 **b.** 1920, 2026

63. a. In 1990, energy consumption in the United States was 87.567
quadrillion BTUs.
b. 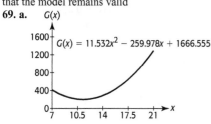   **c.** Yes; energy consumption in
the United States will again be
87.567 when
$x = 78.5$, during 2049.

65. a. Xmin = 0, Xmax = 25, Ymin = 0, Ymax = 40
b. 2012

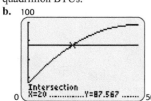

67. a. \$62.85 billion **b.** 2014 **c.** \$264.998 billion; extrapolation;
that the model remains valid
69. a. 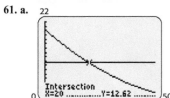 **b.** \$4246.02
c. 2016

71. a. 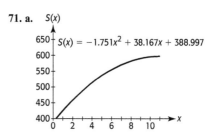 **b.** $x = 5.7187$ and
$x = 16.07$, so in 2006
and 2017 **c.** No;
the model indicates a
maximum in 2011.

73. 2023

3.3 Skills Check

1. 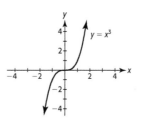 **3.**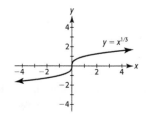

5. 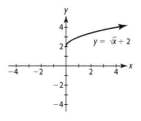 **7.**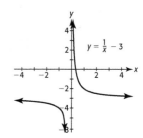

9.

11. a. 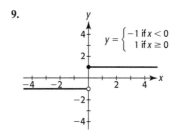 **b.** Piecewise, step function

13. a. 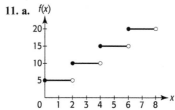 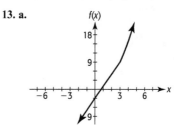 **b.** 5, 16
c. All real numbers

15. a. 0, 3, −3, 5
b.

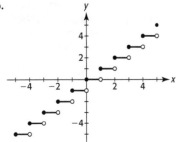

17. a.

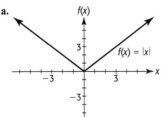

b. 2, 5
c. All real numbers

19. a. 5 **b.** 6 **21. a.** 0 **b.** 29 **23. a.** Increasing **b.** Increasing
25. Concave down **27.** Concave up
29.

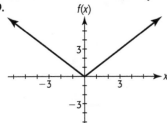

31.

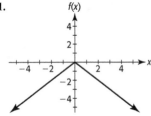

33. They are the same. **35.** $x = \dfrac{7}{2}, x = -\dfrac{5}{2}$ **37.** $x = \dfrac{1}{7}$
39. 16 **41.** 80

3.3 Exercises

43. a. $f(x) = \begin{cases} 7.10 + 0.06747x & \text{if } 0 \le x \le 1200 \\ 88.06 + 0.05788(x - 1200) & \text{if } x > 1200 \end{cases}$

b. $71.87 **c.** $110.05 **45. a.** $P(x) = \begin{cases} 44 & \text{if } 0 < x \le 1 \\ 64 & \text{if } 1 < x \le 2 \\ 84 & \text{if } 2 < x \le 3 \\ 104 & \text{if } 3 < x \le 4 \end{cases}$

b. 64; the postage on a 1.2-oz first-class letter is 64¢.
c. $0 < x \le 4$ **d.** 64, 84 **e.** 64¢, 84¢
47. a. 18,266; 14,832; 7310
b.

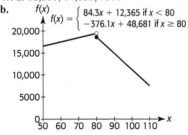

$f(x) = \begin{cases} 84.3x + 12,365 & \text{if } x < 80 \\ -376.1x + 48,681 & \text{if } x \ge 80 \end{cases}$

49. a. Power **b.** 155.196; in 1995, there were 155,196 female physicians. **c.** 358,244
51. a. 61.925%, 67.763%
b. $f(t)$ **c.** Yes, but far in the future

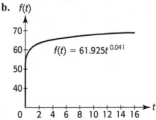

$f(t) = 61.925t^{0.041}$

53. a. Increasing **b.** Concave down **c.** 2022
55. a. Increase **b.** 23.0% **c.** 2012
57. a. y **b.** $0.367, or 37 cents

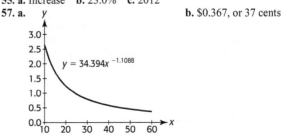

$y = 34.394x^{-1.1088}$

59. 115.2 pounds **61.** 6% **63.** $13,854

3.4 Skills Check

1. $y = 2x^2 - 3x + 1$ **3.** $y = x^2 - 0.5x - 3$

5. $y = \dfrac{x^2}{2} - \dfrac{x}{3} + 6$ **7.** $y = -16x^2 + 32x + 48$

9. $y = 2x^2 - 3x + 1$ **11.** $f(x) = 3x^2 - 2x$
13. The x-values are not equally spaced.
15. $y = 99.9x^2 + 0.64x - 0.75$ **17. a.** $y = 3.545x^{1.323}$
b. $y = 8.114x - 8.067$ **c.** Power function
19. a. $y = 1.292x^{1.178}$ **b.** $y = 2.065x - 1.565$ **c.** They are both
good fits. **21.** $y = 2.98x^{0.614}$

3.4 Exercises

23. a. $y = 3.688x^2 + 459.716x + 2985.640$ **b.** $25,044; $38,564
c. Median income fell in 2009, so this extrapolation may not be
reliable. **25. a.** $y = -54.966x^2 + 4910.104x - 43,958.85$
b. The model appears to fit the data points very well.
c. (44.665, 65,696.78); a 45-year-old earns the maximum median
income of $65,696.78. **27. a.** $y = 0.0052x^2 - 0.62x + 15.0$
b. 50 mph **c.** No; the model estimates the wind chill temperature is a
minimum near 59.4 mph, so the model predicts warmer temperatures
for winds greater than 60 mph.
29. a. y **b.** Yes
                              c. $y = 0.002x^2 -$
                              $0.043x + 0.253$
                              d. 2016

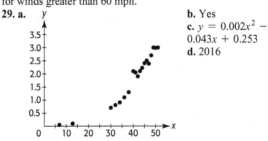

31. a. $y = 511.873x^{-0.046}$ **b.** 454.7 crimes per 100,000 residents

33. a. 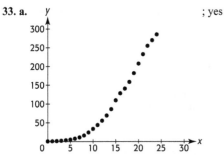 ; yes

b. $y = 0.632x^2 - 2.651x + 1.209$ **c.** 359.57 million
d. over 100%   **35. a.** $y = 0.011x^2 - 0.957x + 24.346$
b. (43.5, 3.531); in 2004, the savings rate was 3.5%, the minimum
during 1980–2009.   **c.** 2019   **37. a.** ii, the quadratic model
b. i, $y = 17\sqrt[3]{x}$ **39. a.** $y = 2.471x^{2.075}$ **b.** \$12.9603 trillion
c. $y = 5.469x^2 - 130.424x + 1217.554$ **d.** Quadratic
41. a. $y = 1.172x^2 + 35.571x + 103.745$ **b.** When $x \approx 65.7$,
during 2026
43. a. $y = 222.434x^{0.3943}$ **b.** $y = 1.689x^2 - 0.923x + 324.097$
c. Power   Quadratic

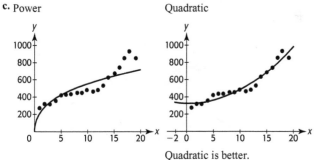

Quadratic is better.

45. a. $y = 3.993x^2 - 432.497x + 12,862.212$
b. $y = 0.046x^{2.628}$ **c.** Quadratic
47. a.

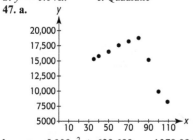

b. $y = -5.009x^2 + 629.699x - 1378.094$ **c.** 8649   **d.** 2013

Chapter 3 **Skills Check**

1. a. (5, 3)   **b.**

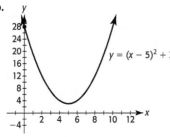

2. a. (−7, −2)   **b.**

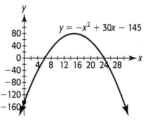

3. a. (1, −27)   **b.**

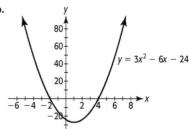

4. a. (−2, −18)   **b.**

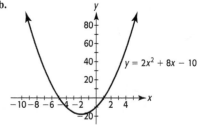

5. a. (15, 80)   **b.**

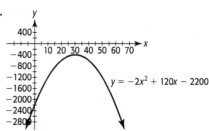

6. a. (30, −400)   **b.**

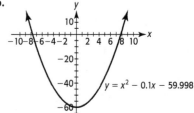

7. a. (0.05, −60.0005)   **b.**

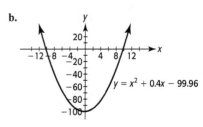

8. a. (−0.2, −100)   **b.**

9. $x = 4, x = 1$ **10.** $x = \frac{1}{2}, x = -\frac{2}{3}$ **11.** $x = -\frac{4}{5}, x = 1$

12. $x = -2, x = \frac{2}{3}$ **13.** $x = 3, x = 1$ **14.** $x = \frac{1}{2}, x = -\frac{3}{2}$

15. a. 4, −2   **b.** 4, −2   **16. a.** −5, 1   **b.** −5, 1
17. 2, −2   **18.** 9, −1   **19.** $z = 2 \pm i\sqrt{2}$ **20.** $w = 2 \pm i$
21. $x = \dfrac{5 \pm i\sqrt{23}}{8}$ **22.** $x = \dfrac{-1 \pm i\sqrt{3}}{4}$

23.

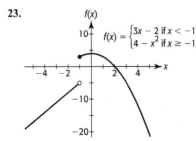

$$f(x) = \begin{cases} 3x - 2 & \text{if } x < -1 \\ 4 - x^2 & \text{if } x \geq -1 \end{cases}$$

24.

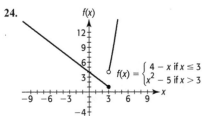

$$f(x) = \begin{cases} 4 - x & \text{if } x \leq 3 \\ x^2 - 5 & \text{if } x > 3 \end{cases}$$

25.

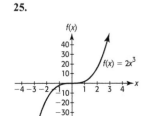

$f(x) = 2x^3$

26.

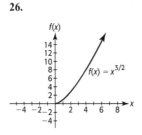

$f(x) = x^{3/2}$

27.

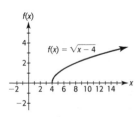

$f(x) = \sqrt{x - 4}$

28.

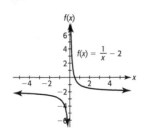

$f(x) = \dfrac{1}{x} - 2$

29.

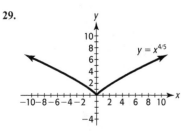

$y = x^{4/5}$

30.

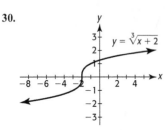

$y = \sqrt[3]{x + 2}$

31. a. Increasing   **b.** Decreasing   **32. a.** Concave up
b. Concave down   **33.** $x = 10, x = -6$ **34.** $x = 5, x = -8$
35. $y = 2x^2 - 3x - 2$ **36.** $y = -2x^2 + 4x + 7$

37. $y = 3.545x^{1.323}$ **38.** $y = 1.043x^2 - 0.513x + 0.977$
39. 128   **40.** $-26, -4, -5$

Chapter 3 **Review**

41. a. 3100   **b.** $84,100   **42. a.** 4050   **b.** $312,050
43. a. 2 sec   **b.** 256 ft   **44. a.** 1.5 sec   **b.** 82.05 m
45. a. 2004   **b.** 137 thousand   **c.** 2008
46. 30 or 120   **47.** 5 sec   **48.** 400 or 3700
49. a.

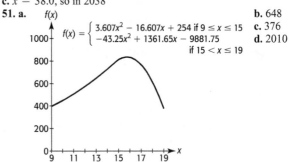

$$y = \begin{cases} 0.08x^2 - 2.64x + 22.35 & \text{if } 15 \leq x \leq 45 \\ -0.525x + 89.82 & \text{if } 45 < x \leq 110 \end{cases}$$

b. 42.57%   **c.** Yes; the model gives 32.07%.
50. a. $y = -0.00482x^2 + 0.754x + 8.512$ **b.** 22.8%
c. $x = 38.0$, so in 2038
51. a.

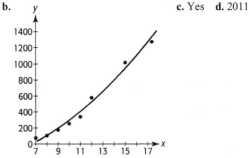

$$f(x) = \begin{cases} 3.607x^2 - 16.607x + 254 & \text{if } 9 \leq x \leq 15 \\ -43.25x^2 + 1361.65x - 9881.75 & \\ & \text{if } 15 < x \leq 19 \end{cases}$$

b. 648   **c.** 376   **d.** 2010

52. 2506 ha
53. a. $y = 5.126x^2 + 0.370x - 221.289$
b.   **c.** Yes   **d.** 2011

54. a. $y = 3.383x^2 + 268.507x + 1812.701$ **b.** 1992   **c.** 2031
55. $y = 18.624x^2 - 440.198x + 20{,}823.439$ thousand people in the
United States in year x
56. a. $y = 0.05143x^2 + 3.18286x + 65.40000$ **b.** 16 yr
57. a. $y = -2.7x + 113$ **b.** $y = 28.369x^{0.443}$

c. $f(x) = \begin{cases} -2.7x + 113 & \text{if } 0 \leq x \leq 10 \\ 28.39x^{0.443} & \text{if } 10 < x \leq 19 \end{cases}$

d. i. 99.5 thousand people employed   **ii.** 1999, 2004   **iii.** 2016
58. a. $y = 780.23x^{0.153}$ **b.** 2053   **59. a.** $y = 1.92x + 8.147$, good
fit   **b.** $y = 0.037x^2 + 1.018x + 12.721$, better fit than linear model
c. linear: $56.2 billion; quadratic: $61.4 billion
60. a. $y = -0.288x^2 + 21.340x - 31.025$ **b.** $y = 14.747x^{1.002}$
c. The power model, as the quadratic model will turn and begin
decreasing

Chapter 4 Additional Topics with Functions

Toolbox Exercises

1. $(-\infty, 0) \cup (0, \infty); (-\infty, 0) \cup (0, \infty)$ **2.** $(-\infty, \infty); \{k\}$
3. $(-\infty, 0) \cup (0, \infty)$ **4.** $(0, \infty); (-\infty, 0)$ **5.** $[0, \infty)$
6. $(-\infty, \infty)$ **7.** Increasing **8.** Increasing **9.** Decreasing
10. Increasing **11.** Decreasing **12.** Increasing
13. Cubing **14.** Square root

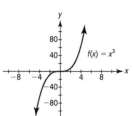

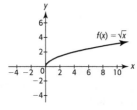

15. Power function, with power $1/6$ **16.** Power function, with power -3

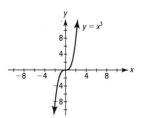

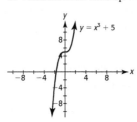

17. Not a function **18.** Not a function **19.** Function
20. Function **21.** Function **22.** Function

4.1 Skills Check

1. a. **b.** Vertical shift 5 units up

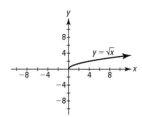

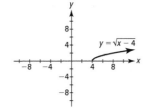

3. a. **b.** Horizontal shift 4 units right

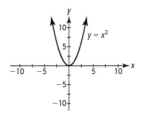

5. a.

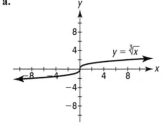

b. Horizontal shift 2 units left, vertical shift 1 unit down

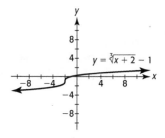

7. a. **b.** Horizontal shift 2 units right, vertical shift 1 unit up

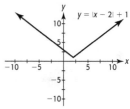

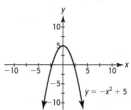

9. a. **b.** Reflection across x-axis, vertical shift 5 units up

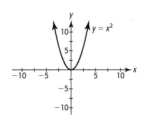

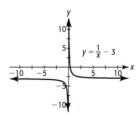

11. a. **b.** Vertical shift 3 units down

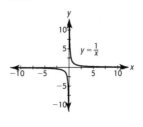

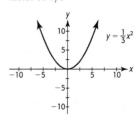

13. a. **b.** Vertical compression using a factor of $1/3$

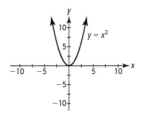

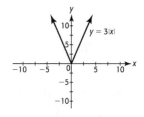

15. a. **b.** Vertical stretch using a factor of 3

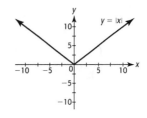

17. Shifted 2 units right and 3 units up   **19.** $y = (x + 4)^{3/2}$
21. $y = 3x^{3/2} + 5$ **23.** $g(x) = -x^2 + 2$
25. $g(x) = |x + 3| - 2$ **27.** y-axis   **29.** x-axis   **31.** Yes
33. y-axis   **35.** Origin   **37.** Origin   **39.** x-axis, y-axis, origin
41. Even   **43.** Even   **45.** Odd   **47.** Even
49.

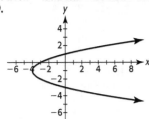

4.1 Exercises

51. a. $y = x^2$ **b.** $M(3) = 25.199$; in 2003, 25,199,000 people 12
and older in the United States used marijuana.
c.

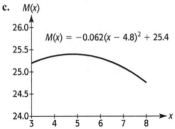

$$M(x) = -0.062(x - 4.8)^2 + 25.4$$

53. a. Linear   **b.** Reciprocal; vertical stretch using a factor of 30,000
and shift down 20 units   **55. a.** Reciprocal function, shift 1 unit left

b.   **c.** Decrease

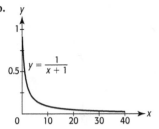

$$y = \frac{1}{x + 1}$$

57. a. Shift 10 units left, shift 1 unit down, reflect about the x-axis,
and vertical stretch using a factor of 1000.
b.

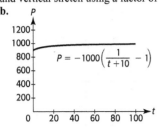

$$P = -1000\left(\frac{1}{t + 10} - 1\right)$$

59. $C(x) = 306.472(x + 5)^{0.464}$
61. $T(x) = -13.898(x + 10)^2 + 255.467(x + 10) + 5425.618$
63. a. 218.75 million   **b.** $C(t) = 0.00056(t + 5)^4$ **c.** $t = 20$; yes
65. a. 11,780   **b.** $S(t) = 638.57(t + 20)^{0.775}$
c. $t = 23$; yes

4.2 Skills Check

1. a. $2x - 1$ **b.** $4x - 9$ **c.** $-3x^2 + 17x - 20$ **d.** $\dfrac{3x - 5}{4 - x}$
e. All real numbers except 4   **3. a.** $x^2 - x + 1$

b. $x^2 - 3x - 1$ **c.** $x^3 - x^2 - 2x$ **d.** $\dfrac{x^2 - 2x}{1 + x}$ **e.** All real

numbers except -1

5. a. $\dfrac{x^2 + x + 5}{5x}$ **b.** $\dfrac{-x^2 - x + 5}{5x}$ **c.** $\dfrac{x + 1}{5x}$ **d.** $\dfrac{5}{x(x + 1)}$

e. All real numbers except 0 and -1 **7. a.** $\sqrt{x + 1} - x^2$

b. $\sqrt{x} - 1 + x^2$ **c.** $\sqrt{x}(1 - x^2)$ **d.** $\dfrac{\sqrt{x}}{1 - x^2}$

e. All real numbers $x \geq 0$ except 1   **9. a.** -8 **b.** 1   **c.** 196   **d.** 3.5

11. a. $6x - 8$ **b.** $6x - 19$ **13. a.** $\dfrac{1}{x^2}$ **b.** $\dfrac{1}{x^2}$

15. a. $\sqrt{2x - 8}$ **b.** $2\sqrt{x - 1} - 7$ **17. a.** $|4x - 3|$ **b.** $4|x - 3|$
19. a. x **b.** x **21. a.** 2   **b.** 1   **23. a.** -2 **b.** -1 **c.** -3
d. -3 **e.** 2

4.2 Exercises

25. a. $P = 66x - 3420$ **b.** \$6480   **27. a.** Cost is quadratic; revenue
is linear.   **b.** $P = 1020x - x^2 - 10,000$ **c.** Quadratic
29. a. $P = 520x - x^2 - 10,000$ **b.** 260   **c.** \$57,600
31. a. $\overline{C}(x)$ is $C(x) = 50,000 + 105x$ divided by $f(x) = x$; that is,
$\overline{C}(x) = \dfrac{C(x)}{x}$.   **b.** \$121.67   **33. a.** $\overline{C}(x) = \dfrac{3000 + 72x}{x}$
b. \$102 per printer
35. a. $S(p) + N(p) = 0.5p^2 + 78p + 12,900$ **b.** $0 \leq p \leq 100$
c. 23,970   **37. a.** $B(8) = 162$ **b.** $P(8) = \$7.54$
c. $(B \cdot P)(8) = \$1221.48$
d. $W(x) = (B \cdot P)(x) = 6(x + 1)^{3/2}(8.5 - 0.12x)$
39. a. $P(x) = 160x - 32,000$ **b.** \$64,000   **c.** \$160 per unit
41. a. No. Adding the percentages is not valid since the percentages
are based on different populations of people. Adding the number of
males and number of females completing college and then dividing by
the total is a legitimate approach.   **b.** 64.78%; 67.76%   **c.** 65.25%;
68.75%; the percentages are relatively close but not the same.
43. a. Meat put in container   **b.** Meat ground   **c.** Meat ground and
ground again   **d.** Meat ground and put in container   **e.** Meat put in
container, then both ground   **f.** d   **45.** $p(s(x)) = x - 18.5$
47. 6.00 Russian rubles   **49.** $100 \cdot \dfrac{g(x)}{f(x)}$ **51.** $f(x) + g(x)$ **53.** 40%

4.3 Skills Check

1. Yes; Domain   Range   **3.** No inverse

5. a. x **b.** Yes   **7.** Yes
9.

x	$f(x)$		x	$f^{-1}(x)$
-1	-7		-7	-1
0	-4		-4	0
1	-1		-1	1
2	2		2	2
3	5		5	3

11. No   **13.** Yes   **15.** Yes   **17.** No   **19. a.** $f^{-1}(x) = \dfrac{x + 4}{3}$
b. Yes   **21.** -2 **23.** $g^{-1}(x) = \dfrac{x - 1}{4}$
25. $g^{-1}(x) = \sqrt{x + 3}$

27.

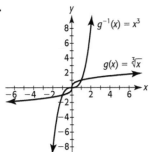

29. Yes; yes

31.

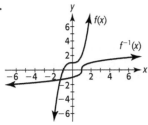

4.3 Exercises

33. a. $t^{-1}(x) = x - 34.5$ **b.** $8\frac{1}{2}$ **35.** $f^{-1}(t) = \dfrac{82.074 - t}{2.087}$; 2010

37. a. $f^{-1}(x) = \dfrac{2.97 - x}{0.085}$; the years after 2000 when the percentage taking antidepressants is $x\%$ **b.** 2008 **39. a.** $W^{-1}(x) = \sqrt[3]{500x}$

b. Given the weight, the inverse function calculates the length.
c. 10 in. **d.** Domain: $x > 0$; range: $W^{-1}(x) > 0$ **41.** Make my day
43. No; two checks could be written for the same amount.

45. a. Yes **b.** $f^{-1}(x) = \sqrt[3]{\dfrac{3x}{4\pi}}$ **c.** Domain: $x > 0$; range: $f^{-1}(x) > 0$

d. To calculate the radius of a sphere given the volume **e.** 25 in.

47. a. $f^{-1}(x) = \dfrac{x}{1.0136}$; dividing Canadian dollars by 1.0136 gives U.S. dollars **b.** \$500 **49. a.** No **b.** $x > 0$; it is a distance, $x \neq 0$.

c. Yes **d.** $I^{-1}(x) = \sqrt{\dfrac{300{,}000}{x}}$; 2 ft **51. a.** $f^{-1}(x) = 0.6154x$;
converts U.S. dollars to U.K. pounds **b.** \$1000

4.4 Skills Check

1. $x = 1$ **3.** $x = -7$ **5.** $x = 6$ **7.** $x = -2, x = 6$
9. $x = 9$ **11.** $x = 23, x = -31$ **13.** $-4 < x < 0$
15. $-3 \leq x \leq 3$ **17.** $4 < x < 5$ **19.** $x \leq -2$ or $x \geq 6$
21. $-1 < x < 7$ **23.** $x \leq 0.314$ or $x \geq 3.186$
25. $x \leq -0.914$ or $x \geq 1.314$ **27.** $x < 0.587$ **29.** $x < 5$
31. $-1 < x < 2$ **33.** $x \leq 4$ or $x \geq 8$ **35. a.** $x \leq -2$ or $x \geq 3$
b. $-2 < x < 3$ **37. a.** No solution **b.** All real numbers
39. $-3 \leq x \leq 2.5$

4.4 Exercises

41. $100 < x < 4000$ units
43. $100 < x < 8000$ units
45. $2 < t < 8.83$ sec
47. 1983 to 1995
49. 1920 to 2026 **51.** 50 to 69.3 mph
53. 1995 to 2005 **55.** 1990 to 2007
57. 2004 to 2018 **59.** 1985 to 2012

Chapter 4 Skills Check

1. Shift 8 units right and 7 units up
2. Shift 1 unit left, vertical stretch using a factor of 2 units, reflect about the x-axis
3. a.

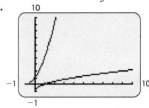

b. The graph of $g(x)$ can be obtained from the graph of $f(x)$ by shifting 2 units left and 3 units down.

4. $[-2, \infty)$ **5.** $y = (x - 6)^{1/3} + 4$ **6.** $y = 3x^{1/3} - 5$
7. f **8.** c **9.** e **10.** Origin **11.** y-axis **12.** Odd
13. $3x^2 + x - 4$ **14.** $9 - x^3 - 6x$ **15.** $18x^3 - 42x^2 + 20x$
16. $\dfrac{5 - x^3}{6x - 4}$, $x \neq \dfrac{2}{3}$ **17.** 38 **18.** $108x^2 - 174x + 68$
19. $18x^2 - 30x - 4$ **20.** 188 **21. a.** x; x
b. They are inverse functions. **22.** $f^{-1}(x) = \dfrac{x + 2}{3}$
23. $g^{-1}(x) = x^3 + 1$ **24.**

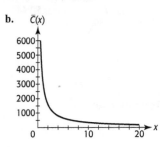

25. Yes **26.** No **27.** $x = -\dfrac{3}{8}$ **28.** $x = -2$
29. $[-2, 9]$ **30.** $x \leq -3$ or $x \geq \dfrac{1}{2}$ **31.** $-2 \leq x \leq 6$
32. $x \geq 4.5$ or $x \leq -3$ **33.** $x < 20$
34. $x \geq 16\sqrt{2} - 2$ or $x \leq -16\sqrt{2} - 2$

Chapter 4 Review

35. a.

b. 64 in.

36. a. 4 sec **b.** 380 ft **c.** Stretched vertically using a factor of 16, and reflected across the x-axis, then shifted 4 units right and 380 units up.
37. a. 2007 **b.** 2004 to 2009 **38.** 2007 to 2010
39. 1990 to 2010

40. a. $\overline{C}(x) = \dfrac{30x + 3150}{x}$ **b.**

41. a.

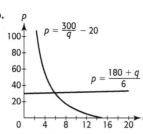

b. The function, and therefore the average cost, is decreasing.   **c.** Vertical stretch using a factor of 50,000 and shift 120 units up

42. a. Demand function   **b.**

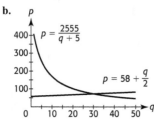

43. a. Demand function   **b.**

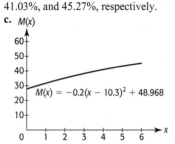

44. a. Reflect across the x-axis, compress vertically using a factor of 0.2, shift right 10.3 units, and shift up 48.968 units
b. 31.67; 35.19; 41.03; 45.27; the percents of high school seniors who tried marijuana in 1991, 1992, 1994, and 1996 were 31.67%, 35.19%, 41.03%, and 45.27%, respectively.
c.

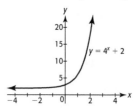

d. $\{1, 2, 3, 4, 5, \ldots, 16\}$

45. a. $R(E(t)) = 0.002805t^2 + 0.35706t + 1.104065$; the function calculates the revenue for Southwest Airlines given the number of years after 1990.   **b.** $R(E(3)) \approx 2.2$; in 1993, Southwest Airlines had revenue of $2.2 billion.   **c.** $E(7) = 24.042$; in 1997, Southwest Airlines had 24,042 employees.   **d.** $3.7 billion
46. a. $f^{-1}(x) = \dfrac{x + 2.886}{0.554}$ **b.** The mean sentence length can be found as a function of the mean time in prison.
47. a. $f^{-1}(x) = \dfrac{x - 28.093}{5.582}$ **b.** Given the number of billions of dollars spent on higher education, the inverse function computes the number of years after 1990.   **c.** 10   **48. a.** Reciprocal function
b. Function P, which gives $0.449   **49.** More than 200 and less than 6000 units   **50.** 1960 to 1996

Chapter 5 Exponential and Logarithmic Functions

Toolbox Exercises

1. a. x^7 **b.** x^5 **c.** $256a^4y^4$ **d.** $\dfrac{81}{z^4}$ **e.** $2^5 = 32$
f. x^8 **2. a.** y^6 **b.** w^6 **c.** $216b^3x^3$ **d.** $\dfrac{125z^3}{8}$

e. 3^5, or 243   **f.** $16y^{12}$ **3.** 10   **4.** 256   **5.** $\dfrac{1}{x^7}$ **6.** $\dfrac{1}{y^8}$
7. $\dfrac{1}{c^{18}}$ **8.** $\dfrac{1}{x^8}$ **9.** a **10.** b^2 **11.** $x^{1/6}$ **12.** $y^{1/15}$
13. $\dfrac{6}{ab^2}$ **14.** $\dfrac{-8a^2}{b^2}$ **15.** $\dfrac{x^{10}}{4}$ **16.** $\dfrac{8y^{18}}{27}$ **17.** $\dfrac{-7}{a^2b}$
18. $\dfrac{-6y^2}{x}$ **19.** 4.6×10^7 **20.** 8.62×10^{11}
21. 9.4×10^{-5} **22.** 2.78×10^{-6} **23.** 437,200
24. 7,910,000   **25.** 0.00056294   **26.** 0.0063478
27. 3.708125×10^6 **28.** 6.460833515×10^1 **29.** $x^{4/3}$
30. $y^{13/20}$ **31.** $c^{5/3}$ **32.** $x^{9/8}$ **33.** $x^{1/4}$ **34.** $y^{1/8}$

5.1 Skills Check

1. c, d, e
3. a.

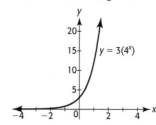

b. 2.718; 0.368; 54.598
c. x-axis   **d.** 1

5.

7.

9.

11.

13. B   **15.** A   **17.** E
19. Vertical shift, 2 units up   **21.** Reflection across the y-axis

23. Vertical stretch using a factor of 3

25. The graph of $y = 3 \cdot 4^{(x-2)} - 3$ is the graph of $y = 3(4^x)$ shifted to the right 2 units and shifted down 3 units.

27. a.

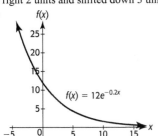

$f(x) = 12e^{-0.2x}$

b. 1.624; 88.669
c. Decay

5.1 Exercises

29. a. \$12,000   **b.** \$8603.73   **c.** No

31. a.

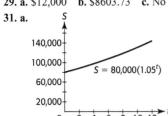

$S = 80,000(1.05^t)$

b. \$130,311.57

33. a.

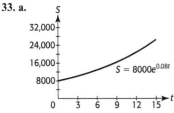

$S = 8000e^{0.08t}$

b. In about 11.45 yr

c.

t (year)	S (\$)
10	17,804.33
20	39,624.26
22	46,499.50

35. a. 376.84 g   **b.**
c. About 24.5 yr

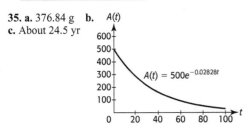

$A(t) = 500e^{-0.02828t}$

37. a.

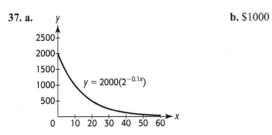

$y = 2000(2^{-0.1x})$

b. \$1000

c. Sales declined drastically after the end of the ad campaign. (Answers may vary.)   **39. a.** \$14,339.44   **b.** Because of future inflation, they should plan to save money. (Answers may vary.)
41. a. \$122,140.28   **b.** In about 14 yr   **43. a.** Increasing
b. 57,128   **c.** 61,577   **d.** Approximately 858 people per year
45. a. About 88.6 g   **b.** About 19,034 yr

47. a.

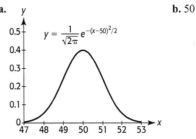

$y = \dfrac{1}{\sqrt{2\pi}}e^{-(x-50)^2/2}$

b. 50

5.2 Skills Check

1. $3^y = x$ **3.** $e^y = 2x$
5. $\log_4 x = y$ **7.** $\log_2 32 = 5$
9. a. 0.845   **b.** 4.454   **c.** 4.806
11. a. 5   **b.** 2   **c.** 3   **d.** 3   **e.** 4
13. a. -3 **b.** 0   **c.** 1   **d.** -4
15.   **17.**

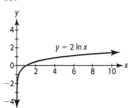

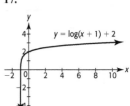

19. a. $y = \log_4 x$ **b.** Graphs are symmetric about the line $y = x$.

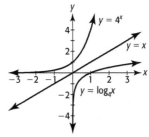

21. $a^x = a; x = 1$ **23.** 14   **25.** 12   **27.** 2.2146   **29.** 2.322
31. $\ln(3x - 2) - \ln(x + 1)$
33. $\dfrac{1}{3}\log_3(4x + 1) - \log_3(4) - 2\log_3(x)$
35. $\log_2 x^3 y$ **37.** $\ln \dfrac{(2a)^4}{b}$

5.2 Exercises

39. a. 57; 78   **b.** Improved health care and better diet. (Answers may vary.)   **41. a.** \$11,293; \$11,835   **b.** Increasing
c.

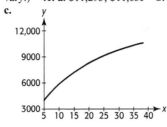

43. \$75.50   **45. a.** 47.1%; 47.5%   **b.** Increasing   **47.** ≈ 6.9 yr
49. 35 quarters, or $8\dfrac{3}{4}$ yr   **51.** 9 yr   **53.** 4.4   **55.** $10^{6.4}I_0 \approx 2,511,886I_0$
57. $10^{7.1}I_0 \approx 12,589,254I_0$ **59.** About 14.1 times as intense
61. 158.5 times as intense   **63.** Decibel reading of higher intensity sound is 20 more than the other.   **65.** $10^{14}I_0$ **67.** Intensity of painful sound is 100 times the intensity of other.
69. $[H^+] = 10^{-7.79} \approx 0.0000000162$ **71.** 251.2 times as acidic

5.3 Skills Check

1. $x \approx 3.204$ **3.** $x = 7.824$ **5.** $x \approx 1.819$ **7.** 0.374 **9.** $x \approx 1.204$
11. 1.6131 **13.** 0.1667 **15.** $x = \dfrac{10}{3}$ **17.** $x = 2$ **19.** $x \approx 3.5$
21. $x \approx 0.769$ **23.** $x = 8$ **25.** $x = e^{1.5} \approx 4.482$
27. $x = \dfrac{e^6}{2} \approx 201.71$ **29.** $x = 50$ **31.** $x = \dfrac{1}{3}$ **33.** $x = 40$
35. $x = 5$ **37.** $x < 5$ **39.** $x \geq 9$

5.3 Exercises

41. 5 **43. a.** $\ln \dfrac{S}{25{,}000} = -0.072x$ **b.** 6 weeks   **45. a.** \$3200
b. 9 days   **47. a.** \$1,457,837   **b.** 2013   **49.** 13.86 yr   **51.** 23.11 yr
53. a. 500 g   **b.** ≈ 24.51 yr   **55.** 3 hours

57. a.

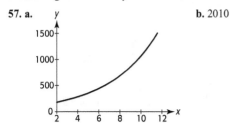

b. 2010

c. No; the model increases and reaches 1500 when $x = 11$. This is not reasonable since the price of gold has historically fluctuated.

59. 7 units   **61.** $t = \dfrac{\ln 2}{\ln (1.07)}$ **63.** 13 yr   **65.** 5 yr   **67. a.** 2008

b. 2008; yes   **69. a.** \$174.08 million   **b.** \$8160.69 million

c. 4588%   **71.** 1996

73. a.

Annual Interest Rate	Rule of 72 Years	Exact Years
2%	36	34.66
3%	24	23.10
4%	18	17.33
5%	14.4	13.86
6%	12	11.55
7%	10.29	9.90
8%	9	8.66
9%	8	7.70
10%	7.2	6.93
11%	6.55	6.30

b. As interest rate increases, the estimate gets closer to actual value.
75. a. ≈ 11.9 **b.** 6 yr
77. 16 yr   **79.** 46 months
81. ≈ 2075 yr

83. a. 16 weeks   **b.**

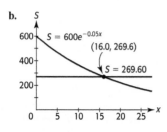

5.4 Skills Check

1. $y = 2(3^x)$ **3.** Not exponential   **5.** $f(x) = 4^x$

7. a.

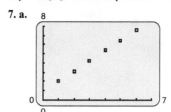

b. Linear model

9. Exponential function   **11.** $y = 0.876(2.494^x)$

13. a.

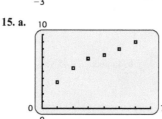

b. Logarithmic model

15. a.

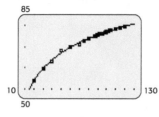

b. $y = 3.671x^{0.505}$

c. $y = -0.125x^2 + 1.886x + 1.960$ **d.** $y = 3.468 + 2.917 \ln x$

5.4 Exercises

17. a. $y = 30{,}000(1.04^t)$ **b.** \$54,028
19. a. $y = 20{,}000(0.98^x)$ **b.** \$18,078   **21. a.** $y = 492.439(1.070^x)$
b. \$20,100.8 billion   **c.** 2014   **23. a.** $y = 1.756(1.085^x)$
b. \$17,749 billion   **c.** 2018
d.

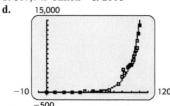

War on terror; Iraq war

25. a. $y = 2.919(1.041^x)$ **b.** 269.6   **c.** 2016
27. a. $y = 11.027 + 14.304 \ln x$
b. $y = -0.0018x^2 + 0.488x + 46.249$ **c.** Logarithmic model

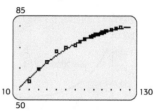

d. Logarithmic: 79.0; quadratic: 78.6
29. a. $y = 27.496 + 4.929 \ln x$ **b.** 2024
31. a. $y = 2400.492(1.062^x)$ **b.** 21,244; extrapolation
33. a. $y = -681.976 + 251.829 \ln x$ **b.** 31.5%

c. $y = 0.627x^2 - 7.400x - 26.675$ **d.** Quadratic function
35. a. $y = 0.028(1.381^x)$ **b.** 89.5 million

5.5 Skills Check

1. 49,801.75   **3.** 17,230.47   **5.** 26,445.08   **7.** 12,311.80
9. 1,723,331.03   **11.** 1123.60; 1191.00; 1191.00

13. $P = S\left(1 + \dfrac{r}{k}\right)^{-kn}$

5.5 Exercises

15. a. \$16,288.19   **b.** \$88,551.38
17. a.   **b.** 8 yr

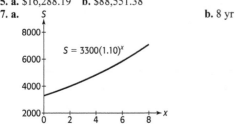
$S = 3300(1.10)^x$

19. \$32,620.38   **21. a.** \$33,194.62   **b.** More; compounded more
frequently   **23.** \$49,958.02   **25. a.** \$20,544.33   **b.** \$29,446.80
27. a. \$28,543.39   **b.** Continuous compounding yields \$903.41 more.
29. a. ≈ 7.27 yr   **b.** ≈ 6.93 yr   **31. a.** \$2954.91   **b.** \$4813.24
33. \$6152.25
35. a.

Years	0	7	14	21	28
Future Value (\$)	1000	2000	4000	8000	16,000

b. $y = 1000(1.104)^x$ **c.** \$1640.01; \$2826.02   **37.** \$29,303.36
39. \$5583.95   **41.** \$4996.09   **43.** \$24,215.65   **45.** $t \approx 17.39$;
17 yr, 5 mo   **47.** 12 yr and after

49.
$$2 = \left(1 + \frac{r}{m}\right)^{mt}$$
$$\ln 2 = mt \ln\left(1 + \frac{r}{m}\right)$$
$$\frac{\ln 2}{m \ln(1 + r/m)} = t$$

5.6 Skills Check

1. $P = \dfrac{S}{(1 + i)^n}$ **3.** $A = R\left(\dfrac{1 - (1 + i)^{-n}}{i}\right)$

5. $R = A\left(\dfrac{i}{1 - (1 + i)^{-n}}\right)$

5.6 Exercises

7. \$52,723.18   **9.** \$21,824.53   **11.** \$486,043.02   **13.** \$9549.11
15. \$7023.58   **17.** \$530,179.96   **19.** \$372,845.60
21. a. \$702,600.91   **b.** \$121,548.38   **c.** \$100,000 plus the annuity
has a larger present value than \$800,000.   **23. a.** \$198,850.99
b. \$576,000   **c.** \$377,149.01   **25. a.** 2%   **b.** 16   **c.** \$736.50
27. a. \$1498.88   **b.** \$639,596.80   **c.** \$289,596.80

5.7 Skills Check

1. 73.83   **3.** 995.51; 999.82

5. a.

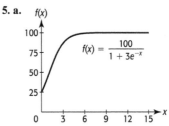
$f(x) = \dfrac{100}{1 + 3e^{-x}}$

b. 25; 99.99
c. Increasing
d. 100

7. a.

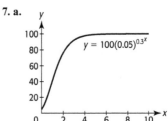
$y = 100(0.05)^{0.3^x}$

b. 5   **c.** 100

5.7 Exercises

9. a.

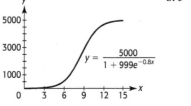
$y = \dfrac{5000}{1 + 999e^{-0.8x}}$

b. 5   **c.** 5000

11. a.

$y = \dfrac{89.786}{1 + 4.6531e^{-0.8256x}}$

b. 29.56%
c. 86.93%
d. 89.786%

13. a. Approximately 218   **b.** 1970   **c.** Seventh day
15. a. $y = \dfrac{89.786}{1 + 4.6531e^{-0.8256x}}$ **b.** Yes

c. $y = 14.137x + 17.624$ **d.** Logistic model

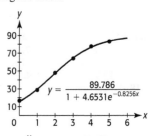
$y = \dfrac{89.786}{1 + 4.6531e^{-0.8256x}}$

17. a. $y = \dfrac{44.742}{1 + 6.870\,e^{-0.0782x}}$ **b.** $y = \dfrac{44.742}{1 + 6.870e^{-0.0782x}}$

c. 40.9%

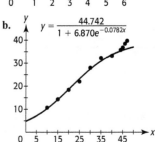

19. a. $y = \dfrac{82.488}{1 + 0.816e^{-0.024x}}$ **b.** 67.9; 77.7   **c.** 82.5

21. a. 4000   **b.** Approximately 9985
c. 10,000

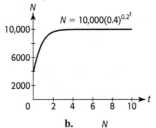

$N = 10{,}000(0.4)^{0.2^t}$

23. a. 21,012
c. 40,000   **b.**

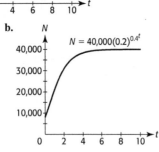

$N = 40{,}000(0.2)^{0.4^t}$

25. a. 10   **b.** 100   **c.** 1000   **d.** 6 yr after the company was formed
27. 10 days   **29.** 11 yr

33. a.

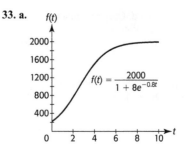

$f(t) = \dfrac{2000}{1 + 8e^{-0.8t}}$

b. 222.22; 1973.76   **c.** 2000
34. a.

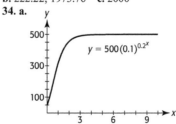

$y = 500(0.1)^{0.2^x}$

b. 50   **c.** 500

Chapter 5 **Skills Check**

1. a.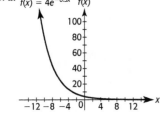

$f(x) = 4e^{-0.3x}$ **b.** ≈ 80.342; ≈ 0.19915

2. Decreasing
3. 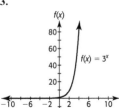   **4.**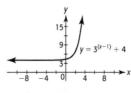

$f(x) = 3^x$ $y = 3^{(x-1)} + 4$

5. Graph of $y = 3^{(x-1)} + 4$ is graph of $f(x) = 3^x$ shifted right 1 unit
and up 4 units.   **6.** Increasing   **7. a.** $y = 500$ **b.** $x = 20$
8. $y = \log_6 x$ **9.** $3x = \log_7 y$ **10.** $x = 4^y$ **11.** $x = 10^y$
12. $x = e^y$ **13.** $y = \log_4 x$ **14.** 1.3424   **15.** 4.0254   **16.** 1
17. 4   **18.** 4   **19.** -3 **20.** 3.6309   **21.** 1.9358
22.   **23.**

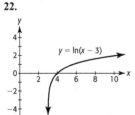

$y = \ln(x - 3)$ $y = \log_3 x$

24. 5.8289   **25.** 0.2012   **26.** 2.8813   **27.** 2
28. $\ln(2x - 5)^3 - \ln(x - 3) = 3\ln(2x - 5) - \ln(x - 3)$

29. $\log_4 \dfrac{x^6}{y^2}$ **30.** Exponential function; $y = 0.810(2.470^x)$

31. 4926.80   **32.** 32,373.02

Chapter 5 **Review**

35. 325 million   **36.** $1515.72   **37.** 1988 and after
38. a. 3 on Richter scale   **b.** $3{,}162{,}278 I_0$
39. Approximately 1259 times as intense

40. 10 yr   **41. a.** $x = 7 \log_2 \dfrac{S}{1000}$ **b.** 30 yr

42. 10 weeks   **43. a.** Exponential growth
b.

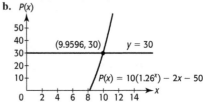

44. a. $P(x) = 10(1.26^x) - 2x - 50$
b.

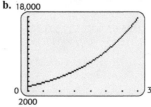

$(9.9596, 30)$ $y = 30$
$P(x) = 10(1.26^x) - 2x - 50$

45. 7 weeks   **46. a.** $y = 54.62$ g   **b.** 8445 years ago
47. 14 yr   **48.** $\approx \$4451.08$ **49.** 15 yr   **50.** $y = 108.319(1.315^x)$
51. a. $y = 98.221(0.870^x)$ **b.** Exponential decay   **c.** 1.5
52. a. $y = 220.936(1.347^x)$ **b.** 686,377 thousand; no;
twice U.S. population   **53.** $y = 165.893(1.055^x)$; $1065 billion
54. a. $y = 0.940 + 28.672 \ln x$ **b.** 86.8%   **c.** 2027; predicted
percent >100 **55.** $y = 4.337 + 40.890 \ln x$

56. a. $N = \dfrac{129.679}{1 + 0.106 e^{-0.0786x}}$ **b.** Excellent fit

57. $20,609.02   **58.** $30,072.61   **59.** $34,426.47
60. $274,419.05   **61.** $209,281.18   **62.** $6899.32
63. $66.43   **64.** $773.16   **65. a.** 26.989%; 31.699%
b. 44.742%   **66. a.** Approximately 1184
b. 16 days   **67. a.** 1298   **b.** 3997
c. 4000   **68. a.** Approximately 17,993

b. 18,000   **69. a.** $y = \dfrac{627.044}{1 + 268.609 e^{-0.324x}}$ **b.** 626   **c.** 2022

Index

Note: Page numbers followed by f indicate figures; those followed by t indicate tables.

Index of Applications

Index of Applications, continued